Guidelines for Drinking-water Quality

THIRD EDITION

Volume 1
Recommendations

WORLD HEALTH ORGANIZATION
Geneva
2004

WHO Library Cataloguing-in-Publication Data

World Health Organization.
  Guidelines for Drinking-water Quality. Vol. 1 : 3rd ed.

  1.Potable water – standards   2.Water – standards   3.Water quality – standards
  4.Guidelines   I.Title.

  ISBN 92 4 154638 7   (NLM Classification: WA 675)

© **World Health Organization 2004**

All rights reserved. Publications of the World Health Organization can be obtained from Marketing and Dissemination, World Health Organization, 20 Avenue Appia, 1211 Geneva 27, Switzerland (tel: +41 22 791 2476; fax: +41 22 791 4857; email: bookorders@who.int). Requests for permission to reproduce or translate WHO publications – whether for sale or for noncommercial distribution – should be addressed to Publications, at the above address (fax: +41 22 791 4806; email: permissions@who.int).

The designations employed and the presentation of the material in this publication do not imply the expression of any opinion whatsoever on the part of the World Health Organization concerning the legal status of any country, territory, city or area or of its authorities, or concerning the delimitation of its frontiers or boundaries. Dotted lines on maps represent approximate border lines for which there may not yet be full agreement.

The mention of specific companies or of certain manufacturers' products does not imply that they are endorsed or recommended by the World Health Organization in preference to others of a similar nature that are not mentioned. Errors and omissions excepted, the names of proprietary products are distinguished by initial capital letters.

The World Health Organization does not warrant that the information contained in this publication is complete and correct and shall not be liable for any damages incurred as a result of its use.

Designed by minimum graphics
Typeset by SNP Best-set Typesetter Ltd., Hong Kong
Printed in China by Sun Fung

Contents

Preface		xv
Acknowledgements		xviii
Acronyms and abbreviations used in text		xx

1. Introduction — 1
- 1.16 General considerations and principles — 1
  - 1.1.1 Microbial aspects — 3
  - 1.1.2 Disinfection — 5
  - 1.1.3 Chemical aspects — 6
  - 1.1.4 Radiological aspects — 7
  - 1.1.5 Acceptability aspects — 7
- 1.2 Roles and responsibilities in drinking-water safety management — 8
  - 1.2.1 Surveillance and quality control — 8
  - 1.2.2 Public health authorities — 10
  - 1.2.3 Local authorities — 11
  - 1.2.4 Water resource management — 12
  - 1.2.5 Drinking-water supply agencies — 13
  - 1.2.6 Community management — 14
  - 1.2.7 Water vendors — 15
  - 1.2.8 Individual consumers — 15
  - 1.2.9 Certification agencies — 16
  - 1.2.10 Plumbing — 17
- 1.3 Supporting documentation to the Guidelines — 18

2. The Guidelines: a framework for safe drinking-water — 22
- 2.1 Framework for safe drinking-water: requirements — 22
  - 2.1.1 Health-based targets — 24
  - 2.1.2 System assessment and design — 25
  - 2.1.3 Operational monitoring — 26
  - 2.1.4 Management plans, documentation and communication — 27
  - 2.1.5 Surveillance of drinking-water quality — 28

	2.2	Guidelines for verification	29
		2.2.1 Microbial water quality	29
		2.2.2 Chemical water quality	30
	2.3	National drinking-water policy	31
		2.3.1 Laws, regulations and standards	31
		2.3.2 Setting national standards	32
	2.4	Identifying priority drinking-water quality concerns	34
		2.4.1 Assessing microbial priorities	35
		2.4.2 Assessing chemical priorities	35
3.	**Health-based targets**		**37**
	3.1	Role and purpose of health-based targets	37
	3.2	Types of health-based targets	39
		3.2.1 Specified technology targets	41
		3.2.2 Performance targets	41
		3.2.3 Water quality targets	42
		3.2.4 Health outcome targets	43
	3.3	General considerations in establishing health-based targets	43
		3.3.1 Assessment of risk in the framework for safe drinking-water	44
		3.3.2 Reference level of risk	44
		3.3.3 Disability-adjusted life-years (DALYs)	45
4.	**Water safety plans**		**48**
	4.1	System assessment and design	51
		4.1.1 New systems	52
		4.1.2 Collecting and evaluating available data	53
		4.1.3 Resource and source protection	56
		4.1.4 Treatment	59
		4.1.5 Piped distribution systems	61
		4.1.6 Non-piped, community and household systems	64
		4.1.7 Validation	67
		4.1.8 Upgrade and improvement	67
	4.2	Operational monitoring and maintaining control	68
		4.2.1 Determining system control measures	68
		4.2.2 Selecting operational monitoring parameters	68
		4.2.3 Establishing operational and critical limits	70
		4.2.4 Non-piped, community and household systems	71
	4.3	Verification	71
		4.3.1 Verification of microbial quality	72
		4.3.2 Verification of chemical quality	73
		4.3.3 Water sources	73
		4.3.4 Piped distribution systems	74

		4.3.5	Verification for community-managed supplies	74
		4.3.6	Quality assurance and quality control	75
	4.4	Management procedures for piped distribution systems		76
		4.4.1	Predictable incidents ("deviations")	77
		4.4.2	Unforeseen events	77
		4.4.3	Emergencies	78
		4.4.4	Closing supply, water avoidance and "boil water" orders	79
		4.4.5	Preparing a monitoring plan	80
		4.4.6	Supporting programmes	80
	4.5	Management of community and household water supplies		81
	4.6	Documentation and communication		82

5. Surveillance   84

	5.1	Types of approaches		85
		5.1.1	Audit	86
		5.1.2	Direct assessment	87
	5.2	Adapting approaches to specific circumstances		88
		5.2.1	Urban areas in developing countries	88
		5.2.2	Surveillance of community drinking-water supplies	88
		5.2.3	Surveillance of household treatment and storage systems	89
	5.3	Adequacy of supply		90
		5.3.1	Quantity (service level)	90
		5.3.2	Accessibility	91
		5.3.3	Affordability	92
		5.3.4	Continuity	92
	5.4	Planning and implementation		93
	5.5	Reporting and communicating		95
		5.5.1	Interaction with community and consumers	96
		5.5.2	Regional use of data	96

6. Application of the guidelines in specific circumstances   99

	6.1	Large buildings		99
		6.1.1	Health risk assessment	100
		6.1.2	System assessment	100
		6.1.3	Management	101
		6.1.4	Monitoring	101
		6.1.5	Independent surveillance and supporting programmes	102
		6.1.6	Drinking-water quality in health care facilities	102
		6.1.7	Drinking-water quality in schools and day care centres	103
	6.2	Emergencies and disasters		104
		6.2.1	Practical considerations	105
		6.2.2	Monitoring	106
		6.2.3	Microbial guidelines	107

	6.2.4	Sanitary inspections and catchment mapping	108
	6.2.5	Chemical and radiological guidelines	108
	6.2.6	Testing kits and laboratories	109
6.3	Safe drinking-water for travellers		109
6.4	Desalination systems		111
6.5	Packaged drinking-water		113
	6.5.1	Safety of packaged drinking-water	113
	6.5.2	Potential health benefits of bottled drinking-water	114
	6.5.3	International standards for bottled drinking-water	114
6.6	Food production and processing		115
6.7	Aircraft and airports		116
	6.7.1	Health risks	116
	6.7.2	System risk assessment	116
	6.7.3	Operational monitoring	116
	6.7.4	Management	117
	6.7.5	Surveillance	117
6.8	Ships		117
	6.8.1	Health risks	117
	6.8.2	System risk assessment	118
	6.8.3	Operational monitoring	119
	6.8.4	Management	119
	6.8.5	Surveillance	120

7. Microbial aspects — 121

7.1	Microbial hazards associated with drinking-water		121
	7.1.1	Waterborne infections	121
	7.1.2	Persistence and growth in water	124
	7.1.3	Public health aspects	125
7.2	Health-based target setting		126
	7.2.1	Health-based targets applied to microbial hazards	126
	7.2.2	Risk assessment approach	126
	7.2.3	Risk-based performance target setting	131
	7.2.4	Presenting the outcome of performance target development	133
	7.2.5	Issues in adapting risk-based performance target setting to national/local circumstances	133
	7.2.6	Health outcome targets	134
7.3	Occurrence and treatment of pathogens		135
	7.3.1	Occurrence	136
	7.3.2	Treatment	137
7.4	Verification of microbial safety and quality		142
7.5	Methods of detection of faecal indicator bacteria		143

CONTENTS

8.	**Chemical aspects**		**145**
8.1	Chemical hazards in drinking-water		145
8.2	Derivation of chemical guideline values		147
	8.2.1	Approaches taken	148
	8.2.2	Threshold chemicals	149
	8.2.3	Alternative approaches	152
	8.2.4	Non-threshold chemicals	154
	8.2.5	Data quality	154
	8.2.6	Provisional guideline values	155
	8.2.7	Chemicals with effects on acceptability	156
	8.2.8	Non-guideline chemicals	156
	8.2.9	Mixtures	156
8.3	Analytical aspects		157
	8.3.1	Analytical achievability	157
	8.3.2	Analytical methods	158
8.4	Treatment		166
	8.4.1	Treatment achievability	166
	8.4.2	Chlorination	171
	8.4.3	Ozonation	172
	8.4.4	Other disinfection processes	172
	8.4.5	Filtration	173
	8.4.6	Aeration	175
	8.4.7	Chemical coagulation	175
	8.4.8	Activated carbon adsorption	176
	8.4.9	Ion exchange	177
	8.4.10	Membrane processes	178
	8.4.11	Other treatment processes	178
	8.4.12	Disinfection by-products – process control measures	179
	8.4.13	Treatment for corrosion control	180
8.5	Guideline values for individual chemicals, by source category		184
	8.5.1	Naturally occurring chemicals	184
	8.5.2	Chemicals from industrial sources and human dwellings	185
	8.5.3	Chemicals from agricultural activities	187
	8.5.4	Chemicals used in water treatment or from materials in contact with drinking-water	188
	8.5.5	Pesticides used in water for public health purposes	190
	8.5.6	Cyanobacterial toxins	192
9.	**Radiological aspects**		**197**
9.1	Sources and health effects of radiation exposure		198
	9.1.1	Radiation exposure through drinking-water	200
	9.1.2	Radiation-induced health effects through drinking-water	200

9.2		Units of radioactivity and radiation dose	201
9.3		Guidance levels for radionuclides in drinking-water	202
9.4		Monitoring and assessment for dissolved radionuclides	204
	9.4.1	Screening of drinking-water supplies	204
	9.4.2	Strategy for assessing drinking-water	205
	9.4.3	Remedial measures	205
9.5	Radon		206
	9.5.1	Radon in air and water	206
	9.5.2	Risk	207
	9.5.3	Guidance on radon in drinking-water supplies	207
9.6		Sampling, analysis and reporting	207
	9.6.1	Measuring gross alpha and gross beta activity concentrations	207
	9.6.2	Measuring potassium-40	208
	9.6.3	Measuring radon	208
	9.6.4	Sampling	209
	9.6.5	Reporting of results	209

10. Acceptability aspects — 210

10.1	Taste, odour and appearance	211
	10.1.1 Biologically derived contaminants	211
	10.1.2 Chemically derived contaminants	213
	10.1.3 Treatment of taste, odour and appearance problems	219
10.2	Temperature	220

11. Microbial fact sheets — 221

11.1	Bacterial pathogens	222
	11.1.1 *Acinetobacter*	222
	11.1.2 *Aeromonas*	224
	11.1.3 *Bacillus*	225
	11.1.4 *Burkholderia pseudomallei*	226
	11.1.5 *Campylobacter*	228
	11.1.6 *Escherichia coli* pathogenic strains	229
	11.1.7 *Helicobacter pylori*	231
	11.1.8 *Klebsiella*	232
	11.1.9 *Legionella*	233
	11.1.10 *Mycobacterium*	235
	11.1.11 *Pseudomonas aeruginosa*	237
	11.1.12 *Salmonella*	239
	11.1.13 *Shigella*	240
	11.1.14 *Staphylococcus aureus*	242
	11.1.15 *Tsukamurella*	243

		11.1.16 *Vibrio*	244
		11.1.17 *Yersinia*	246
	11.2	Viral pathogens	247
		11.2.1 Adenoviruses	248
		11.2.2 Astroviruses	250
		11.2.3 Caliciviruses	251
		11.2.4 Enteroviruses	253
		11.2.5 Hepatitis A virus	254
		11.2.6 Hepatitis E virus	256
		11.2.7 Rotaviruses and orthoreoviruses	257
	11.3	Protozoan pathogens	259
		11.3.1 *Acanthamoeba*	259
		11.3.2 *Balantidium coli*	261
		11.3.3 *Cryptosporidium*	262
		11.3.4 *Cyclospora cayetanensis*	264
		11.3.5 *Entamoeba histolytica*	265
		11.3.6 *Giardia intestinalis*	267
		11.3.7 *Isospora belli*	268
		11.3.8 Microsporidia	270
		11.3.9 *Naegleria fowleri*	272
		11.3.10 *Toxoplasma gondii*	274
	11.4	Helminth pathogens	275
		11.4.1 *Dracunculus medinensis*	276
		11.4.2 *Fasciola* spp.	278
	11.5	Toxic cyanobacteria	279
	11.6	Indicator and index organisms	281
		11.6.1 Total coliform bacteria	282
		11.6.2 *Escherichia coli* and thermotolerant coliform bacteria	284
		11.6.3 Heterotrophic plate counts	285
		11.6.4 Intestinal enterococci	287
		11.6.5 *Clostridium perfringens*	288
		11.6.6 Coliphages	289
		11.6.7 *Bacteroides fragilis* phages	292
		11.6.8 Enteric viruses	294
12.	**Chemical fact sheets**		**296**
	12.1	Acrylamide	296
	12.2	Alachlor	297
	12.3	Aldicarb	298
	12.4	Aldrin and dieldrin	300
	12.5	Aluminium	301
	12.6	Ammonia	303

12.7	Antimony	304
12.8	Arsenic	306
12.9	Asbestos	308
12.10	Atrazine	308
12.11	Barium	310
12.12	Bentazone	311
12.13	Benzene	312
12.14	Boron	313
12.15	Bromate	315
12.16	Brominated acetic acids	316
12.17	Cadmium	317
12.18	Carbofuran	319
12.19	Carbon tetrachloride	320
12.20	Chloral hydrate (trichloroacetaldehyde)	321
12.21	Chlordane	323
12.22	Chloride	324
12.23	Chlorine	325
12.24	Chlorite and chlorate	326
12.25	Chloroacetones	329
12.26	Chlorophenols (2-chlorophenol, 2,4-dichlorophenol, 2,4,6-trichlorophenol)	329
12.27	Chloropicrin	331
12.28	Chlorotoluron	332
12.29	Chlorpyrifos	333
12.30	Chromium	334
12.31	Copper	335
12.32	Cyanazine	337
12.33	Cyanide	339
12.34	Cyanogen chloride	340
12.35	2,4-D (2,4-dichlorophenoxyacetic acid)	340
12.36	2,4-DB	342
12.37	DDT and metabolites	343
12.38	Dialkyltins	345
12.39	1,2-Dibromo-3-chloropropane (DBCP)	346
12.40	1,2-Dibromoethane (ethylene dibromide)	347
12.41	Dichloroacetic acid	349
12.42	Dichlorobenzenes (1,2-dichlorobenzene, 1,3-dichlorobenzene, 1,4-dichlorobenzene)	350
12.43	1,1-Dichloroethane	352
12.44	1,2-Dichloroethane	353
12.45	1,1-Dichloroethene	354
12.46	1,2-Dichloroethene	355

12.47	Dichloromethane	357
12.48	1,2-Dichloropropane (1,2-DCP)	358
12.49	1,3-Dichloropropane	359
12.50	1,3-Dichloropropene	360
12.51	Dichlorprop (2,4-DP)	361
12.52	Di(2-ethylhexyl)adipate	362
12.53	Di(2-ethylhexyl)phthalate	363
12.54	Dimethoate	364
12.55	Diquat	366
12.56	Edetic acid (EDTA)	367
12.57	Endosulfan	368
12.58	Endrin	369
12.59	Epichlorohydrin	370
12.60	Ethylbenzene	372
12.61	Fenitrothion	373
12.62	Fenoprop (2,4,5-TP; 2,4,5-trichlorophenoxy propionic acid)	374
12.63	Fluoride	375
12.64	Formaldehyde	377
12.65	Glyphosate and AMPA	379
12.66	Halogenated acetonitriles (dichloroacetonitrile, ibromoacetonitrile, bromochloroacetonitrile, trichloroacetonitrile)	380
12.67	Hardness	382
12.68	Heptachlor and heptachlor epoxide	383
12.69	Hexachlorobenzene (HCB)	385
12.70	Hexachlorobutadiene (HCBD)	386
12.71	Hydrogen sulfide	387
12.72	Inorganic tin	388
12.73	Iodine	389
12.74	Iron	390
12.75	Isoproturon	391
12.76	Lead	392
12.77	Lindane	394
12.78	Malathion	396
12.79	Manganese	397
12.80	MCPA [4-(2-methyl-4-chlorophenoxy)acetic acid]	399
12.81	Mecoprop (MCPP; [2(2-methyl-chlorophenoxy) propionic acid])	401
12.82	Mercury	402
12.83	Methoxychlor	403
12.84	Methyl parathion	404
12.85	Metolachlor	405

12.86	Microcystin-LR	407
12.87	Molinate	408
12.88	Molybdenum	410
12.89	Monochloramine	411
12.90	Monochloroacetic acid	412
12.91	Monochlorobenzene	413
12.92	MX	414
12.93	Nickel	415
12.94	Nitrate and nitrite	417
12.95	Nitrilotriacetic acid (NTA)	420
12.96	Parathion	421
12.97	Pendimethalin	422
12.98	Pentachlorophenol (PCP)	424
12.99	Permethrin	425
12.100	pH	426
12.101	2-Phenylphenol and its sodium salt	427
12.102	Polynuclear aromatic hydrocarbons (PAHs)	428
12.103	Propanil	430
12.104	Pyriproxyfen	431
12.105	Selenium	432
12.106	Silver	434
12.107	Simazine	435
12.108	Sodium	436
12.109	Styrene	437
12.110	Sulfate	438
12.111	2,4,5-T (2,4,5-Trichlorophenoxyacetic acid)	439
12.112	Terbuthylazine (TBA)	440
12.113	Tetrachloroethene	442
12.114	Toluene	443
12.115	Total dissolved solids (TDS)	444
12.116	Trichloroacetic acid	445
12.117	Trichlorobenzenes (total)	446
12.118	1,1,1-Trichloroethane	447
12.119	Trichloroethene	448
12.120	Trifluralin	450
12.121	Trihalomethanes (bromoform, bromodichloromethane, dibromochloromethane, chloroform)	451
12.122	Uranium	454
12.123	Vinyl chloride	456
12.124	Xylenes	458
12.125	Zinc	459

Annex 1 Bibliography	461
Annex 2 Contributors to the development of the Third Edition of the *Guidelines for Drinking-water Quality*	467
Annex 3 Default assumptions	486
Annex 4 Chemical summary tables	488
Index	494

Preface

Access to safe drinking-water is essential to health, a basic human right and a component of effective policy for health protection.

The importance of water, sanitation and hygiene for health and development has been reflected in the outcomes of a series of international policy forums. These have included health-oriented conferences such as the International Conference on Primary Health Care, held in Alma-Ata, Kazakhstan (former Soviet Union), in 1978. They have also included water-oriented conferences such as the 1977 World Water Conference in Mar del Plata, Argentina, which launched the water supply and sanitation decade of 1981–1990, as well as the Millennium Declaration goals adopted by the General Assembly of the United Nations (UN) in 2000 and the outcome of the Johannesburg World Summit for Sustainable Development in 2002. Most recently, the UN General Assembly declared the period from 2005 to 2015 as the International Decade for Action, "Water for Life."

Access to safe drinking-water is important as a health and development issue at a national, regional and local level. In some regions, it has been shown that investments in water supply and sanitation can yield a net economic benefit, since the reductions in adverse health effects and health care costs outweigh the costs of undertaking the interventions. This is true for major water supply infrastructure investments through to water treatment in the home. Experience has also shown that interventions in improving access to safe water favour the poor in particular, whether in rural or urban areas, and can be an effective part of poverty alleviation strategies.

In 1983–1984 and in 1993–1997, the World Health Organization (WHO) published the first and second editions of the *Guidelines for Drinking-water Quality* in three volumes as successors to previous WHO International Standards. In 1995, the decision was made to pursue the further development of the Guidelines through a process of rolling revision. This led to the publication of addenda to the second edition of the Guidelines, on chemical and microbial aspects, in 1998, 1999 and 2002; the publication of a text on *Toxic Cyanobacteria in Water*; and the preparation of expert reviews on key issues preparatory to the development of a third edition of the Guidelines.

In 2000, a detailed plan of work was agreed upon for development of the third edition of the Guidelines. As with previous editions, this work was shared between WHO Headquarters and the WHO Regional Office for Europe (EURO). Leading the process of the development of the third edition were the Programme on Water Sanitation and Health within Headquarters and the European Centre for Environment and Health, Rome, within EURO. Within WHO Headquarters, the Programme on Chemical Safety provided inputs on some chemical hazards, and the Programme on Radiological Safety contributed to the section dealing with radiological aspects. All six WHO Regional Offices participated in the process.

This revised Volume 1 of the Guidelines is accompanied by a series of publications providing information on the assessment and management of risks associated with microbial hazards and by internationally peer-reviewed risk assessments for specific chemicals. These replace the corresponding parts of the previous Volume 2. Volume 3 provides guidance on good practice in surveillance, monitoring and assessment of drinking-water quality in community supplies. The Guidelines are also accompanied by other publications explaining the scientific basis of their development and providing guidance on good practice in implementation.

This volume of the *Guidelines for Drinking-water Quality* explains requirements to ensure drinking-water safety, including minimum procedures and specific guideline values, and how those requirements are intended to be used. The volume also describes the approaches used in deriving the guidelines, including guideline values. It includes fact sheets on significant microbial and chemical hazards. The development of this third edition of the *Guidelines for Drinking-water Quality* includes a substantive revision of approaches to ensuring microbial safety. This takes account of important developments in microbial risk assessment and its linkages to risk management. The development of this orientation and content was led over an extended period by Dr Arie Havelaar (RIVM, Netherlands) and Dr Jamie Bartram (WHO).

Since the second edition of WHO's *Guidelines for Drinking-water Quality*, there have been a number of events that have highlighted the importance and furthered understanding of various aspects of drinking-water quality and health. These are reflected in this third edition of the Guidelines.

These Guidelines supersede those in previous editions (1983–1984, 1993–1997 and addenda in 1998, 1999 and 2002) and previous International Standards (1958, 1963 and 1971). The Guidelines are recognized as representing the position of the UN system on issues of drinking-water quality and health by "UN-Water," the body that coordinates amongst the 24 UN agencies and programmes concerned with water issues. This edition of the Guidelines further develops concepts, approaches and information in previous editions:

- Experience has shown that microbial hazards continue to be the primary concern in both developing and developed countries. Experience has also shown the value of a systematic approach towards securing microbial safety. This edition includes

significantly expanded guidance on ensuring microbial safety of drinking-water, building on principles – such as the multiple-barrier approach and the importance of source protection – considered in previous editions. The Guidelines are accompanied by documentation describing approaches towards fulfilling requirements for microbial safety and providing guidance to good practice in ensuring that safety is achieved.
- Information on many chemicals has been revised. This includes information on chemicals not considered previously; revisions to take account of new scientific information; and, in some cases, lesser coverage where new information suggests a lesser priority.
- Experience has also shown the necessity of recognizing the important roles of many different stakeholders in ensuring drinking-water safety. This edition includes discussion of the roles and responsibilities of key stakeholders in ensuring drinking-water safety.
- The need for different tools and approaches in supporting safe management of large piped supplies versus small community supplies remains relevant, and this edition describes the principal characteristics of the different approaches.
- There has been increasing recognition that only a few key chemicals cause large-scale health effects through drinking-water exposure. These include fluoride and arsenic. Other chemicals, such as lead, selenium and uranium, may also be significant under certain conditions. Interest in chemical hazards in drinking-water was highlighted by recognition of the scale of arsenic exposure through drinking-water in Bangladesh and elsewhere. The revised Guidelines and associated publications provide guidance on identifying local priorities and on management of the chemicals associated with large-scale effects.
- WHO is frequently approached for guidance on the application of the *Guidelines for Drinking-water Quality* to situations other than community supplies or managed utilities. This revised edition includes information on application of the Guidelines to several specific circumstances and is accompanied by texts dealing with some of these in greater detail.

The *Guidelines for Drinking-water Quality* are kept up to date through a process of rolling revision, which leads to periodic release of documents that may add to or supersede information in this volume.

The Guidelines are addressed primarily to water and health regulators, policy-makers and their advisors, to assist in the development of national standards. The Guidelines and associated documents are also used by many others as a source of information on water quality and health and on effective management approaches.

Acknowledgements

The preparation of the current edition of the *Guidelines for Drinking-water Quality* and supporting documentation covered a period of eight years and involved the participation of over 490 experts from 90 developing and developed countries. The contributions of all who participated in the preparation and finalization of the *Guidelines for Drinking-water Quality*, including those individuals listed in Annex 2, are gratefully acknowledged.

The work of the following Working Groups was crucial to the development of the third edition of the *Guidelines for Drinking-water Quality*:

Microbial aspects working group
Ms T. Boonyakarnkul, Department of Health, Thailand (*Surveillance and control*)
Dr D. Cunliffe, SA Department of Human Services, Australia (*Public health*)
Prof. W. Grabow, University of Pretoria, South Africa (*Pathogen-specific information*)
Dr A. Havelaar, RIVM, The Netherlands (Working Group coordinator; *Risk assessment*)
Prof. M. Sobsey, University of North Carolina, USA (*Risk management*)

Chemical aspects working group
Mr J.K. Fawell, United Kingdom (*Organic and inorganic constituents*)
Ms M. Giddings, Health Canada (*Disinfectants and disinfection by-products*)
Prof. Y. Magara, Hokkaido University, Japan (*Analytical achievability*)
Dr E. Ohanian, EPA, USA (*Disinfectants and disinfection by-products*)
Dr P. Toft, Canada (*Pesticides*)

Protection and control working group
Dr I. Chorus, Umweltbundesamt, Germany (*Resource and source protection*)
Dr J. Cotruvo, USA (*Materials and additives*)
Dr G. Howard, DfID, Bangladesh, and formerly Loughborough University, United Kingdom (*Monitoring and assessment*)
Mr P. Jackson, WRc-NSF, United Kingdom (*Treatment achievability*)

ACKNOWLEDGEMENTS

The WHO coordinators were:

Dr J. Bartram, Coordinator, Programme on Water Sanitation and Health, WHO Headquarters, and formerly WHO European Centre for Environmental Health

Mr P. Callan, Programme on Water Sanitation and Health, WHO Headquarters, on secondment from National Health and Medical Research Council, Australia

Ms C. Vickers acted as a liaison between the Working Groups and the International Programme on Chemical Safety, WHO Headquarters.

Ms Marla Sheffer of Ottawa, Canada, was responsible for the editing of the Guidelines. Mr Hiroki Hashizume provided support to the work of the Chemical Aspects Working Group. Ms Mary-Ann Lundby, Ms Grazia Motturi and Ms Penny Ward provided secretarial and administrative support throughout the process and to individual meetings.

The preparation of these Guidelines would not have been possible without the generous support of the following, which is gratefully acknowledged: the Ministry of Health of Italy; the Ministry of Health, Labour and Welfare of Japan; the National Health and Medical Research Council, Australia; the Swedish International Development Cooperation Agency, Sweden and the United States Environmental Protection Agency.

Acronyms and abbreviations used in text

AAS	atomic absorption spectrometry
AD	Alzheimer disease
ADI	acceptable daily intake
AES	atomic emission spectrometry
AIDS	acquired immunodeficiency syndrome
AMPA	aminomethylphosphonic acid
BaP	benzo[a]pyrene
BDCM	bromodichloromethane
BMD	benchmark dose
bw	body weight
CAC	Codex Alimentarius Commission
CAS	Chemical Abstracts Service
CICAD	Concise International Chemical Assessment Document
CSAF	chemical-specific adjustment factor
Ct	product of disinfectant concentration and contact time
DAEC	diffusely adherent *E. coli*
DALY	disability-adjusted life-year
DBCM	dibromochloromethane
DBCP	1,2-dibromo-3-chloropropane
DBP	disinfection by-product
DCB	dichlorobenzene
DCP	dichloropropane
DDT	dichlorodiphenyltrichloroethane
DEHA	di(2-ethylhexyl)adipate
DEHP	di(2-ethylhexyl)phthalate
DNA	deoxyribonucleic acid

EAAS	electrothermal atomic absorption spectrometry
EAEC	enteroaggregative *E. coli*
EBCT	empty bed contact time
EC	electron capture
ECD	electron capture detector
EDTA	edetic acid; ethylenediaminetetraacetic acid
EHC	Environmental Health Criteria monograph
EHEC	enterohaemorrhagic *E. coli*
EIEC	enteroinvasive *E. coli*
ELISA	enzyme-linked immunosorbent assay
EPEC	enteropathogenic *E. coli*
ETEC	enterotoxigenic *E. coli*
EURO	WHO Regional Office for Europe
FAAS	flame atomic absorption spectrometry
FAO	Food and Agriculture Organization of the United Nations
FD	fluorescence detector
FID	flame ionization detector
FPD	flame photodiode detector
GAC	granular activated carbon
GAE	granulomatous amoebic encephalitis
GC	gas chromatography
GL	guidance level (used for radionuclides in drinking-water)
GV	guideline value
HACCP	hazard analysis and critical control points
HAd	human adenovirus
HAstV	human astrovirus
HAV	hepatitis A virus
Hb	haemoglobin
HCB	hexachlorobenzene
HCBD	hexachlorobutadiene
HCH	hexachlorocyclohexane
HEV	hepatitis E virus
HIV	human immunodeficiency virus
HPC	heterotrophic plate count
HPLC	high-performance liquid chromatography
HRV	human rotavirus
HuCV	human calicivirus
HUS	haemolytic uraemic syndrome

IAEA	International Atomic Energy Agency
IARC	International Agency for Research on Cancer
IC	ion chromatography
ICP	inductively coupled plasma
ICRP	International Commission on Radiological Protection
IDC	individual dose criterion
IPCS	International Programme on Chemical Safety
ISO	International Organization for Standardization
JECFA	Joint FAO/WHO Expert Committee on Food Additives
JMPR	Joint FAO/WHO Meeting on Pesticide Residues
K_{ow}	octanol/water partition coefficient
LI	Langelier Index
LOAEL	lowest-observed-adverse-effect level
MCB	monochlorobenzene
MCPA	4-(2-methyl-4-chlorophenoxy)acetic acid
MCPP	2(2-methyl-chlorophenoxy) propionic acid; mecoprop
metHb	methaemoglobin
MMT	methylcyclopentadienyl manganese tricarbonyl
MS	mass spectrometry
MX	3-chloro-4-dichloromethyl-5-hydroxy-2(5H)-furanone
NAS	National Academy of Sciences (USA)
NOAEL	no-observed-adverse-effect level
NOEL	no-observed-effect level
NTA	nitrilotriacetic acid
NTP	National Toxicology Program (USA)
NTU	nephelometric turbidity unit
P/A	presence/absence
PAC	powdered activated carbon
PAH	polynuclear aromatic hydrocarbon
PAM	primary amoebic meningoencephalitis
PCP	pentachlorophenol
PCR	polymerase chain reaction
PD	photoionization detector
PMTDI	provisional maximum tolerable daily intake
PT	purge and trap
PTDI	provisional tolerable daily intake

PTWI	provisional tolerable weekly intake
PVC	polyvinyl chloride
QMRA	quantitative microbial risk assessment
RDL	reference dose level
RIVM	Rijksinstituut voor Volksgezondheid en Milieu (Dutch National Institute of Public Health and Environmental Protection)
RNA	ribonucleic acid
SI	Système international d'unités (International System of Units)
SOP	standard operating procedure
SPADNS	sulfo phenyl azo dihydroxy naphthalene disulfonic acid
TBA	terbuthylazine
TCB	trichlorobenzene
TCU	true colour unit
TD_{05}	tumorigenic dose$_{05}$, the intake or exposure associated with a 5% excess incidence of tumours in experimental studies in animals
TDI	tolerable daily intake
TDS	total dissolved solids
THM	trihalomethane
TID	thermal ionization detector
UF	uncertainty factor
UNICEF	United Nations Children's Fund
UNSCEAR	United Nations Scientific Committee on the Effects of Atomic Radiation
USA	United States of America
US EPA	United States Environmental Protection Agency
UV	ultraviolet
UVPAD	ultraviolet photodiode array detector
WHO	World Health Organization
WHOPES	World Health Organization Pesticide Evaluation Scheme
WQT	water quality target
WSP	water safety plan
YLD	years of healthy life lost in states of less than full health, i.e., years lived with a disability
YLL	years of life lost by premature mortality

1
Introduction

1.1 General considerations and principles

The primary purpose of the *Guidelines for Drinking-water Quality* is the protection of public health.

Water is essential to sustain life, and a satisfactory (adequate, safe and accessible) supply must be available to all. Improving access to safe drinking-water can result in tangible benefits to health. Every effort should be made to achieve a drinking-water quality as safe as practicable.

> Diseases related to contamination of drinking-water constitute a major burden on human health. Interventions to improve the quality of drinking-water provide significant benefits to health.

Safe drinking-water, as defined by the Guidelines, does not represent any significant risk to health over a lifetime of consumption, including different sensitivities that may occur between life stages. Those at greatest risk of waterborne disease are infants and young children, people who are debilitated or living under unsanitary conditions and the elderly. Safe drinking-water is suitable for all usual domestic purposes, including personal hygiene. The Guidelines are applicable to packaged water and ice intended for human consumption. However, water of higher quality may be required for some special purposes, such as renal dialysis and cleaning of contact lenses, or for certain purposes in food production and pharmaceutical use. Those who are severely immunocompromised may need to take additional steps, such as boiling drinking-water, due to their susceptibility to organisms that would not normally be of concern through drinking-water. The Guidelines may not be suitable for the protection of aquatic life, or for some industries.

The Guidelines are intended to support the development and implementation of risk management strategies that will ensure the safety of drinking-water supplies through the control of hazardous constituents of water. These strategies may include national or regional standards developed from the scientific basis provided in the Guidelines. The Guidelines describe reasonable minimum requirements of safe practice to protect the health of consumers and/or derive numerical "guideline values" for constituents of water or indicators of water quality. Neither the minimum safe practices nor the numeric guideline values are mandatory limits. In order to define such limits, it is necessary to consider the guidelines in the context of local or national environmental, social, economic and cultural conditions.

The main reason for not promoting the adoption of international standards for drinking-water quality is the advantage provided by the use of a risk–benefit approach (qualitative or quantitative) in the establishment of national standards and regulations. Further, the Guidelines are best implemented through an integrated preventive management framework for safety applied from catchment to consumer. The Guidelines provide a scientific point of departure for national authorities to develop drinking-water regulations and standards appropriate for the national situation. In developing standards and regulations, care should be taken to ensure that scarce resources are not unnecessarily diverted to the development of standards and the monitoring of substances of relatively minor importance to public health. The approach followed in these Guidelines is intended to lead to national standards and regulations that can be readily implemented and enforced and are protective of public health.

The nature and form of drinking-water standards may vary among countries and regions. There is no single approach that is universally applicable. It is essential in the development and implementation of standards that the current and planned legislation relating to water, health and local government are taken into account and that the capacity to develop and implement regulations is assessed. Approaches that may work in one country or region will not necessarily transfer to other countries or regions. It is essential that each country review its needs and capacities in developing a regulatory framework.

The judgement of safety – or what is an acceptable level of risk in particular circumstances – is a matter in which society as a whole has a role to play. The final judgement as to whether the benefit resulting from the adoption of any of the guidelines and guideline values as national or local standards justifies the cost is for each country to decide.

Although the Guidelines describe a quality of water that is acceptable for lifelong consumption, the establishment of these Guidelines, including guideline values, should not be regarded as implying that the quality of drinking-water may be degraded to the recommended level. Indeed, a continuous effort should be made to maintain drinking-water quality at the highest possible level.

An important concept in the allocation of resources to improving drinking-water safety is that of incremental improvements towards long-term targets. Priorities set to remedy the most urgent problems (e.g., protection from pathogens; see section 1.1.1) may be linked to long-term targets of further water quality improvements (e.g., improvements in the acceptability of drinking-water; see section 1.1.5).

The basic and essential requirements to ensure the safety of drinking-water are a "framework" for safe drinking-water, comprising health-based targets established by a competent health authority; adequate and properly managed systems (adequate infrastructure, proper monitoring and effective planning and management); and a system of independent surveillance.

A holistic approach to drinking-water supply risk assessment and risk management increases confidence in the safety of drinking-water. This approach entails systematic assessment of risks throughout a drinking-water supply – from the catchment and its source water through to the consumer – and identification of the ways in which these risks can be managed, including methods to ensure that control measures are working effectively. It incorporates strategies to deal with day-to-day management of water quality, including upsets and failures.

The Guidelines are applicable to large metropolitan and small community piped drinking-water systems and to non-piped drinking-water systems in communities and in individual dwellings. The Guidelines are also applicable to a range of specific circumstances, including large buildings, travellers and conveyances.

The great majority of evident water-related health problems are the result of microbial (bacteriological, viral, protozoan or other biological) contamination. Nevertheless, an appreciable number of serious health concerns may occur as a result of the chemical contamination of drinking-water.

1.1.1 Microbial aspects

Securing the microbial safety of drinking-water supplies is based on the use of multiple barriers, from catchment to consumer, to prevent the contamination of drinking-water or to reduce contamination to levels not injurious to health. Safety is increased if multiple barriers are in place, including protection of water resources, proper selection and operation of a series of treatment steps and management of distribution systems (piped or otherwise) to maintain and protect treated water quality. The preferred strategy is a management approach that places the primary emphasis on preventing or reducing the entry of pathogens into water sources and reducing reliance on treatment processes for removal of pathogens.

> The potential health consequences of microbial contamination are such that its control must always be of paramount importance and must never be compromised.

In general terms, the greatest microbial risks are associated with ingestion of water that is contaminated with human or animal (including bird) faeces. Faeces can be a source of pathogenic bacteria, viruses, protozoa and helminths.

Faecally derived pathogens are the principal concerns in setting health-based targets for microbial safety. Microbial water quality often varies rapidly and over a wide range. Short-term peaks in pathogen concentration may increase disease risks considerably and may trigger outbreaks of waterborne disease. Furthermore, by the time microbial contamination is detected, many people may have been exposed. For these reasons, reliance cannot be placed solely on end-product testing, even when frequent, to ensure the microbial safety of drinking-water.

Particular attention should be directed to a water safety framework and implementing comprehensive water safety plans (WSPs) to consistently ensure drinking-water safety and thereby protect public health (see chapter 4). Management of microbial drinking-water safety requires a system-wide assessment to determine potential hazards that can affect the system (see section 4.1); identification of the control measures needed to reduce or eliminate the hazards, and operational monitoring to ensure that barriers within the system are functioning efficiently (see section 4.2); and the development of management plans to describe actions taken under both normal and incident conditions. These are the three components of a WSP.

Failure to ensure drinking-water safety may expose the community to the risk of outbreaks of intestinal and other infectious diseases. Drinking-water-borne outbreaks are particularly to be avoided because of their capacity to result in the simultaneous infection of a large number of persons and potentially a high proportion of the community.

In addition to faecally borne pathogens, other microbial hazards (e.g., guinea worm [*Dracunculus medinensis*], toxic cyanobacteria and *Legionella*) may be of public health importance under specific circumstances.

The infective stages of many helminths, such as parasitic roundworms and flatworms, can be transmitted to humans through drinking-water. As a single mature larva or fertilized egg can cause infection, these should be absent from drinking-water. However, the water route is relatively unimportant for helminth infection, except in the case of the guinea worm.

Legionella bacteria are ubiquitous in the environment and can proliferate at the higher temperatures experienced at times in piped drinking-water distribution systems and more commonly in hot and warm water distribution systems. Exposure to *Legionella* from drinking-water is through inhalation and can be controlled through the implementation of basic water quality management measures in buildings and through the maintenance of disinfection residuals throughout the piped distribution system.

Public health concern regarding cyanobacteria relates to their potential to produce a variety of toxins, known as "cyanotoxins." In contrast to pathogenic bacteria, cyanobacteria do not proliferate within the human body after uptake; they proliferate only in the aquatic environment before intake. While the toxic peptides (e.g., microcystins) are usually contained within the cells and thus may be largely eliminated by filtration, toxic alkaloids such as cylindrospermopsin and neurotoxins are also released into the water and may break through filtration systems.

Some microorganisms will grow as biofilms on surfaces in contact with water. With few exceptions, such as *Legionella*, most of these organisms do not cause illness in healthy persons, but they can cause nuisance through generation of tastes and odours or discoloration of drinking-water supplies. Growth following drinking-water treatment is often referred to as "regrowth." It is typically reflected in measurement of increasing heterotrophic plate counts (HPC) in water samples. Elevated HPC occur especially in stagnant parts of piped distribution systems, in domestic plumbing, in some bottled water and in plumbed-in devices such as softeners, carbon filters and vending machines.

While water can be a very significant source of infectious organisms, many of the diseases that may be waterborne may also be transmitted by other routes, including person-to-person contact, droplets and aerosols and food intake. Depending on circumstance and in the absence of waterborne outbreaks, these routes may be more important than waterborne transmission.

Microbial aspects of water quality are considered in more detail in chapter 7, with fact sheets on specific microorganisms provided in chapter 11.

1.1.2 Disinfection

Disinfection is of unquestionable importance in the supply of safe drinking-water. The destruction of microbial pathogens is essential and very commonly involves the use of reactive chemical agents such as chlorine.

Disinfection is an effective barrier to many pathogens (especially bacteria) during drinking-water treatment and should be used for surface waters and for groundwater subject to faecal contamination. Residual disinfection is used to provide a partial safeguard against low-level contamination and growth within the distribution system.

Chemical disinfection of a drinking-water supply that is faecally contaminated will reduce the overall risk of disease but may not necessarily render the supply safe. For example, chlorine disinfection of drinking-water has limitations against the protozoan pathogens – in particular *Cryptosporidium* – and some viruses. Disinfection efficacy may also be unsatisfactory against pathogens within flocs or particles, which protect them from disinfectant action. High levels of turbidity can protect microorganisms from the effects of disinfection, stimulate the growth of bacteria and give rise to a significant chlorine demand. An effective overall management strategy incorporates multiple barriers, including source water protection and appropriate treatment processes, as well as protection during storage and distribution in conjunction with disinfection to prevent or remove microbial contamination.

The use of chemical disinfectants in water treatment usually results in the formation of chemical by-products. However, the risks to health from these by-products are extremely small in comparison with the risks associated with inadequate disinfection, and it is important that disinfection not be compromised in attempting to control such by-products.

> Disinfection should not be compromised in attempting to control disinfection by-products (DBPs).

Some disinfectants such as chlorine can be easily monitored and controlled as a drinking-water disinfectant, and frequent monitoring is recommended wherever chlorination is practised.

Disinfection of drinking-water is considered in more detail in chapter 8, with fact sheets on specific disinfectants and DBPs provided in chapter 12.

1.1.3 Chemical aspects

The health concerns associated with chemical constituents of drinking-water differ from those associated with microbial contamination and arise primarily from the ability of chemical constituents to cause adverse health effects after prolonged periods of exposure. There are few chemical constituents of water that can lead to health problems resulting from a single exposure, except through massive accidental contamination of a drinking-water supply. Moreover, experience shows that in many, but not all, such incidents, the water becomes undrinkable owing to unacceptable taste, odour and appearance.

In situations where short-term exposure is not likely to lead to health impairment, it is often most effective to concentrate the available resources for remedial action on finding and eliminating the source of contamination, rather than on installing expensive drinking-water treatment for the removal of the chemical constituent.

There are many chemicals that may occur in drinking-water; however, only a few are of immediate health concern in any given circumstance. The priority given to both monitoring and remedial action for chemical contaminants in drinking-water should be managed to ensure that scarce resources are not unnecessarily directed towards those of little or no health concern.

Exposure to high levels of fluoride, which occurs naturally, can lead to mottling of teeth and, in severe cases, crippling skeletal fluorosis. Similarly, arsenic may occur naturally, and excess exposure to arsenic in drinking-water may result in a significant risk of cancer and skin lesions. Other naturally occurring chemicals, including uranium and selenium, may also give rise to health concern when they are present in excess.

The presence of nitrate and nitrite in water has been associated with methaemoglobinaemia, especially in bottle-fed infants. Nitrate may arise from the excessive application of fertilizers or from leaching of wastewater or other organic wastes into surface water and groundwater.

Particularly in areas with aggressive or acidic waters, the use of lead pipes and fittings or solder can result in elevated lead levels in drinking-water, which cause adverse neurological effects.

There are few chemicals for which the contribution from drinking-water to overall intake is an important factor in preventing disease. One example is the effect of fluoride in drinking-water in increasing prevention against dental caries. The Guidelines do not attempt to define minimum desirable concentrations for chemicals in drinking-water.

Guideline values are derived for many chemical constituents of drinking-water. A guideline value normally represents the concentration of a constituent that does not result in any significant risk to health over a lifetime of consumption. A number of provisional guideline values have been established based on the practical level of treatment achievability or analytical achievability. In these cases, the guideline value is higher than the calculated health-based value.

The chemical aspects of drinking-water quality are considered in more detail in chapter 8, with fact sheets on specific chemical contaminants provided in chapter 12.

1.1.4 Radiological aspects

The health risk associated with the presence of naturally occurring radionuclides in drinking-water should also be taken into consideration, although the contribution of drinking-water to total exposure to radionuclides is very small under normal circumstances.

Formal guideline values are not set for individual radionuclides in drinking-water. Rather, the approach used is based on screening drinking-water for gross alpha and gross beta radiation activity. While finding levels of activity above screening values does not indicate any immediate risk to health, it should trigger further investigation into determining the radionuclides responsible and the possible risks, taking into account local circumstances.

The guidance values recommended in this volume do not apply to drinking-water supplies contaminated during emergencies arising from accidental releases of radioactive substances to the environment.

Radiological aspects of drinking-water quality are considered in more detail in chapter 9.

1.1.5 Acceptability aspects

Water should be free of tastes and odours that would be objectionable to the majority of consumers.

In assessing the quality of drinking-water, consumers rely principally upon their senses. Microbial, chemical and physical water constituents may affect the appearance, odour or taste of the water, and the consumer will evaluate the quality and acceptability of the water on the basis of these criteria. Although these substances may have no direct health effects, water that is highly turbid, is highly coloured or has an objectionable taste or odour may be regarded by consumers as unsafe and may be rejected. In extreme cases, consumers may avoid aesthetically unacceptable but otherwise safe drinking-water in favour of more pleasant but potentially unsafe sources. It is therefore wise to be aware of consumer perceptions and to take into account both health-related guidelines and aesthetic criteria when assessing drinking-water supplies and developing regulations and standards.

Changes in the normal appearance, odour or taste of a drinking-water supply may signal changes in the quality of the raw water source or deficiencies in the treatment process and should be investigated.

Acceptability aspects of drinking-water quality are considered in more detail in chapter 10.

1.2 Roles and responsibilities in drinking-water safety management

Preventive management is the preferred approach to drinking-water safety and should take account of the characteristics of the drinking-water supply from catchment and source to its use by consumers. As many aspects of drinking-water quality management are often outside the direct responsibility of the water supplier, it is essential that a collaborative multiagency approach be adopted to ensure that agencies with responsibility for specific areas within the water cycle are involved in the management of water quality. One example is where catchments and source waters are beyond the drinking-water supplier's jurisdiction. Consultation with other authorities will generally be necessary for other elements of drinking-water quality management, such as monitoring and reporting requirements, emergency response plans and communication strategies.

> A preventive integrated management approach with collaboration from all relevant agencies is the preferred approach to ensuring drinking-water safety.

Major stakeholders that could affect or be affected by decisions or activities of the drinking-water supplier should be encouraged to coordinate their planning and management activities where appropriate. These could include, for example, health and resource management agencies, consumers, industry and plumbers. Appropriate mechanisms and documentation should be established for stakeholder commitment and involvement.

1.2.1 Surveillance and quality control

In order to protect public health, a dual-role approach, differentiating the roles and responsibilities of service providers from those of an authority responsible for independent oversight protective of public health ("drinking-water supply surveillance"), has proven to be effective.

Organizational arrangements for the maintenance and improvement of drinking-water supply services should take into account the vital and complementary roles of the agency responsible for surveillance and of the water supplier. The two functions of surveillance and quality control are best performed by separate and independent entities because of the conflict of interest that arises when the two are combined. In this:

— national agencies provide a framework of targets, standards and legislation to enable and require suppliers to meet defined obligations;
— agencies involved in supplying water for consumption by any means should be required to ensure and verify that the systems they administer are capable of delivering safe water and that they routinely achieve this; and
— a surveillance agency is responsible for independent (external) surveillance through periodic audit of all aspects of safety and/or verification testing.

1. INTRODUCTION

In practice, there may not always be a clear division of responsibilities between the surveillance and drinking-water supply agencies. In some cases, the range of professional, governmental, nongovernmental and private institutions may be wider and more complex than that discussed above. Whatever the existing framework, it is important that clear strategies and structures be developed for implementing water safety plans, quality control and surveillance, collating and summarizing data, reporting and disseminating the findings and taking remedial action. Clear lines of accountability and communication are essential.

> Surveillance of drinking-water quality can be defined as "the continuous and vigilant public health assessment and review of the safety and acceptability of drinking-water supplies" (WHO, 1976).

Surveillance is an investigative activity undertaken to identify and evaluate potential health risks associated with drinking-water. Surveillance contributes to the protection of public health by promoting improvement of the quality, quantity, accessibility, coverage (i.e., populations with reliable access), affordability and continuity of drinking-water supplies (termed "service indicators"). The surveillance authority must have the authority to determine whether a water supplier is fulfilling its obligations.

In most countries, the agency responsible for the surveillance of drinking-water supply services is the ministry of health (or public health) and its regional or departmental offices. In some countries, it may be an environmental protection agency; in others, the environmental health departments of local government may have some responsibility.

Surveillance requires a systematic programme of surveys, which may include auditing, analysis, sanitary inspection and/or institutional and community aspects. It should cover the whole of the drinking-water system, including sources and activities in the catchment, transmission infrastructure, treatment plants, storage reservoirs and distribution systems (whether piped or unpiped).

Ensuring timely action to prevent problems and ensure the correction of faults should be an aim of a surveillance programme. There may at times be a need for penalties to encourage and ensure compliance. The surveillance agency must therefore be supported by strong and enforceable legislation. However, it is important that the agency develops a positive and supportive relationship with suppliers, with the application of penalties used as a last resort.

> Drinking-water suppliers are responsible at all times for the quality and safety of the water that they produce.

The surveillance agency should be empowered by law to compel water suppliers to recommend the boiling of water or other measures when microbial contamination that could threaten public health is detected.

1.2.2 Public health authorities

In order to effectively support the protection of public health, a national entity with responsibility for public health will normally act in four areas:

- *Surveillance of health status and trends*, including outbreak detection and investigation, generally directly but in some instances through a decentralized body.
- Directly establish drinking-water *norms and standards*. National public health authorities often have the primary responsibility for setting norms on drinking-water supply, which may include the setting of water quality targets (WQTs), performance and safety targets and directly specified requirements (e.g., treatment). Normative activity is not restricted to water quality but also includes, for example, regulation and approval of materials and chemicals used in the production and distribution of drinking-water (see section 8.5.4) and establishing minimum standards in areas such as domestic plumbing (see section 1.2.10). Nor is it a static activity, because as changes occur in drinking-water supply practice, in technologies and in materials available (e.g., in plumbing materials and treatment processes), so health priorities and responses to them will also change.
- Representing health concerns in *wider policy development*, especially health policy and integrated water resource management (see section 1.2.4). Health concerns will often suggest a supportive role towards resource allocation to those concerned with drinking-water supply extension and improvement; will often involve lobbying for the primary requirement to satisfy drinking-water needs above other priorities; and may imply involvement in conflict resolution.
- *Direct action*, generally through subsidiary bodies (e.g., regional and local environmental health administrations) or by providing guidance to other local entities (e.g., local government) in surveillance of drinking-water supplies. These roles vary widely according to national and local structures and responsibilities and frequently include a supportive role to community suppliers, where local authorities often intervene directly.

Public health surveillance (i.e., surveillance of health status and trends) contributes to verifying drinking-water safety. It takes into consideration disease in the entire population, which may be exposed to pathogenic microorganisms from a range of sources, not only drinking-water. National public health authorities may also undertake or direct research to evaluate the role of water as a risk factor in disease – for example, through case–control, cohort or intervention studies. Public health surveillance teams typically operate at national, regional and local levels, as well as in cities and rural health centres. Routine public health surveillance includes:

— ongoing monitoring of reportable diseases, many of which can be caused by waterborne pathogens;
— outbreak detection;
— long-term trend analysis;

— geographic and demographic analysis; and
— feedback to water authorities.

Public health surveillance can be enhanced in a variety of ways to identify possible waterborne outbreaks in response to suspicion about unusual disease incidence or following deterioration of water quality. Epidemiological investigations include:

— outbreak investigations;
— intervention studies to evaluate intervention options; and
— case–control or cohort studies to evaluate the role of water as a risk factor in disease.

However, public health surveillance cannot be relied upon to provide information in a timely manner to enable short-term operational response to control waterborne disease. Limitations include:

— outbreaks of non-reportable disease;
— time delay between exposure and illness;
— time delay between illness and reporting;
— low level of reporting; and
— difficulties in identifying causative pathogens and sources.

The public health authority operates reactively, as well as proactively, against the background of overall public health policy and in interaction with all stakeholders. In accounting for public health context, priority will normally be afforded to disadvantaged groups. This will generally entail balancing drinking-water safety management and improvement with the need to ensure access to reliable supplies of safe drinking-water in adequate quantities.

In order to develop an understanding of the national drinking-water situation, the national public health authority should periodically produce reports outlining the state of national water quality and highlighting public health concerns and priorities in the context of overall public health priorities. This implies the need for effective exchange of information between local, regional and national agencies.

National health authorities should lead or participate in formulation and implementation of policy to ensure access to some form of reliable, safe drinking-water supply. Where this has not been achieved, appropriate tools and education should be made available to implement individual or household-level treatment and safe storage.

1.2.3 Local authorities

Local environmental health authorities often play an important role in managing water resources and drinking-water supplies. This may include catchment inspection and authorization of activities in the catchment that may impact on source water quality. It can also include verifying and auditing (surveillance) of the management of formal drinking-water systems. Local environmental health authorities will also give specific guidance to communities or individuals in designing and implementing community and household drinking-water systems and correcting deficiencies, and they may also be responsible for surveillance of community and household drinking-water supplies. They have an important role to play in educating consumers where household water treatment is necessary.

Management of household and small community drinking-water supplies generally requires education programmes about drinking-water supply and water quality. Such programmes should normally include:

— water hygiene awareness raising;
— basic technical training and technology transfer in drinking-water supply and management;
— consideration of and approaches to overcoming sociocultural barriers to acceptance of water quality interventions;
— motivation, mobilization and social marketing activities; and
— a system of continued support, follow-up and dissemination of the water quality programme to achieve and maintain sustainability.

These programmes can be administered at the community level by local health authorities or other entities, such as nongovernmental organizations and the private sector. If the programme arises from other entities, the involvement of the local health authority in the development and implementation of the water quality education and training programme is strongly encouraged.

Approaches to participatory hygiene and sanitation education and training programmes are described in other WHO documents (see Simpson-Hébert et al., 1996; Sawyer et al., 1998; Brikké, 2000).

1.2.4 Water resource management

Water resource management is an integral aspect of the preventive management of drinking-water quality. Prevention of microbial and chemical contamination of source water is the first barrier against drinking-water contamination of public health concern.

Water resource management and potentially polluting human activity in the catchment will influence water quality downstream and in aquifers. This will impact on treatment steps required to ensure safe water, and preventive action may be preferable to upgrading treatment.

The influence of land use on water quality should be assessed as part of water resource management. This assessment is not normally undertaken by health authorities or drinking-water supply agencies alone and should take into consideration:

— land cover modification;
— extraction activities;
— construction/modification of waterways;
— application of fertilizers, herbicides, pesticides and other chemicals;
— livestock density and application of manure;

— road construction, maintenance and use;
— various forms of recreation;
— urban or rural residential development, with particular attention to excreta disposal, sanitation, landfill and waste disposal; and
— other potentially polluting human activities, such as industry, military sites, etc.

Water resource management may be the responsibility of catchment management agencies and/or other entities controlling or affecting water resources, such as industrial, agricultural, navigation and flood control entities.

The extent to which the responsibilities of health or drinking-water supply agencies include water resource management varies greatly between countries and communities. Regardless of government structures and sector responsibilities, it is important that health authorities liaise and collaborate with sectors managing the water resource and regulating land use in the catchment.

Establishing close collaboration between the public health authority, water supplier and resource management agency assists recognition of the health hazards potentially occurring in the system. It is also important for ensuring that the protection of drinking-water resources is considered in decisions for land use or regulations to control contamination of water resources. Depending on the setting, this may include involvement of further sectors, such as agriculture, traffic, tourism or urban development.

To ensure the adequate protection of drinking-water sources, national authorities will normally interact with other sectors in formulating national policy for integrated water resource management. Regional and local structures for implementing the policy will be set up, and national authorities will guide regional and local authorities by providing tools.

Regional environmental or public health authorities have an important task in participating in the preparation of integrated water resource management plans to ensure the best available drinking-water source quality. For further information, see the supporting documents *Protecting Surface Waters for Health* and *Protecting Groundwaters for Health* (section 1.3).

1.2.5 Drinking-water supply agencies

Drinking-water supplies vary from very large urban systems servicing populations with tens of millions to small community systems providing water to very small populations. In most countries, they include community sources as well as piped means of supply.

Drinking-water supply agencies are responsible for quality assurance and quality control (see section 1.2.1). Their key responsibilities are to prepare and implement WSPs (for more information, see chapter 4).

In many cases, the water supplier is not responsible for the management of the catchment feeding sources of its supplies. The roles of the water supplier with respect to catchments are to participate in interagency water resource management activities; to understand the risks arising from potentially contaminating activities and incidents; and to use this information in assessing risks to the drinking-water supply and developing and applying appropriate management. Although drinking-water suppliers may not undertake catchment surveys and pollution risk assessment alone, their roles include recognizing the need for them and initiating multiagency collaboration – for example, with health and environmental authorities.

Experience has shown that an association of stakeholders in drinking-water supply (e.g., operators, managers and specialist groups such as small suppliers, scientists, sociologists, legislators, politicians, etc.) can provide a valuable non-threatening forum for interchange of ideas.

For further information, see the supporting document *Water Safety Plans* (section 1.3).

1.2.6 Community management

Community-managed drinking-water systems, with both piped and non-piped distribution, are common worldwide in both developed and developing countries. The precise definition of a community drinking-water system will vary. While a definition based on population size or the type of supply may be appropriate under many conditions, approaches to administration and management provide a distinction between the drinking-water systems of small communities and those of larger towns and cities. This includes the increased reliance on often untrained and sometimes unpaid community members in the administration and operation of community drinking-water systems. Drinking-water systems in periurban areas in developing countries – the communities surrounding major towns and cities – may also have the characteristics of community systems.

Effective and sustainable programmes for the management of community drinking-water quality require the active support and involvement of local communities. These communities should be involved at all stages of such programmes, including initial surveys; decisions on siting of wells, siting of off-takes or establishing protection zones; monitoring and surveillance of drinking-water supplies; reporting faults, carrying out maintenance and taking remedial action; and supportive actions, including sanitation and hygiene practices.

A community may already be highly organized and taking action on health or drinking-water supply issues. Alternatively, it may lack a well developed drinking-water system; some sectors of the community, such as women, may be poorly represented; and there may be disagreements or factional conflicts. In this situation, achieving community participation will take more time and effort to bring people together, resolve differences, agree on common aims and take action. Visits, possibly over several years, will often be needed to provide support and encouragement and to ensure that the structures created for safe drinking-water supply continue to operate. This may involve setting up hygiene and health educational programmes to ensure that the community:

— is aware of the importance of drinking-water quality and its relation to health and of the need for safe drinking-water in sufficient quantities for domestic use for drinking, cooking and hygiene;
— recognizes the importance of surveillance and the need for a community response;
— understands and is prepared to play its role in the surveillance process;
— has the necessary skills to perform that role; and
— is aware of requirements for the protection of drinking-water supplies from pollution.

For further information, see WHO *Guidelines for Drinking-water Quality*, second edition, Volume 3; the supporting document *Water Safety Plans* (section 1.3); Simpson-Hébert et al. (1996); Sawyer et al. (1998); and Brikké (2000).

1.2.7 Water vendors

Vendors selling water to households or at collection points are common in many parts of the world where scarcity of water or faults in or lack of infrastructure limits access to suitable quantities of drinking-water. Water vendors use a range of modes of transport to carry drinking-water for sale directly to the consumer, including tanker trucks and wheelbarrows/trolleys. In the context of these Guidelines, water vending does not include bottled or packaged water (which is considered in section 6.5) or water sold through vending machines.

There are a number of health concerns associated with water supplied to consumers by water vendors. These include access to adequate volumes and concern regarding inadequate treatment or transport in inappropriate containers, which can result in contamination.

Where the source of water is uncertain or the quality of the water is unknown, water can be treated or re-treated in small quantities to significantly improve its quality and safety. The simplest and most important treatment for microbially contaminated water is disinfection. If bulk supplies in tankers are used, sufficient chlorine should be added to ensure that a free residual chlorine concentration of at least 0.5 mg/litre after a contact time of at least 30 min is present at the delivery point. Tankers should normally be reserved for potable water use. Before use, tankers should be either chemically disinfected or steam-cleaned.

Local authorities should implement surveillance programmes for water provided by vendors and, where necessary, develop education programmes to improve the collection, treatment and distribution of water to prevent contamination.

1.2.8 Individual consumers

Everyone consumes water from one source or another, and consumers often play important roles in the collection, treatment and storage of water. Consumer actions may help to ensure the safety of the water they consume and may also contribute to improvement or contamination of the water consumed by others. Consumers have the responsibility for ensuring that their actions do not impact adversely on water quality. Installation and maintenance of household plumbing systems should be undertaken preferably by qualified and authorized plumbers (see section 1.2.10) or other persons with appropriate expertise to ensure that cross-connection or backflow events do not result in contamination of local water supplies.

In most countries, there are populations whose water is derived from household sources, such as private wells and rainwater. In households using non-piped water supplies, appropriate efforts are needed to ensure safe collection, storage and perhaps treatment of their drinking-water. In some circumstances, households and individuals may wish to treat water in the home to increase their confidence in its safety, not only where community supplies are absent, but also where community supplies are known to be contaminated or causing waterborne disease (see chapter 7). Public health, surveillance and/or other local authorities may provide guidance to support households and individual consumers in ensuring the safety of their drinking-water (see section 6.3). Such guidance is best provided in the context of a community education and training programme.

1.2.9 Certification agencies

Certification is used to verify that devices and materials used in the drinking-water supply meet a given level of quality and safety. Certification is a process in which an independent organization validates the claims of the manufacturers against a formal standard or criterion or provides an independent assessment of possible risks of contamination from a material or process. The certification agency may be responsible for seeking data from manufacturers, generating test results, conducting inspections and audits and possibly making recommendations on product performance.

Certification has been applied to technologies used at household and community levels, such as hand pumps; materials used by water supplies, such as treatment chemicals; and devices used in the household for collection, treatment and storage.

Certification of products or processes involved in the collection, treatment, storage and distribution of water can be overseen by government agencies or private organizations. Certification procedures will depend on the standards against which the products are certified, certification criteria and the party that performs the certification.

National, local government or private (third-party auditing) certification programmes have a number of possible objectives:

— certification of products to ensure that their use does not threaten the safety of the user or the general public, such as by causing contamination of drinking-water with toxic substances, substances that could affect consumer acceptability or substances that support the growth of microorganisms;
— product testing, to avoid retesting at local levels or prior to each procurement;

1. INTRODUCTION

— ensuring uniform quality and condition of products;
— certification and accreditation of analytical and other testing laboratories; and
— control of materials and chemicals used for the treatment of drinking-water, including the performance of devices for household use.

An important step in any certification procedure is the establishment of standards, which must form the basis of assessment of the products. These standards should also – as far as possible – contain the criteria for approval. In procedures for certification on technical aspects, these standards are generally developed in cooperation with the manufacturers, the certifying agency and the consumers. The national public health authorities should have responsibility for developing the parts of the approval process or criteria relating directly to public health. For further information, see section 8.5.4.

1.2.10 Plumbing

Significant adverse health effects have been associated with inadequate plumbing systems within public and private buildings arising from poor design, incorrect installation, alterations and inadequate maintenance.

Numerous factors influence the quality of water within a building's piped distribution system and may result in microbial or chemical contamination of drinking-water. Outbreaks of gastrointestinal disease can occur through faecal contamination of drinking-water within buildings arising from deficiencies in roof storage tanks and cross-connections with wastewater pipes, for example. Poorly designed plumbing systems can cause stagnation of water and provide a suitable environment for the proliferation of *Legionella*. Plumbing materials, pipes, fittings and coatings can result in elevated heavy metal (e.g., lead) concentrations in drinking-water, and inappropriate materials can be conducive to bacterial growth. Potential adverse health effects may not be confined to the individual building. Exposure of other consumers to contaminants is possible through contamination of the local public distribution system, beyond the particular building, through cross-contamination of drinking-water and backflow.

The delivery of water that complies with relevant standards within buildings generally relies on a plumbing system that is not directly managed by the water supplier. Reliance is therefore placed on proper installation and servicing of plumbing and, for larger buildings, on building-specific WSPs (see section 6.1).

To ensure the safety of drinking-water supplies within the building system, plumbing practices must prevent the introduction of hazards to health. This can be achieved by ensuring that:

— pipes carrying either water or wastes are watertight, durable, of smooth and unobstructed interior and protected against anticipated stresses;
— cross-connections between the drinking-water supply and the wastewater removal systems do not occur;

— water storage systems are intact and not subject to intrusion of microbial and chemical contaminants;
— hot and cold water systems are designed to minimize the proliferation of *Legionella* (see also sections 6.1 and 11.1.9);
— appropriate protection is in place to prevent backflow;
— the system design of multistorey buildings minimizes pressure fluctuations;
— waste is discharged without contaminating drinking-water; and
— plumbing systems function efficiently.

It is important that plumbers are appropriately qualified, have the competence to undertake necessary installation and servicing of plumbing systems to ensure compliance with local regulations and use only materials approved as safe for use with drinking-water.

Design of the plumbing systems of new buildings should normally be approved prior to construction and be inspected by an appropriate regulatory body during construction and prior to commissioning of the buildings.

1.3 Supporting documentation to the guidelines

These Guidelines are accompanied by separate texts that provide background information substantiating the derivation of the guidelines and providing guidance on good practice towards effective implementation. These are available as published texts and electronically through the Internet (http://www.who.int/water_sanitation_health/dwq/en/) and CD-ROM. Reference details are provided in Annex 1.

Assessing Microbial Safety of Drinking Water: Improving Approaches and Methods
This book provides a state-of-the-art review of approaches and methods used in assessing the microbial safety of drinking-water. It offers guidance on the selection and use of indicators alongside operational monitoring to meet specific information needs and looks at potential applications of "new" technologies and emerging methods.

Chemical Safety of Drinking-water: Assessing Priorities for Risk Management
This document provides tools that assist users to undertake a systematic assessment of their water supply system(s) locally, regionally or nationally; to prioritize the chemicals likely to be of greatest significance; to consider how these might be controlled or eliminated; and to review or develop standards that are appropriate.

Domestic Water Quantity, Service Level and Health
This paper reviews the requirements for water for health-related purposes to determine acceptable minimum needs for consumption (hydration and food preparation) and basic hygiene.

1. INTRODUCTION

Evaluation of the H$_2$S Method for Detection of Fecal Contamination of Drinking Water
This report critically reviews the scientific basis, validity, available data and other information concerning the use of "H$_2$S tests" as measures or indicators of faecal contamination in drinking-water.

Hazard Characterization for Pathogens in Food and Water: Guidelines
This document provides a practical framework and structured approach for the characterization of microbial hazards, to assist governmental and research scientists.

Heterotrophic Plate Counts and Drinking-water Safety: The Significance of HPCs for Water Quality and Human Health
This document provides a critical assessment of the role of the HPC measurement in drinking-water safety management.

Managing Water in the Home: Accelerated Health Gains from Improved Water Supply
This report describes and critically reviews the various methods and systems for household water collection, treatment and storage. It assesses the ability of household water treatment and storage methods to provide water with improved microbial quality.

Pathogenic Mycobacteria in Water: A Guide to Public Health Consequences, Monitoring and Management
This book describes the current knowledge about the distribution of pathogenic environmental mycobacteria (PEM) in water and other parts of the environment. Included are discussions of the routes of transmission that lead to human infection, the most significant disease symptoms that can follow infection and the classical and modern methods of analysis of PEM species. The book concludes with a discussion of the issues surrounding the control of PEM in drinking-water and the assessment and management of risks.

Quantifying Public Health Risk in the WHO Guidelines for Drinking-water Quality: A Burden of Disease Approach
This report provides a discussion paper on the concepts and methodology of Disability Adjusted Life Years (DALYs) as a common public health metric and its usefulness for drinking-water quality and illustrates the approach for several drinking-water contaminants already examined using the burden of disease approach.

Safe Piped Water: Managing Microbial Water Quality in Piped Distribution Systems
The development of pressurized pipe networks for supplying drinking-water to individual dwellings, buildings and communal taps is an important component in

the continuing development and health of many communities. This publication considers the introduction of microbial contaminants and growth of microorganisms in distribution networks and the practices that contribute to ensuring drinking-water safety in piped distribution systems.

Toxic Cyanobacteria in Water: A Guide to their Public Health Consequences, Monitoring and Management
This book describes the state of knowledge regarding the impact of cyanobacteria on health through the use of water. It considers aspects of risk management and details the information needed for protecting drinking-water sources and recreational water bodies from the health hazards caused by cyanobacteria and their toxins. It also outlines the state of knowledge regarding the principal considerations in the design of programmes and studies for monitoring water resources and supplies and describes the approaches and procedures used.

Upgrading Water Treatment Plants
This book provides a practical guide to improving the performance of water treatment plants. It will be an invaluable source of information for those who are responsible for designing, operating, maintaining or upgrading water treatment plants.

Water Safety Plans
The improvement of water quality control strategies, in conjunction with improvements in excreta disposal and personal hygiene, can be expected to deliver substantial health gains in the population. This document provides information on improved strategies for the control and monitoring of drinking-water quality.

Water Treatment and Pathogen Control: Process Efficiency in Achieving Safe Drinking-water
This publication provides a critical analysis of the literature on removal and inactivation of pathogenic microbes in water to aid the water quality specialist and design engineer in making decisions regarding microbial water quality.

Texts in preparation or in revision:
Arsenic in Drinking-water: Assessing and managing health risks (in preparation)
Desalination for Safe Drinking-water Supply (in preparation)
Guide to Hygiene and Sanitation in Aviation (in revision)
Guide to Ship Sanitation (in revision)
Health Aspects of Plumbing (in preparation)
Legionella and the Prevention of Legionellosis (in finalization)
Protecting Groundwaters for Health – Managing the Quality of Drinking-water Sources (in preparation)

1. INTRODUCTION

Protecting Surface Waters for Health – Managing the Quality of Drinking-water Sources (in preparation)

Rapid Assessment of Drinking-water Quality: A Handbook for Implementation (in preparation)

2
The Guidelines: a framework for safe drinking-water

The quality of drinking-water may be controlled through a combination of protection of water sources, control of treatment processes and management of the distribution and handling of the water. Guidelines must be appropriate for national, regional and local circumstances, which requires adaptation to environmental, social, economic and cultural circumstances and priority setting.

2.1 Framework for safe drinking-water: requirements

The Guidelines outline a preventive management "framework for safe drinking-water" that comprises five key components:

— health-based targets based on an evaluation of health concerns (chapter 3);
— system assessment to determine whether the drinking-water supply (from source through treatment to the point of consumption) as a whole can deliver water that meets the health-based targets (section 4.1);
— operational monitoring of the control measures in the drinking-water supply that are of particular importance in securing drinking-water safety (section 4.2);
— management plans documenting the system assessment and monitoring plans and describing actions to be taken in normal operation and incident conditions, including upgrade and improvement, documentation and communication (sections 4.4–4.6); and
— a system of independent surveillance that verifies that the above are operating properly (chapter 5).

In support of the framework for safe drinking-water, the Guidelines provide a range of supporting information, including microbial aspects (chapters 7 and 11), chemical aspects (chapters 8 and 12), radiological aspects (chapter 9) and acceptability aspects (chapter 10). Figure 2.1 provides an overview of the interrelationship of the individual chapters of the Guidelines in ensuring drinking-water safety.

There is a wide range of microbial and chemical constituents of drinking-water that can cause adverse human health effects. The detection of these constituents in both raw water and water delivered to consumers is often slow, complex and costly,

2. THE GUIDELINES: A FRAMEWORK FOR SAFE DRINKING-WATER

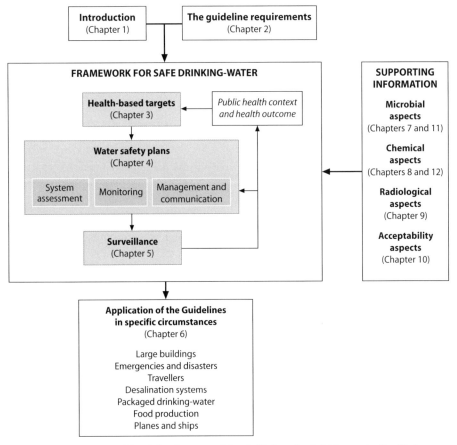

Figure 2.1 Interrelationship of the chapters of the *Guidelines for Drinking-water Quality* in ensuring drinking-water safety

which limits early warning capability and affordability. Reliance on water quality determination alone is insufficient to protect public health. As it is neither physically nor economically feasible to test for all drinking-water quality parameters, the use of monitoring effort and resources should be carefully planned and directed at significant or key characteristics.

Some characteristics not related to health, such as those with significant impacts on acceptability of water, may also be of importance. Where water has unacceptable aesthetic characteristics (e.g., appearance, taste and odour), further investigation may be required to determine whether there are problems with significance for health.

The control of the microbial and chemical quality of drinking-water requires the development of management plans, which, when implemented, provide the basis for system protection and process control to ensure that numbers of pathogens and concentrations of chemicals present a negligible risk to public health and that water is acceptable to consumers. The management plans developed by water suppliers are best termed "water safety plans" (WSPs). A WSP comprises system assessment and design, operational monitoring and management plans, including documentation and communication. The elements of a WSP build on the multiple-barrier principle, the principles of hazard analysis and critical control points (HACCP) and other systematic management approaches. The plans should address all aspects of the drinking-water supply and focus on the control of abstraction, treatment and delivery of drinking-water.

Many drinking-water supplies provide adequate safe drinking-water in the absence of formalized WSPs. Major benefits of developing and implementing a WSP for these supplies include the systematic and detailed assessment and prioritization of hazards and the operational monitoring of barriers or control measures. In addition, a WSP provides for an organized and structured system to minimize the chance of failure through oversight or lapse of management and for contingency plans to respond to system failures or unforeseen hazardous events.

2.1.1 Health-based targets

Health-based targets are an essential component of the drinking-water safety framework. They should be established by a high-level authority responsible for health in consultation with others, including water suppliers and affected communities. They should take account of the overall public health situation and contribution of drinking-water quality to disease due to waterborne microbes and chemicals, as a part of overall water and health policy. They must also take account of the importance of ensuring access to water, especially among those who are not served.

Health-based targets provide the basis for the application of the Guidelines to all types of drinking-water supply. Constituents of drinking-water may cause adverse health effects from single exposures (e.g., microbial pathogens) or long-term exposures (e.g., many chemicals). Due to the range of constituents in water, their mode of action and the nature of fluctuations in their concentration, there are four principal types of health-based targets used as a basis for identifying safety requirements:

- *Health outcome targets:* In some circumstances, especially where waterborne disease contributes to a measurable burden, reducing exposure through drinking-water has the potential to appreciably reduce overall risks of disease. In such circumstances, it is possible to establish a health-based target in terms of a quantifiable reduction in the overall level of disease. This is most applicable where adverse effects follow shortly after exposure, where such effects are readily and reliably monitored and where changes in exposure can also be readily and reliably monitored. This type of health outcome target is primarily applicable to some microbial hazards in developing countries and chemical hazards with clearly defined health effects largely attributable to water (e.g., fluoride). In other circumstances, health outcome targets may be the basis for evaluation of results through quantitative risk assessment models. In these cases, health outcomes are estimated based on information concerning exposure and dose–response relationships. The results may be employed directly as a basis for the specification of water quality targets or provide the basis for development of the other types of health-based targets. Health outcome targets based on information on the impact of tested interventions on the health of real populations are ideal but rarely available. More common are health outcome targets based on defined levels of tolerable risk, either absolute or fractions of total disease burden, preferably based on epidemiological evidence or, alternatively, risk assessment studies.

- *Water quality targets (WQTs):* WQTs are established for individual drinking-water constituents that represent a health risk from long-term exposure and where fluctuations in concentration are small or occur over long periods. They are typically expressed as guideline values (concentrations) of the substances or chemicals of concern.
- *Performance targets:* Performance targets are employed for constituents where short-term exposure represents a public health risk or where large fluctuations in numbers or concentration can occur over short periods with significant health implications. They are typically expressed in terms of required reductions of the substance of concern or effectiveness in preventing contamination.
- *Specified technology targets:* National regulatory agencies may establish targets for specific actions for smaller municipal, community and household drinking-water supplies. Such targets may identify specific permissible devices or processes for given situations and/or for generic drinking-water system types.

It is important that health-based targets are realistic under local operating conditions and are set to protect and improve public health. Health-based targets underpin development of WSPs, provide information with which to evaluate the adequacy of existing installations and assist in identifying the level and type of inspection and analytical verifications that are appropriate.

Most countries apply several types of targets for different types of supply and different contaminants. In order to ensure that they are relevant and supportive, representative scenarios should be developed, including description of assumptions, management options, control measures and indicator systems for verification, where appropriate. These should be supported by general guidance addressing the identification of national, regional or local priorities and progressive implementation, thereby helping to ensure that best use is made of available resources.

Health-based targets are considered in more detail in chapter 3.

2.1.2 System assessment and design

Assessment of the drinking-water system is equally applicable to large utilities with piped distribution systems, piped and non-piped community supplies, including hand pumps, and individual domestic supplies. Assessment can be of existing infrastructure or of plans for new supplies or for upgrading of existing supplies. As drinking-water quality varies throughout the system, the assessment should aim to determine whether the final quality of water delivered to the consumer will routinely meet established health-based targets. Understanding source quality and changes through the system requires expert input. The assessment of systems should be reviewed periodically.

The system assessment needs to take into consideration the behaviour of selected constituents or groups of constituents that may influence water quality. Having identified and documented actual and potential hazards, including potentially hazardous events and scenarios that may affect water quality, the level of risk for each hazard can then be estimated and ranked, based on the likelihood and severity of the consequences.

Validation is an element of system assessment. It is undertaken to ensure that the information supporting the plan is correct and is concerned with the assessment of the scientific and technical inputs into the WSP. Evidence to support the WSP can come from a wide variety of sources, including scientific literature, trade associations, regulation and legislation departments, historical data, professional bodies and supplier knowledge.

If the system is theoretically capable of meeting the health-based targets, the WSP is the management tool that will assist in actually meeting the health-based targets, and it should be developed following the steps outlined in subsequent sections. If the system is unlikely to be capable of meeting the health-based targets, a programme of upgrading (which may include capital investment or training) should be initiated to ensure that the drinking-water supply would meet the targets. In the interim, every effort should be made to supply water of the highest achievable quality. Where a significant risk to public health exists, additional measures may be appropriate.

Assessment and design are considered in more detail in section 4.1 (see also the supporting document *Upgrading Water Treatment Plants*; section 1.3).

2.1.3 Operational monitoring

Control measures are actions implemented in the drinking-water system that prevent, reduce or eliminate contamination and are identified in system assessment. They include, for example, catchment management actions, the plinth surrounding a well, filters and disinfection infrastructure and piped distribution systems. If collectively operating properly, they would ensure that health-based targets are met.

Operational monitoring is the conduct of planned observations or measurements to assess whether the control measures in a drinking-water system are operating properly. It is possible to set limits for control measures, monitor those limits and take corrective action in response to a detected deviation before the water becomes unsafe. Examples of limits are that the plinth surrounding a hand pump is complete and not damaged, the turbidity of water following filtration is below a certain value or the chlorine residual after disinfection plants or at the far point of the distribution system is above an agreed value.

The frequency of operational monitoring varies with the nature of the control measure – for example, checking plinth integrity monthly to yearly, monitoring turbidity on-line or very frequently and monitoring disinfection residual at multiple points daily or continuously on-line. If monitoring shows that a limit does not meet specifications, then there is the potential for water to be, or to become, unsafe. The objective is timely monitoring of control measures, with a logically based sampling plan, to prevent the delivery of potentially unsafe water.

In most cases, operational monitoring will be based on simple and rapid observations or tests, such as turbidity or structural integrity, rather than complex microbial or chemical tests. The complex tests are generally applied as part of validation and verification activities (discussed in sections 4.1.7 and 4.3, respectively) rather than as part of operational monitoring.

In order not only to have confidence that the chain of supply is operating properly, but to confirm that water quality is being maintained and achieved, it is necessary to carry out verification, as outlined in section 2.2.

The use of indicator bacteria in monitoring of water quality is discussed in the supporting document *Assessing Microbial Safety of Drinking Water* (section 1.3), and operational monitoring is considered in more detail in section 4.2.

2.1.4 *Management plans, documentation and communication*

A management plan documents system assessment and operational monitoring and verification plans and describes actions in both normal operation and during "incidents" where a loss of control of the system may occur. The management plan should also outline procedures and other supporting programmes required to ensure optimal operation of the drinking-water system.

As the management of some aspects of the drinking-water system often falls outside the responsibility of a single agency, it is essential that the roles, accountabilities and responsibilities of the various agencies involved be defined in order to coordinate their planning and management. Appropriate mechanisms and documentation should therefore be established for ensuring stakeholder involvement and commitment. This may include establishing working groups, committees or task forces, with appropriate representatives, and developing partnership agreements, including for example signed memoranda of understanding (see also section 1.2).

Documentation of all aspects of drinking-water quality management is essential. Documents should describe activities that are undertaken and how procedures are performed. They should also include detailed information on:

— assessment of the drinking-water system (including flow diagrams and potential hazards and the outcome of validation);
— control measures and operational monitoring and verification plan;
— routine operation and management procedures;

— incident and emergency response plans; and
— supporting measures, including:
   — training programmes
   — research and development
   — procedures for evaluating results and reporting
   — performance evaluations, audits and reviews
   — communication protocols
   — community consultation.

Documentation and record systems should be kept as simple and focused as possible. The level of detail in the documentation of procedures should be sufficient to provide assurance of operational control when coupled with a suitably qualified and competent operator.

Mechanisms should be established to periodically review and, where necessary, revise documents to reflect changing circumstances. Documents should be assembled in a manner that will enable any necessary modifications to be made easily. A document control system should be developed to ensure that current versions are in use and obsolete documents are discarded.

Appropriate documentation and reporting of incidents or emergencies should also be established. The organization should learn as much as possible from an incident to improve preparedness and planning for future events. Review of an incident may indicate necessary amendments to existing protocols.

Effective communication to increase community awareness and knowledge of drinking-water quality issues and the various areas of responsibility helps consumers to understand and contribute to decisions about the service provided by a drinking-water supplier or land use constraints imposed in catchment areas. A thorough understanding of the diversity of views held by individuals or groups in the community is necessary to satisfy community expectations.

Management, documentation and communication are considered in more detail in sections 4.4, 4.5 and 4.6.

2.1.5 Surveillance of drinking-water quality

The surveillance agency is responsible for an independent (external) and periodic review of all aspects of safety, whereas the water supplier is responsible at all times for regular quality control, for operational monitoring and for ensuring good operating practice.

Surveillance contributes to the protection of public health by assessing compliance with WSPs and promoting improvement of the quality, quantity, accessibility, coverage, affordability and continuity of drinking-water supplies.

Surveillance requires a systematic programme of surveys that may include auditing of WSPs, analysis, sanitary inspection and institutional and community aspects. It should cover the whole of the drinking-water system, including sources and activities in the catchment, transmission infrastructure, whether piped or unpiped, treatment plants, storage reservoirs and distribution systems.

Since incremental improvement and prioritizing action in systems presenting greatest overall risk to public health are important, there are advantages to adopting a grading scheme for the relative safety of drinking-water supplies (see chapter 4). More sophisticated grading schemes may be of particular use in community supplies where the frequency of testing is low and exclusive reliance on analytical results is particularly inappropriate. Such schemes will typically take account of both analytical findings and sanitary inspection through approaches such as those presented in section 4.1.2.

The role of surveillance is discussed in section 1.2.1 and chapter 5.

2.2 Guidelines for verification

Drinking-water safety is secured by application of a WSP, which includes monitoring the efficiency of control measures using appropriately selected determinants. In addition to this operational monitoring, a final verification of quality is required.

Verification is the use of methods, procedures or tests in addition to those used in operational monitoring to determine if the performance of the drinking-water supply is in compliance with the stated objectives outlined by the health based targets and/or whether the WSP needs modification and revalidation.

2.2.1 Microbial water quality

For microbial water quality, verification is likely to include microbiological testing. In most cases, it will involve the analysis of faecal indicator microorganisms, but in some circumstances it may also include assessment of specific pathogen densities. Verification of the microbial quality of drinking-water may be undertaken by the supplier, surveillance agencies or a combination of the two (see sections 4.3.1 and 7.4).

Approaches to verification include testing of source water, water immediately after treatment, water in distribution systems or stored household water. Verification of the microbial quality of drinking-water includes testing for *Escherichia coli* as an indicator of faecal pollution. *E. coli* provides conclusive evidence of recent faecal pollution and should not be present in drinking-water. In practice, testing for thermotolerant coliform bacteria can be an acceptable alternative in many circumstances. While *E. coli* is a useful indicator, it has limitations. Enteric viruses and protozoa are more resistant to disinfection; consequently, the absence of *E. coli* will not necessarily indicate freedom from these organisms. Under certain circumstances, it may be desirable to include more resistant microorganisms, such as bacteriophages and/or bacterial spores. Such circumstances could include the use of source water known to be contaminated with enteric viruses and parasites or high levels of viral and parasitic diseases in the community.

Water quality can vary rapidly, and all systems are subject to occasional failure. For example, rainfall can greatly increase the levels of microbial contamination in source waters, and waterborne outbreaks often occur following rainfall. Results of analytical testing must be interpreted taking this into account.

2.2.2 Chemical water quality

Assessment of the adequacy of the chemical quality of drinking-water relies on comparison of the results of water quality analysis with guideline values.

For additives (i.e., chemicals deriving primarily from materials and chemicals used in the production and distribution of drinking-water), emphasis is placed on the direct control of the quality of these products. In controlling drinking-water additives, testing procedures typically assess the contribution of the additive to drinking-water and take account of variations over time in deriving a value that can be compared with the guideline value (see section 8.5.4).

As indicated in chapter 1, most chemicals are of concern only with long-term exposure; however, some hazardous chemicals that occur in drinking-water are of concern because of effects arising from sequences of exposures over a short period. Where the concentration of the chemical of interest varies widely, even a series of analytical results may fail to fully identify and describe the public health risk (e.g., nitrate, which is associated with methaemoglobinaemia in bottle-fed infants). In controlling such hazards, attention must be given to both knowledge of causal factors such as fertilizer use in agriculture and trends in detected concentrations, since these will indicate whether a significant problem may arise in the future. Other hazards may arise intermittently, often associated with seasonal activity or seasonal conditions. One example is the occurrence of blooms of toxic cyanobacteria in surface water.

A *guideline value* represents the concentration of a constituent that does not exceed tolerable risk to the health of the consumer over a lifetime of consumption. Guidelines for some chemical contaminants (e.g., lead, nitrate) are set to be protective for susceptible subpopulations. These guidelines are also protective of the general population over a lifetime.

The exceedance of a guideline value does not necessarily result in a significant risk to health. Therefore, deviations above the guideline values in either the short or long term do not necessarily mean that the water is unsuitable for consumption. The amount by which, and the period for which, any guideline value can be exceeded without affecting public health depends upon the specific substance involved. However, exceedance should be a signal:

— as a minimum, to investigate the cause with a view to taking remedial action as necessary; and

— to consult with, and seek advice from, the authority responsible for public health.

When a guideline value is exceeded, it is recommended that the authority responsible for public health be consulted for advice on suitable action, taking into account the intake of the substance from sources other than drinking-water, the toxicity of the substance, the likelihood and nature of any adverse effects and the practicality of remedial measures. In applying the guideline values, an important consideration is that unless there are appropriate alternative supplies available, maintenance of adequate quantities of water is a high priority. The use of the Guidelines in emergencies is considered in more detail in section 6.2.

It is important that recommended guideline values are both practical and feasible to implement as well as protective of public health. Guideline values are not normally set at concentrations lower than the detection limits achievable under routine laboratory operating conditions. Moreover, guideline values are established taking into account available techniques for controlling, removing or reducing the concentration of the contaminant to the desired level. In some instances, therefore, *provisional* guideline values have been set for contaminants for which there is some uncertainty in available information or calculated guideline values are not practically achievable.

2.3 National drinking-water policy
2.3.1 *Laws, regulations and standards*
The aim of national drinking-water laws and standards should be to ensure that the consumer enjoys safe potable water, not to shut down deficient water supplies.

Effective control of drinking-water quality is supported ideally by adequate legislation, standards and codes and their enforcement. The precise nature of the legislation in each country will depend on national, constitutional and other considerations. It will generally outline the responsibility and authority of a number of agencies and describe the relationship between them, as well as establish basic policy principles (e.g., water supplied for drinking-water should be safe). The national regulations, adjusted as necessary, should be applicable to all water supplies. This would normally embody different approaches to situations where formal responsibility for drinking-water quality is assigned to a defined entity and situations where community management prevails.

Legislation should make provision for the establishment and amendment of drinking-water quality standards and guidelines, as well as for the establishment of regulations for the development and protection of drinking-water sources and the treatment, maintenance and distribution of safe drinking-water.

Legislation should establish the legal functions and responsibilities of the water supplier and would generally specify that the water supplier is legally responsible at all times for the quality of the water sold and/or supplied to the consumer and for the proper supervision, inspection, maintenance and safe operation of the drinking-water system. It is the water supplier that actually provides water to the public – the "consumer" – and that should be legally responsible for its quality and safety. The supplier is responsible for continuous and effective quality assurance and quality control of water supplies, including inspection, supervision, preventive maintenance, routine testing of water quality and remedial actions as required. However, the supplier is normally responsible for the quality of the water only up to a defined point in the distribution system and may not have responsibility for deterioration of water quality as a result of poor plumbing or unsatisfactory storage tanks in households and buildings.

Where consecutive agencies manage water – for example, a drinking-water wholesaler, a municipal water supplier and a local water distribution company – each agency should carry responsibility for the quality of the water arising from its actions.

Legal and organizational arrangements aimed at ensuring compliance with the legislation, standards or codes of practice for drinking-water quality will normally provide for an independent surveillance agency, as outlined in section 1.2.1 and chapter 5. The legislation should define the duties, obligations and powers of the water surveillance agency. The surveillance agency should preferably be represented at the national level and should operate at national, regional and local levels. The surveillance agency should be given the necessary powers to administer and enforce laws, regulations, standards and codes concerned with water quality. It should also be able to delegate those powers to other specified agencies, such as municipal councils, local health departments, regional authorities and qualified, government-authorized private audit or testing services. Its responsibilities should include the surveillance of water quality to ensure that water delivered to the consumer, through either piped or non-piped distribution systems, meets drinking-water supply service standards; approving sources of drinking-water; and surveying the provision of drinking-water to the population as a whole. There needs to be a high level of knowledge, training and understanding in such an agency in order that drinking-water supply does not suffer from inappropriate regulatory action. The surveillance agency should be empowered by law to compel water suppliers to recommend the boiling of water or other measures when microbial contamination that could threaten public health is detected.

Implementation of programmes to provide safe drinking-water should not be delayed because of a lack of appropriate legislation. Even where legally binding guidelines or standards for drinking-water have yet to be promulgated, it may be possible to encourage, and even enforce, the supply of safe drinking-water through educational efforts or commercial, contractual arrangements between consumer and supplier (e.g., based on civil law) or through interim measures, including health, food or welfare legislation, for example.

Drinking-water quality legislation may usefully provide for interim standards, permitted deviations and exemptions as part of a national or regional policy, rather than as a result of local initiatives. This can take the form of temporary exemptions for certain communities or areas for defined periods of time. Short- and medium-term targets should be set so that the most significant risks to human health are controlled first.

2.3.2 Setting national standards

In countries where universal access to safe drinking-water at an acceptable level of service has not been achieved, policy should refer to expressed targets for increases in access. Such policy statements should be consistent with achievement of the Millennium Development Goals (http://www.developmentgoals.org/) of the United Nations (UN) Millennium Declaration and should take account of levels of acceptable access outlined in General Comment 15 on the Right to Water of the UN Committee on Economic, Social and Cultural Rights (http://www.unhchr.ch/html/menu2/6/cescr.htm) and associated documents.

In developing national drinking-water standards based on these Guidelines, it will be necessary to take account of a variety of environmental, social, cultural, economic, dietary and other conditions affecting potential exposure. This may lead to national standards that differ appreciably from these Guidelines. A programme based on modest but realistic goals – including fewer water quality parameters of priority health concern at attainable levels consistent with providing a reasonable degree of public health protection in terms of reduction of disease or reduced risk of disease within the population – may achieve more than an overambitious one, especially if targets are upgraded periodically.

The authority to establish and revise drinking-water standards, codes of practice and other technical regulations should be delegated to the appropriate government minister – preferably the minister of health – who is responsible for ensuring the safety of water supplies and the protection of public health. The authority to establish and enforce quality standards and regulations may be vested in a ministry other than the one usually responsible for public and/or environmental health. Consideration should then be given to requiring that regulations and standards are promulgated only after approval by the public health or environmental health authority so as to ensure their conformity with health protection principles.

Drinking-water supply policy should normally outline the requirements for protection of water sources and resources, the need for appropriate treatment, preventive maintenance within distribution systems and requirements to support maintaining water safety after collection from communal sources.

The basic water legislation should not specify sampling frequencies but should give the administration the power to establish a list of parameters to be measured and the frequency and location of such measurements.

Standards and codes should normally specify the quality of the water to be supplied to the consumer, the practices to be followed in selecting and developing water sources and in treatment processes and distribution or household storage systems, and procedures for approving water systems in terms of water quality.

Setting national standards should ideally involve consideration of the quality of the water, the quality of service, "target setting" and the quality of infrastructure and systems, as well as enforcement action. For example, national standards should define protection zones around water sources, minimum standard specifications for operating systems, hygiene practice standards in construction and minimum standards for health protection. Some countries include these details in a "sanitary code" or "code of good practice." It is preferable to include in regulations the requirement to consult with drinking-water supply agencies and appropriate professional bodies, since doing so makes it more likely that drinking-water controls will be implemented effectively.

The costs associated with drinking-water quality surveillance and control should be taken into account in developing national legislation and standards.

To ensure that standards are acceptable to consumers, communities served, together with the major water users, should be involved in the standards-setting process. Public health agencies may be closer to the community than those responsible for its drinking-water supply. At a local level, they also interact with other sectors (e.g., education), and their combined action is essential to ensure active community involvement.

Other ministries, such as those responsible for public works, housing, natural resources or the environment, may administer normative and regulatory functions concerned with the design of drinking-water supply and waste disposal systems, equipment standards, plumbing codes and rules, water allocation, natural resource protection and conservation and waste collection, treatment and disposal.

2.4 Identifying priority drinking-water quality concerns

These Guidelines cover a large number of potential constituents in drinking-water in order to meet the varied needs of countries worldwide. Generally, only a few constituents will be of concern under any given circumstances. It is essential that the national regulatory agency and local water authorities determine and respond to the constituents of relevance. This will ensure that efforts and investments can be directed to those constituents that are of public health significance.

Guidelines are established for potentially hazardous water constituents and provide a basis for assessing drinking-water quality. Different parameters may require different priorities for management to improve and protect public health. In general, the order of priority is to:

— ensure an adequate supply of microbiologically safe water and maintain acceptability to discourage consumers from using potentially less microbiologically safe water;
— manage key chemical contaminants known to cause adverse health effects; and
— address other chemical contaminants.

Priority setting should be undertaken on the basis of a systematic assessment based on collaborative effort among all relevant agencies and may be applied at national and system-specific levels. It may require the formation of a broad-based interagency committee including authorities such as health, water resources, drinking-water supply, environment, agriculture and geological services/mining to establish a mechanism for sharing information and reaching consensus on drinking-water quality issues.

Sources of information that should be considered in determining priorities include catchment type (protected, unprotected), geology, topography, agricultural land use, industrial activities, sanitary surveys, records of previous monitoring, inspections and local and community knowledge. The wider the range of data sources used, the more useful the results of the process will be. In many situations, authorities or consumers may have already identified a number of drinking-water quality problems, particularly where they cause obvious health effects or acceptability problems. These existing problems would normally be assigned a high priority.

2.4.1 Assessing microbial priorities

The most common and widespread health risk associated with drinking-water is microbial contamination, the consequences of which mean that its control must always be of paramount importance. Priority needs to be given to improving and developing the drinking-water supplies that represent the greatest public health risk.

> The most common and widespread health risk associated with drinking-water is microbial contamination, the consequences of which mean that its control must always be of paramount importance.

Microbial contamination of major urban systems has the potential to cause large outbreaks of waterborne disease. Ensuring quality in such systems is therefore a priority. Nevertheless, the majority (around 80%) of the global population without access to improved drinking-water supplies resides in rural areas. Similarly, small and community supplies in most countries contribute disproportionately to overall drinking-water quality concerns. Identifying local and national priorities should take factors such as these into account.

Health-based targets for microbial contaminants are discussed in section 3.2, and a comprehensive consideration of microbial aspects of drinking-water quality is contained in chapter 7.

2.4.2 Assessing chemical priorities

Not all of the chemicals with guideline values will be present in all water supplies or, indeed, all countries. If they do exist, they may not be found at levels of concern. Conversely, some chemicals without guideline values or not addressed in the Guidelines may nevertheless be of legitimate local concern under special circumstances.

Risk management strategies (as reflected in national standards and monitoring activities) and commitment of resources should give priority to those chemicals that pose a risk to human health or to those with significant impacts on acceptability of water.

Only a few chemicals have been shown to cause widespread health effects in humans as a consequence of exposure through drinking-water when they are present in excessive quantities. These include fluoride and arsenic. Human health effects have also been demonstrated in some areas associated with lead (from domestic plumbing), and there is concern because of the potential extent of exposure to selenium and uranium in some areas at concentrations of human health significance. Iron and manganese are of widespread significance because of their effects on acceptability. These constituents should be taken into consideration as part of any priority-setting process. In some cases, assessment will indicate that no risk of significant exposure exists at the national, regional or system level.

Drinking-water may be only a minor contributor to the overall intake of a particular chemical, and in some circumstances controlling the levels in drinking-water, at potentially considerable expense, may have little impact on overall exposure. Drinking-water risk management strategies should therefore be considered in conjunction with other potential sources of human exposure.

The process of "short-listing" chemicals of concern may initially be a simple classification of high and low risk to identify broad issues. This may be refined using data from more detailed assessments and analysis and may take into consideration rare events, variability and uncertainty.

Guidance is provided in the supporting document *Chemical Safety of Drinking-water* (section 1.3) on how to undertake prioritization of chemicals in drinking-water. This deals with issues including:

— the probability of exposure (including the period of exposure) of the consumer to the chemical;
— the concentration of the chemical that is likely to give rise to health effects (see also section 8.5); and
— the evidence of health effects or exposure arising through drinking-water, as opposed to other sources, and relative ease of control of the different sources of exposure.

Additional information on the hazards and risks of many chemicals not included in these Guidelines is available from several sources, including WHO Environmental Health Criteria monographs (EHCs) and Concise International Chemical Assessment Documents (CICADs) (http://www.who.int/pcs/index.htm), reports by the Joint FAO/WHO Meeting on Pesticide Residues (JMPR) and Joint FAO/WHO Expert Committee on Food Additives (JECFA) and information from competent national authorities, such as the US Environmental Protection Agency (US EPA) (www.epa.gov/waterscience). These information sources have been peer reviewed and provide readily accessible information on toxicology, hazards and risks of many less common contaminants. They can help water suppliers and health officials to decide upon the significance (if any) of a detected chemical and on the response that might be appropriate.

3
Health-based targets

3.1 Role and purpose of health-based targets

Health-based targets should be part of overall public health policy, taking into account status and trends and the contribution of drinking-water to the transmission of infectious disease and to overall exposure to hazardous chemicals both in individual settings and within overall health management. The purpose of setting targets is to mark out milestones to guide and chart progress towards a predetermined health and/or water safety goal. To ensure effective health protection and improvement, targets need to be realistic and relevant to local conditions (including economic, environmental, social and cultural conditions) and financial, technical and institutional resources. This normally implies periodic review and updating of priorities and targets and, in turn, that norms and standards should be periodically updated to take account of these factors and the changes in available information (see section 2.3).

Health-based targets provide a "benchmark" for water suppliers. They provide information with which to evaluate the adequacy of existing installations and policies and assist in identifying the level and type of inspection and analytical verification that are appropriate and in developing auditing schemes. Health-based targets underpin the development of WSPs and verification of their successful implementation. They should lead to improvements in public health outcomes.

Health-based targets should assist in determining specific interventions appropriate to delivering safe drinking-water, including control measures such as source protection and treatment processes.

> The judgement of safety – or what is a tolerable risk in particular circumstances – is a matter in which society as a whole has a role to play. The final judgement as to whether the benefit resulting from the adoption of any of the health-based targets justifies the cost is for each country to decide.

The use of health-based targets is applicable in countries at all levels of development. Different types of target will be applicable for different purposes, so that in most countries several types of target may be used for various purposes. Care must be taken to develop targets that account for the exposures that contribute most to disease. Care must also be taken to reflect the advantages of progressive, incremental improvement, which will often be based on categorization of public health risk (see section 4.1.2).

Health-based targets are typically national in character. Using information and approaches in these Guidelines, national authorities should be able to establish health-based targets that will protect and improve drinking-water quality and, consequently, human health and also support the best use of available resources in specific national and local circumstances.

In order to minimize the likelihood of outbreaks of disease, care is required to account properly for drinking-water supply performance both in steady state and during maintenance and periods of short-term water quality deterioration. Performance of the drinking-water system during short-term events (such as variation in source water quality, system challenges and process problems) must therefore be considered in the development of health-based targets. Both short-term and catastrophic events can result in periods of very degraded source water quality and greatly decreased efficiency in many processes, both of which provide a logical and sound justification for the long-established "multiple-barrier principle" in water safety.

The processes of formulating, implementing and evaluating health-based targets provide benefits to the overall preventive management of drinking-water quality. These benefits are outlined in Table 3.1.

Targets can be a helpful tool both for encouraging and for measuring incremental progress in improving drinking-water quality management. Improvements can relate to the scientific basis for target setting, progressive evolution to target types that more precisely reflect the health protection goals and the use of targets in defining and promoting categorization for progressive improvement, especially of existing water supplies. Water quality managers, be they suppliers or legislators, should aim at continuously improving water quality management. An example of phased improvement

Table 3.1 Benefits of health-based targets

Target development stage	Benefit
Formulation	Provides insight into the health of the population
	Reveals gaps in knowledge
	Supports priority setting
	Increases the transparency of health policy
	Promotes consistency among national health programmes
	Stimulates debate
Implementation	Inspires and motivates collaborating authorities to take action
	Improves commitment
	Fosters accountability
	Guides the rational allocation of resources
Evaluation	Supplies established milestones for incremental improvements
	Provides opportunity to take action to correct deficiencies and/or deviations
	Identifies data needs and discrepancies

is given in section 5.4. The degree of improvement may be large, as in moving from the initial phase to the intermediate phase, or relatively small.

Ideally, health-based targets should be set using quantitative risk assessment and should take into account local conditions and hazards. In practice, however, they may evolve from epidemiological evidence of waterborne disease based on surveillance, intervention studies or historical precedent or be adapted from international practice and guidance.

3.2 Types of health-based targets

The approaches presented here for developing health-based targets are based on a consistent framework applicable to all types of hazards and for all types of water supplies (see Table 3.2 and below). This offers flexibility to account for national priorities and to support a risk–benefit approach. The framework includes different types of health-based targets. They differ considerably with respect to the amount of resources needed to develop and implement the targets and in relation to the precision with which the public health benefits of risk management actions can be defined. Target types at the bottom of Table 3.2 require least interpretation by practitioners in implementation but depend on a number of assumptions. The targets towards the top of the table require considerably greater scientific and technical underpinning in order to overcome the need to make assumptions and are therefore more precisely related to the level of health protection. The framework is forward looking, in that currently critical data for developing the next stage of target setting may not be available, and a need to collect additional data may become obvious.

Establishing health-based targets should take account not only of "steady-state" conditions but also the possibility of short-term events (such as variation in environmental water quality, system challenges and process problems) that may lead to significant risk to public health.

For microbial pathogens, health-based targets will employ groups of selected pathogens that combine both control challenges and health significance in terms of health hazard and other relevant data. More than one pathogen is required in order to assess the diverse range of challenges to the safeguards available. Where the burden of waterborne microbial disease is high, health-based targets can be based on achieving a measurable reduction in the existing levels of community disease, such as diarrhoea or cholera, as an incremental step in public health improvement of drinking-water quality. While health-based targets may be expressed in terms of tolerable exposure to specific pathogens (i.e., WQTs), care is required in relating this to overall population exposure, which may be focused on short periods of time, and such targets are inappropriate for direct pathogen monitoring. These conditions relate to the recognized phenomenon of short periods of decreased efficiency in many processes and provide a logical justification for the long-established multiple-barrier principle in water safety. Targets must also account for background rates of disease during normal conditions of drinking-water supply performance and efficiency.

Table 3.2 **Nature, application and assessment of health-based targets**

Type of target	Nature of target	Typical applications	Assessment
Health outcome			
• epidemiology based	Reduction in detected disease incidence or prevalence	Microbial or chemical hazards with high measurable disease burden largely water-associated	Public health surveillance and analytical epidemiology
• risk assessment based	Tolerable level of risk from contaminants in drinking-water, absolute or as a fraction of the total burden by all exposures	Microbial or chemical hazards in situations where disease burden is low or cannot be measured directly	Quantitative risk assessment
Water quality	Guideline values applied to water quality	Chemical constituents found in source waters	Periodic measurement of key chemical constituents to assess compliance with relevant guideline values (see section 8.5)
	Guideline values applied in testing procedures for materials and chemicals	Chemical additives and by-products	Testing procedures applied to the materials and chemicals to assess their contribution to drinking-water exposure taking account of variations over time (see section 8.5)
Performance	Generic performance target for removal of groups of microbes	Microbial contaminants	Compliance assessment through system assessment (see section 4.1) and operational monitoring (see section 4.2)
	Customized performance targets for removal of groups of microbes	Microbial contaminants	Individually reviewed by public health authority; assessment would then proceed as above
	Guideline values applied to water quality	Threshold chemicals with effects on health that vary widely (e.g., nitrate and cyanobacterial toxins)	Compliance assessment through system assessment (see section 4.1) and operational monitoring (see section 4.2)
Specified technology	National authorities specify specific processes to adequately address constituents with health effects (e.g., generic WSPs for an unprotected catchment)	Constituents with health effect in small municipalities and community supplies	Compliance assessment through system assessment (see section 4.1) and operational monitoring (see section 4.2)

Note: Each target type is based on those above it in this table, and assumptions with default values are introduced in moving down between target types. These assumptions simplify the application of the target and reduce potential inconsistencies.

3. HEALTH-BASED TARGETS

For chemical constituents of drinking-water, health-based targets can be developed using the guideline values outlined in section 8.5. These have been established on the basis of the health effect of the chemical in water. In developing national drinking-water standards (or health-based targets) based on these guideline values, it will be necessary to take into consideration a variety of environmental, social, cultural, economic, dietary and other conditions affecting potential exposure. This may lead to national targets that differ appreciably from the guideline values.

3.2.1 Specified technology targets

Specified technology targets are most frequently applied to small community supplies and to devices used at household level. They may take the form of recommendations concerning technologies applicable in certain circumstances and/or licensing programmes to restrict access to certain technologies or provide guidance for their application.

Smaller municipal and community drinking-water suppliers often have limited resources and ability to develop individual system assessments and/or management plans. National regulatory agencies may therefore directly specify requirements or approved options. This may imply, for example, providing guidance notes for protection of well heads, specific and approved treatment processes in relation to source types and requirements for protection of drinking-water quality in distribution.

In some circumstances, national or regional authorities may wish to establish model WSPs to be used by local suppliers either directly or with limited adaptation. This may be of particular importance when supplies are community managed. In these circumstances, an approach focusing on ensuring that operators receive adequate training and support to overcome management weaknesses is likely to be more effective than enforcement of compliance.

3.2.2 Performance targets

Performance targets are most frequently applied to the control of microbial hazards in piped supplies varying from small to large.

In situations where short-term exposure is relevant to public health, because water quality varies rapidly or it is not possible to detect hazards between production and consumption, it is necessary to ensure that control measures are in place and operating optimally and to verify their effectiveness in order to secure safe drinking-water.

Performance targets assist in the selection and use of control measures that are capable of preventing pathogens from breaching the barriers of source protection, treatment and distribution systems or preventing growth within the distribution system.

Performance targets should define requirements in relation to source water quality with prime emphasis on processes and practices that will ensure that the targets can be routinely achieved. Most commonly, targets for removal of pathogen groups through water treatment processes will be specified in relation to broad categories of source water quality or source water type and less frequently in relation to specific data on source water quality. The derivation of performance targets requires the integration of factors such as tolerable disease burden (tolerable risk), including severity of disease outcomes and dose–response relationships for specific pathogens (target microbes) (see section 7.3).

Performance targets should be developed for target microbes representing groups of pathogens that combine both control challenges and health significance. In practice, more than one target microbe will normally be required in order to properly reflect diverse challenges to the safeguards available. While performance targets may be derived in relation to exposure to specific pathogens, care is required in relating this to overall population exposure and risk, which may be concentrated into short periods of time.

The principal practical application of performance targets for pathogen control is in assessing the adequacy of drinking-water treatment infrastructure. This is achieved by using information on performance targets with either specific information on treatment performance or assumptions regarding performance of technology types concerning pathogen removal. Examples of performance targets and of treatment effects on pathogens are given in chapter 7.

Performance requirements are also important in certification of devices for drinking-water treatment and for pipe installation that prevents ingress. Certification of devices and materials is discussed elsewhere (see section 1.2.9).

3.2.3 Water quality targets

Adverse health consequences may arise from exposure to chemicals following long-term and, in some cases, short-term exposure. Furthermore, concentrations of most chemicals in drinking-water do not normally fluctuate widely over short periods of time. Management through periodic analysis of drinking-water quality and comparison with WQTs such as guideline values is therefore commonly applied to many chemicals in drinking-water where health effects arise from long-term exposure. While a preventive management approach to water quality should be applied to all drinking-water systems, the guideline values for individual chemicals described in section 8.5 provide health-based targets for chemicals in drinking-water.

Where water treatment processes have been put in place to remove specific chemicals (see section 8.4), WQTs should be used to determine appropriate treatment requirements.

It is important that WQTs are established only for those chemicals that, following rigorous assessment, have been determined to be of health concern or of concern for the acceptability of the drinking-water to consumers. There is little value in undertaking measurements for chemicals that are unlikely to be in the system, that will be present only at concentrations much lower than the guideline value or that have no human health effects or effects on drinking-water acceptability.

3. HEALTH-BASED TARGETS

WQTs are also used in the certification process for chemicals that occur in water as a result of treatment processes or from materials in contact with water. In such applications, assumptions are made in order to derive standards for materials and chemicals that can be employed in their certification. Generally, allowance must be made for the incremental increase over levels found in water sources. For some materials (e.g., domestic plumbing), assumptions must also account for the relatively high release of some substances for a short period following installation.

For microbial hazards, WQTs in terms of pathogens serve primarily as a step in the development of performance targets and have no direct application. In some circumstances, especially where non-conventional technologies are employed in large facilities, it may be appropriate to establish WQTs for microbial contaminants.

3.2.4 Health outcome targets

In some circumstances, especially where there is a measurable burden of water-related disease, it is possible to establish a health-based target in terms of a quantifiable reduction in the overall level of disease. This is most applicable where adverse effects soon follow exposure and are readily and reliably monitored and where changes in exposure can also be readily and reliably monitored. This type of health outcome target is therefore primarily applicable to microbial hazards in both developing and developed countries and to chemical hazards with clearly defined health effects largely attributable to water (e.g., fluoride).

In other circumstances, health-based targets may be based on the results of quantitative risk assessment. In these cases, health outcomes are estimated based on information concerning exposure and dose–response relationships. The results may be employed directly as a basis to define WQTs or may provide the basis for development of performance targets.

There are limitations in the available data and models for quantitative microbial risk assessment (QMRA). Short-term fluctuations in water quality may have a major impact on overall health risks – including those associated with background rates of disease and outbreaks – and are a particular focus of concern in expanding application of QMRA. Further developments in these fields will significantly enhance the applicability and usefulness of this approach.

3.3 General considerations in establishing health-based targets

While water can be a major source of enteric pathogens and hazardous chemicals, it is by no means the only source. In setting targets, consideration needs to be given to other sources of hazards, including food, air and person-to-person contact, as well as the impact of poor sanitation and personal hygiene. There is limited value in establishing a strict target concentration for a chemical if drinking-water provides only a small proportion of total exposure. The cost of meeting such targets could unnecessarily divert funding from other, more pressing health interventions. It is important to take account of the impact of the proposed intervention on overall rates of disease. For some pathogens and their associated diseases, interventions in water quality may be ineffective and may therefore not be justified. This may be the case where other routes of exposure dominate. For others, long experience has shown the effectiveness of drinking-water supply and quality management (e.g., typhoid, dysentery caused by *Shigella*).

Health-based targets and water quality improvement programmes in general should also be viewed in the context of a broader public health policy, including initiatives to improve sanitation, waste disposal, personal hygiene and public education on mechanisms for reducing both personal exposure to hazards and the impact of personal activity on water quality. Improved public health, reduced carriage of pathogens and reduced human impacts on water resources all contribute to drinking-water safety (see Howard et al., 2002).

3.3.1 Assessment of risk in the framework for safe drinking-water

In the framework for safe drinking-water, assessment of risk is not a goal in its own right but is part of an iterative cycle that uses the assessment of risk to derive management decisions that, when implemented, result in incremental improvements in water quality. For the purposes of these Guidelines, the emphasis of incremental improvement is on health. However, in applying the Guidelines to specific circumstances, non-health factors should be taken into account, as they may have a considerable impact upon both costs and benefits.

3.3.2 Reference level of risk

Descriptions of a "reference level of risk" in relation to water are typically expressed in terms of specific health outcomes – for example, a maximum frequency of diarrhoeal disease or cancer incidence or maximum frequency of infection (but not necessarily disease) with a specific pathogen.

There is a range of water-related illnesses with differing severities, including acute, delayed and chronic effects and both morbidity and mortality. Effects may be as diverse as adverse birth outcomes, cancer, cholera, dysentery, infectious hepatitis, intestinal worms, skeletal fluorosis, typhoid and Guillain-Barré syndrome.

Decisions about risk acceptance are highly complex and need to take account of different dimensions of risk. In addition to the "objective" dimensions of probability, severity and duration of an effect, there are important environmental, social, cultural, economic and political dimensions that play important roles in decision-making. Negotiations play an important role in these processes, and the outcome may very well be unique in each situation. Notwithstanding the complexity of decisions about risk, there is a need for a baseline definition of tolerable risk for the development of guidelines and as a departure point for decisions in specific situations.

A reference level of risk enables the comparison of water-related diseases with one another and a consistent approach for dealing with each hazard. For the purposes of these Guidelines, a reference level of risk is used for broad equivalence between the levels of protection afforded to toxic chemicals and those afforded to microbial pathogens. For these purposes, only the health effects of waterborne diseases are taken into account. The reference level of risk is 10^{-6} disability-adjusted life-years (DALYs) per person per year, which is approximately equivalent to a lifetime excess cancer risk of 10^{-5} (i.e., 1 excess case of cancer per 100 000 of the population ingesting drinking-water containing the substance at the guideline value over a life span) (see section 3.3.3 for further details). For a pathogen causing watery diarrhoea with a low case fatality rate (e.g., 1 in 100 000), this reference level of risk would be equivalent to 1/1000 annual risk of disease to an individual (approximately 1/10 over a lifetime). The reference level of risk can be adapted to local circumstances on the basis of a risk–benefit approach. In particular, account should be taken of the fraction of the burden of a particular disease that is likely to be associated with drinking-water. Public health prioritization would normally indicate that major contributors should be dealt with preferentially, taking account of the costs and impacts of potential interventions. This is also the rationale underlying the incremental development and application of standards. The application of DALYs for setting a reference level of risk is a new and evolving approach. A particular challenge is to define human health effects associated with exposure to non-threshold chemicals.

3.3.3 Disability-adjusted life-years (DALYs)

The diverse hazards that may be present in water are associated with very diverse adverse health outcomes. Some outcomes are acute (diarrhoea, methaemoglobinaemia), and others are delayed (cancer by years, infectious hepatitis by weeks); some are potentially severe (cancer, adverse birth outcomes, typhoid), and others are typically mild (diarrhoea and dental fluorosis); some especially affect certain age ranges (skeletal fluorosis in older adults often arises from exposure in childhood; infection with hepatitis E virus [HEV] has a very high mortality rate among pregnant women), and some have very specific concern for certain vulnerable subpopulations (cryptosporidiosis is mild and self-limiting for the population at large but has a high mortality rate among those who test positive for human immunodeficiency virus [HIV]). In addition, any one hazard may cause multiple effects (e.g., gastroenteritis, Gullain-Barré syndrome, reactive arthritis and mortality associated with *Campylobacter*).

In order to be able to objectively compare water-related hazards and the different outcomes with which they are associated, a common "metric" that can take account of differing probabilities, severities and duration of effects is needed. Such a metric should also be applicable regardless of the type of hazard, applying to microbial, chemical and radiological hazards. The metric used in the *Guidelines for Drinking-water Quality* is the DALY. WHO has quite extensively used DALYs to evaluate public health priorities and to assess the disease burden associated with environmental exposures.

The basic principle of the DALY is to weight each health effect for its severity from 0 (normal good health) to 1 (death). This weight is multiplied by the duration of the effect – the time in which disease is apparent (when the outcome is death, the "duration" is the remaining life expectancy) – and by the number of people affected by a particular outcome. It is then possible to sum the effects of all different outcomes due to a particular agent.

Thus, the DALY is the sum of years of life lost by premature mortality (YLL) and years of healthy life lost in states of less than full health, i.e., years lived with a disability (YLD), which are standardized by means of severity weights. Thus:

$$DALY = YLL + YLD$$

Key advantages of using DALYs are its "aggregation" of different effects and its combining of quality and quantity of life. In addition – and because the approaches taken require explicit recognition of assumptions made – it is possible to discuss these and assess the impact of their variation. The use of an outcome metric also focuses attention on actual rather than potential hazards and thereby promotes and enables rational public health priority setting. Most of the difficulties in using DALYs relate to availability of data – for example, on exposure and on epidemiological associations.

DALYs can also be used to compare the health impact of different agents in water. For example, ozone is a chemical disinfectant that produces bromate as a by-product. DALYs have been used to compare the risks from *Cryptosporidium parvum* and bromate and to assess the net health benefits of ozonation in drinking-water treatment.

In previous editions of the *Guidelines for Drinking-water Quality* and in many national drinking-water standards, a "tolerable" risk of cancer has been used to derive guideline values for non-threshold chemicals such as genotoxic carcinogens. This is necessary because there is some (theoretical) risk at any level of exposure. In this and previous editions of the Guidelines, an upper-bound excess lifetime risk of cancer of 10^{-5} has been used, while accepting that this is a conservative position and almost certainly overestimates the true risk.

Different cancers have different severities, manifested mainly by different mortality rates. A typical example is renal cell cancer, associated with exposure to bromate in drinking-water. The theoretical disease burden of renal cell cancer, taking into account an average case:fatality ratio of 0.6 and average age at onset of 65 years, is 11.4 DALYs per case (Havelaar et al., 2000). These data can be used to assess tolerable lifetime cancer risk and a tolerable annual loss of DALYs. Here, we account for the lifelong exposure to carcinogens by dividing the tolerable risk over a life span of 70 years and multiplying by the disease burden per case: (10^{-5} cancer cases / 70 years of life) × 11.4 DALYs per case = 1.6×10^{-6} DALYs per person-year or a tolerable loss of 1.6 healthy life-years in a population of a million over a year.

For guideline derivation, the preferred option is to define an upper level of tolerable risk that is the same for exposure to each hazard (contaminant or constituent in water). As noted above, for the purposes of these Guidelines, the reference level of risk employed is 10^{-6} DALYs per person-year. This is approximately equivalent to the 10^{-5} excess lifetime risk of cancer used in this and previous editions of the Guidelines to determine guideline values for genotoxic carcinogens. For countries that use a stricter definition of the level of acceptable risk of carcinogens (such as 10^{-6}), the tolerable loss will be proportionately lower (such as 10^{-7} DALYs per person-year).

Further information on the use of DALYs in establishing health-based targets is included in the supporting document *Quantifying Public Health Risk in the WHO Guidelines for Drinking-water Quality* (see section 1.3).

4
Water safety plans

The most effective means of consistently ensuring the safety of a drinking-water supply is through the use of a comprehensive risk assessment and risk management approach that encompasses all steps in water supply from catchment to consumer. In these Guidelines, such approaches are termed *water safety plans* (WSPs). The WSP approach has been developed to organize and systematize a long history of management practices applied to drinking-water and to ensure the applicability of these practices to the management of drinking-water quality. It draws on many of the principles and concepts from other risk management approaches, in particular the multiple-barrier approach and HACCP (as used in the food industry).

This chapter focuses on the principles of WSPs and is not a comprehensive guide to the application of these practices. Further information on how to develop a WSP is available in the supporting document *Water Safety Plans* (section 1.3).

Some elements of a WSP will often be implemented as part of a drinking-water supplier's usual practice or as part of benchmarked good practice without consolidation into a comprehensive WSP. This may include quality assurance systems (e.g., ISO 9001:2000). Existing good management practices provide a suitable platform for integrating WSP principles. However, existing practices may not include system-tailored hazard identification and risk assessment as a starting point for system management.

WSPs can vary in complexity, as appropriate for the situation. In many cases, they will be quite simple, focusing on the key hazards identified for the specific system. The wide range of examples of control measures given in the following text does not imply that all of these are appropriate in all cases. WSPs are a powerful tool for the drinking-water supplier to manage the supply safely. They also assist surveillance by public health authorities.

WSPs should, by preference, be developed for individual drinking-water systems. However, for small systems, this may not be realistic, and either specified technology WSPs or model WSPs with guides for their development are prepared. For smaller systems, the WSP is likely to be developed by a statutory body or accredited third-party organization. In these settings, guidance on household water storage, handling and use may also be required. Plans dealing with household water should be linked to a hygiene education programme and advice to households in maintaining water safety.

A WSP has three key components, which are guided by health-based targets (see chapter 3) and overseen through drinking-water supply surveillance (see chapter 5). They are:

> A WSP comprises, as a minimum, the three essential actions that are the responsibility of the drinking-water supplier in order to ensure that drinking-water is safe. These are:
>
> ■ a system assessment;
> ■ effective operational monitoring; and
> ■ management.

— *system assessment* to determine whether the drinking-water supply chain (up to the point of consumption) as a whole can deliver water of a quality that meets health-based targets. This also includes the assessment of design criteria of new systems;
— identifying control measures in a drinking-water system that will collectively control identified risks and ensure that the health-based targets are met. For each control measure identified, an appropriate means of *operational monitoring* should be defined that will ensure that any deviation from required performance is rapidly detected in a timely manner; and
— *management* plans describing actions to be taken during normal operation or incident conditions and documenting the system assessment (including upgrade and improvement), monitoring and communication plans and supporting programmes.

The primary objectives of a WSP in ensuring good drinking-water supply practice are the minimization of contamination of source waters, the reduction or removal of contamination through treatment processes and the prevention of contamination during storage, distribution and handling of drinking-water. These objectives are equally applicable to large piped drinking-water supplies, small community supplies and household systems and are achieved through:

— development of an understanding of the specific system and its capability to supply water that meets health-based targets;
— identification of potential sources of contamination and how they can be controlled;
— validation of control measures employed to control hazards;
— implementation of a system for monitoring the control measures within the water system;
— timely corrective actions to ensure that safe water is consistently supplied; and
— undertaking verification of drinking-water quality to ensure that the WSP is being implemented correctly and is achieving the performance required to meet relevant national, regional and local water quality standards or objectives.

For the WSP to be relied on for controlling the hazards and hazardous events for which it was set in place, it needs to be supported by accurate and reliable technical

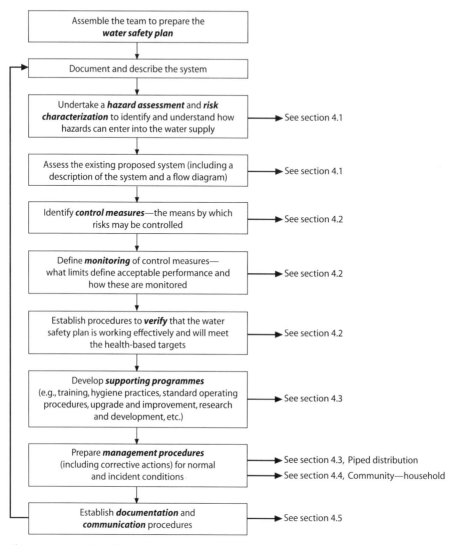

Figure 4.1 Overview of the key steps in developing a water safety plan (WSP)

information. This process of obtaining evidence that the WSP is effective is known as *validation*. Such information could be obtained from relevant industry bodies, from partnering and benchmarking with larger authorities (to optimize resource sharing), from scientific and technical literature and from expert judgement. Assumptions and manufacturer specifications for each piece of equipment and each barrier need to be validated for each system being studied to ensure that the equipment or barrier is effective in that system. System-specific validation is essential, as variabilities in water composition, for instance, may have a large impact on the efficacy of certain removal processes.

Validation normally includes more extensive and intensive monitoring than routine operational monitoring, in order to determine whether system units are performing as assumed in the system assessment. This process often leads to improvements in operating performance through the identification of the most effective and robust operating modes. Additional benefits of the validation process may include identification of more suitable operational monitoring parameters for unit performance.

Verification of drinking-water quality provides an indication of the overall performance of the drinking-water system and the ultimate quality of drinking-water being supplied to consumers. This incorporates monitoring of drinking-water quality as well as assessment of consumer satisfaction.

Where a defined entity is responsible for a drinking-water supply, its responsibility should include the preparation and implementation of a WSP. This plan should normally be reviewed and agreed upon with the authority responsible for protection of public health to ensure that it will deliver water of a quality consistent with the health-based targets.

Where there is no formal service provider, the competent national or regional authority should act as a source of information and guidance on the adequacy of appropriate management of community and individual drinking-water supplies. This will include defining requirements for operational monitoring and management. Approaches to verification in these circumstances will depend on the capacity of local authorities and communities and should be defined in national policy.

4.1 System assessment and design

The first stage in developing a WSP is to form a multidisciplinary team of experts with a thorough understanding of the drinking-water system involved. Typically, such a team would include individuals involved in each stage of the supply of drinking-water, such as engineers, catchment and water managers, water quality specialists, environmental or public health or hygienist professionals, operational staff and representatives of consumers. In most settings, the team will include members from several institutions, and there should be some independent members, such as from professional organizations or universities.

Effective management of the drinking-water system requires a comprehensive understanding of the system, the range and magnitude of hazards that may be present and the ability of existing processes and infrastructure to manage actual or potential risks. It also requires an assessment of capabilities to meet targets. When a new system or an upgrade of an existing system is being planned, the first step in developing a WSP is the collection and evaluation of all available relevant information and consideration of what risks may arise during delivery of water to the consumer.

> Effective risk management requires the identification of potential hazards, their sources and potential hazardous events and an assessment of the level of risk presented by each. In this context:
>
> - a **hazard** is a biological, chemical, physical or radiological agent that has the potential to cause harm;
> - a **hazardous event** is an incident or situation that can lead to the presence of a hazard (what can happen and how); and
> - **risk** is the likelihood of identified hazards causing harm in exposed populations in a specified time frame, including the magnitude of that harm and/or the consequences.

Assessment of the drinking-water system supports subsequent steps in the WSP in which effective strategies for control of hazards are planned and implemented.

The assessment and evaluation of a drinking-water system are enhanced through the development of a flow diagram. Diagrams provide an overview description of the drinking-water system, including characterization of the source, identification of potential pollution sources in the catchment, measures for resource and source protection, treatment processes, storage and distribution infrastructure. It is essential that the representation of the drinking-water system is conceptually accurate. If the flow diagram is not correct, it is possible to overlook potential hazards that may be significant. To ensure accuracy, the flow diagram should be validated by visually checking the diagram against features observed on the ground.

Data on the occurrence of pathogens and chemicals in source waters combined with information concerning the effectiveness of existing controls enable an assessment of whether health-based targets can be achieved with the existing infrastructure. They also assist in identifying catchment management measures, treatment processes and distribution system operating conditions that would reasonably be expected to achieve those targets if improvements are required.

> It may often be more efficient to invest in preventive processes within the catchment than to invest in major treatment infrastructure to manage a hazard.

To ensure the accuracy of the assessment, it is essential that all elements of the drinking-water system (resource and source protection, treatment and distribution) are considered concurrently and that interactions and influences between each element and their overall effect are taken into consideration.

4.1.1 New systems

When drinking-water supply sources are being investigated or developed, it is prudent to undertake a wide range of analyses in order to establish overall safety and to determine potential sources of contamination of the drinking-water supply source. These would normally include hydrological analysis, geological assessment and land use inventories to determine potential chemical and radiological contaminants.

When designing new systems, all water quality factors should be taken into account in selecting technologies for abstraction and treatment of new resources. Variations in the turbidity and other parameters of raw surface waters can be very great, and allowance must be made for this. Treatment plants should be designed to take account of variations known or expected to occur with significant frequency rather than for average water quality; otherwise, filters may rapidly become blocked or sedimentation tanks overloaded. The chemical aggressiveness of some groundwaters may affect the integrity of borehole casings and pumps, leading to unacceptably high levels of iron in the supply, eventual breakdown and expensive repair work. Both the quality and availability of drinking-water may be reduced and public health endangered.

4.1.2 Collecting and evaluating available data

Table 4.1 provides examples of areas that should normally be taken into consideration as part of the assessment of the drinking-water system. In most cases, consultation with public health and other sectors, including land and water users and all those who regulate activities in the catchment, will be required for the analysis of catchments. A structured approach is important to ensure that significant issues are not overlooked and that areas of greatest risk are identified.

The overall assessment of the drinking-water system should take into consideration any historical water quality data that assist in understanding source water characteristics and drinking-water system performance both over time and following specific events (e.g., heavy rainfall).

Prioritizing hazards for control

Once potential hazards and their sources have been identified, the risk associated with each hazard or hazardous event should be compared so that priorities for risk management can be established and documented. Although there are numerous contaminants that can compromise drinking-water quality, not every hazard will require the same degree of attention.

The risk associated with each hazard or hazardous event may be described by identifying the likelihood of occurrence (e.g., certain, possible, rare) and evaluating the severity of consequences if the hazard occurred (e.g., insignificant, major, catastrophic). The aim should be to distinguish between important and less important hazards or hazardous events. The approach used typically involves a semiquantitative matrix.

Simple scoring matrices typically apply technical information from guidelines, scientific literature and industry practice with well informed "expert" judgement supported by peer review or benchmarking. Scoring is specific for each drinking-water system, since each system is unique. Where generic WSPs are developed for technologies used by small drinking-water systems, the scoring will be specific to the technology rather than the individual drinking-water system.

By using a semiquantitative scoring, control measures can be ranked in relation to the most significant hazards. A variety of approaches to ranking risk can be applied.

Table 4.1 Examples of information useful in assessing a drinking-water system

Component of drinking-water system	Information to consider in assessing component of drinking-water system
Catchments	• Geology and hydrology • Meteorology and weather patterns • General catchment and river health • Wildlife • Competing water uses • Nature and intensity of development and land use • Other activities in the catchment that potentially release contaminants into source water • Planned future activities
Surface water	• Description of water body type (e.g., river, reservoir, dam) • Physical characteristics (e.g., size, depth, thermal stratification, altitude) • Flow and reliability of source water • Retention times • Water constituents (physical, chemical, microbial) • Protection (e.g., enclosures, access) • Recreational and other human activity • Bulk water transport
Groundwater	• Confined or unconfined aquifer • Aquifer hydrogeology • Flow rate and direction • Dilution characteristics • Recharge area • Wellhead protection • Depth of casing • Bulk water transport
Treatment	• Treatment processes (including optional processes) • Equipment design • Monitoring equipment and automation • Water treatment chemicals used • Treatment efficiencies • Disinfection removals of pathogens • Disinfectant residual / contact time
Service reservoirs and distribution	• Reservoir design • Retention times • Seasonal variations • Protection (e.g., covers, enclosures, access) • Distribution system design • Hydraulic conditions (e.g., water age, pressures, flows) • Backflow protection • Disinfectant residuals

An example of an approach is given in Table 4.2. Application of this matrix relies to a significant extent on expert opinion to make judgements on the health risk posed by hazards or hazardous events.

An example of descriptors that can be used to rate the likelihood of occurrence and severity of consequences is given in Table 4.3. A "cut-off" point must be deter-

4. WATER SAFETY PLANS

Table 4.2 Example of a simple risk scoring matrix for ranking risks

Likelihood	Severity of consequences				
	Insignificant	Minor	Moderate	Major	Catastrophic
Almost certain					
Likely					
Moderately likely					
Unlikely					
Rare					

Table 4.3 Examples of definitions of likelihood and severity categories that can be used in risk scoring

Item	Definition
Likelihood categories	
Almost certain	Once per day
Likely	Once per week
Moderately likely	Once per month
Unlikely	Once per year
Rare	Once every 5 years
Severity categories	
Catastrophic	Potentially lethal to large population
Major	Potentially lethal to small population
Moderate	Potentially harmful to large population
Minor	Potentially harmful to small population
Insignificant	No impact or not detectable

mined, above which all hazards will require immediate attention. There is little value in expending large amounts of effort to consider very small risks.

Control measures

The assessment and planning of control measures should ensure that health-based targets will be met and should be based on hazard identification and assessment. The level of control applied to a hazard should be proportional to the associated ranking. Assessment of control measures involves:

> Control measures are those steps in drinking-water supply that directly affect drinking-water quality and that collectively ensure that drinking-water consistently meets health-based targets. They are activities and processes applied to prevent hazard occurrence.

— identifying existing control measures for each significant hazard or hazardous event from catchment to consumer;
— evaluating whether the control measures, when considered together, are effective in controlling risk to acceptable levels; and
— if improvement is required, evaluating alternative and additional control measures that could be applied.

Identification and implementation of control measures should be based on the multiple-barrier principle. The strength of this approach is that a failure of one barrier may be compensated by effective operation of the remaining barriers, thus minimizing the likelihood of contaminants passing through the entire system and being present in sufficient amounts to cause harm to consumers. Many control measures may contribute to control more than one hazard, while some hazards may require more than one control measure for effective control. Examples of control measures are provided in the following sections.

All control measures are important and should be afforded ongoing attention. They should be subject to operational monitoring and control, with the means of monitoring and frequency of data collection based on the nature of the control measure and the rapidity with which change may occur (see section 4.4.3).

4.1.3 Resource and source protection

Effective catchment management has many benefits. By decreasing the contamination of the source water, the amount of treatment required is reduced. This may reduce the production of treatment by-products and minimize operational costs.

Hazard identification

Understanding the reasons for variations in raw water quality is important, as it will influence the requirements for treatment, treatment efficiency and the resulting health risk associated with the finished water. In general, raw water quality is influenced by both natural and human use factors. Important natural factors include wildlife, climate, topography, geology and vegetation. Human use factors include point sources (e.g., municipal and industrial wastewater discharges) and non-point sources (e.g., urban and agricultural runoff, including agrochemicals, livestock or recreational use). For example, discharges of municipal wastewater can be a major source of pathogens; urban runoff and livestock can contribute substantial microbial load; body contact recreation can be a source of faecal contamination; and agricultural runoff can lead to increased challenges to treatment.

Whether water is drawn from surface or underground sources, it is important that the characteristics of the local catchment or aquifer are understood and that the scenarios that could lead to water pollution are identified and managed. The extent to which potentially polluting activities in the catchment can be reduced may appear to be limited by competition for water and pressure for increased development in the catchment. However, introducing good practice in containment of hazards is often possible without substantially restricting activities, and collaboration between stakeholders may be a powerful tool to reduce pollution without reducing beneficial development.

Resource protection and source protection provide the first barriers in protection of drinking-water quality. Where catchment management is beyond the jurisdiction of the drinking-water supplier, the planning and implementation of control measures will require coordination with other agencies. These may include planning authorities, catchment boards, environmental and water resource regulators, road authorities, emergency services and agricultural, industrial and other commercial entities whose activities have an impact on water quality. It may not be possible to apply all aspects of resource and source protection initially; nevertheless, priority should be given to catchment management. This will contribute to a sense of ownership and joint responsibility for drinking-water resources through multistakeholder bodies that assess pollution risks and develop plans for improving management practices for reducing these risks.

Groundwater from depth and confined aquifers is usually microbially safe and chemically stable in the absence of direct contamination; however, shallow or unconfined aquifers can be subject to contamination from discharges or seepages associated with agricultural practices (e.g., pathogens, nitrates and pesticides), on-site sanitation and sewerage (pathogens and nitrates) and industrial wastes. Hazards and hazardous events that can have an impact on catchments and that should be taken into consideration as part of a hazard assessment include:

— rapid variations in raw water quality;
— sewage and septic system discharges;
— industrial discharges;
— chemical use in catchment areas (e.g., use of fertilizers and agricultural pesticides);
— major spills (including relationship to public roads and transport routes), both accidental and deliberate;
— human access (e.g., recreational activity);
— wildlife and livestock;
— land use (e.g., animal husbandry, agriculture, forestry, industrial area, waste disposal, mining) and changes in land use;
— inadequate buffer zones and vegetation, soil erosion and failure of sediment traps;
— stormwater flows and discharges;
— active or closed waste disposal or mining sites / contaminated sites / hazardous wastes;
— geology (naturally occurring chemicals);
— unconfined and shallow aquifer (including groundwater under direct influence of surface water);
— inadequate wellhead protection, uncased or inadequately cased bores and unhygienic practices; and
— climatic and seasonal variations (e.g., heavy rainfalls, droughts) and natural disasters.

Further hazards and hazardous situations that can have an impact on storage reservoirs and intakes and that should be taken into consideration as part of a hazard assessment include:

- human access / absence of exclusion areas;
- short circuiting of reservoir;
- depletion of reservoir storage;
- lack of selective withdrawal;
- lack of alternative water sources;
- unsuitable intake location;
- cyanobacterial blooms;
- stratification; and
- failure of alarms and monitoring equipment.

Control measures
Effective resource and source protection includes the following elements:

- developing and implementing a catchment management plan, which includes control measures to protect surface water and groundwater sources;
- ensuring that planning regulations include the protection of water resources (land use planning and watershed management) from potentially polluting activities and are enforced; and
- promoting awareness in the community of the impact of human activity on water quality.

Examples of control measures for effective protection of source water and catchments include:

- designated and limited uses;
- registration of chemicals used in catchments;
- specific protective requirements (e.g., containment) for chemical industry or refuelling stations;
- reservoir mixing/destratification to reduce growth of cyanobacteria or to reduce anoxic hypolimnion and solubilization of sedimentary manganese and iron;
- pH adjustment of reservoir water;
- control of human activities within catchment boundaries;
- control of wastewater effluents;
- land use planning procedures, use of planning and environmental regulations to regulate potential water-polluting developments;
- regular inspections of catchment areas;
- diversion of local stormwater flows;
- protection of waterways;
- runoff interception; and
- security to prevent tampering.

Similarly, control measures for effective protection of water extraction and storage systems include:

4. WATER SAFETY PLANS

— use of available water storage during and after periods of heavy rainfall;
— appropriate location and protection of intake;
— appropriate choice of off-take depth from reservoirs;
— proper well construction, including casing, sealing and wellhead security;
— proper location of wells;
— water storage systems to maximize retention times;
— storages and reservoirs with appropriate stormwater collection and drainage;
— security from access by animals; and
— security to prevent unauthorized access and tampering.

Where a number of water sources are available, there may be flexibility in the selection of water for treatment and supply. It may be possible to avoid taking water from rivers and streams when water quality is poor (e.g., following heavy rainfall) in order to reduce risk and prevent potential problems in subsequent treatment processes.

Retention of water in reservoirs can reduce the number of faecal microorganisms through settling and inactivation, including solar (ultraviolet [UV]) disinfection but also provides opportunities for contamination to be introduced. Most pathogenic microorganisms of faecal origin (enteric pathogens) do not survive indefinitely in the environment. Substantial die-off of enteric bacteria will occur over a period of weeks. Enteric viruses and protozoa will often survive for longer periods (weeks to months) but are often removed by settling and antagonism from indigenous microbes. Retention also allows suspended material to settle, which makes subsequent disinfection more effective and reduces the formation of DBPs.

Control measures for groundwater sources should include protecting the aquifer and the local area around the borehead from contamination and ensuring the physical integrity of the bore (surface sealed, casing intact, etc.).

Further information on the use of indicators in catchment characterization is available in chapter 4 of the supporting document *Assessing Microbial Safety of Drinking Water* (section 1.3).

4.1.4 Treatment

After source water protection, the next barriers to contamination of the drinking-water system are those of water treatment processes, including disinfection and physical removal of contaminants.

Hazard identification

Hazards may be introduced during treatment, or hazardous circumstances may allow contaminants to pass through treatment in significant concentrations. Constituents of drinking-water can be introduced through the treatment process, including chemical additives used in the treatment process or products in contact with drinking-water. Sporadic high turbidity in source water can overwhelm treatment processes, allowing enteric pathogens into treated water and the distribution system. Similarly, suboptimal filtration following filter backwashing can lead to the introduction of pathogens into the distribution system.

Examples of potential hazards and hazardous events that can have an impact on the performance of drinking-water treatment include the following:

— flow variations outside design limits;
— inappropriate or insufficient treatment processes, including disinfection;
— inadequate backup (infrastructure, human resources);
— process control failure and malfunction or poor reliability of equipment;
— use of unapproved or contaminated water treatment chemicals and materials;
— chemical dosing failures;
— inadequate mixing;
— failure of alarms and monitoring equipment;
— power failures;
— accidental and deliberate pollution;
— natural disasters;
— formation of DBPs; and
— cross-connections to contaminated water/wastewater, internal short circuiting.

Control measures

Control measures may include pretreatment, coagulation/flocculation/sedimentation, filtration and disinfection.

Pretreatment includes processes such as roughing filters, microstrainers, off-stream storage and bankside filtration. Pretreatment options may be compatible with a variety of treatment processes ranging in complexity from simple disinfection to membrane processes. Pretreatment can reduce and/or stabilize the microbial, natural organic matter and particulate load.

Coagulation, flocculation, sedimentation (or flotation) and filtration remove particles, including microorganisms (bacteria, viruses and protozoa). It is important that processes are optimized and controlled to achieve consistent and reliable performance. Chemical coagulation is the most important step in determining the removal efficiency of coagulation/flocculation/clarification processes. It also directly affects the removal efficiency of granular media filtration units and has indirect impacts on the efficiency of the disinfection process. While it is unlikely that the coagulation process itself introduces any new microbial hazards to finished water, a failure or inefficiency in the coagulation process could result in an increased microbial load entering drinking-water distribution.

Various filtration processes are used in drinking-water treatment, including granular, slow sand, precoat and membrane (microfiltration, ultrafiltration, nanofiltration and reverse osmosis) filtration. With proper design and operation, filtration can act as a consistent and effective barrier for microbial pathogens and may in some cases be the only treatment barrier (e.g., for removing *Cryptosporidium* oocysts by direct filtration when chlorine is used as the sole disinfectant).

Application of an adequate level of disinfection is an essential element for most treatment systems to achieve the necessary level of microbial risk reduction. Taking account of the level of microbial inactivation required for the more resistant microbial pathogens through the application of the Ct concept (product of disinfectant concentration and contact time) for a particular pH and temperature ensures that other more sensitive microbes are also effectively controlled. Where disinfection is used, measures to minimize DBP formation should be taken into consideration.

The most commonly used disinfection process is chlorination. Ozonation, UV irradiation, chloramination and application of chlorine dioxide are also used. These methods are very effective in killing bacteria and can be reasonably effective in inactivating viruses (depending on type) and many protozoa, including *Giardia* and *Cryptosporidium*. For effective removal or inactivation of protozoal cysts and oocysts, filtration with the aid of coagulation/flocculation (to reduce particles and turbidity) followed by disinfection (by one or a combination of disinfectants) is the most practical method.

Examples of treatment control measures include:

— coagulation/flocculation and sedimentation;
— use of approved water treatment chemicals and materials;
— control of water treatment chemicals;
— process controls;
— availability of backup systems;
— water treatment process optimization, including
    — chemical dosing
    — filter backwashing
    — flow rate
— use of water in storage in periods of poor-quality raw water; and
— security to prevent unauthorized access and tampering.

Storage of water after disinfection and before supply to consumers can improve disinfection by increasing disinfectant contact times. This can be particularly important for more resistant microorganisms, such as *Giardia* and some viruses.

Further information can be found in the supporting document *Water Treatment and Pathogen Control* (section 1.3).

4.1.5 Piped distribution systems

Water treatment should be optimized to prevent microbial growth, corrosion of pipe materials and the formation of deposits through measures such as:

— continuous and reliable elimination of particles and the production of water of low turbidity;

- precipitation and removal of dissolved (and particulate) iron and manganese;
- minimizing the carry-over of residual coagulant (dissolved, colloidal or particulate), which may precipitate in reservoirs and pipework;
- reducing as far as possible the dissolved organic matter and especially easily biodegradable organic carbon, which provides nutrients for microorganisms; and
- maintaining the corrosion potential within limits that avoid damage to the structural materials and consumption of disinfectant.

Maintaining good water quality in the distribution system will depend on the design and operation of the system and on maintenance and survey procedures to prevent contamination and to prevent and remove accumulation of internal deposits.

Further information is available in the supporting document *Safe, Piped Water* (section 1.3).

Hazard identification

The protection of the distribution system is essential for providing safe drinking-water. Because of the nature of the distribution system, which may include many kilometres of pipe, storage tanks, interconnections with industrial users and the potential for tampering and vandalism, opportunities for microbial and chemical contamination exist.

Contamination can occur within the distribution system:

- when contaminated water in the subsurface material and especially nearby sewers surrounding the distribution system enters because of low internal pipe pressure or through the effect of a "pressure wave" within the system (infiltration/ingress);
- when contaminated water is drawn into the distribution system or storage reservoir through backflow resulting from a reduction in line pressure and a physical link between contaminated water and the storage or distribution system;
- through open or insecure treated water storage reservoirs and aqueducts, which are potentially vulnerable to surface runoff from the land and to attracting animals and waterfowl as faecal contamination sources and may be insecure against vandalism and tampering;
- through pipe bursts when existing mains are repaired or replaced or when new water mains are installed, potentially leading to the introduction of contaminated soil or debris into the system;
- through human error resulting in the unintentional cross-connection of wastewater or stormwater pipes to the distribution system or through illegal or unauthorized connections;
- through leaching of chemicals and heavy metals from materials such as pipes, solders / jointing compounds, taps and chemicals used in cleaning and disinfection of distribution systems; and
- when petrol or oil diffuses through plastic pipes.

4. WATER SAFETY PLANS

In each case, if the contaminated water contains pathogens or hazardous chemicals, it is likely that consumers will be exposed to them.

Even where disinfectant residuals are employed to limit microbial occurrence, they may be inadequate to overcome the contamination or may be ineffective against some or all of the pathogen types introduced. As a result, pathogens may occur in concentrations that could lead to infection and illness.

Where water is supplied intermittently, the resulting low water pressure will allow the ingress of contaminated water into the system through breaks, cracks, joints and pinholes. Intermittent supplies are not desirable but are very common in many countries and are frequently associated with contamination. The control of water quality in intermittent supplies represents a significant challenge, as the risks of infiltration and backflow increase significantly. The risks may be elevated seasonally as soil moisture conditions increase the likelihood of a pressure gradient developing from the soil to the pipe. Where contaminants enter the pipes in an intermittent supply, the charging of the system when supply is restored may increase risks to consumers, as a concentrated "slug" of contaminated water can be expected to flow through the system. Where household storage is used to overcome intermittent supply, localized use of disinfectants to reduce microbial proliferation may be warranted.

Drinking-water entering the distribution system may contain free-living amoebae and environmental strains of various heterotrophic bacterial and fungal species. Under favourable conditions, amoebae and heterotrophs, including strains of *Citrobacter*, *Enterobacter* and *Klebsiella*, may colonize distribution systems and form biofilms. There is no evidence to implicate the occurrence of most microorganisms from biofilms (excepting, for example, *Legionella*, which can colonize water systems in buildings) with adverse health effects in the general population through drinking-water, with the possible exception of severely immunocompromised people (see the supporting document *Heterotrophic Plate Counts and Drinking-water Safety*; section 1.3).

Water temperatures and nutrient concentrations are not generally elevated enough within the distribution system to support the growth of *E. coli* (or enteric pathogenic bacteria) in biofilms. Thus, the presence of *E. coli* should be considered as evidence of recent faecal contamination.

Natural disasters, including flood, drought and earth tremors, may significantly affect piped water distribution systems.

Control measures

Water entering the distribution system must be microbially safe and ideally should also be biologically stable. The distribution system itself must provide a secure barrier to contamination as the water is transported to the user. Maintaining a disinfectant residual throughout the distribution system can provide some protection against contamination and limit microbial growth problems. Chloramination has proved successful in controlling *Naegleria fowleri* in water and sediments in long pipelines and may reduce regrowth of *Legionella* within buildings.

Residual disinfectant will provide partial protection against microbial contamination, but may also mask the detection of contamination through conventional faecal indicator bacteria such as *E. coli*, particularly by resistant organisms. Where a disinfectant residual is used within a distribution system, measures to minimize DBP production should be taken into consideration.

Water distribution systems should be fully enclosed, and storage reservoirs and tanks should be securely roofed with external drainage to prevent contamination. Control of short circuiting and prevention of stagnation in both storage and distribution contribute to prevention of microbial growth. A number of strategies can be adopted to maintain the quality of water within the distribution system, including use of backflow prevention devices, maintaining positive pressure throughout the system and implementation of efficient maintenance procedures. It is also important that appropriate security measures be put in place to prevent unauthorized access to or interference with the drinking-water system infrastructure.

Control measures may include using a more stable secondary disinfecting chemical (e.g., chloramines instead of free chlorine), undertaking a programme of pipe replacement, flushing and relining and maintaining positive pressure in the distribution system. Reducing the time that water is in the system by avoiding stagnation in storage tanks, loops and dead-end sections will also contribute to maintaining drinking-water quality.

Other examples of distribution system control measures include the following:

— distribution system maintenance;
— availability of backup systems (power supply);
— maintaining an adequate disinfectant residual;
— implementing cross-connection and backflow prevention devices;
— fully enclosed distribution system and storages;
— appropriate repair procedures, including subsequent disinfection of water mains;
— maintaining adequate system pressure; and
— maintaining security to prevent sabotage, illegal tapping and tampering.

Further information is available in the supporting document *Safe, Piped Water* (section 1.3).

4.1.6 Non-piped, community and household systems
Hazard identification

Hazard identification would ideally be on a case-by-case basis. In practice, however, for non-piped, community and household drinking-water systems, reliance is typically placed on general assumptions of hazardous conditions that are relevant for technologies or system types and that may be defined at a national or regional level.

Examples of hazards and hazardous situations potentially associated with various non-piped sources of water include the following:

- tubewell fitted with a hand pump
  — ingress of contaminated surface water directly into borehole
  — ingress of contaminants due to poor construction or damage to the lining
  — leaching of microbial contaminants into aquifer
- simple protected spring
  — contamination directly through "backfill" area
  — contaminated surface water causes rapid recharge
- simple dug well
  — ingress of contaminants due to poor construction or damage to the lining
  — contamination introduced by buckets
- rainwater collection
  — bird and other animal droppings found on roof or in guttering
  — first flush of water can enter storage tank.

Further guidance is provided in the supporting document *Water Safety Plans* (section 1.3) and in Volume 3 of the *Guidelines for Drinking-Water Quality*.

Control measures

The control measures required ideally depend on the characteristics of the source water and the associated catchment; in practice, standard approaches may be applied for each of these, rather than customized assessment of each system.

Examples of control measures for various non-piped sources include the following:

- tubewell fitted with a hand pump
  — proper wellhead completion measures
  — provide adequate set-back distances for contaminant sources such as latrines or animal husbandry, ideally based on travel time
- simple protected spring
  — maintain effective spring protection measures
  — establish set-back distance based on travel time
- simple dug well
  — proper construction and use of a mortar seal on lining
  — install and maintain hand pump or other sanitary means of abstraction
- rainwater collection
  — cleaning of roof and gutters
  — first-flush diversion unit.

In most cases, contamination of groundwater supplies can be controlled by a combination of simple measures. In the absence of fractures or fissures, which may allow rapid transport of contaminants to the source, groundwater in confined or deep aquifers will generally be free of pathogenic microorganisms. Bores should be encased to a reasonable depth, and boreheads should be sealed to prevent ingress of surface water or shallow groundwater.

Rainwater systems, particularly those involving storage in above-ground tanks, can be a relatively safe supply of water. The principal sources of contamination are birds, small mammals and debris collected on roofs. The impact of these sources can be minimized by simple measures: guttering should be cleared regularly; overhanging branches should be kept to a minimum (because they can be a source of debris and can increase access to roof catchment areas by birds and small mammals); and inlet pipes to tanks should include leaf litter strainers. First-flush diverters, which prevent the initial roof-cleaning wash of water (20–25 litres) from entering tanks, are recommended. If first-flush diverters are not available, a detachable downpipe can be used manually to provide the same result.

In general, surface waters will require at least disinfection, and usually also filtration, to ensure microbial safety. The first barrier is based on minimizing contamination from human waste, livestock and other hazards at the source.

The greater the protection of the water source, the less the reliance on treatment or disinfection. Water should be protected during storage and delivery to consumers by ensuring that the distribution and storage systems are enclosed.

This applies to both piped systems (section 4.1.5) and vendor-supplied water (section 6.5). For water stored in the home, protection from contamination can be achieved by use of enclosed or otherwise safely designed storage containers that prevent the introduction of hands, dippers or other extraneous sources of contamination.

For control of chemical hazards, reliance may be placed primarily on initial screening of sources and on ensuring the quality and performance of treatment chemicals, materials and devices available for this use, including water storage systems.

Model WSPs are available in the supporting document *Water Safety Plans* (section 1.3) for the following types of water supply:

— groundwater from protected boreholes / wells with mechanized pumping;
— conventional treatment of water;
— multistage filtration;
— storage and distribution through supplier-managed piped systems;
— storage and distribution through community-managed piped systems;
— water vendors;
— water on conveyances (planes, ships and trains);
— tubewell from which water is collected by hand;
— springs from which water is collected by hand;
— simple protected dug wells; and
— rainwater catchments.

Guidance is also available regarding how water safety may be assured for household water collection, transport and storage (see the supporting document *Managing Water*

4. WATER SAFETY PLANS

in the Home; section 1.3). This should be used in conjunction with hygiene education programmes to support health promotion in order to reduce water-related disease.

4.1.7 Validation

Validation is concerned with obtaining evidence on the performance of control measures. It should ensure that the information supporting the WSP is correct, thus enabling achievement of health-based targets.

> Validation is an investigative activity to identify the effectiveness of a control measure. It is typically an intensive activity when a system is initially constructed or rehabilitated. It provides information on reliably achievable quality improvement or maintenance to be used in system assessment in preference to assumed values and also to define the operational criteria required to ensure that the control measure contributes to effective control of hazards.

Validation of treatment processes is required to show that treatment processes can operate as required. It can be undertaken during pilot stage studies and/or during initial implementation of a new or modified water treatment system. It is also a useful tool in the optimization of existing treatment processes.

The first stage of validation is to consider data that already exist. These will include data from the scientific literature, trade associations, regulation and legislation departments and professional bodies, historical data and supplier knowledge. This will inform the testing requirements. Validation is not used for day-to-day management of drinking-water supplies; as a result, microbial parameters that may be inappropriate for operational monitoring can be used, and the lag time for return of results and additional costs from pathogen measurements can often be tolerated.

4.1.8 Upgrade and improvement

The assessment of the drinking-water system may indicate that existing practices and technologies may not ensure drinking-water safety. In some instances, all that may be needed is to review, document and formalize these practices and address any areas where improvements are required; in others, major infrastructure changes may be needed. The assessment of the system should be used as a basis to develop a plan to address identified needs for full implementation of a WSP.

Improvement of the drinking-water system may encompass a wide range of issues, such as:

— capital works;
— training;
— enhanced operational procedures;
— community consultation programmes;
— research and development;
— developing incident protocols; and
— communication and reporting.

Upgrade and improvement plans can include short-term (e.g., 1 year) or long-term programmes. Short-term improvements might include, for example, improvements to community consultation and the development of community awareness programmes. Long-term capital works projects could include covering of water storages or enhanced coagulation and filtration.

Implementation of improvement plans may have significant budgetary implications and therefore may require detailed analysis and careful prioritization in accord with the outcomes of risk assessment. Implementation of plans should be monitored to confirm that improvements have been made and are effective. Control measures often require considerable expenditure, and decisions about water quality improvements cannot be made in isolation from other aspects of drinking-water supply that compete for limited financial resources. Priorities will need to be established, and improvements may need to be phased in over a period of time.

4.2 Operational monitoring and maintaining control

> Operational monitoring assesses the performance of control measures at appropriate time intervals. The intervals may vary widely – for example, from on-line control of residual chlorine to quarterly verification of the integrity of the plinth surrounding a well.

The objectives of operational monitoring are for the drinking-water supplier to monitor each control measure in a timely manner to enable effective system management and to ensure that health-based targets are achieved.

4.2.1 Determining system control measures

The identity and number of control measures are system specific and will be determined by the number and nature of hazards and magnitude of associated risks.

Control measures should reflect the likelihood and consequences of loss of control. Control measures have a number of operational requirements, including the following:

— operational monitoring parameters that can be measured and for which limits can be set to define the operational effectiveness of the activity;
— operational monitoring parameters that can be monitored with sufficient frequency to reveal failures in a timely fashion; and
— procedures for corrective action that can be implemented in response to deviation from limits.

4.2.2 Selecting operational monitoring parameters

The parameters selected for operational monitoring should reflect the effectiveness of each control measure, provide a timely indication of performance, be readily measured and provide opportunity for an appropriate response. Examples include measurable variables, such as chlorine residuals, pH and turbidity, or observable factors, such as the integrity of vermin-proofing screens.

Enteric pathogens and indicator bacteria are of limited use for operational monitoring, because the time taken to process and analyse water samples does not allow operational adjustments to be made prior to supply.

A range of parameters can be used in operational monitoring:

- For source waters, these include turbidity, UV absorbency, algal growth, flow and retention time, colour, conductivity and local meteorological events (see the supporting documents *Protecting Surface Waters for Health* and *Protecting Groundwaters for Health*; section 1.3).
- For treatment, parameters may include disinfectant concentration and contact time, UV intensity, pH, light absorbency, membrane integrity, turbidity and colour (see the supporting document *Water Treatment and Pathogen Control*; section 1.3).
- In piped distribution systems, operational monitoring parameters may include the following:
  — *Chlorine residual monitoring* provides a rapid indication of problems that will direct measurement of microbial parameters. A sudden disappearance of an otherwise stable residual can indicate ingress of contamination. Alternatively, difficulties in maintaining residuals at points in a distribution system or a gradual disappearance of residual may indicate that the water or pipework has a high oxidant demand due to growth of bacteria.
  — The presence or absence of *faecal indicator bacteria* is another commonly used operational monitoring parameter. However, there are pathogens that are more resistant to chlorine disinfection than the most commonly used indicator – *E. coli* or thermotolerant coliforms. Therefore, the presence of more resistant faecal indicator bacteria (e.g., intestinal enterococci), *Clostridium perfringens* spores or coliphages as an operational monitoring parameter may be more appropriate in certain circumstances.
  — *Heterotrophic bacteria* present in a supply can be a useful indicator of changes, such as increased microbial growth potential, increased biofilm activity, extended retention times or stagnation and a breakdown of integrity of the system. The numbers of heterotrophic bacteria present in a supply may reflect the presence of large contact surfaces within the treatment system, such as in-line filters, and may not be a direct indicator of the condition within the distribution system (see the supporting document *Heterotrophic Plate Counts and Drinking-water Safety*; section 1.3).
  — *Pressure measurement* and *turbidity* are also useful operational monitoring parameters in piped distribution systems.

Guidance for management of distribution system operation and maintenance is available (see the supporting document *Safe, Piped Water*; section 1.3) and includes the

Table 4.4 **Examples of operational monitoring parameters that can be used to monitor control measures**

Operational parameter	Raw water	Coagulation	Sedimentation	Filtration	Disinfection	Distribution system
pH		✓	✓		✓	✓
Turbidity (or particle count)	✓	✓	✓	✓	✓	✓
Dissolved oxygen	✓					
Stream/river flow	✓					
Rainfall	✓					
Colour	✓					
Conductivity (total dissolved solids, or TDS)	✓					
Organic carbon	✓		✓			
Algae, algal toxins and metabolites	✓					✓
Chemical dosage		✓			✓	
Flow rate		✓	✓	✓	✓	
Net charge		✓				
Streaming current value		✓				
Headloss				✓		
Ct[a]					✓	
Disinfectant residual					✓	✓
DBPs					✓	✓
Hydraulic pressure						✓

[a] Ct = Disinfectant concentration × contact time.

development of a monitoring programme for water quality and other parameters such as pressure.

Examples of operational monitoring parameters are provided in Table 4.4.

4.2.3 Establishing operational and critical limits

Control measures need to have defined limits for operational acceptability – termed operational limits – that can be applied to operational monitoring parameters. Operational limits should be defined for parameters applying to each control measure. If monitoring shows that an operational limit has been exceeded, then predetermined corrective actions (see section 4.4) need to be applied. The detection of the deviation and implementation of corrective action(s) should be possible in a time frame adequate to maintain performance and water safety.

For some control measures, a second series of "critical limits" may also be defined, outside of which confidence in water safety would be lost. Deviations from critical limits will usually require urgent action, including immediate notification of the appropriate health authority.

Operational and critical limits can be upper limits, lower limits, a range or an "envelope" of performance measures.

4. WATER SAFETY PLANS

4.2.4 Non-piped, community and household systems

Generally, surface water or shallow groundwater should not be used as a source of drinking-water without sanitary protection or treatment.

Monitoring of water sources (including rainwater tanks) by community operators or households will typically involve periodic sanitary inspection. The sanitary inspection forms used should be comprehensible and easy to use; for instance, the forms may be pictorial. The risk factors included should be preferably related to activities that are under the control of the operator and that may affect water quality. The links to action from the results of operational monitoring should be clear, and training will be required.

Operators should also undertake regular physical assessments of the water, especially after heavy rains, to monitor whether any obvious changes in water quality occur (e.g., changes in colour, odour or turbidity).

Treatment of water from community sources (such as boreholes, wells and springs) as well as household rainwater collection is rarely practised; however, if treatment is applied, then operational monitoring is advisable.

Collection, transportation and storage of water in the home

Maintaining the quality of water during collection and manual transport is the responsibility of the household. Good hygiene practices are required and should be supported through hygiene education. Hygiene education programmes should provide households and communities with skills to monitor and manage their water hygiene.

Household treatment of water has proven to be effective in delivery of public health gains. Monitoring of treatment processes will be specific to the technology. When household treatment is introduced, it is essential that information (and, where appropriate, training) be provided to users to ensure that they understand basic operational monitoring requirements.

4.3 Verification

> In addition to operational monitoring of the performance of the individual components of a drinking-water system, it is necessary to undertake final **verification** for reassurance that the system as a whole is operating safely. Verification may be undertaken by the supplier, by an independent authority or by a combination of these, depending on the administrative regime in a given country. It typically includes testing for faecal indicator organisms and hazardous chemicals.

Verification provides a final check on the overall safety of the drinking-water supply chain. Verification may be undertaken by the surveillance agency and/or can be a component of supplier quality control.

For microbial verification, testing is typically for faecal indicator bacteria in treated water and water in distribution. For verification of chemical safety, testing for chemicals of concern may be at the end of treatment, in distribution or at the point of consumption (depending on whether the concentrations are likely to change in distribution).

Frequencies of sampling should reflect the need to balance the benefits and costs of obtaining more information. Sampling frequencies are usually based on the population served or on the volume of water supplied, to reflect the increased population risk. Frequency of testing for individual characteristics will also depend on variability. Sampling and analysis are required most frequently for microbial and less often for chemical constituents. This is because even brief episodes of microbial contamination can lead directly to illness in consumers, whereas episodes of chemical contamination that would constitute an acute health concern, in the absence of a specific event (e.g., chemical overdosing at a treatment plant), are rare. Sampling frequencies for water leaving treatment depend on the quality of the water source and the type of treatment.

4.3.1 Verification of microbial quality

Verification of microbial quality of water in supply must be designed to ensure the best possible chance of detecting contamination. Sampling should therefore account for potential variations of water quality in distribution. This will normally mean taking account of locations and of times of increased likelihood of contamination.

Faecal contamination will not be distributed evenly throughout a piped distribution system. In systems where water quality is good, this significantly reduces the probability of detecting faecal indicator bacteria in the relatively few samples collected.

The chances of detecting contamination in systems reporting predominantly negative results for faecal indicator bacteria can be increased by using more frequent presence/absence (P/A) testing. P/A testing can be simpler, faster and less expensive than quantitative methods. Comparative studies of the P/A and quantitative methods demonstrate that the P/A methods can maximize the detection of faecal indicator bacteria. However, P/A testing is appropriate only in a system where the majority of tests for indicators provide negative results.

The more frequently the water is examined for faecal indicators, the more likely it is that contamination will be detected. Frequent examination by a simple method is more valuable than less frequent examination by a complex test or series of tests.

The nature and likelihood of contamination can vary seasonally, with rainfall and with other local conditions. Sampling should normally be random but should be increased at times of epidemics, flooding or emergency operations or following interruptions of supply or repair work.

4. WATER SAFETY PLANS

4.3.2 Verification of chemical quality

Issues that need to be addressed in developing chemical verification include the availability of appropriate analytical facilities, the cost of analyses, the possible deterioration of samples, the stability of the contaminant, the likely occurrence of the contaminant in various supplies, the most suitable point for monitoring and the frequency of sampling.

For a given chemical, the location and frequency of sampling will be determined by its principal sources (see chapter 8) and variability. Substances that do not change significantly in concentration over time require less frequent sampling than those that might vary significantly.

In many cases, source water sampling once per year, or even less, may be adequate, particularly in stable groundwaters, where the naturally occurring substances of concern will vary very slowly over time. Surface waters are likely to be more variable and require a greater number of samples, depending on the contaminant and its importance.

Sampling locations will depend on the water quality characteristic being examined. Sampling at the treatment plant or at the head of the distribution system may be sufficient for constituents where concentrations do not change during delivery. However, for those constituents that can change during distribution, sampling should be undertaken following consideration of the behaviour and/or source of the specific substance. Samples should include points near the extremities of the distribution system and taps connected directly to the mains in houses and large multi-occupancy buildings. Lead, for example, should be sampled at consumers' taps, since the source of lead is usually service connections or plumbing in buildings.

For further information, see the supporting document *Chemical Safety of Drinking-water* (section 1.3).

4.3.3 Water sources

Testing source waters is particularly important where there is no water treatment. It will also be useful following failure of the treatment process or as part of an investigation of a waterborne disease outbreak. The frequency of testing will depend on the reason that the sampling is being carried out. Testing frequency may be:

— on a regular basis (the frequency of verification testing will depend on several factors, including the size of the community supplied, the reliability of the quality of the drinking-water / degree of treatment and the presence of local risk factors);
— on an occasional basis (e.g., random or during visits to community-managed drinking-water supplies); and
— increased following degradation of source water quality resulting from predictable incidents, emergencies or unplanned events considered likely to increase the potential for a breakthrough in contamination (e.g., following a flood, upstream spills).

Prior to commissioning a new drinking-water supply, a wider range of analyses should be carried out, including parameters identified as potentially being present from a review of data from similar supplies or from a risk assessment of the source.

4.3.4 Piped distribution systems

The choice of sampling points will be dependent on the individual water supply. The nature of the public health risk posed by pathogens and the contamination potential throughout distribution systems mean that collection of samples for microbial analysis (and associated parameters, such as chlorine residual) will typically be done frequently and from dispersed sampling sites. Careful consideration of sampling points and frequency is required for chemical constituents that arise from piping and plumbing materials and that are not controlled through their direct regulation and for constituents that change in distribution, such as trihalomethanes (THMs).

Recommended minimum sample numbers for verification of the microbial quality of drinking-water are shown in Table 4.5.

The use of stratified random sampling in distribution systems has proven to be effective.

4.3.5 Verification for community-managed supplies

If the performance of a community drinking-water system is to be properly evaluated, a number of factors must be considered. Some countries that have developed national strategies for the surveillance and quality control of drinking-water systems have adopted *quantitative service indicators* (i.e., quality, quantity, accessibility, coverage, affordability and continuity) for application at community, regional and national levels. Usual practice would be to include the critical parameters for microbial quality (normally *E. coli*, chlorine, turbidity and pH) and for a sanitary inspection to be carried out. Methods for these tests must be standardized and approved. It is recommended that field test kits be validated for performance against reference or standard methods and approved for use in verification testing.

Together, service indicators provide a basis for setting targets for community drinking-water supplies. They serve as a quantitative guide to the adequacy of drink-

Table 4.5 Recommended minimum sample numbers for faecal indicator testing in distribution systems[a]

Population	Total number of samples per year
Point sources	Progressive sampling of all sources over 3- to 5-year cycles (maximum)
Piped supplies	
<5000	12
5000–100 000	12 per 5000 head of population
>100 000–500 000	12 per 10 000 head of population plus an additional 120 samples
>500 000	12 per 100 000 head of population plus an additional 180 samples

[a] Parameters such as chlorine, turbidity and pH should be tested more frequently as part of operational and verification monitoring.

ing-water supplies and provide consumers with an objective measure of the quality of the overall service and thus the degree of public health protection afforded.

Periodic testing and sanitary inspection of community drinking-water supplies should typically be undertaken by the surveillance agency and should assess microbial hazards and known problem chemicals (see also chapter 5). Frequent sampling is unlikely to be possible, and one approach is therefore a rolling programme of visits to ensure that each supply is visited once every 3–5 years. The primary purpose is to inform strategic planning and policy rather than to assess compliance of individual drinking-water supplies. Comprehensive analysis of chemical quality of all sources is recommended prior to commissioning as a minimum and preferably every 3–5 years thereafter.

Advice on the design of sampling programmes and on the frequency of sampling is given in ISO standards (Table 4.6).

4.3.6 Quality assurance and quality control

Appropriate quality assurance and analytical quality control procedures should be implemented for all activities linked to the production of drinking-water quality data. These procedures will ensure that the data are fit for purpose – in other words, that the results produced are of adequate accuracy. Fit for purpose, or adequate accuracy, will be defined in the water quality monitoring programme, which will include a statement about accuracy and precision of the data. Because of the wide range of substances, methods, equipment and accuracy requirements likely to be involved in the monitoring of drinking-water, many detailed, practical aspects of analytical quality control are concerned. These are beyond the scope of this publication.

The design and implementation of a quality assurance programme for analytical laboratories are described in detail in *Water Quality Monitoring* (Bartram & Ballance,

Table 4.6 International Organization for Standardization (ISO) standards for water quality giving guidance on sampling

ISO standard no.	Title (water quality)
5667–1:1980	Sampling – Part 1: Guidance on the design of sampling programmes
5667–2:1991	Sampling – Part 2: Guidance on sampling techniques
5667–3:1994	Sampling – Part 3: Guidance on the preservation and handling of samples
5667–4:1987	Sampling – Part 4: Guidance on sampling from lakes, natural and man-made
5667–5:1991	Sampling – Part 5: Guidance on sampling of drinking-water and water used for food and beverage processing
5667–6:1990	Sampling – Part 6: Guidance on sampling of rivers and streams
5667–13:1997	Sampling – Part 13: Guidance on sampling of sludges from sewage and water-treatment works
5667–14:1998	Sampling – Part 14: Guidance on quality assurance of environmental water sampling and handling
5667–16:1998	Sampling – Part 16: Guidance on biotesting of samples
5668–17:2000	Sampling – Part 17: Guidance on sampling of suspended sediments
13530:1997	Water quality – Guide to analytical control for water analysis

1996). The relevant chapter draws upon the standard ISO 17025:2000 *General requirements for the competence of testing and calibration laboratories*, which provides a framework for the management of quality in analytical laboratories.

4.4 Management procedures for piped distribution systems

> Effective management implies definition of actions to be taken in response to variations that occur during normal operational conditions; of actions to be taken in specific "incident" situations where a loss of control of the system may occur; and of procedures to be followed in unforeseen and emergency situations. Management procedures should be documented alongside system assessment, monitoring plans, supporting programmes and communication required to ensure safe operation of the system.

Much of a management plan will describe actions to be taken in response to "normal" variation in operational monitoring parameters in order to maintain optimal operation in response to operational monitoring parameters reaching operational limits.

A significant deviation in operational monitoring where a critical limit is exceeded (or in verification) is often referred to as an "incident." An incident is any situation in which there is reason to suspect that water being supplied for drinking may be, or may become, unsafe (i.e., confidence in water safety is lost). As part of a WSP, management procedures should be defined for response to predictable incidents as well as unpredictable incidents and emergencies. Incident triggers could include:

— non-compliance with operational monitoring criteria;
— inadequate performance of a sewage treatment plant discharging to source water;
— spillage of a hazardous substance into source water;
— failure of the power supply to an essential control measure;
— extreme rainfall in a catchment;
— detection of unusually high turbidity (source or treated water);
— unusual taste, odour or appearance of water;
— detection of microbial indicator parameters, including unusually high faecal indicator densities (source or treated water) and unusually high pathogen densities (source water); and
— public health indicators or a disease outbreak for which water is a suspect vector.

Incident response plans can have a range of alert levels. These can be minor early warning, necessitating no more than additional investigation, through to emergency. Emergencies are likely to require the resources of organizations beyond the drinking-water supplier, particularly the public health authorities.

Incident response plans typically comprise:

— accountabilities and contact details for key personnel, often including several organizations and individuals;

- lists of measurable indicators and limit values/conditions that would trigger incidents, along with a scale of alert levels;
- clear description of the actions required in response to alerts;
- location and identity of the standard operating procedures (SOPs) and required equipment;
- location of backup equipment;
- relevant logistical and technical information; and
- checklists and quick reference guides.

The plan may need to be followed at very short notice, so standby rosters, effective communication systems and up-to-date training and documentation are required.

Staff should be trained in response to ensure that they can manage incidents and/or emergencies effectively. Incident and emergency response plans should be periodically reviewed and practised. This improves preparedness and provides opportunities to improve the effectiveness of plans before an emergency occurs.

Following any incident or emergency, an investigation should be undertaken involving all concerned staff. The investigation should consider factors such as:

- What was the cause of the problem?
- How was the problem first identified or recognized?
- What were the most essential actions required?
- What communication problems arose, and how were they addressed?
- What were the immediate and longer-term consequences?
- How well did the emergency response plan function?

Appropriate documentation and reporting of the incident or emergency should also be established. The organization should learn as much as possible from the incident or emergency to improve preparedness and planning for future incidents. Review of the incident or emergency may indicate necessary amendments to existing protocols.

The preparation of clear procedures, definition of accountability and provision of equipment for the sampling and storing of water in the event of an incident can be valuable for follow-up epidemiological or other investigations, and the sampling and storage of water from early on during a suspected incident should be part of the response plan.

4.4.1 Predictable incidents ("deviations")

Many incidents (e.g., exceedance of a critical limit) can be foreseen, and management plans can specify resulting actions. Actions may include, for example, temporary change of water sources (if possible), increasing coagulation dose, use of backup disinfection or increasing disinfectant concentrations in distribution systems.

4.4.2 Unforeseen events

Some scenarios that lead to water being considered potentially unsafe might not be specifically identified within incident response plans. This may be either because the events were unforeseen or because they were considered too unlikely to justify preparing detailed corrective action plans. To allow for such events, a general incident response plan should be developed. The plan would be used to provide general guidance on identifying and handling of incidents along with specific guidance on responses that would be applied to many different types of incident.

A protocol for situation assessment and declaring incidents would be provided in a general incident response plan that includes personal accountabilities and categorical selection criteria. The selection criteria may include:

— time to effect;
— population affected; and
— nature of the suspected hazard.

The success of general incident responses depends on the experience, judgement and skill of the personnel operating and managing the drinking-water systems. However, generic activities that are common in response to many incidents can be incorporated within general incident response plans. For example, for piped systems, emergency flushing SOPs can be prepared and tested for use in the event that contaminated water needs to be flushed from a piped system. Similarly, SOPs for rapidly changing or bypassing reservoirs can be prepared, tested and incorporated. The development of such a "toolkit" of supporting material limits the likelihood of error and speeds up responses during incidents.

4.4.3 Emergencies

Water suppliers should develop plans to be invoked in the event of an emergency. These plans should consider potential natural disasters (e.g., earthquakes, floods, damage to electrical equipment by lightning strikes), accidents (e.g., spills in the watershed), damage to treatment plant and distribution system and human actions (e.g., strikes, sabotage). Emergency plans should clearly specify responsibilities for coordinating measures to be taken, a communication plan to alert and inform users of the drinking-water supply and plans for providing and distributing emergency supplies of drinking-water.

Plans should be developed in consultation with relevant regulatory authorities and other key agencies and should be consistent with national and local emergency response arrangements. Key areas to be addressed in emergency response plans include:

— response actions, including increased monitoring;
— responsibilities and authorities internal and external to the organization;
— plans for emergency drinking-water supplies;
— communication protocols and strategies, including notification procedures (internal, regulatory body, media and public); and
— mechanisms for increased public health surveillance.

4. WATER SAFETY PLANS

During an emergency in which there is evidence of faecal contamination of the drinking-water supply, it may be necessary either to modify the treatment of existing sources or to temporarily use alternative sources of drinking-water. It may be necessary to increase disinfection at source or to rechlorinate during distribution.

If microbial quality cannot be maintained, it may be necessary to advise consumers to boil the water during the emergency (see section 4.4.4). Boiling water itself has health risks (e.g., scalding), and initiating superchlorination and undertaking immediate corrective measures may be preferable.

In emergencies, such as during outbreaks of potentially waterborne disease or when faecal contamination of a drinking-water supply is detected, the concentration of free chlorine should be increased to greater than 0.5 mg/litre throughout the system as a minimum immediate response.

It is impossible to give general guidance concerning emergencies in which chemicals cause massive contamination of the drinking-water supply, caused either by accident or by deliberate action. The guideline values recommended in these Guidelines (see section 8.5 and Annex 4) relate to a level of exposure that is regarded as tolerable throughout life; acute toxic effects are not normally considered. The length of time during which exposure to a chemical far in excess of the guideline value would be toxicologically detrimental will depend upon factors that vary from contaminant to contaminant. In an emergency situation, the public health authorities should be consulted about appropriate action.

4.4.4 Closing supply, water avoidance and "boil water" orders

Incident response plans for emergencies and unplanned events should include an evaluation of the basis for issuing water avoidance and boil water orders. The objective of the order should be taken in the public interest, and the order will typically be managed by public health authorities.

A decision to close a drinking-water supply carries an obligation to provide an alternative safe supply and is very rarely justifiable because of the adverse effects, especially to health, of restricting access to water.

Issuing a boil water order is a serious measure that should be undertaken only when the public health authority, having consulted with the incident response team, is convinced of an ongoing risk to health from drinking water, which outweighs any risk from the boil water order itself. The public interest is not always best served by boil water orders, which can have negative public health consequences through scalds and anxiety. In addition, if boil water notices are issued frequently or are left in place for long periods, the public response will decrease. If a notice is issued, advice must be clear and easy to understand, or it may be ignored because of confusion over what to do. When issuing a boil water notice, it is good practice to establish criteria for removing the notice.

4.4.5 Preparing a monitoring plan

Programs should be developed for operational and verification monitoring and documented as part of a WSP, detailing the strategies and procedures to follow for monitoring the various aspects of the drinking-water system. The monitoring plans should be fully documented and should include the following information:

- parameters to be monitored;
- sampling or assessment location and frequency;
- sampling or assessment methods and equipment;
- schedules for sampling or assessment;
- methods for quality assurance and validation of results;
- requirements for checking and interpreting results;
- responsibilities and necessary qualifications of staff;
- requirements for documentation and management of records, including how monitoring results will be recorded and stored; and
- requirements for reporting and communication of results.

4.4.6 Supporting programmes

Many actions are important in ensuring drinking-water safety but do not directly affect drinking-water quality and are therefore not control measures. These are referred to as "supporting programmes" and should also be documented in a WSP.

> Actions that are important in ensuring drinking-water safety but do not directly affect drinking-water quality are referred to as supporting programmes.

Supporting programmes could involve:

- controlling access to treatment plants, catchments and reservoirs, and implementing the appropriate security measures to prevent transfer of hazards from people when they do enter source water;
- developing verification protocols for the use of chemicals and materials in the drinking-water supply – for instance, to ensure the use of suppliers that participate in quality assurance programmes;
- using designated equipment for attending to incidents such as mains bursts (e.g., equipment should be designated for potable water work only and not for sewage work); and
- training and educational programmes for personnel involved in activities that could influence drinking-water safety; training should be implemented as part of induction programmes and frequently updated.

Supporting programmes will consist almost entirely of items that drinking-water suppliers and handlers will ordinarily have in place as part of their normal operation. For most, the implementation of supporting programmes will involve:

4. WATER SAFETY PLANS

— collation of existing operational and management practices;
— initial and, thereafter, periodic review and updating to continually improve practices;
— promotion of good practices to encourage their use; and
— audit of practices to check that they are being used, including taking corrective actions in case of non-conformance.

Codes of good operating and management practice and hygienic working practice are essential elements of supporting programmes. These are often captured within SOPs. They include, but are not limited to:

— hygienic working practices documented in maintenance SOPs;
— attention to personal hygiene;
— training and competence of personnel involved in drinking-water supply;
— tools for managing the actions of staff, such as quality assurance systems;
— securing stakeholder commitment, at all levels, to the provision of safe drinking-water;
— education of communities whose activities may influence drinking-water quality;
— calibration of monitoring equipment; and
— record keeping.

Comparison of one set of supporting programmes with the supporting programmes of other suppliers, through peer review, benchmarking and personnel or document exchange, can stimulate ideas for improved practice.

Supporting programmes can be extensive, be varied and involve multiple organizations and individuals. Many supporting programmes involve water resource protection measures and typically include aspects of land use control. Some water resource protection measures are engineered, such as effluent treatment processes and stormwater management practices that may be used as control measures.

4.5 Management of community and household water supplies

Community drinking-water supplies worldwide are more frequently contaminated than larger drinking-water supplies, may be more prone to operating discontinuously (or intermittently) and break down or fail more frequently.

To ensure safe drinking-water, the focus in small supplies should be on:

— informing the public;
— assessing the water supply to determine whether it is able to meet identified health-based targets (see section 4.1);
— monitoring identified control measures and training operators to ensure that all likely hazards can be controlled and that risks are maintained at a tolerable level (see section 4.2);
— operational monitoring of the drinking-water system (see section 4.2);

— implementing systematic water quality management procedures (see section 4.4.1), including documentation and communication (see section 4.6);
— establishing appropriate incident response protocols (usually encompassing actions at the individual supply, backed by training of operators, and actions required by local or national authorities) (see sections 4.4.2, 4.4.3 and 4.4.4); and
— developing programmes to upgrade and improve existing water delivery (usually defined at a national or regional level rather than at the level of individual supplies) (see section 4.1.8).

For point sources serving communities or individual households, the emphasis should be on selecting the best available quality source water and on protecting its quality by the use of multiple barriers (usually within source protection) and maintenance programmes. Whatever the source (groundwater, surface water or rainwater tanks), communities and householders should assure themselves that the water is safe to drink. Generally, surface water and shallow groundwater under the direct influence of surface water (which includes shallow groundwater with preferential flow paths) should receive treatment.

The parameters recommended for the minimum monitoring of community supplies are those that best establish the hygienic state of the water and thus the risk of waterborne disease. The essential parameters of water quality are *E. coli* – thermotolerant (faecal) coliforms are accepted as suitable substitutes – and chlorine residual (if chlorination is practised).

These should be supplemented, where appropriate, by pH adjustment (if chlorination is practised) and measurement of turbidity.

These parameters may be measured on site using relatively unsophisticated testing equipment. On-site testing is essential for the determination of turbidity and chlorine residual, which change rapidly during transport and storage; it is also important for the other parameters where laboratory support is lacking or where transportation problems would render conventional sampling and analysis impractical.

Other health-related parameters of local significance should also be measured. The overall approach to control of chemical contamination is outlined in chapter 8.

4.6 Documentation and communication

Documentation of a WSP should include:

— description and assessment of the drinking-water system (see section 4.1), including programmes to upgrade and improve existing water delivery (see section 4.1.8);
— the plan for operational monitoring and verification of the drinking-water system (see section 4.2);

— water safety management procedures for normal operation, incidents (specific and unforeseen) and emergency situations (see sections 4.4.1, 4.4.2 and 4.4.3), including communication plans; and
— description of supporting programmes (see section 4.4.6).

Records are essential to review the adequacy of the WSP and to demonstrate the adherence of the drinking-water system to the WSP. Five types of records are generally kept:

— supporting documentation for developing the WSP including validation;
— records and results generated through operational monitoring and verification;
— outcomes of incident investigations;
— documentation of methods and procedures used; and
— records of employee training programmes.

By tracking records generated through operational monitoring and verification, an operator or manager can detect that a process is approaching its operational or critical limit. Review of records can be instrumental in identifying trends and in making operational adjustments. Periodic review of WSP records is recommended so that trends can be noted and appropriate actions decided upon and implemented. Records are also essential when surveillance is implemented through auditing-based approaches.

Communication strategies should include:

— procedures for promptly advising of any significant incidents within the drinking-water supply, including notification of the public health authority;
— summary information to be made available to consumers – for example, through annual reports and on the Internet; and
— establishment of mechanisms to receive and actively address community complaints in a timely fashion.

The right of consumers to health-related information on the water supplied to them for domestic purposes is fundamental. However, in many communities, the simple right of access to information will not ensure that individuals are aware of the quality of the water supplied to them; furthermore, the probability of consuming unsafe water may be relatively high. The agencies responsible for monitoring should therefore develop strategies for disseminating and explaining the significance of health-related information. Further information on communication is provided in section 5.5.

5
Surveillance

Drinking-water supply surveillance is "the continuous and vigilant public health assessment and review of the safety and acceptability of drinking-water supplies" (WHO, 1976). This surveillance contributes to the protection of public health by promoting improvement of the quality, quantity, accessibility, coverage, affordability and continuity of water supplies (known as service indicators) and is complementary to the quality control function of the drinking-water supplier. Drinking-water supply surveillance does not remove or replace the responsibility of the drinking-water supplier to ensure that a drinking-water supply is of acceptable quality and meets predetermined health-based and other performance targets.

All members of the population receive drinking-water by some means – including the use of piped supplies with or without treatment and with or without pumping (supplied via domestic connection or public standpipe), delivery by tanker truck or carriage by beasts of burden or collection from groundwater sources (springs or wells) or surface sources (lakes, rivers and streams). It is important for the surveillance agency to build up a picture of the frequency of use of the different types of supply, especially as a preliminary step in the planning of a surveillance programme. There is little to be gained from surveillance of piped water supplies alone if these are available to only a small proportion of the population or if they represent a minority of supplies.

Information alone does not lead to improvement. Instead, the effective management and use of the information generated by surveillance make possible the rational improvement of water supplies – where "rational" implies that available resources are used for maximum public health benefit.

Surveillance is an important element in the development of strategies for incremental improvement of the quality of drinking-water supply services. It is important that strategies be developed for implementing surveillance, collating, analysing and summarizing data and reporting and disseminating the findings and are accompanied by recommendations for remedial action. Follow-up will be required to ensure that remedial action is taken.

Surveillance extends beyond drinking-water supplies operated by a discrete drinking-water supplier to include drinking-water supplies that are managed by communities and includes assurance of good hygiene in the collection and storage of household water.

The surveillance agency must have, or have access to, legal expertise in addition to expertise on drinking-water and water quality (see section 2.3.1). Drinking-water supply surveillance is also used to ensure that any transgressions that may occur are appropriately investigated and resolved. In many cases, it will be more appropriate to use surveillance as a mechanism for collaboration between public health agencies and drinking-water suppliers to improve drinking-water supply than to resort to enforcement, particularly where the problem lies mainly with community-managed drinking-water supplies.

The authorities responsible for drinking-water supply surveillance may be the public health ministry or other agency (see section 1.2.1), and their roles encompass four areas of activity:

— public health oversight of organized drinking-water supplies;
— public health oversight and information support to populations without access to organized drinking-water supplies, including communities and households;
— consolidation of information from diverse sources to enable understanding of the overall drinking-water supply situation for a country or region as a whole as an input to the development of coherent public health-centred policies and practices; and
— participation in the investigation, reporting and compilation of outbreaks of waterborne disease.

A drinking-water supply surveillance programme should normally include processes for approval of WSPs. This approval will normally involve review of the system assessment, of the identification of appropriate control measures and supporting programmes and of operational monitoring and management plans. It should ensure that the WSP covers normal operating conditions and predictable incidents (deviations) and has contingency plans in case of an emergency or unforeseen event.

The surveillance agency may also support or undertake the development of WSPs for community-managed drinking-water supplies and household water storage. Such plans may be generic for particular technologies rather than specific for individual systems.

5.1 Types of approaches

There are two types of approaches to surveillance of drinking-water quality: audit-based approaches and approaches relying on direct assessment. Implementation of surveillance will generally include a mixture of these approaches according to supply type and may involve using rolling programmes whereby systems are addressed progressively. Often it is not possible to undertake extensive surveillance of all community or household supplies. In these cases, well designed surveys should be undertaken in order to understand the situation at the national or regional level.

5.1.1 Audit

In the audit approach to surveillance, assessment activities, including verification testing, are undertaken largely by the supplier, with third-party auditing to verify compliance. It is increasingly common that analytical services are procured from accredited external laboratories. Some authorities are also experimenting with the use of such arrangements for services such as sanitary inspection, sampling and audit reviews.

An audit approach requires the existence of a stable source of expertise and capacity within the surveillance agency in order to:

— review and approve new WSPs;
— undertake or oversee auditing of the implementation of individual WSPs as a programmed routine activity; and
— respond to, investigate and provide advice on receipt of reports on significant incidents.

Periodic audit of implementation of WSPs is required:

— at intervals (the frequency of routine audits will be dependent on factors such as the size of the population served and the nature and quality of source water / treatment facilities);
— following substantial changes to the source, the distribution or storage system or treatment process; and
— following significant incidents.

Periodic audit would normally include the following elements, in addition to review of the WSP:

— examination of records to ensure that system management is being carried out as described in the WSP;
— ensuring that operational monitoring parameters are kept within operational limits and that compliance is being maintained;
— ensuring that verification programmes are operated by the water supplier (either through in-house expertise or through a third-party arrangement);
— assessment of supporting programmes and of strategies for improvement and updating of the WSP; and
— in some circumstances, sanitary inspection, which may cover the whole of the drinking-water system, including sources, transmission infrastructure, treatment plants, storage reservoirs and distribution systems.

In response to reports of significant incidents, it is necessary to ensure that:

— the event is investigated promptly and appropriately;
— the cause of the event is determined and corrected;

— the incident and corrective action are documented and reported to appropriate authorities; and
— the WSP is reassessed to avoid the occurrence of a similar situation.

The implementation of an audit-based approach places responsibility on the drinking-water supplier to provide the surveillance agency with information regarding system performance against agreed indicators. In addition, a programme of announced and unannounced visits by auditors to drinking-water suppliers should be implemented to review documentation and records of operational practice in order to ensure that data submitted are reliable. Such an approach does not necessarily imply that water suppliers are likely to falsify records, but it does provide an important means of reassuring consumers that there is true independent verification of the activities of the water supplier. The surveillance agency will normally retain the authority to undertake some analysis of drinking-water quality to verify performance or enter into a third-party arrangement for such analysis.

5.1.2 Direct assessment

It may be appropriate for the drinking-water supply surveillance agency to carry out independent testing of water supplies. Such an approach often implies that the agency has access to analytical facilities of its own, with staff trained to carry out sampling, analysis and sanitary inspection.

Direct assessment also implies that surveillance agencies have the capacity to assess findings and to report to and advise suppliers and communities.

A surveillance programme based on direct assessment would normally include:

— specified approaches to large municipality / small municipality / community supplies and individual household supplies;
— sanitary inspections to be carried out by qualified personnel;
— sampling to be carried out by qualified personnel;
— tests to be conducted using suitable methods by accredited laboratories or using approved field testing equipment and qualified personnel; and
— procedures on reporting findings and follow-up to ensure that they have been acted on.

For community-managed drinking-water supplies and where the development of in-house verification or third-party arrangements is limited, direct assessment may be used as the principal system of surveillance. This may apply to drinking-water supplies in small towns by small-scale private sector operators or local government. Direct assessment may lead to the identification of requirements to amend or update the WSP, and the process to be followed when undertaking such amendments should be clearly identified.

Where direct assessment is carried out by the surveillance agency, it complements other verification testing. General guidance on verification testing, which is also applicable to surveillance through direct assessment, is provided in section 4.3.

5.2 Adapting approaches to specific circumstances

5.2.1 Urban areas in developing countries

Drinking-water supply arrangements in urban areas of developing countries are typically complex. There will often be a large piped supply with household and public connections and a range of alternative drinking-water supplies, including point sources and vended water. In these situations, the surveillance programme should take account of the different sources of drinking-water and the potential for deterioration in quality during collection, storage and use. Furthermore, the population will vary in terms of socioeconomic status and vulnerability to water-related disease.

In many situations, zoning the urban area on the basis of vulnerability and drinking-water supply arrangements is required. The zoning system should include all populations within the urban area, including informal and periurban settlements, regardless of their legal status, in order to direct resources to where greatest improvements (or benefits) to public health will be achieved. This provides a mechanism to ensure that non-piped drinking-water sources are also included within drinking-water supply surveillance activities.

Experience has shown that zoning can be developed using qualitative and quantitative methods and is useful in identifying vulnerable groups and priority communities where drinking-water supply improvements are required.

5.2.2 Surveillance of community drinking-water supplies

Small community-managed drinking-water supplies are found in most countries and may be the predominant form of drinking-water supply for large sections of the population. The precise definition of a "community drinking-water supply" will vary, but administration and management arrangements are often what set community supplies apart. Community-managed supplies may include simple piped water systems or a range of point sources, such as boreholes with hand pumps, dug wells and protected springs.

The control of water safety and implementation of surveillance programmes for such supplies often face significant constraints. These typically include:

— limited capacity and skills within the community to undertake process control and verification; this may increase the need both for surveillance to assess the state of drinking-water supplies and for surveillance staff to provide training and support to community members; and
— the very large number of widely dispersed supplies, which significantly increases overall costs in undertaking surveillance activities.

Furthermore, it is often these supplies that present the greatest water quality problems.

Experience from both developing and developed countries has shown that surveillance of community-managed drinking-water supplies can be effective when well designed and when the objectives are geared more towards a supportive role to enhance community management and evaluation of overall strategies to their support than towards enforcement of compliance.

Surveillance of community drinking-water supplies requires a systematic programme of surveys that encompass all aspects of the drinking-water supply to the population as a whole, including sanitary inspection (including catchments) and institutional and community aspects. Surveillance should address variability in source water quality, treatment process efficacy and the quality of distributed or household-treated and household-stored water.

Experience has also shown that the role of surveillance may include health education and health promotion activities to improve healthy behaviour and management of drinking-water supply and sanitation. Participatory activities can include sanitary inspection by communities and, where appropriate, community-based testing of drinking-water quality using affordable field test kits and other accessible testing resources.

In the evaluation of overall strategies, the principal aim should be to derive overall lessons for improving water safety for all community supplies, rather than relying on monitoring the performance of individual supplies.

Frequent visits to every individual supply may be impractical because of the very large numbers of such supplies and the limitations of resources for such visits. However, surveillance of large numbers of community supplies can be achieved through a rolling programme of visits. Commonly, the aim will be to visit each supply periodically (once every 3–5 years at a minimum) using either stratified random sampling or cluster sampling to select specific supplies to be visited. During each visit, sanitary inspection and water quality analysis will normally be done to provide insight to contamination and its causes.

During each visit, testing of water stored in the home may be undertaken in a sample of households. The objective for such testing is to determine whether contamination occurs primarily at the source or within the home. This will allow evaluation of the need for investment in supply improvement or education on good hygiene practices for household treatment and safe storage. Household testing may also be used to evaluate the impact of a specific hygiene education programme.

5.2.3 Surveillance of household treatment and storage systems

Where water is handled during storage in households, it may be vulnerable to contamination, and sampling of household-stored water is of interest in independent surveillance. It is often undertaken on a "survey" basis to develop insights into the extent and nature of prevailing problems.

Surveillance systems managed by public health authorities for drinking-water supplies using household treatment and household storage containers are therefore recommended. The principal focus of surveillance of household-based interventions will be assessment of their acceptance and impact through sample surveys so as to evaluate and inform overall strategy development and refinement.

5.3 Adequacy of supply

As the drinking-water supply surveillance agency has an interest in the health of the population at large, its interest extends beyond water quality to include all aspects of the adequacy of drinking-water supply for the protection of public health.

In undertaking an assessment of the adequacy of the drinking-water supply, the following basic service parameters of a drinking-water supply should normally be taken into consideration:

- *Quality*: whether the supply has an approved WSP (see chapter 4) that has been validated and is subject to periodic audit to demonstrate compliance (see chapter 3);
- *Quantity (service level)*: the proportion of the population using water from different levels of drinking-water supply (e.g., no access, basic access, intermediate access and optimal access)
- *Accessibility*: the percentage of the population that has reasonable access to an improved drinking-water supply;
- *Affordability*: the tariff paid by domestic consumers; and
- *Continuity*: the percentage of the time during which drinking-water is available (daily, weekly and seasonally).

5.3.1 Quantity (service level)

The quantity of water collected and used by households has an important influence on health. There is a basic human physiological requirement for water to maintain adequate hydration and an additional requirement for food preparation. There is a further requirement for water to support hygiene, which is necessary for health.

Estimates of the volume of water needed for health purposes vary widely. In deriving WHO guideline values, it is assumed that the daily per capita consumption of drinking-water is approximately 2 litres for adults, although actual consumption varies according to climate, activity level and diet. Based on currently available data, a minimum volume of 7.5 litres per capita per day will provide sufficient water for hydration and incorporation into food for most people under most conditions. In addition, adequate domestic water is needed for food preparation, laundry and personal and domestic hygiene, which are also important for health. Water may also be important in income generation and amenity uses.

The quantities of water collected and used by households are primarily a function of the distance to the water supply or total collection time required. This broadly equates to the level of service. Four levels of service can be defined, as shown in Table 5.1.

Service level is a useful and easily measured indicator that provides a valid surrogate for the quantity of water collected by households and is the preferred indicator for surveillance. Available evidence indicates that health gains accrue from improving

Table 5.1 Service level and quantity of water collected

Service level	Distance/time	Likely volumes of water collected	Public health risk from poor hygiene	Intervention priority and actions
No access	More than 1 km / more than 30 min round-trip	Very low – 5 litres per capita per day	**Very high** Hygiene practice compromised Basic consumption may be compromised	**Very high** Provision of basic level of service Hygiene education
Basic access	Within 1 km / within 30 min round-trip	Average approximately 20 litres per capita per day	**High** Hygiene may be compromised Laundry may occur off-plot	**High** Hygiene education Provision of improved level of service
Intermediate access	Water provided on-plot through at least one tap (yard level)	Average approximately 50 litres per capita per day	**Low** Hygiene should not be compromised Laundry likely to occur on-plot	**Low** Hygiene promotion still yields health gains Encourage optimal access
Optimal access	Supply of water through multiple taps within the house	Average 100–200 litres per capita per day	**Very low** Hygiene should not be compromised Laundry will occur on-plot	**Very low** Hygiene promotion still yields health gains

Source: Howard & Bartram (2003).

service level in two key stages: the delivery of water within 1 km or 30 min total collection time; and when supplied to a yard level of service. Further health gains are likely to occur once water is supplied through multiple taps, as this will increase water availability for diverse hygiene practices. The volume of water collected may also depend on the reliability and cost of water. Therefore, collection of data on these indicators is important.

5.3.2 Accessibility

From the public health standpoint, the proportion of the population with reliable access to safe drinking-water is the most important single indicator of the overall success of a drinking-water supply programme.

There are a number of definitions of access (or coverage), many with qualifications regarding safety or adequacy. The preferred definition is that used by WHO and UNICEF in their "Joint Monitoring Programme," which defines "reasonable access" to improved sources as being "availability of at least 20 litres per person per day within one kilometre of the user's dwelling." Improved and unimproved water supply technologies in the WHO/UNICEF Joint Monitoring Programme have been defined in terms of providing "reasonable access," as summarized below:

- **Improved water supply technologies:**
  - Household connection
  - Public standpipe
  - Borehole
  - Protected dug well
  - Protected spring
  - Rainwater collection
- **Unimproved water supply technologies:**
  - Unprotected well
  - Unprotected spring
  - Vendor-provided water
  - Bottled water
  - Tanker truck provision of water.

5.3.3 Affordability

The affordability of water has a significant influence on the use of water and selection of water sources. Households with the lowest levels of access to safe water supply frequently pay more for their water than do households connected to a piped water system. The high cost of water may force households to use alternative sources of water of poorer quality that represent a greater risk to health. Furthermore, high costs of water may reduce the volumes of water used by households, which in turn may influence hygiene practices and increase risks of disease transmission.

When assessing affordability, it is important to collect data on the price at the point of purchase. Where households are connected to the drinking-water supplier, this will be the tariff applied. Where water is purchased from public standpipes or from neighbours, the price at the point of purchase may be very different from the drinking-water supplier tariff. Many alternative water sources (notably vendors) also involve costs, and these costs should be included in evaluations of affordability. In addition to recurrent costs, the costs for initial acquisition of a connection should also be considered when evaluating affordability.

5.3.4 Continuity

Interruptions to drinking-water supply either through intermittent sources or resulting from engineering inefficiencies are a major determinant of the access to and quality of drinking-water. Analysis of data on continuity of supply requires the consideration of several components. Continuity can be classified as follows:

- year-round service from a reliable source with no interruption of flow at the tap or source;
- year-round service with frequent (daily or weekly) interruptions, of which the most common causes are:

- —restricted pumping regimes in pumped systems, whether planned or due to power failure or sporadic failure;
  - —peak demand exceeding the flow capacity of the transmission mains or the capacity of the reservoir;
  - —excessive leakage within the distribution systems;
  - —excessive demands on community-managed point sources;
- seasonal service variation resulting from source fluctuation, which typically has three causes:
  - —natural variation in source volume during the year;
  - —volume limitation because of competition with other uses such as irrigation;
  - —periods of high turbidity when the source water may be untreatable; and
- compounded frequent and seasonal discontinuity.

This classification reflects broad categories of continuity, which are likely to affect hygiene in different ways. Daily or weekly discontinuity results in low supply pressure and a consequent risk of in-pipe recontamination. Other consequences include reduced availability and lower volume use, which adversely affect hygiene. Household water storage may be necessary, and this may lead to an increase in the risk of contamination during such storage and associated handling. Seasonal discontinuity often forces users to obtain water from inferior and distant sources. As a consequence, in addition to the obvious reduction in quality and quantity, time is lost in water collection.

5.4 Planning and implementation

For drinking-water supply surveillance to lead to improvements in drinking-water supply, it is vital that the mechanisms for promoting improvement are recognized and used.

The focus of drinking-water supply improvement (whether as investment priority at regional or national levels, development of hygiene education programmes or enforcement of compliance) will depend on the nature of the drinking-water supplies and the types of problems identified. A checklist of mechanisms for drinking-water supply improvement based on the output of surveillance is given below:

- **Establishing national priorities** – When the most common problems and shortcomings in drinking-water systems have been identified, national strategies can be formulated for improvements and remedial measures; these might include changes in training (of managers, administrators, engineers or field staff), rolling programmes for rehabilitation or improvement or changes in funding strategies to target specific needs.
- **Establishing regional priorities** – Regional offices of drinking-water supply agencies can decide which communities to work in and which remedial activities are priorities; public health criteria should be considered when priorities are set.

- **Establishing hygiene education programmes** – Not all of the problems revealed by surveillance are technical in nature, and not all are solved by drinking-water suppliers; surveillance also looks at problems involving community and household supplies, water collection and transport and household treatment and storage. The solutions to many of these problems are likely to require educational and promotional activities.
- **Auditing of WSPs and upgrading** – The information generated by surveillance can be used to audit WSPs and to assess whether these are in compliance. Systems and their associated WSPs should be upgraded where they are found to be deficient, although feasibility must be considered, and enforcement of upgrading should be linked to strategies for progressive improvement.
- **Ensuring community operation and maintenance** – Support should be provided by a designated authority to enable community members to be trained so that they are able to assume responsibility for the operation and maintenance of community drinking-water supplies.
- **Establishing public awareness and information channels** – Publication of information on public health aspects of drinking-water supplies, water quality and the performance of suppliers can encourage suppliers to follow good practices, mobilize public opinion and response and reduce the need for regulatory enforcement, which should be an option of last resort.

In order to make best use of limited resources where surveillance is not yet practised, it is advisable to start with a basic programme that develops in a planned manner. Activities in the early stages should generate enough useful data to demonstrate the value of surveillance. Thereafter, the objective should be to progress to more advanced surveillance as resources and conditions permit.

The activities normally undertaken in the initial, intermediate and advanced stages of development of drinking-water supply surveillance are summarized as follows:

- **Initial phase:**
  - Establish requirements for institutional development.
  - Provide training for staff involved in programme.
  - Define the role of participants, e.g., quality assurance / quality control by supplier, surveillance by public health authority.
  - Develop methodologies suitable for the area.
  - Commence routine surveillance in priority areas (including inventories).
  - Limit verification to essential parameters and known problem substances.
  - Establish reporting, filing and communication systems.
  - Advocate improvements according to identified priorities.
  - Establish reporting to local suppliers, communities, media and regional authorities.
  - Establish liaison with communities; identify community roles in surveillance and means of promoting community participation.

- **Intermediate phase:**
  — Train staff involved in programme.
  — Establish and expand systematic routine surveillance.
  — Expand access to analytical capability (often by means of regional laboratories, national laboratories being largely responsible for analytical quality control and training of regional laboratory staff).
  — Undertake surveys for chemical contaminants using wider range of analytical methods.
  — Evaluate all methodologies (sampling, analysis, etc.).
  — Use appropriate standard methods (e.g., analytical methods, fieldwork procedures).
  — Develop capacity for statistical analysis of data.
  — Establish national database.
  — Identify common problems, promote activities to address them at regional and national levels.
  — Expand reporting to include interpretation at national level.
  — Draft or revise health-based targets as part of framework for safe drinking-water.
  — Use legal enforcement where necessary.
  — Involve communities routinely in surveillance implementation.
- **Advanced phase:**
  — Train staff involved in programme.
  — Establish routine testing for all health and acceptability parameters at defined frequencies.
  — Use full network of national, regional and local laboratories (including analytical quality control).
  — Use national framework for drinking-water safety.
  — Improve water services on the basis of national and local priorities, hygiene education and enforcement of standards.
  — Establish regional database archives compatible with national database.
  — Disseminate data at all levels (local, regional and national).
  — Involve communities routinely in surveillance implementation.

5.5 Reporting and communicating

An essential element of a successful surveillance programme is the reporting of results to stakeholders. It is important to establish appropriate systems of reporting to all relevant bodies. Proper reporting and feedback will support the development of effective remedial strategies. The ability of the surveillance programme to identify and advocate interventions to improve water supply is highly dependent on the ability to analyse and present information in a meaningful way to different target audiences. The target audiences for surveillance information will typically include:

— public health officials at local, regional and national levels;
— water suppliers;
— local administrations;
— communities and water users; and
— local, regional and national authorities responsible for development planning and investment.

5.5.1 Interaction with community and consumers

Community participation is a desirable component of surveillance, particularly for community and household drinking-water supplies. As primary beneficiaries of improved drinking-water supplies, community members have a right to take part in decision-making. The community represents a resource that can be drawn upon for local knowledge and experience. They are the people who are likely to first notice problems in the drinking-water supply and therefore can provide an indication of when immediate remedial action is required. Communication strategies should include:

— provision of summary information to consumers (e.g., through annual reports or the Internet); and
— establishment and involvement of consumer associations at local, regional and national levels.

> The right of consumers to information on the safety of the water supplied to them for domestic purposes is fundamental.

However, in many communities, the simple right of access to information will not ensure that individuals are aware of the quality or safety of the water supplied to them. The agencies responsible for surveillance should develop strategies for disseminating and explaining the significance of results obtained.

It may not be feasible for the surveillance agency to provide feedback information directly to the entire community. Thus, it may be appropriate to use community organizations, where these exist, to provide an effective channel for providing feedback information to users. Some local organizations (e.g., local councils and community-based organizations, such as women's groups, religious groups and schools) have regular meetings in the communities that they serve and can therefore provide a mechanism of relaying important information to a large number of people within the community. Furthermore, by using local organizations, it is often easier to initiate a process of discussion and decision-making within the community concerning water quality. The most important elements in working with local organizations are to ensure that the organization selected can access the whole community and can initiate discussion on the results of surveillance.

5.5.2 Regional use of data

Strategies for regional prioritization are typically of a medium-term nature and have specific data requirements. While the management of information at a national level is aimed at highlighting common or recurrent problems, the objective at a regional level is to assign a degree of priority to individual interventions. It is therefore important to derive a relative measure of health risk. While this information cannot be used on its own to determine which systems should be given immediate attention (which would also require the analysis of economic, social, environmental and cultural factors), it provides an extremely important tool for determining regional priorities. It should be a declared objective to ensure that remedial action is carried out each year on a predetermined proportion of the systems classified as high risk.

At the regional level, it is also important to monitor the improvement in (or deterioration of) both individual drinking-water supplies and the supplies as a whole. In this context, simple measures, such as the mean sanitary inspection score of all systems, the proportion of systems with given degrees of faecal contamination, the population with different levels of service and the mean cost of domestic consumption, should be calculated yearly and changes monitored.

In many developing and developed countries, a high proportion of small-community drinking-water systems fail to meet requirements for water safety. In such circumstances, it is important that realistic goals for progressive improvement are agreed upon and implemented. It is practical to classify water quality results in terms of an overall grading for water safety linked to priority for action, as illustrated in Table 5.2.

Grading schemes may be of particular use in community supplies where the frequency of testing is low and reliance on analytical results alone is especially inappropriate. Such schemes will typically take account of both analytical findings and results of the sanitary inspection through schema such as illustrated in Figure 5.1.

Combined analysis of sanitary inspection and water quality data can be used to identify the most important causes of and control measures for contamination. This is important to support effective and rational decision-making. For instance, it will be important to know whether on-site or off-site sanitation could be associated with contamination of drinking-water, as the remedial actions required to address either source of contamination will be very different. This analysis may also identify other factors associated with contamination, such as heavy rainfall. As the data will be non-parametric, suitable methods for analysis include chi-square, odds ratios and logistic regression models.

Table 5.2 Categorization of drinking-water systems based on compliance with performance and safety targets (see also table 7.7)

	Proportion (%) of samples negative for *E. coli*		
	Population size:		
Quality of water system	<5000	5000–100 000	>100 000
Excellent	90	95	99
Good	80	90	95
Fair	70	85	90
Poor	60	80	85

GUIDELINES FOR DRINKING-WATER QUALITY

Figure 5.1 Example of assessment of priority of remedial actions of community drinking-water supplies based on a grading system of microbial quality and sanitary inspection rating or score

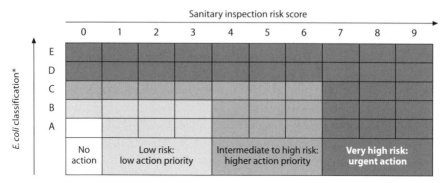

* Based on frequency of *E. coli* positivity in drinking-water and/or *E. coli* concentrations in drinking-water.

Grading	Description
A	Completely satisfactory, extremely low level of risk
B	Satisfactory, very low level of risk
C	Marginally satisfactory, low level of microbial risk when water leaves the plant, but may not be satisfactory chemically
D	Unsatisfactory level of risk
E	Unacceptable level of risk

Source: Lloyd & Bartram (1991)

6
Application of the Guidelines in specific circumstances

These Guidelines provide a generally applicable approach to drinking-water safety. In chapters 2–5, approaches and, where appropriate, aspects of their application to drinking-water supply through piped distribution and through community supplies are described. In applying the Guidelines in specific circumstances, additional factors may be important. This chapter describes the application of the Guidelines in some commonly encountered specific circumstances and issues that should be taken into account in each.

6.1 Large buildings

Responsibility for many actions essential to the control of drinking-water quality in large buildings may be outside the responsibility of the drinking-water supplier. Significant contamination can occur because of factors within the built environment, and specific requirements in the large building environment (including hospitals and health care facilities) are distinct from those in the domestic environment.

General drinking-water safety is assured by maintenance protocols, regular cleaning, temperature management and maintenance of a disinfectant residual. For these reasons, authorities responsible for building safety should be responsible for developing and implementing WSPs. Regulatory or other appropriate authorities may provide guidance on the development and application of WSPs for large building drinking-water systems, which should be implemented by managers.

WSPs for large buildings may usefully address not only drinking-water systems but also other water systems, such as cooling towers and evaporative condensers of air conditioning devices.

The regulator can specify compliance requirements for buildings in general or for individual buildings. Compliance may require that maintenance and monitoring programmes be carried out through a building-specific WSP. It may be appropriate to display maintenance and monitoring programmes and certification of compliance at a conspicuous location within the building. Compliance could be verified and certified by an independent auditor.

6.1.1 Health risk assessment

The principal hazards that may accrue in the drinking-water systems of large buildings are ingress of microbial contamination (which may affect only the building or also the wider supply), proliferation and dispersal of bacteria growing on water contact surfaces (especially *Legionella*) and addition of chemical substances from piping, jointing and plumbing materials.

Faecal contamination may occur through cross-connection and backflow and from buried/immersed tanks and pipes, especially if not maintained with positive internal water pressure.

Legionella bacteria are the cause of legionellosis, including legionnaires' disease. They are ubiquitous in the environment and can proliferate at temperatures experienced at times in piped distribution systems. The route of infection is by inhalation of droplets or aerosols; however, exposure from piped water systems is preventable through the implementation of basic water quality management measures, including maintaining water temperature outside the range at which *Legionella* proliferates (25–50 °C) and maintaining disinfectant residuals throughout the piped distribution system.

Devices such as cooling towers and hot or warm water systems, if not appropriately maintained, can provide suitable conditions for the survival and growth of *Legionella*. In large buildings, there is increased potential for growth of *Legionella* in long water distribution systems, and maintenance of these systems needs particular attention. In addition to supporting the growth of *Legionella*, devices such as cooling towers and hot or warm water systems can disseminate contaminated water in aerosols.

For further information on *Legionella* in drinking-water, see section 11.1.9 and the supporting document *Legionella and the Prevention of Legionellosis* (see section 1.3).

Hospitals, nursing care homes, other health care facilities, schools, hotels and some other large buildings are high-risk environments, because of both the complex nature of their drinking-water systems and the sensitivities of their occupants. Requirements similar to those outlined above for other large buildings apply, but heightened vigilance in control measure monitoring and verification is generally justified.

6.1.2 System assessment

Because WSPs for large buildings are limited to the building environment and since dose–response is not easily described for bacteria arising from growth, adequate control measures are defined in terms of practices that have been shown to be effective.

In undertaking an assessment of the building's distribution system, a range of specific issues must be taken into consideration. These factors relate to ingress and proliferation of contaminants and include:

— pressure of water within the system;
— intermittent supplies;

6. APPLICATION OF THE GUIDELINES IN SPECIFIC CIRCUMSTANCES

— temperature of water;
— cross-connections, especially in mixed systems;
— backflow prevention; and
— system design to minimize dead/blind ends (i.e., a length of pipe, closed at one end, through which no water passes) and other areas of potential stagnation.

6.1.3 Management

The aim of a distribution system within a large building is to supply safe drinking-water at adequate pressure and flow. Pressure is influenced by the action of friction at the pipe wall, flow rate and pipe length, gradient and diameter. For the purposes of maintaining drinking-water quality, it is important to minimize transit times and avoid low flows and pressures. Pressure at any point in the system should be maintained within a range whereby the maximum pressure avoids pipe bursts and the minimum pressure ensures that water is supplied at adequate flow rates for all expected demands. In some buildings, this may require pressure boosting in the network.

Where piped water is stored in tanks to reduce the effect of intermittent supplies, and particularly where water is supplied directly to equipment, the potential for backflow of water into the mains network exists. This may be driven by high pressures generated in equipment connected to mains water supplies or by low pressures in the mains. Water quality in intermittent systems may deteriorate on recharging, where surges may lead to leakage and dislodgement of biofilm and acceptability problems.

A backflow event will be a sanitary problem if there is cross-connection between the potable supply and a source of contamination. Positive pressure should be maintained throughout the piped distribution system. Effective maintenance procedures should be implemented to prevent backflow. In situations where backflow is of particular concern, backflow prevention devices may be used in addition to the primary objective of reducing or eliminating backflow. Situations presenting a potentially high public health risk (e.g., dental chairs, laboratories) should receive special attention.

Significant points of risk exist in areas where pipes carrying drinking-water pass through drains or other places where stagnant water pools. The risk associated with ingress of contamination in these situations may be controlled by reducing the formation of such stagnant pools and by routing pipework to avoid such areas. The design and management of piped water systems in buildings must also take into account the impact of slow flows and dead ends.

Wherever possible, drinking-water taps should be situated in areas where the pipes are well flushed to minimize leaching from pipes, materials and plumbing fittings.

6.1.4 Monitoring

Monitoring of control measures includes:

— temperature, including frequent (e.g., weekly) monitoring of remote areas;
— disinfectants and pH, when employed (e.g., weekly to monthly); and
— microbial quality of water, particularly following system maintenance or repairs.

Daily monitoring may be necessary in the presence of suspected water-related cases of illness.

Monitoring of drinking-water quality is required to be more frequent when the building is new or recently commissioned or following maintenance of the system. When the building's drinking-water system has not stabilized, monitoring should be more frequent until the water quality has stabilized.

6.1.5 Independent surveillance and supporting programmes

Independent surveillance is a desirable element in ensuring continued water safety within a large building and should be undertaken by the relevant health agency or other independent authority.

In order to ensure safety of drinking-water within buildings, supportive activities of national regulatory agencies include the following:

— specific attention to application of codes of good practice (e.g., at commissioning and in contracting construction and rehabilitation);
— suitable training for engineers and plumbers;
— regulation of the plumbing community;
— effective certification of materials and devices in the marketplace; and
— inclusion of WSPs as an essential component of building safety provision.

A WSP would normally document its use of and reliance on such measures – for instance, in using only approved professionals to conduct maintenance and in insisting on their use of certified materials.

6.1.6 Drinking-water quality in health care facilities

Health care facilities include hospitals, health centres and hospices, residential care, dental offices and dialysis units. Drinking-water should be suitable for human consumption and for all usual domestic purposes, including personal hygiene. However, it may not be suitable for all uses or for some patients within health care facilities, and further processing or treatment or other safeguards may be required.

Drinking-water can contain a range of microorganisms, including *Pseudomonas aeruginosa*, non-tuberculous *Mycobacterium* spp., *Acinetobacter* spp., *Aeromonas* spp. and *Aspergillus*. There is no evidence that these microorganisms represent a health concern through water consumption by the general population, including most patients in health care facilities. However, additional processing may be required to ensure safety for consumption by severely immunosuppressed persons, such as those with neutrophil counts below 500 per µl (see the supporting document *Heterotrophic Plate Counts and Drinking-water Safety*; section 1.3).

6. APPLICATION OF THE GUIDELINES IN SPECIFIC CIRCUMSTANCES

Microorganisms in drinking-water also have the potential to cause infections if drinking-water is used to wash burns or to wash medical devices such as endoscopes and catheters. Water used for such purposes needs to be of a higher quality than described in these Guidelines and may require additional processing, such as microfiltration or sterilization, depending on use.

Health care facilities may include environments that support the proliferation and dissemination of *Legionella* (see section 11.1.9 and the supporting document *Legionella and the Prevention of Legionellosis*; section 1.3).

Renal dialysis requires large volumes of water that exceed the chemical and microbial quality requirements for drinking-water. Water used for dialysis requires special processing to minimize the presence of microorganisms, endotoxins, toxins and chemical contaminants. The vulnerability of renal dialysis patients was demonstrated in 1996 by the death of 50 such patients after exposure to water contaminated by high levels of microcystin (Jochimsen et al., 1998; Pouria et al., 1998). Dialysis patients are also sensitive to chloramines, and this needs to be considered when chloramination is used to disinfect drinking-water supplies, particularly in areas where there are home dialysis patients.

All health care facilities should have specific WSPs as part of their infection control programme. These plans should address issues such as water quality and treatment requirements, cleaning of specialized equipment and control of microbial growth in water systems and ancillary equipment.

6.1.7 Drinking-water quality in schools and day care centres

A long-term approach to improving hygiene in the community includes working with children in schools. This enables the concept of good hygiene, of which drinking-water safety is a part, to become part of a general understanding of health and the influence of the environment. Schoolchildren can relay hygiene concepts to family and households. As young children learn from what they see around them, the school environment itself should meet the requirements of good hygiene – for example, by providing toilets or latrines, water for hand-washing, generally clean surroundings and hygienic facilities for the preparation and serving of school meals. Visual demonstration of the presence of bacteria on unwashed hands has been shown to be valuable (e.g., using UV fluorescence of bacteria or the hydrogen sulfide paper strip method).

One of the most important characteristics of effective health education is that it builds on concepts, ideas and practices that people already have. Hygiene education programmes should be based on an understanding of the factors that influence behaviour at the community level. These might include:

— enabling factors, such as money, materials and time to carry out appropriate patterns of behaviour;
— pressure from particular members of the family and community (e.g., elders, traditional healers, opinion leaders);

— beliefs and attitudes among community members with respect to hygienic behaviour, especially the perceived benefits and disadvantages of taking action; and
— the understanding of the relationship between health and hygiene.

An understanding of the factors that influence hygiene-related behaviours will help in identifying the resources (e.g., soap, storage containers), the key individuals in the home and community and the important beliefs that should be taken into account. This will help to ensure that the content of the hygiene education is relevant to the community. Good advice should:

— result in improved health;
— be affordable;
— require a minimum of effort and time to put into practice;
— be realistic;
— be culturally acceptable;
— meet a perceived need; and
— be easy to understand.

6.2 Emergencies and disasters

Drinking-water safety is one of the most important public health issues in most emergencies and disasters. The greatest waterborne risk to health in most emergencies is the transmission of faecal pathogens, due to inadequate sanitation, hygiene and protection of water sources. Some disasters, including those caused by or involving damage to chemical and nuclear industrial installations or spillage in transport or volcanic activity, may create acute problems from chemical or radiological water pollution.

Different types of disaster affect water quality in different ways. When people are displaced by conflict and natural disaster, they may move to an area where unprotected water sources are contaminated. When population density is high and sanitation is inadequate, unprotected water sources in and around the temporary settlement are highly likely to become contaminated. If there is a significant prevalence of disease cases and carriers in a population of people with low immunity due to malnutrition or the burden of other diseases, then the risk of an outbreak of waterborne disease is increased. The quality of urban drinking-water supplies is particularly at risk following earthquakes, mudslides and other structurally damaging disasters. Water treatment works may be damaged, causing untreated or partially treated water to be distributed, and sewers and water transmission pipes may be broken, causing contamination of drinking-water in the distribution system. Floods may contaminate wells, boreholes and surface water sources with faecal matter washed from the ground surface or from overflowing latrines and sewers. During droughts, people may be forced to use unprotected water supplies when normal supplies dry up; as more people and animals use fewer water sources, the risk of contamination is increased.

Emergency situations that are appropriately managed tend to stabilize after a matter of days or weeks. Many develop into long-term situations that can last for several years before a permanent solution is found. Water quality concerns may change over that time, and water quality parameters that pose long-term risks to health may become more important.

6.2.1 Practical considerations

Available sources of water are very limited in most emergency situations, and providing a sufficient quantity of water for personal and domestic hygiene as well as for drinking and cooking is important. Guidelines and national drinking-water quality standards should therefore be flexible, taking into consideration the risks and benefits to health in the short and long term, and should not excessively restrict water availability for hygiene, as this would often result in an increased overall risk of disease transmission.

There are a number of factors to take into consideration when providing drinking-water for a population affected by a disaster, including the following:

- *The quantity of water available and the reliability of supply* – This is likely to be the overriding concern in most emergency situations, as it is usually easier to improve water quality than to increase its availability or to move the affected population closer to another water source.
- *The equitability of access to water* – Even if sufficient water is available to meet minimum needs, additional measures may be needed to ensure that access is equitable. Unless water points are sufficiently close to their dwellings, people will not be able to collect enough water for their needs. Water may need to be rationed to ensure that everyone's basic needs are met.
- *The quality of the raw water* – It is preferable to choose a source of water that can be supplied with little or no treatment, provided it is available in sufficient quantity.
- *Sources of contamination and the possibility of protecting the water source* – This should always be a priority in emergencies, whether or not disinfection of the water supply is considered necessary.
- *The treatment processes required for rapidly providing a sufficient quantity of potable water* – As surface water sources are commonly used to provide water to large populations in emergencies, clarification of the raw water – for example, by flocculation and sedimentation and/or by filtration – is commonly required before disinfection.
- *The treatment processes appropriate for post-emergency situations* – The affordability, simplicity and reliability of water treatment processes in the longer term should be considered early on in the emergency response.
- *The need to disinfect drinking-water supplies* – In emergencies, hygiene conditions are normally poor and the risk of disease outbreaks is high, particularly in populations with low immunity. It is therefore crucial to disinfect the water supplies, ensuring a residual disinfection capacity in the water. This practice would

considerably reduce the likelihood of disease transmission through contamination of water in the home.
- *Acceptability* – It is important to ensure that drinking-water provided in emergencies is acceptable to the consumers, or they may resort to water from unprotected or untreated supplies.
- *The need for vessels to collect and store water* – Vessels that are hygienic and appropriate to local needs and habits are needed for the collection and storage of water to be used for washing, cooking and bathing.
- *Epidemiological considerations* – Contamination of water may occur during collection, storage and use in the home, as a result of lack of sanitation or poor hygiene due to an insufficient quantity of water. Other transmission routes for major waterborne and sanitation-related diseases in emergencies and disasters include person-to-person contact, aerosols and food intake. The importance of all routes should be considered when applying the Guidelines, selecting and protecting water sources and choosing options for water treatment.

In many emergency situations, water is collected from central water collection points, stored in containers and then transferred to cooking and drinking vessels by the affected people. This process provides many opportunities for contamination of the water after it leaves the supply system. It is therefore important that people are aware of the risks to health from contamination of water from the point of collection to the moment of consumption and have the means to reduce or eliminate these risks. When water sources are close to dwelling areas, they may easily be contaminated through indiscriminate defecation, which should be strongly discouraged. Establishing and maintaining water quality in emergencies require the rapid recruitment, training and management of operations staff and the establishment of systems for maintenance and repairs, consumable supplies and monitoring. Communication with the affected population is extremely important for reducing health problems due to poor water quality. Detailed information may be found in Wisner & Adams (2003).

6.2.2 Monitoring
Water safety should be monitored during emergencies. Monitoring may involve sanitary inspection and one or more of:

— sanitary inspection and water sampling and analysis;
— monitoring of water treatment processes, including disinfection;
— monitoring of water quality at all water collection points and in a sample of homes; and
— water quality assessment in the investigation of disease outbreaks or the evaluation of hygiene promotion activities, as required.

Monitoring and reporting systems should be designed and managed to ensure that action is swiftly taken to protect health. Health information should also be monitored

6. APPLICATION OF THE GUIDELINES IN SPECIFIC CIRCUMSTANCES

to ensure that water quality can be rapidly investigated when there is a possibility that water quality might contribute to a health problem and that treatment processes – particularly disinfection – can be modified as required.

6.2.3 Microbial guidelines

The objective of zero *E. coli* per 100 ml of water is the goal for all water supplies and should be the target even in emergencies; however, it may be difficult to achieve in the immediate post-disaster period. This highlights the need for appropriate disinfection.

An indication of a certain level of faecal indicator bacteria *alone* is not a reliable guide to microbial water safety. Some faecal pathogens, including many viruses and protozoal cysts and oocysts, may be more resistant to treatment (e.g., by chlorine) than common faecal indicator bacteria. More generally, if a sanitary survey suggests the risk of faecal contamination, then even a very low level of faecal contamination may be considered to present a risk, especially during an outbreak of a potentially waterborne disease, such as cholera.

Drinking-water should be disinfected in emergency situations, and an adequate disinfectant residual (e.g., chlorine) should be maintained in the system. Turbid water should be clarified wherever possible to enable disinfection to be effective. Minimum target concentrations for chlorine at point of delivery are 0.2 mg/litre in normal circumstances and 0.5 mg/litre in high-risk circumstances.

Where there is a concern about the quality of drinking-water in an emergency situation that cannot be addressed through central services, then the appropriateness of household-level treatment should be evaluated, including, for example:

— bringing water to a rolling boil and cooling before consumption;
— adding sodium or calcium hypochlorite solution, such as household bleach, to a bucket of water, mixing thoroughly and allowing to stand for about 30 min prior to consumption; turbid water should be clarified by settling and/or filtration before disinfection;
— vigorously shaking small volumes of water in a clean, transparent container, such as a soft drink bottle, for 20 s and exposing the container to sunlight for at least 6 h;
— applying products such as tablets or other dosing techniques to disinfect the water, with or without clarification by flocculation or filtration; and
— end-use units and devices for field treatment of drinking-water.

Emergency decontamination processes may not always accomplish the level of disinfection recommended for optimal conditions, particularly with regard to resistant pathogens. However, implementation of emergency procedures may reduce numbers of pathogens to levels at which the risk of waterborne disease is largely controlled.

The parameters most commonly measured to assess microbial safety are as follows:

- *E. coli (see above):* Thermotolerant coliforms may provide a simpler surrogate.
- *Residual chlorine:* Taste does not give a reliable indication of chlorine concentration. Chlorine content should be tested in the field with, for example, a colour comparator, generally used in the range of 0.2–1 mg/litre.
- *pH:* It is necessary to know the pH of water, because more alkaline water requires a longer contact time or a higher free residual chlorine level at the end of the contact time for adequate disinfection (0.4–0.5 mg/litre at pH 6–8, rising to 0.6 mg/litre at pH 8–9; chlorination may be ineffective above pH 9).
- *Turbidity:* Turbidity adversely affects the efficiency of disinfection. Turbidity is also measured to determine what type and level of treatment are needed. It can be carried out with a simple turbidity tube that allows a direct reading in nephelometric turbidity units (NTU).

6.2.4 Sanitary inspections and catchment mapping

It is possible to assess the likelihood of faecal contamination of water sources through a sanitary inspection. Sanitary inspection and water quality testing are complementary activities; the findings of each assists the interpretation of the other. Where water quality analysis cannot be performed, sanitary inspection can still provide valuable information to support effective decision-making. A sanitary inspection makes it possible to see what needs to be done to protect the water source. This procedure can be combined with bacteriological, physical and chemical testing to enable field teams to assess and act on risks from contamination and to provide the basis for monitoring water supplies in the post-disaster period.

Even when it is possible to carry out testing of microbial quality, results are not instantly available. Thus, the immediate assessment of contamination risk may be based on gross indicators such as proximity to sources of faecal contamination (human or animal), colour and smell, the presence of dead fish or animals, the presence of foreign matter such as ash or debris or the presence of a chemical or radiation hazard or wastewater discharge point upstream. Catchment mapping involving the identification of sources and pathways of pollution can be an important tool for assessing the likelihood of contamination of a water source.

It is important to use a standard reporting format for sanitary inspections and catchment mapping to ensure that information gathered by different staff is reliable and that information gathered on different water sources may be compared. For an example format, see WHO (1997) and Davis & Lambert (2002). For more information on catchment mapping, see House & Reed (1997).

6.2.5 Chemical and radiological guidelines

Many chemicals in drinking-water are of concern only after extended periods of exposure. Thus, to reduce the risk of outbreaks of waterborne and water-washed (e.g., trachoma, scabies, skin infections) disease, it is preferable to supply water in an emergency, even if it significantly exceeds the guideline values for some chemical parameters, rather than restrict access to water, provided the water can be treated to kill pathogens and can be supplied rapidly to the affected population. Where water sources are likely to be used for long periods, chemical and radiological contaminants of more long-term health concern should be given greater attention. In some situations, this may entail adding treatment processes or seeking alternative sources.

Water from sources that are considered to have a significant risk of chemical or radiological contamination should be avoided, even as a temporary measure. In the long term, achieving the guidelines should be the aim of emergency drinking-water supply programmes based on the progressive improvement of water quality. Procedures for identifying priority chemicals in drinking-water are outlined in the supporting document *Chemical Safety of Drinking-water* (section 1.3).

6.2.6 Testing kits and laboratories

Portable testing kits allow the determination in the field of key water quality parameters, such as thermotolerant coliform count, free residual chlorine, pH, turbidity and filterability.

Where large numbers of water samples need testing or a broad range of parameters is of interest, laboratory analysis is usually most appropriate. If the drinking-water supplier's laboratories or laboratories at environmental health offices and universities no longer function because of the disaster, then a temporary laboratory may need to be set up. Where samples are transported to laboratories, handling is important. Poor handling may lead to meaningless or misleading results.

Workers should be trained in the correct procedures for collecting, labelling, packing and transporting samples and in supplying supporting information from the sanitary survey to help interpret laboratory results. For guidance on methods of water sampling and testing, see WHO (1997) and Bartram & Ballance (1996).

6.3 Safe drinking-water for travellers

Diarrhoea is the most common cause of ill health for travellers; up to 80% of all travellers are affected in high-risk areas. In localities where the quality of potable water and sanitation and food hygiene practices are questionable, the numbers of parasites, bacteria and viruses in water and food can be substantial, and numerous infections can occur. Cases occur among people staying in resorts and hotels in all categories. No vaccine is capable of conferring general protection against diarrhoea, which is caused by many different pathogens. It is important that travellers are aware of possible risks and take appropriate steps to minimize these.

Contaminated food, water and drinks are the most common sources of infections. Careful selection of drinking-water sources and appropriate water treatment offer significant protection. Preventive measures while living or travelling in areas with unsafe drinking-water include the following:

- Always avoid consumption or use of unsafe water (even when brushing teeth) if you are unsure about water quality.
- Avoid unpasteurized juices and ice made from untreated water.
- Avoid salads or other uncooked meals that may have been washed or prepared with unsafe water.
- Drink water that you have boiled, filtered and/or treated with chlorine or iodine and stored in clean containers.
- Consume ice only if it is known to be of drinking-water quality.
- Drink bottled water if it is known to be safe, carbonated bottled beverages (water and sodas) only from sealed, tamper-proof containers and pasteurized/canned juices and pasteurized milk.
- Drink coffee and tea made from boiled water and served and stored in clean containers.

The greatest health risk from drinking-water for travellers is associated with microbial constituents of water. Water can be treated or re-treated in small quantities to significantly improve its safety. The simplest and most important beneficial treatments for microbially contaminated water are boiling, disinfection and filtration to inactivate or remove pathogenic microorganisms. These treatments will generally not reduce most chemical constituents in drinking-water. However, most chemicals are of health concern only after long-term exposure. Numerous simple treatment approaches and commercially available technologies are also available to travellers to treat drinking-water for single-person use.

Bringing water to a rolling boil is the most effective way to kill disease-causing pathogens, even at high altitudes and even for turbid water. The hot water should be allowed to cool down on its own without the addition of ice. If water for boiling is to be clarified, this should be done before boiling.

Chemical disinfection is effective for killing bacteria, some viruses and some protozoa (but not, for example, *Cryptosporidium* oocysts). Some form of chlorine and iodine are the chemicals most widely used for disinfection by travellers. After chlorination, a carbon (charcoal) filter may be used to remove excess chlorine taste and, in the case of iodine treatment, to remove excess iodine. Silver is not very effective for eliminating disease-causing microorganisms, since silver by itself is slow acting. If water is turbid (not clear or with suspended solid matter), it should be clarified before disinfection; clarification includes filtration, settling and decanting. Portable filtration devices that have been tested and rated to remove protozoa and some bacteria are also available; ceramic filters and some carbon block filters are the most common types. The filter's pore size rating must be 1 µm (absolute) or less to ensure removal of *Cryptosporidium* oocysts (these very fine filters may require a pre-filter to remove larger particles in order to avoid clogging the final filter). A combination of technologies (filtration followed by chemical disinfection or boiling) is recommended, as most filtering devices do not remove viruses.

For people with weakened immune systems, extra precautions are recommended to reduce the risk of infection from contaminated water. While drinking boiled water is safest, certified bottled or mineral water may also be acceptable. Iodine as a water disinfectant is not recommended for pregnant women, those with a history of thyroid disease and those with known hypersensitivity to iodine, unless there is also an effective post-treatment iodine removal system such as granular carbon in use.

6.4 Desalination systems

The principal purpose of desalination is to enable sources of brackish or salty water, otherwise unacceptable for human consumption, to be used for this purpose.

The use of desalination to provide drinking-water is increasing and is likely to continue to increase because of water scarcity driven by pressures arising from population growth, over-exploitation of water resources and pollution of other water sources. While most (around 60%) of currently constructed capacity is in the eastern Mediterranean region, desalination facilities exist all over the world, and their use is likely to increase in all continents.

Most present applications of desalination are for estuarine water, coastal water and seawater. Desalination may also be applied to brackish inland waters (both surface water and groundwater) and may be used on board vessels. Small-scale desalination units also exist for household and community use and present specific challenges to effective operation and maintenance.

Further guidance on desalination for safe drinking-water supply is available in the supporting document *Desalination for Safe Drinking-water Supply* (section 1.3).

In applying the Guidelines to desalinated water supply systems, account should be taken of certain major differences between these and systems abstracting water from freshwater sources. These differences include the factors described in section 6.4.1. Once taken into account, the general requirements of these Guidelines for securing microbial, chemical and radiological safety should apply.

Brackish water, coastal water and seawater sources may contain hazards not encountered in freshwater systems. These include diverse harmful algal events associated with micro- and macroalgae and cyanobacteria; certain free-living bacteria (including *Vibrio* spp., such as *V. parahaemolyticus* and *V. cholerae*); and some chemicals, such as boron and bromide, that are more abundant in seawater.

Harmful algal events may be associated with exo- and endotoxins that may not be destroyed by heating, are inside algal cells or are free in the water. They are usually non-volatile, and, where they are destroyed by chlorination, this usually requires extremely long contact times. Although a number of toxins have been identified, it is possible that there are other unrecognized toxins. Minimizing of the potential for abstracting water containing toxic algae through location/siting and intake design plus effective monitoring and intake management is an important control measure.

Other chemical issues, such as control of "additives," DBPs and pesticides, are similar to those encountered in fresh waters (see chapter 8), except that a larger variety and greater quantities may be involved in desalination. Due to the presence of bromide in seawater, the distribution of DBPs will likely be dominated by brominated organics.

Approaches to monitoring and assessing the quality of freshwater sources may not be directly applicable to sources subject to desalination. For example, many faecal indicator bacteria die off more rapidly than pathogens (especially viruses) in saline than in fresh water.

The effectiveness of some of the processes employed in desalination to remove some substances of health concern remains inadequately understood. Examples of inefficiencies include imperfect membrane and/or membrane seal integrity (membrane treatment); bacterial growth through membranes/biofilm development on membranes (in membrane treatment systems); and carry-over, especially of volatile substances (with vapour).

Because of the apparently high effectiveness of some of the processes used in removal of both microorganisms and chemical constituents (especially distillation and reverse osmosis), these processes may be employed as single-stage treatments or combined with only a low level of residual disinfectant. The absence of multiple barriers places great stress on the continuously safe operation of that process and implies that even a short-term decrease in effectiveness may present an increased risk to human health. This, in turn, implies the need for on-line monitoring linked to rapid management intervention. For further information, see the supporting document *Water Treatment and Pathogen Control* (section 1.3).

Water produced by desalination is "aggressive" towards materials used, for example, in water supply and domestic plumbing and pipes. Special consideration should be given to the quality of such materials, and normal procedures for certification of materials as suitable for potable water use may not be adequate for water that has not been "stabilized."

Because of the aggressivity of desalinated water and because desalinated water may be considered bland, flavourless and unacceptable, desalinated water is commonly treated by adding chemical constituents such as calcium and magnesium carbonate with carbon dioxide. Once such treatment has been applied, desalinated waters should be no more aggressive than waters normally encountered in the drinking-water supply. Chemicals used in such treatment should be subject to normal procedures for certification.

Desalinated waters are commonly blended with small volumes of more mineral-rich waters to improve their acceptability and particularly to reduce their aggressivity to materials. Blending waters should be fully potable, as described here and elsewhere in the Guidelines. Where seawater is used for this purpose, the major ions added are sodium and chloride. This does not contribute to improving hardness or ion balance, and only small amounts (e.g., 1–3%) can be added without leading to problems of acceptability. Blended waters from coastal and estuarine areas may be more susceptible to contamination with petroleum hydrocarbons, which could give rise to taste and odour problems. Some groundwaters or surface waters, after suitable treatment, may be employed for blending in higher proportions and may improve hardness and ion balance.

Desalinated water is a manufactured product. Concern has been expressed about the impact of extremes of major ion composition or ratios for human health. There is limited evidence to describe the health risk associated with long-term consumption of such water, although concerns regarding mineral content may be limited by the stabilization processes outlined above (see WHO, 2003b).

Desalinated water, by virtue of its manufacture, often contains lower than usual concentrations of other ions commonly found in water, some of which are essential elements. Water typically contributes a small proportion of these, and most intake is through food. Exceptions include fluoride, and declining dental health has been reported from populations consuming desalinated water with very low fluoride content where there is a moderate to high risk of dental caries (WHO, 2003b).

Desalinated water may be more subject to "microbial growth" problems than other waters as a result of one or more of the following: higher initial temperature (from treatment process), higher temperature (application in hot climates) and/or the effect of aggressivity on materials (thereby releasing nutrients). The direct health significance of such growth (see the supporting document *Heterotrophic Plate Counts and Drinking-water Safety*; section 1.3), with the exception of *Legionella* (see chapter 11), is inadequately understood. Nitrite formation by organisms in biofilms may prove problematic where chloramination is practised and excess ammonia is present. Precaution implies that preventive management should be applied as part of good management practice.

6.5 Packaged drinking-water

Bottled water and ice are widely available in both industrialized and developing countries. Consumers may have various reasons for purchasing packaged drinking-water, such as taste, convenience or fashion; for many consumers, however, safety and potential health benefits are important considerations.

6.5.1 Safety of packaged drinking-water

Water is packaged for consumption in a range of vessels, including cans, laminated boxes and plastic bags, and as ice prepared for consumption. However, it is most commonly prepared in glass or plastic bottles. Bottled water also comes in various sizes, from single servings to large carboys holding up to 80 litres.

In applying the Guidelines to bottled waters, certain chemical constituents may be more readily controlled than in piped distribution systems, and stricter standards may therefore be preferred in order to reduce overall population exposure. Similarly, when flexibility exists regarding the source of the water, stricter standards for certain naturally occurring substances of health concern, such as arsenic, may be more readily achieved than in piped distribution systems.

However, some substances may prove to be more difficult to manage in bottled water than in tap water. Some hazards may be associated with the nature of the product (e.g., glass chips and metal fragments). Other problems may arise because bottled water is stored for longer periods and at higher temperatures than water distributed in piped distribution systems or because containers and bottles are reused without adequate cleaning or disinfection. Control of materials used in containers and closures for bottled water is, therefore, of special concern. Some microorganisms that are normally of little or no public health significance may grow to higher levels in bottled water. This growth appears to occur less frequently in gasified water and in water bottled in glass containers than in still water and water bottled in plastic containers. The public health significance of this microbial growth remains uncertain, especially for vulnerable individuals, such as bottle-fed infants and immunocompromised individuals. In regard to bottle-fed infants, as bottled water is not sterile, it should be disinfected – for example, by boiling – prior to its use in the preparation of infant formula. For further information, see the supporting document *Heterotrophic Plate Counts and Drinking-water Safety* (section 1.3).

6.5.2 Potential health benefits of bottled drinking-water

There is a belief by some consumers that natural mineral waters have medicinal properties or offer other health benefits. Such waters are typically of high mineral content, sometimes significantly higher than concentrations normally accepted in drinking-water. Such waters often have a long tradition of use and are often accepted on the basis that they are considered foods rather than drinking-water *per se*. Although certain mineral waters may be useful in providing essential micro-nutrients, such as calcium, these Guidelines do not make recommendations regarding minimum concentrations of essential compounds, because of the uncertainties surrounding mineral nutrition from drinking-water.

Packaged waters with very low mineral content, such as distilled or demineralized waters, are also consumed. Rainwater, which is similarly low in minerals, is consumed by some populations without apparent adverse health effects. There is insufficient scientific information on the benefits or hazards of regularly consuming these types of bottled waters (see WHO, 2003b).

6.5.3 International standards for bottled drinking-water

The *Guidelines for Drinking-water Quality* provide a basis for derivation of standards for all packaged waters. As with other sources of drinking-water, safety is pursued through a combination of safety management and end product quality standards and testing. The international framework for packaged water regulation is provided by the Codex Alimentarius Commission (CAC) of WHO and the FAO. CAC has developed a *Standard for Natural Mineral Waters* and an associated Code of Practice. The Standard describes the product and its compositional and quality factors, including limits for certain chemicals, hygiene, packaging and labelling. The CAC has also developed a *Standard for Bottled/Packaged Waters* to cover packaged drinking-water other than natural mineral waters. Both relevant CAC standards refer directly to these Guidelines.

The CAC *Code of Practice for Collecting, Processing and Marketing of Natural Mineral Waters* provides guidance on a range of good manufacturing practices and provides a generic WSP applied to packaged drinking-water.

Under the existing CAC *Standard for Natural Mineral Waters* and associated Code of Practice, natural mineral waters must conform to strict requirements, including collection and bottling without further treatment from a natural source, such as a spring or well. In comparison, the CAC *Standard for Bottled/Packaged Waters* includes waters from other sources, in addition to springs and wells, and treatment to improve their safety and quality. The distinctions between these standards are especially relevant in regions where natural mineral waters have a long cultural history.

For further information on CAC, its Codex Committee on Natural Mineral Waters, the CAC *Standard for Natural Mineral Waters* and its companion Code of Practice, readers are referred to the CAC website (http://www.codexalimentarius.net/).

6.6 Food production and processing

The quality of water defined by the Guidelines is such that it is suitable for all normal uses in the food industry. Some processes have special water quality requirements in order to secure the desired characteristics of the product, and the Guidelines do not necessarily guarantee that such special requirements are met.

Deterioration in drinking-water quality may have severe impacts on food processing facilities and potentially upon public health. The consequences of a failure to use water of potable quality will depend on the use of the water and the subsequent processing of potentially contaminated materials. Variations in water quality that may be tolerated occasionally in drinking-water supply may be unacceptable for some uses in the food industry. These variations may result in a significant financial impact on food production – for example, through product recalls.

The diverse uses of water in food production and processing have different water quality requirements. Uses include:

— irrigation and livestock watering;
— those in which water may be incorporated in or adhere to a product (e.g., as an ingredient, or where used in washing or "refreshing" of foods);
— misting of salad vegetables in grocery stores; and
— those in which contact between the water and foodstuff should be minimal (as in heating and cooling and cleaning water).

To reduce microbial contamination, specific treatments (e.g., heat) capable of removing a range of pathogenic organisms of public health concern may be used. The effect of these treatments should be taken into account when assessing the impacts of deterioration in drinking-water quality on a food production or processing facility.

Information on deterioration of the quality of a drinking-water supply should be promptly communicated to vulnerable food production facilities.

6.7 Aircraft and airports
6.7.1 Health risks
The importance of water as a potential vehicle for infectious disease transmission on aircraft has been well documented. In general terms, the greatest microbial risks are those associated with ingestion of water that is contaminated with human and animal excreta.

If the source of water used to replenish aircraft supplies is contaminated, and unless adequate precautions are taken, disease can be spread through the aircraft water. It is thus imperative that airports comply with Article 14.2 (Part III – Health Organization) of the International Health Regulations (1969) and be provided with potable drinking-water from a source approved by the appropriate regulatory agency (WHO, 1983).

A potable water source is not a safeguard if the water is subsequently contaminated during transfer, storage or distribution in aircraft. Airports usually have special arrangements for managing water after it has entered the airport. Water may be delivered to aircraft by water servicing vehicles or water bowsers. Transfer of water from the water carriers to the aircraft provides the opportunity for microbial or chemical contamination (e.g., from water hoses).

A WSP covering water management within airports from receipt of the water through to its transfer to the aircraft, complemented by measures (e.g., safe materials and good practices in design, construction, operation and maintenance of aircraft systems) to ensure that water quality is maintained on the aircraft, provides a framework for water safety in aviation.

6.7.2 System risk assessment
In undertaking an assessment of the general airport/aircraft water distribution system, a range of specific issues must be taken into consideration, including:

— quality of source water;
— design and construction of airport storage tanks and pipes;
— design and construction of water servicing vehicles;
— water loading techniques;
— any treatment systems on aircraft;
— maintenance of on-board plumbing; and
— prevention of cross-connections, including backflow prevention.

6.7.3 Operational monitoring
The airport authority has responsibility for safe drinking-water supply, including for operational monitoring, until water is transferred to the aircraft operator. The primary

emphasis of monitoring is as a verification of management processes. Monitoring of control measures includes:

— quality of source water;
— hydrants, hoses and bowsers for cleanliness and repair;
— disinfectant residuals and pH;
— backflow preventers;
— filters; and
— microbial quality of water, particularly after maintenance or repairs.

6.7.4 Management

Even if potable water is supplied to the airport, it is necessary to introduce precautions to prevent contamination during the transfer of water to the aircraft and in the aircraft drinking-water system itself. Staff employed in drinking-water supply must not be engaged in activities related to aircraft toilet servicing without first taking all necessary precautions (e.g., thorough handwashing, change of outer garments).

All water servicing vehicles must be cleansed and disinfected frequently.

Supporting programmes that should be documented as part of a WSP for airports include:

— suitable training for crews dealing with water transfer and treatment; and
— effective certification of materials used on aircraft for storage tanks and pipes.

6.7.5 Surveillance

Independent surveillance resembles that described in chapter 5 and is an essential element in ensuring drinking-water safety in aviation. This implies:

— periodic audit and direct assessment;
— review and approval of WSPs;
— specific attention to the aircraft industry's codes of practice, the supporting document *Guide to Hygiene and Sanitation in Aviation* (section 1.3) and airport health or airline regulations; and
— responding, investigating and providing advice on receipt of report on significant incidents.

6.8 Ships

6.8.1 Health risks

The importance of water as a vehicle for infectious disease transmission on ships has been clearly documented. In general terms, the greatest microbial risks are associated with ingestion of water that is contaminated with human and animal excreta. Waterborne transmission of the enterotoxigenic *E. coli*, Norovirus, *Vibrio* spp., *Salmonella typhi*, *Salmonella* spp. (non-typhi), *Shigella* spp., *Cryptosporidium* spp., *Giardia lamblia* and *Legionella* spp. on ships has been confirmed (see Rooney et al., in press).

Chemical water poisoning can also occur on ships. For example, one outbreak of acute chemical poisoning implicated hydroquinone, an ingredient of photo developer, as the disease-causing agent in the ship's potable water supply. Chronic chemical poisoning on a ship could also occur if crew or passengers were exposed to small doses of harmful chemicals over long periods of time.

The supporting document *Guide to Ship Sanitation* (section 1.3) describes the factors that can be encountered during water treatment, transfer, production, storage or distribution in ships. This revised Guide includes description of specific features of the organization of the supply and the regulatory framework.

The organization of water supply systems covering shore facilities and ships differs considerably from conventional water transfer on land. Even though a port authority may receive potable water from a municipal or private supply, it usually has special arrangements for managing the water after it has entered the port. Water is delivered to ships by hoses or transferred to the ship via water boats or barges. Transfer of water from shore to ships can provide possibilities for microbial or chemical contamination.

In contrast to a shore facility, plumbing aboard ships consists of numerous piping systems, carrying potable water, seawater, sewage and fuel, fitted into a relatively confined space. Piping systems are normally extensive and complex, making them difficult to inspect, repair and maintain. A number of waterborne outbreaks on ships have been caused by contamination of potable water after it had been loaded onto the ship – for example, by sewage or bilge when the water storage systems were not adequately designed and constructed. During distribution, it may be difficult to prevent water quality deterioration due to stagnant water and dead ends.

Water distribution on ships may also provide greater opportunities for contamination to occur than onshore, because ship movement increases the possibility of surge and backflow.

6.8.2 *System risk assessment*

In undertaking an assessment of the ship's drinking-water system, a range of specific issues must be taken into consideration, including:

— quality of source water;
— water loading equipment;
— water loading techniques;
— design and construction of storage tanks and pipes;
— filtration systems and other treatment systems on board the ship;
— backflow prevention;
— pressure of water within the system;
— system design to minimize dead ends and areas of stagnation; and
— residual disinfection.

6. APPLICATION OF THE GUIDELINES IN SPECIFIC CIRCUMSTANCES

6.8.3 Operational monitoring

The ship's master is responsible for operational monitoring. The primary emphasis of monitoring is as a verification of management processes. Monitoring of control measures includes:

— quality of source water;
— hydrants and hoses for cleanliness and repair;
— disinfectant residuals and pH (e.g., daily);
— backflow prevention devices (e.g., monthly to yearly);
— filters (before and during each use); and
— microbial quality of treated water, particularly after maintenance or repairs.

The frequency of monitoring should reflect the probable rate of change in water quality. For example, monitoring of drinking-water on ships may be more frequent when the ship is new or recently commissioned, with frequencies decreasing in the light of review of results. Similarly, if the ship's water system has been out of control, monitoring following restoration of the system would be more frequent until it is verified that the system is clearly under control.

6.8.4 Management

The port authority has responsibility for providing safe potable water for loading onto vessels. The ship's master will not normally have direct control of pollution of water supplied at port. If water is suspected to have come from an unsafe source, the ship's master may have to decide if any additional treatment (e.g., hyperchlorination and/or filtration) is necessary. When treatment on board or prior to boarding is necessary, the treatment selected should be that which is best suited to the water and which is most easily operated and maintained by the ship's officers and crew.

During transfer from shore to ship and on board, water must be provided with sanitary safeguards through the shore distribution system, including connections to the ship system, and throughout the ship system, to prevent contamination of the water.

Potable water should be stored in one or more tanks that are constructed, located and protected so as to be safe against contamination. Potable water lines should be protected and located so that they will not be submerged in bilge water or pass through tanks storing non-potable liquids.

The ship's master should ensure that crew and passengers receive a sufficient and uninterrupted drinking-water supply and that contamination is not introduced in the distribution system. The distribution systems on ships are especially vulnerable to contamination when the pressure falls. Backflow prevention devices should be installed to prevent contamination of water where loss of pressure could result in backflow.

The potable water distribution lines should not be cross-connected with the piping or storage tanks of any non-potable water system.

Water safety is secured through repair and maintenance protocols, including the ability to contain potential contamination by valving and the cleanliness of personnel, their working practices and the materials employed.

Current practice on many ships is to use disinfectant residuals to control the growth of microorganisms in the distribution system. Residual disinfection alone should not be relied on to "treat" contaminated water, since the disinfection can be readily overwhelmed by contamination.

Supporting programmes that should be documented as part of the WSP for ships include:

— suitable training for crew dealing with water transfer and treatment; and
— effective certification of materials used on ships for storage tanks and pipes.

6.8.5 Surveillance

Independent surveillance is a desirable element in ensuring drinking-water safety on ships. This implies:

— periodic audit and direct assessment;
— review and approval of WSPs;
— specific attention to the shipping industry's codes of practice, the supporting document *Guide to Ship Sanitation* (section 1.3) and port health or shipping regulations; and
— responding, investigating and providing advice on receipt of report on significant incidents.

7
Microbial aspects

The greatest risk from microbes in water is associated with consumption of drinking-water that is contaminated with human and animal excreta, although other sources and routes of exposure may also be significant.

This chapter focuses on organisms for which there is evidence, from outbreak studies or from prospective studies in non-outbreak situations, of disease being caused by ingestion of drinking-water, inhalation of droplets or contact with drinking-water; and their control.

7.1 Microbial hazards associated with drinking-water

Infectious diseases caused by pathogenic bacteria, viruses and parasites (e.g., protozoa and helminths) are the most common and widespread health risk associated with drinking-water. The public health burden is determined by the severity of the illness(es) associated with pathogens, their infectivity and the population exposed.

Breakdown in water supply safety may lead to large-scale contamination and potentially to detectable disease outbreaks. Other breakdowns and low-level, potentially repeated contamination may lead to significant sporadic disease, but is unlikely to be associated with the drinking-water source by public health surveillance.

Quantified risk assessment can assist in understanding and managing risks, especially those associated with sporadic disease.

7.1.1 Waterborne infections

The pathogens that may be transmitted through contaminated drinking-water are diverse. Table 7.1 and Figure 7.1 provide general information on pathogens that are of relevance for drinking-water supply management. The spectrum changes in response to variables such as increases in human and animal populations, escalating use of wastewater, changes in lifestyles and medical interventions, population movement and travel and selective pressures for new pathogens and mutants or recombinations of existing pathogens. The immunity of individuals also varies considerably, whether acquired by contact with a pathogen or influenced by such factors as age, sex, state of health and living conditions.

Table 7.1 Waterborne pathogens and their significance in water supplies

Pathogen	Health significance	Persistence in water supplies[a]	Resistance to chlorine[b]	Relative infectivity[c]	Important animal source
Bacteria					
Burkholderia pseudomallei	Low	May multiply	Low	Low	No
Campylobacter jejuni, C. coli	High	Moderate	Low	Moderate	Yes
Escherichia coli – Pathogenic[d]	High	Moderate	Low	Low	Yes
E. coli – Enterohaemorrhagic	High	Moderate	Low	High	Yes
Legionella spp.	High	Multiply	Low	Moderate	No
Non-tuberculous mycobacteria	Low	Multiply	High	Low	No
Pseudomonas aeruginosa[e]	Moderate	May multiply	Moderate	Low	No
Salmonella typhi	High	Moderate	Low	Low	No
Other salmonellae	High	May multiply	Low	Low	Yes
Shigella spp.	High	Short	Low	Moderate	No
Vibrio cholerae	High	Short	Low	Low	No
Yersinia enterocolitica	High	Long	Low	Low	Yes
Viruses					
Adenoviruses	High	Long	Moderate	High	No
Enteroviruses	High	Long	Moderate	High	No
Hepatitis A	High	Long	Moderate	High	No
Hepatitis E	High	Long	Moderate	High	Potentially
Noroviruses and Sapoviruses	High	Long	Moderate	High	Potentially
Rotavirus	High	Long	Moderate	High	No
Protozoa					No
Acanthamoeba spp.	High	Long	High	High	No
Cryptosporidium parvum	High	Long	High	High	Yes
Cyclospora cayetanensis	High	Long	High	High	No
Entamoeba histolytica	High	Moderate	High	High	No
Giardia intestinalis	High	Moderate	High	High	Yes
Naegleria fowleri	High	May multiply[f]	High	High	No
Toxoplasma gondii	High	Long	High	High	Yes
Helminths					
Dracunculus medinensis	High	Moderate	Moderate	High	No
Schistosoma spp.	High	Short	Moderate	High	Yes

Note: Waterborne transmission of the pathogens listed has been confirmed by epidemiological studies and case histories. Part of the demonstration of pathogenicity involves reproducing the disease in suitable hosts. Experimental studies in which volunteers are exposed to known numbers of pathogens provide relative information. As most studies are done with healthy adult volunteers, such data are applicable to only a part of the exposed population, and extrapolation to more sensitive groups is an issue that remains to be studied in more detail.

[a] Detection period for infective stage in water at 20 °C: short, up to 1 week; moderate, 1 week to 1 month; long, over 1 month.
[b] When the infective stage is freely suspended in water treated at conventional doses and contact times. Resistance moderate, agent may not be completely destroyed.
[c] From experiments with human volunteers or from epidemiological evidence.
[d] Includes enteropathogenic, enterotoxigenic and enteroinvasive.
[e] Main route of infection is by skin contact, but can infect immunosuppressed or cancer patients orally.
[f] In warm water.

For pathogens transmitted by the faecal–oral route, drinking-water is only one vehicle of transmission. Contamination of food, hands, utensils and clothing can also play a role, particularly when domestic sanitation and hygiene are poor. Improvements in the quality and availability of water, in excreta disposal and in general hygiene are all important in reducing faecal–oral disease transmission.

7. MICROBIAL ASPECTS

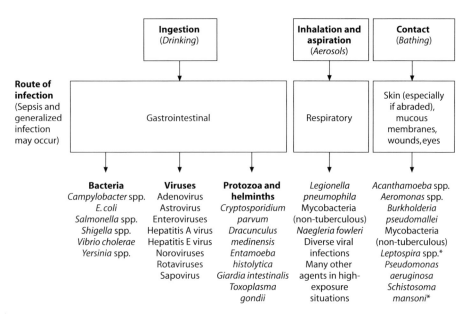

* Primarily from contact with highly contaminated surface waters.

Figure 7.1 Transmission pathways for and examples of water-related pathogens

Drinking-water safety is not related only to faecal contamination. Some organisms grow in piped water distribution systems (e.g., *Legionella*), whereas others occur in source waters (guinea worm *Dracunculus medinensis*) and may cause outbreaks and individual cases. Some other microbes (e.g., toxic cyanobacteria) require specific management approaches, which are covered elsewhere in these Guidelines (see section 11.5).

> Infectious diseases caused by pathogenic bacteria, viruses, protozoa and helminths are the most common and widespread health risk associated with drinking-water.

Certain serious illnesses result from inhalation of water droplets (aerosols) in which the causative organisms have multiplied because of warm temperatures and the presence of nutrients. These include legionellosis and Legionnaires' disease, caused by *Legionella* spp., and those caused by the amoebae *Naegleria fowleri* (primary amoebic meningoencephalitis [PAM]) and *Acanthamoeba* spp. (amoebic meningitis, pulmonary infections).

Schistosomiasis (bilharziasis) is a major parasitic disease of tropical and subtropical regions that is transmitted when the larval stage (cercariae), which is released by infected aquatic snails, penetrates the skin. It is primarily spread by contact with water. Ready availability of safe drinking-water contributes to disease prevention by reducing the need for contact with contaminated water resources – for example, when collecting water to carry to the home or when using water for bathing or laundry.

It is conceivable that unsafe drinking-water contaminated with soil or faeces could act as a carrier of other parasitic infections, such as balantidiasis (*Balantidium coli*) and certain helminths (species of *Fasciola, Fasciolopsis, Echinococcus, Spirometra, Ascaris, Trichuris, Toxocara, Necator, Ancylostoma, Strongyloides* and *Taenia solium*). However, in most of these, the normal mode of transmission is ingestion of the eggs in food contaminated with faeces or faecally contaminated soil (in the case of *Taenia solium*, ingestion of the larval cysticercus stage in uncooked pork) rather than ingestion of contaminated drinking-water.

Other pathogens that may be naturally present in the environment may be able to cause disease in people with impaired local or general immune defence mechanisms, such as the elderly or the very young, patients with burns or extensive wounds, those undergoing immunosuppressive therapy or those with acquired immunodeficiency syndrome (AIDS). If water used by such persons for drinking or bathing contains sufficient numbers of these organisms, they can produce various infections of the skin and the mucous membranes of the eye, ear, nose and throat. Examples of such agents are *Pseudomonas aeruginosa* and species of *Flavobacterium, Acinetobacter, Klebsiella, Serratia, Aeromonas* and certain "slow-growing" (non-tuberculous) mycobacteria (see the supporting document *Pathogenic Mycobacteria in Water*; section 1.3).

Most of the human pathogens listed in Table 7.1 (which are described in more detail in chapter 11) are distributed worldwide; some, however, such as those causing outbreaks of cholera or guinea worm disease, are regional. Eradication of *D. medinensis* is a recognized target of the World Health Assembly (1991).

It is likely that there are pathogens not shown in Table 7.1 that are also transmitted by water. This is because the number of known pathogens for which water is a transmission route continues to increase as new or previously unrecognized pathogens continue to be discovered (see WHO, 2003a).

7.1.2 Persistence and growth in water

While typical waterborne pathogens are able to persist in drinking-water, most do not grow or proliferate in water. Microorganisms like *E. coli* and *Campylobacter* can accumulate in sediments and are mobilized when water flow increases.

After leaving the body of their host, most pathogens gradually lose viability and the ability to infect. The rate of decay is usually exponential, and a pathogen will become undetectable after a certain period. Pathogens with low persistence must rapidly find new hosts and are more likely to be spread by person-to-person contact or poor personal hygiene than by drinking-water. Persistence is affected by several factors, of which temperature is the most important. Decay is usually faster at higher temperatures and may be mediated by the lethal effects of UV radiation in sunlight acting near the water surface.

The most common waterborne pathogens and parasites are those that have high infectivity and either can proliferate in water or possess high resistance to decay outside the body.

Viruses and the resting stages of parasites (cysts, oocysts, ova) are unable to multiply in water. Conversely, relatively high amounts of biodegradable organic carbon, together with warm temperatures and low residual concentrations of chlorine, can permit growth of *Legionella*, *V. cholerae*, *Naegleria fowleri*, *Acanthamoeba* and nuisance organisms in some surface waters and during water distribution (see also the supporting document *Heterotrophic Plate Counts and Drinking-water Safety*; section 1.3).

Microbial water quality may vary rapidly and widely. Short-term peaks in pathogen concentration may increase disease risks considerably and may also trigger outbreaks of waterborne disease. Results of water quality testing for microbes are not normally available in time to inform management action and prevent the supply of unsafe water.

7.1.3 Public health aspects

Outbreaks of waterborne disease may affect large numbers of persons, and the first priority in developing and applying controls on drinking-water quality should be the control of such outbreaks. Available evidence also suggests that drinking-water can contribute to background rates of disease in non-outbreak situations, and control of drinking-water quality should therefore also address waterborne disease in the general community.

Experience has shown that systems for the detection of waterborne disease outbreaks are typically inefficient in countries at all levels of socioeconomic development, and failure to detect outbreaks is not a guarantee that they do not occur; nor does it suggest that drinking-water should necessarily be considered safe.

Some of the pathogens that are known to be transmitted through contaminated drinking-water lead to severe and sometimes life-threatening disease. Examples include typhoid, cholera, infectious hepatitis (caused by hepatitis A virus [HAV] or HEV) and disease caused by *Shigella* spp. and *E. coli* O157. Others are typically associated with less severe outcomes, such as self-limiting diarrhoeal disease (e.g., Norovirus, *Cryptosporidium*).

The effects of exposure to pathogens are not the same for all individuals or, as a consequence, for all populations. Repeated exposure to a pathogen may be associated with a lower probability or severity of illness because of the effects of acquired immunity. For some pathogens (e.g., HAV), immunity is lifelong, whereas for others (e.g., *Campylobacter*), the protective effects may be restricted to a few months to years. On the other hand, sensitive subgroups (e.g., the young, the elderly, pregnant women and the immunocompromised) in the population may have a greater probability of illness or the illness may be more severe, including mortality. Not all pathogens have greater effects in all sensitive subgroups.

Not all infected individuals will develop symptomatic disease. The proportion of the infected population that is asymptomatic (including carriers) differs between pathogens and also depends on population characteristics, such as prevalence of immunity. Carriers and those with asymptomatic infections as well as individuals developing symptoms may all contribute to secondary spread of pathogens.

7.2 Health-based target setting
7.2.1 Health-based targets applied to microbial hazards
General approaches to health-based target setting are described in section 2.1.1 and chapter 3.

Sources of information on health risks may be from both epidemiology and risk assessment, and typically both are employed as complementary sources.

Health-based targets may also be set using a health outcome approach, where the waterborne disease burden is believed to be sufficiently high to allow measurement of the impact of interventions – i.e., to measure reductions in disease that can be attributed to drinking-water.

Risk assessment is especially valuable where the fraction of disease that can be attributed to drinking-water is low or difficult to measure directly through public health surveillance or analytical epidemiological studies.

Data – from both epidemiology and risk assessment – with which to develop health-based targets for many pathogens are limited, but are increasingly being produced. Locally generated data will always be of great value in setting national targets.

For the control of microbial hazards, the most frequent form of health-based target applied is performance targets (see section 3.2.2), which are anchored to a tolerable burden of disease. WQTs (see section 3.2.3) are typically not developed for pathogens, because monitoring finished water for pathogens is not considered a feasible or cost-effective option.

7.2.2 Risk assessment approach
In many circumstances, estimating the effects of improved drinking-water quality on health risks in the population is possible through constructing and applying risk assessment models.

QMRA is a rapidly evolving field that systematically combines available information on exposure and dose–response to produce estimates of the disease burden associated with exposure to pathogens. Mathematical modelling is used to estimate the effects of low doses of pathogens in drinking-water on populations and subpopulations.

Interpreting and applying information from analytical epidemiological studies to derive health-based targets for application at a national or local level require consideration of a number of factors, including the following:
- Are specific estimates of disease reduction or indicative ranges of expected reductions to be provided?
- How representative of the target population was the study sample in order to ensure confidence in the reliability of the results across a wider group?

- To what extent will minor differences in demographic or socioeconomic conditions affect expected outcomes?

Risk assessment commences with problem formulation to identify all possible hazards and their pathways from source(s) to recipient(s). Human exposure to the pathogens (environmental concentrations and volumes ingested) and dose–responses of these selected organisms are then combined to characterize the risks. With the use of additional information (social, cultural, political, economic, environmental, etc.), management options can be prioritized. To encourage stakeholder support and participation, a transparent procedure and active risk communication at each stage of the process are important. An example of a risk assessment approach is described in Table 7.2 and outlined below.

Problem formulation and hazard identification

All potential hazards, sources and events that can lead to the presence of these hazards (i.e., what can happen and how) should be identified and documented for each component of the drinking-water system, regardless of whether or not the component is under the direct control of the drinking-water supplier. This includes point sources of pollution (e.g., human and industrial waste discharge) as well as diffuse sources (e.g., those arising from agricultural and animal husbandry activities). Continuous, intermittent or seasonal pollution patterns should also be considered, as well as extreme and infrequent events, such as droughts and floods.

The broader sense of hazards focuses on hazardous scenarios, which are events that may lead to exposure of consumers to specific pathogenic microorganisms. In this, the hazardous event (e.g., peak contamination of source water with domestic wastewater) may be referred to as the hazard.

Representative organisms are selected that, if controlled, would ensure control of all pathogens of concern. Typically, this implies inclusion of at least one bacterial pathogen, virus and protozoan.

Table 7.2 Risk assessment paradigm for pathogen health risks

Step	Aim
1. Problem formulation and hazard identification	To identify all possible hazards associated with drinking-water that would have an adverse public health consequence, as well as their pathways from source(s) to consumer(s)
2. Exposure assessment	To determine the size and nature of the population exposed and the route, amount and duration of the exposure
3. Dose–response assessment	To characterize the relationship between exposure and the incidence of the health effect
4. Risk characterization	To integrate the information from exposure, dose–response and health interventions in order to estimate the magnitude of the public health problem and to evaluate variability and uncertainty

Source: Adapted from Haas et al. (1999).

Exposure assessment

Exposure assessment involves estimation of the number of pathogenic microbes to which an individual is exposed, principally through ingestion. Exposure assessment is a predictive activity that often involves subjective judgement. It inevitably contains uncertainty and must account for variability of factors such as concentrations of microorganisms over time, volumes ingested, etc.

Exposure can be considered as a single dose of pathogens that a consumer ingests at a certain point of time or the total amount over several exposures (e.g., over a year). Exposure is determined by the concentration of microbes in drinking-water and the volume of water consumed.

It is rarely possible or appropriate to directly measure pathogens in drinking-water on a regular basis. More often, concentrations in source waters are assumed or measured, and estimated reductions – for example, through treatment – are applied to estimate the concentration in the water consumed. Pathogen measurement, when performed, is generally best carried out at the location where the pathogens are at highest concentration (generally source waters). Estimation of their removal by sequential control measures is generally achieved by the use of surrogates (such as *E. coli* for enteric bacterial pathogens).

The other component of exposure assessment, which is common to all pathogens, is the volume of unboiled water consumed by the population, including person-to-person variation in consumption behaviour and especially consumption behaviour of at-risk groups. For microbial hazards, it is important that the unboiled volume of drinking-water, both consumed directly and used in food preparation, is used in the risk assessment, as heating will rapidly inactivate pathogens. This amount is lower than that used for deriving chemical guideline values and WQTs.

The daily exposure of a consumer can be assessed by multiplying the concentration of pathogens in drinking-water by the volume of drinking-water consumed. For the purposes of the Guidelines, unboiled drinking-water consumption is assumed to be 1 litre of water per day.

Dose–response assessment

The probability of an adverse health effect following exposure to one or more pathogenic organisms is derived from a dose–response model. Available dose–response data have been obtained mainly from studies using healthy adult volunteers. Several subgroups in the population, such as children, the elderly and immunocompromised persons, are more sensitive to infectious disease; currently, however, adequate data are lacking to account for this.

The conceptual basis for the infection model is the observation that exposure to the described dose leads to the probability of infection as a conditional event. For infection to occur, one or more viable pathogens must have been ingested. Furthermore, one or more of these ingested pathogens must have survived in the host's body. An important concept is the single-hit principle (i.e., that even a single organism may be able to cause infection and disease, possibly with a low probability). This concept supersedes the concept of (minimum) infectious dose that is frequently used in older literature (see the supporting document *Hazard Characterization for Pathogens in Food and Water*; section 1.3).

In general, well dispersed pathogens in water are considered to be Poisson distributed. When the individual probability of any organism to survive and start infection is the same, the dose–response relation simplifies to an exponential function. If, however, there is heterogeneity in this individual probability, this leads to the beta-Poisson dose–response relation, where the "beta" stands for the distribution of the individual probabilities among pathogens (and hosts). At low exposures, such as would typically occur in drinking-water, the dose–response model is approximately linear and can be represented simply as the probability of infection resulting from exposure to a single organism (see the supporting document *Hazard Characterization for Pathogens in Food and Water*; section 1.3).

Risk characterization

Risk characterization brings together the data collected on pathogen exposure, dose–response, severity and disease burden.

The probability of infection can be estimated as the product of the exposure by drinking-water and the probability that exposure to one organism would result in infection. The probability of infection per day is multiplied by 365 to calculate the probability of infection per year. In doing so, it is assumed that different exposure events are independent, in that no protective immunity is built up. This simplification is justified for low risks only.

Not all infected individuals will develop clinical illness; asymptomatic infection is common for most pathogens. The percentage of infected persons that will develop clinical illness depends on the pathogen, but also on other factors, such as the immune status of the host. Risk of illness per year is obtained by multiplying the probability of infection by the probability of illness given infection.

The low numbers in Table 7.3 can be interpreted to represent the probability that a single individual will develop illness in a given year. For example, a risk of illness for *Campylobacter* of 2.5×10^{-4} per year indicates that, on average, 1 out of 4000 consumers would contract campylobacteriosis from drinking-water.

To translate the risk of developing a specific illness to disease burden per case, the metric DALYs is used. This should reflect not only the effects of acute end-points (e.g., diarrhoeal illness) but also mortality and the effects of more serious end-points (e.g., Guillain-Barré syndrome associated with *Campylobacter*). Disease burden per case varies widely. For example, the disease burden per 1000 cases of rotavirus diarrhoea is 480 DALYs in low-income regions, where child mortality frequently occurs. However, it is only 14 DALYs per 1000 cases in high-income regions, where hospital facilities are accessible to the great majority of the population (see the supporting document *Quantifying Public Health Risk in the WHO Guidelines for Drinking-water*

Table 7.3 **Linking tolerable disease burden and source water quality for reference pathogens: example calculation**

River water (human and animal pollution)		Cryptosporidium	Campylobacter	Rotavirus[a]
Raw water quality (C_R)	Organisms per litre	10	100	10
Treatment effect needed to reach tolerable risk (PT)	Percent reduction	99.994%	99.99987%	99.99968%
Drinking-water quality (C_D)	Organisms per litre	6.3×10^{-4}	1.3×10^{-4}	3.2×10^{-5}
Consumption of unheated drinking-water (V)	Litres per day	1	1	1
Exposure by drinking-water (E)	Organisms per day	6.3×10^{-4}	1.3×10^{-4}	3.2×10^{-5}
Dose–response (r)	Probability of infection per organism	4.0×10^{-3}	1.8×10^{-2}	2.7×10^{-1}
Risk of infection ($P_{inf,d}$)	Per day	2.5×10^{-6}	2.3×10^{-6}	8.5×10^{-6}
Risk of infection ($P_{inf,y}$)	Per year	9.2×10^{-4}	8.3×10^{-4}	3.1×10^{-3}
Risk of (diarrhoeal) illness given infection ($P_{ill|inf}$)		0.7	0.3	0.5
Risk of (diarrhoeal) illness (P_{ill})	Per year	6.4×10^{-4}	2.5×10^{-4}	1.6×10^{-3}
Disease burden (db)	DALYs per case	1.5×10^{-3}	4.6×10^{-3}	1.4×10^{-2}
Susceptible fraction (f_s)	Percentage of population	100%	100%	6%
Disease burden (DB)	DALYs per year	1×10^{-6}	1×10^{-6}	1×10^{-6}
Formulas:	$C_D = C_R \times (1 - PT)$			
	$E = C_D \times V$			
	$P_{inf,d} = E \times r$			

[a] Data from high-income regions. In low-income regions, severity is typically higher, but drinking-water transmission is unlikely to dominate.

Quality; section 1.3). This considerable difference in disease burden results in far stricter treatment requirements in low-income regions for the same source water quality in order to obtain the same risk (expressed as DALYs per year). Ideally, the default disease burden estimates in Table 7.3 should be adapted to specific national situations. In Table 7.3, no accounting is made for effects on immunocompromised persons (e.g., cryptosporidiosis in HIV/AIDS patients), which is significant in some countries. Section 3.3.3 gives more information on the DALY metric and how it is applied to derive a reference level of risk.

Only a proportion of the population may be susceptible to some pathogens, because immunity developed after an initial episode of infection or illness may provide lifelong protection. Examples include HAV and rotaviruses. It is estimated that in developing countries, all children above the age of 5 years are immune to rotaviruses because of repeated exposure in the first years of life. This translates to an

7. MICROBIAL ASPECTS

average of 17% of the population being susceptible to rotavirus illness. In developed countries, rotavirus infection is also common in the first years of life, and the illness is diagnosed mainly in young children, but the percentage of young children as part of the total population is lower. This translates to an average of 6% of the population in developed countries being susceptible.

The uncertainty of the risk estimate is the result of the uncertainty and variability of the data collected in the various steps of the risk assessment. Risk assessment models should ideally account for this variability and uncertainty, although here we present only point estimates (see below).

It is important to choose the most appropriate point estimate for each of the variables. Theoretical considerations show that risks are directly proportional to the arithmetic mean of the ingested dose. Hence, arithmetic means of variables such as concentration in raw water, removal by treatment and consumption of drinking-water are recommended. This recommendation is different from the usual practice among microbiologists and engineers of converting concentrations and treatment effects to log-values and making calculations or specifications on the log-scale. Such calculations result in estimates of the geometric mean rather than the arithmetic mean, and these may significantly underestimate risk. Analysing site-specific data may therefore require going back to the raw data rather than relying on reported log-transformed values.

7.2.3 Risk-based performance target setting

The process outlined above enables estimation of risk on a population level, taking account of source water quality and impact of control. This can be compared with the reference level of risk (see section 3.3.2) or a locally developed tolerable risk. The calculations enable quantification of the degree of source protection or treatment that is needed to achieve a specified level of acceptable risk and analysis of the estimated impact of changes in control measures.

Performance targets are most frequently applied to treatment performance – i.e., to determine the microbial reduction necessary to ensure water safety. A performance target may be applied to a specific system (i.e., allow account to be taken of specific source water characteristics) or generalized (e.g., impose source water quality assumptions on all systems of a certain type or abstracting water from a certain type of source).

Figure 7.2 illustrates the targets for treatment performance for a range of pathogens occurring in the raw water. For example, 10 microorganisms per litre of source water will lead to a performance target of 4.2 logs (or 99.994%) for *Cryptosporidium* or of 5.5 logs (99.99968%) for rotavirus in high-income regions (see also Table 7.4 below). The difference in performance targets for rotavirus in high- and low-income countries (5.5 and 7.6 logs; Figure 7.2) is related to the difference in disease severity by this organism. In low-income countries, the child case fatality rate is relatively high, and, as a consequence, the disease burden is higher. Also, a larger proportion of the

Figure 7.2 Performance targets for selected bacterial, viral and protozoan pathogens in relation to raw water quality (to achieve 10^{-6} DALYs per person per year)

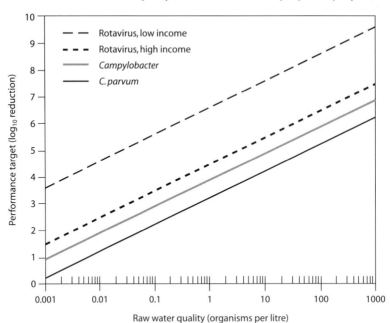

Table 7.4 Health-based targets derived from example calculation in Table 7.3

	Cryptosporidium	Campylobacter	Rotavirus[a]
Organisms per litre in source water	10	100	10
Health outcome target	10^{-6} DALYs per person per year	10^{-6} DALYs per person per year	10^{-6} DALYs per person per year
Risk of diarrhoeal illness[b]	1 per 1600 per year	1 per 4000 per year	1 per 11 000 per year
Drinking-water quality	1 per 1600 litres	1 per 8000 litres	1 per 32 000 litres
Performance target[c]	4.2 $\log_{10}$ units	5.9 $\log_{10}$ units	5.5 $\log_{10}$ units

[a] Data from high-income regions. In low-income regions, severity is typically higher, but drinking-water transmission is unlikely to dominate.
[b] For the susceptible population.
[c] Performance target is a measure of log reduction of pathogens based on source water quality.

population in low-income countries is under the age of 5 and at risk for rotavirus infection.

The derivation of these performance targets is described in Table 7.4, which provides an example of the data and calculations that would normally be used to construct a risk assessment model for waterborne pathogens. The table presents data for representatives of the three major groups of pathogens (bacteria, viruses and protozoa) from a range of sources. These example calculations aim at achieving the reference level of risk of 10^{-6} DALYs per person per year, as described in section 3.3.3. The data in the table illustrate the calculations needed to arrive at a risk estimate and are not guideline values.

7.2.4 Presenting the outcome of performance target development

Table 7.4 presents some data from Table 7.3 in a format that is more meaningful to risk managers. The average concentration of pathogens in drinking-water is included for information. It is not a WQT, nor is it intended to encourage pathogen monitoring in finished water. As an example, a concentration of 6.3×10^{-4} *Cryptosporidium* per litre (see Table 7.3) corresponds to 1 oocyst per 1600 litres (see Table 7.4). The performance target (in the row "Treatment effect" in Table 7.3), expressed as a percent reduction, is the most important management information in the risk assessment table. It can also be expressed as a log-reduction value. For example, 99.99968% reduction for rotavirus corresponds to $5.5 \log_{10}$ units.

7.2.5 Issues in adapting risk-based performance target setting to national/local circumstances

The choice of pathogens in Table 7.4 was based mainly on availability of data on resistance to water treatment, infectivity and disease burden. The pathogens illustrated may not be priority pathogens in all regions of the world, although amending pathogen selection would normally have a small impact on the overall conclusions derived from applying the model.

Wherever possible, country- or site-specific information should be used in assessments of this type. If no specific data are available, an approximate risk estimate can be based on default values (see Table 7.5 below).

Table 7.4 accounts only for changes in water quality derived from treatment and not source protection measures, which are often important contributors to overall safety, impacting on pathogen concentration and/or variability. The risk estimates presented in Table 7.3 also assume that there is no degradation of water quality in the distribution network. These may not be realistic assumptions under all circumstances, and it is advisable to take these factors into account wherever possible.

Table 7.4 presents point estimates only and does not account for variability and uncertainty. Full risk assessment models would incorporate such factors by representing the input variables by statistical distributions rather than by point estimates. However, such models are currently beyond the means of many countries, and data to define such distributions are scarce. Producing such data may involve considerable efforts in terms of time and resources, but will lead to much improved insight into the actual source water quality and treatment performance.

The necessary degree of treatment also depends on the values assumed for variables (e.g., drinking-water consumption, fraction of the population that is susceptible) that can be taken into account in the risk assessment model. Figure 7.3 shows the effect of variation in the consumption of unboiled drinking-water on the performance targets for *Cryptosporidium parvum*. For example, if the raw water concentration

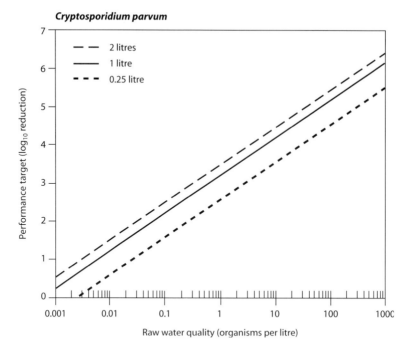

Figure 7.3 Performance targets for *Cryptosporidium parvum* in relation to the daily consumption of unboiled drinking-water (to achieve 10^{-6} DALYs per person per year)

is 1 oocyst per litre, the performance target varies between 2.6 and 3.5 $\log_{10}$ units if consumption values vary between 0.25 and 2 litres per day. Some outbreak data suggest that in developed countries, a significant proportion of the population above 5 years of age may not be immune to rotavirus illness. Figure 7.4 shows the effect of variation in the susceptible fraction of the population. For example, if the raw water concentration is 10 virus particles per litre, the performance target increases from 5.5 to 6.7 if the susceptible fraction increases from 6 to 100%.

7.2.6 Health outcome targets

Health outcome targets that identify disease reductions in a community may be applied to the WSPs developed for specified water quality interventions at community and household levels. These targets would identify expected disease reductions in communities receiving the interventions.

The prioritization of water quality interventions should focus on those aspects that are estimated to contribute more than e.g. 5% of the burden of a given disease (e.g., 5% of total diarrhoea). In many parts of the world, the implementation of a water quality intervention that results in an estimated health gain of more than 5% would

7. MICROBIAL ASPECTS

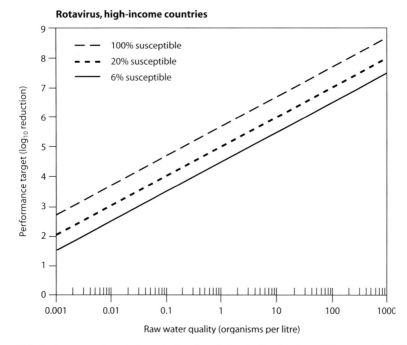

Figure 7.4 Performance targets for rotavirus in relation to the fraction of the population that is susceptible to illness (to achieve 10^{-6} DALYs per person per year)

be considered extremely worthwhile. Directly demonstrating the health gains arising from improving water quality – as assessed, for example, by reduced *E. coli* counts at the point of consumption – may be possible where disease burden is high and effective interventions are applied and can be a powerful tool to demonstrate a first step in incremental water safety improvement.

Where a specified quantified disease reduction is identified as a health outcome target, it may be advisable to undertake ongoing proactive public health surveillance among representative communities rather than through passive surveillance.

7.3 Occurrence and treatment of pathogens

As discussed in section 4.1, system assessment involves determining whether the drinking-water supply chain as a whole can deliver drinking-water quality that meets identified targets. This requires an understanding of the quality of source water and the efficacy of control measures.

An understanding of pathogen occurrence in source waters is essential, because it facilitates selection of the highest-quality source for drinking-water supply, determines pathogen loads and concentrations in source waters and provides a basis for establishing treatment requirements to meet health-based targets within a WSP.

Understanding the efficacy of control measures includes validation (see sections 2.1.2 and 4.1.7). Validation is important both in ensuring that treatment will achieve the desired goals (performance targets) and in assessing areas in which efficacy may be improved (e.g., by comparing performance achieved with that shown to be achievable through well run processes).

7.3.1 Occurrence

The occurrence of pathogens and indicator organisms in groundwater and surface water sources depends on a number of factors, including intrinsic physical and chemical characteristics of the catchment area and the magnitude and range of human activities and animal sources that release pathogens to the environment.

In surface waters, potential pathogen sources include point sources, such as municipal sewerage and urban stormwater overflows, as well as non-point sources, such as contaminated runoff from agricultural areas and areas with sanitation through on-site septic systems and latrines. Other sources are wildlife and direct access of livestock to surface water bodies. Many pathogens in surface water bodies will reduce in concentration due to dilution, settling and die-off due to environmental effects (thermal, sunlight, predation, etc.).

Groundwater is often less vulnerable to the immediate influence of contamination sources due to the barrier effects provided by the overlying soil and its unsaturated zone. Groundwater contamination is more frequent where these protective barriers are breached, allowing direct contamination. This may occur through contaminated or abandoned wells or underground pollution sources, such as latrines and sewer lines. However, a number of studies have demonstrated pathogens and indicator organisms in groundwater, even at depth in the absence of such hazardous circumstances, especially where surface contamination is intense, as with land application of manures or other faecal impacts from intensive animal husbandry (e.g., feedlots). Impacts of these contamination sources can be greatly reduced by, for example, aquifer protection measures and proper well design and construction.

For more detailed discussion on both pathogen sources and key factors determining their fate, refer to the supporting documents *Protecting Surface Waters for Health* and *Protecting Groundwaters for Health* (section 1.3).

Table 7.5 presents estimates of high concentrations of enteric pathogens and microbial indicators in different types of surface waters and groundwaters, derived primarily from a review of published data. High values have been presented because they represent higher-risk situations and, therefore, greater degrees of vulnerability. The table includes two categories of data for rivers and streams: one for impacted sources and one for less impacted sources. More detailed information about these data is published in a variety of references, including several papers cited in Dangendorf et al. (2003).

The data in Table 7.5 provide a useful guide to the concentrations of enteric pathogens and indicator microorganisms in a variety of sources. However, there are a number of limitations and sources of uncertainty in these data, including:

7. MICROBIAL ASPECTS

Table 7.5 Examples of high detectable concentrations (per litre) of enteric pathogens and faecal indicators in different types of source waters from the scientific literature

Pathogen or indicator group	Lakes and reservoirs	Impacted rivers and streams	Wilderness rivers and streams	Groundwater
Campylobacter	20–500	90–2500	0–1100	0–10
Salmonella	—	3–58 000 (3–1000)[a]	1–4	—
E. coli (generic)	10 000–1 000 000	30 000–1 000 000	6000–30 000	0–1000
Viruses	1–10	30–60	0–3	0–2
Cryptosporidium	4–290	2–480	2–240	0–1
Giardia	2–30	1–470	1–2	0–1

[a] Lower range is a more recent measurement.

— the lack of knowledge on sampling locations in relation to pollution sources;
— concerns about the sensitivity of analytical techniques, particularly for viruses and protozoa; and
— the lack of knowledge about the viability and human infectivity of *Cryptosporidium* oocysts, *Giardia* cysts and viruses detected in the different studies, because the various methods used are based upon non-culture methods (e.g., microscopy or molecular/nucleic acid analysis).

While the table provides an indication of concentrations that might be present in water sources, by far the most accurate way of determining pathogen loads and concentrations in specific catchments and other water sources is by analysing water quality over a period of time, taking care to include consideration of seasonal variation and peak events such as storms. Direct measurement of pathogens and indicators in the specific source waters for which a WSP and its target pathogens are being established is recommended wherever possible, because this provides the best estimates of microbial concentrations and loads.

7.3.2 Treatment

Waters of very high quality – for example, groundwater from confined aquifers – may rely on source water and distribution system protection as the principal control measures for provision of safe water. More typically, water treatment is required to remove or destroy pathogenic microorganisms. In many cases (e.g., poor-quality surface water), multiple treatment stages are required, including, for example, coagulation, flocculation, sedimentation, filtration and disinfection. Table 7.6 provides a summary of treatment processes that are commonly used individually or in combination to achieve microbial reductions.

The microbial reductions presented in Table 7.6 are for broad groups or categories of microbes: bacteria, viruses and protozoa. This is because it is generally the case that treatment efficacy for microbial reduction differs among these microbial groups due to the inherently different properties of the microbes (e.g., size, nature of protective outer layers, physicochemical surface properties, etc.). Within these microbial groups,

Table 7.6 **Reductions of bacteria, viruses and protozoa achieved by typical and enhanced water treatment processes**

Treatment process	Enteric pathogen group	Baseline removal	Maximum removal possible
Pretreatment			
Roughing filters	Bacteria	50%	Up to 95% if protected from turbidity spikes by dynamic filter or if used only when ripened
	Viruses	No data available	
	Protozoa	No data available, some removal likely	Performance for protozoan removal likely to correspond to turbidity removal
Microstraining	Bacteria, viruses, protozoa	Zero	Generally ineffective
Off-stream/ bankside storage	All	Recontamination may be significant and add to pollution levels in incoming water; growth of algae may cause deterioration in quality	Avoiding intake at periods of peak turbidity equivalent to 90% removal; compartmentalized storages provide 15–230 times rates of removal
	Bacteria	Zero (assumes short circuiting)	90% removal in 10–40 days actual detention time
	Viruses	Zero (assumes short circuiting)	93% removal in 100 days actual detention time
	Protozoa	Zero (assumes short circuiting)	99% removal in 3 weeks actual detention time
Bankside infiltration	Bacteria	99.9% after 2 m 99.99% after 4 m (minimum based on virus removal)	
	Viruses	99.9% after 2 m 99.99% after 4 m	
	Protozoa	99.99%	
Coagulation/flocculation/sedimentation			
Conventional clarification	Bacteria	30%	90% (depending on the coagulant, pH, temperature, alkalinity, turbidity)
	Viruses	30%	70% (as above)
	Protozoa	30%	90% (as above)
High-rate clarification	Bacteria	At least 30%	
	Viruses	At least 30%	
	Protozoa	95%	99.99% (depending on use of appropriate blanket polymer)
Dissolved air flotation	Bacteria	No data available	
	Viruses	No data available	
	Protozoa	95%	99.9% (depending on pH, coagulant dose, flocculation time, recycle ratio)

Table 7.6 *Continued*

Treatment process	Enteric pathogen group	Baseline removal	Maximum removal possible
Lime softening	Bacteria	20% at pH 9.5 for 6 h at 2–8 °C	99% at pH 11.5 for 6 h at 2–8 °C
	Viruses	90% at pH < 11 for 6 h	99.99% at pH > 11, depending on the virus and on settling time
	Protozoa	Low inactivation	99% through precipitative sedimentation and inactivation at pH 11.5
Ion exchange			
	Bacteria	Zero	
	Viruses	Zero	
	Protozoa	Zero	
Filtration			
Granular high-rate filtration	Bacteria	No data available	99% under optimum coagulation conditions
	Viruses	No data available	99.9% under optimum coagulation conditions
	Protozoa	70%	99.9% under optimum coagulation conditions
Slow sand filtration	Bacteria	50%	99.5% under optimum ripening, cleaning and refilling and in the absence of short circuiting
	Viruses	20%	99.99% under optimum ripening, cleaning and refilling and in the absence of short circuiting
	Protozoa	50%	99% under optimum ripening, cleaning and refilling and in the absence of short circuiting
Precoat filtration, including diatomaceous earth and perlite	Bacteria	30–50%	96–99.9% using chemical pretreatment with coagulants or polymers
	Viruses	90%	98% using chemical pretreatment with coagulants or polymers
	Protozoa	99.9%	99.99%, depending on media grade and filtration rate
Membrane filtration – microfiltration	Bacteria	99.9–99.99%, providing adequate pretreatment and membrane integrity conserved	
	Viruses	<90%	
	Protozoa	99.9–99.99%, providing adequate pretreatment and membrane integrity conserved	
Membrane filtration – ultrafiltration,	Bacteria	Complete removal, providing adequate pretreatment and membrane integrity conserved	

continued

GUIDELINES FOR DRINKING-WATER QUALITY

Table 7.6 *Continued*

Treatment process	Enteric pathogen group	Baseline removal	Maximum removal possible
nanofiltration and reverse osmosis	Viruses	Complete removal with nanofilters, with reverse osmosis and at lower pore sizes of ultrafilters, providing adequate pretreatment and membrane integrity conserved	
	Protozoa	Complete removal, providing adequate pretreatment and membrane integrity conserved	
Disinfection			
Chlorine	Bacteria	Ct_{99}: 0.08 mg·min/litre at 1–2 °C, pH 7; 3.3 mg·min/litre at 1–2 °C, pH 8.5	
	Viruses	Ct_{99}: 12 mg·min/litre at 0–5 °C; 8 mg·min/litre at 10 °C; both at pH 7–7.5	
	Protozoa	*Giardia* Ct_{99}: 230 mg·min/litre at 0.5 °C; 100 mg·min/litre at 10 °C; 41 mg·min/litre at 25 °C; all at pH 7–7.5 *Cryptosporidium* not killed	
Monochloramine	Bacteria	Ct_{99}: 94 mg·min/litre at 1–2 °C, pH 7; 278 mg·min/litre at 1–2 °C, pH 8.5	
	Viruses	Ct_{99}: 1240 mg·min/litre at 1 °C; 430 mg·min/litre at 15 °C; both at pH 6–9	
	Protozoa	*Giardia* Ct_{99}: 2550 mg·min/litre at 1 °C; 1000 mg·min/litre at 15 °C; both at pH 6–9 *Cryptosporidium* not inactivated	
Chlorine dioxide	Bacteria	Ct_{99}: 0.13 mg·min/litre at 1–2 °C, pH 7; 0.19 mg·min/litre at 1–2 °C, pH 8.5	
	Viruses	Ct_{99}: 8.4 mg·min/litre at 1 °C; 2.8 mg·min/litre at 15 °C; both at pH 6–9	
	Protozoa	*Giardia* Ct_{99}: 42 mg·min/litre at 1 °C; 15 mg·min/litre at 10 °C; 7.3 mg·min/litre at 25 °C; all at pH 6–9 *Cryptosporidium* Ct_{99}: 40 mg·min/litre at 22 °C, pH 8	

Table 7.6 *Continued*

Treatment process	Enteric pathogen group	Baseline removal	Maximum removal possible
Ozone	Bacteria	Ct_{99}: 0.02 mg·min/litre at 5 °C, pH 6–7	
	Viruses	Ct_{99}: 0.9 mg·min/litre at 1 °C, 0.3 mg·min/litre at 15 °C	
	Protozoa	*Giardia* Ct_{99}: 1.9 mg·min/litre at 1 °C; 0.63 mg·min/litre at 15 °C, pH 6–9 *Cryptosporidium* Ct_{99}: 40 mg·min/litre at 1 °C; 4.4 mg·min/litre at 22 °C	
UV irradiation	Bacteria	99% inactivation: 7 mJ/cm^2	
	Viruses	99% inactivation: 59 mJ/cm^2	
	Protozoa	*Giardia* 99% inactivation: 5 mJ/cm^2 *Cryptosporidium* 99.9% inactivation: 10 mJ/cm^2	

Note: Ct and UV apply to microorganisms in suspension, not embedded in particles or in biofilm.

differences in treatment process efficiencies are smaller among the specific species, types or strains of microbes. Such differences do occur, however, and the table presents conservative estimates of microbial reductions based on the more resistant or persistent pathogenic members of that microbial group. Where differences in removal by treatment between specific members of a microbial group are great, the results for the individual microbes are presented separately in the table.

Non-piped water supplies such as roof catchments (rainwater harvesting) and water collected from wells or springs may often be contaminated with pathogens. Such sources often require treatment and protected storage to achieve safe water. Many of the processes used for water treatment in households are the same as those used for community-managed and other piped water supplies (Table 7.6). The performance of these treatment processes at the household level is likely to be similar to that for baseline removal of microbes, as shown in Table 7.6. However, there are additional water treatment technologies recommended for use in non-piped water supplies at the household level that typically are not used for piped supplies.

Further information about these water treatment processes, their operations and their performance for pathogen reduction is provided in more detail in supporting documents (for piped water supplies: *Water Treatment and Pathogen Control*; for non-piped [primarily household] water supplies: *Managing Water in the Home*; see section 1.3).

7.4 Verification of microbial safety and quality

Pathogenic agents have several properties that distinguish them from other drinking-water contaminants:

- Pathogens are discrete and not in solution.
- Pathogens are often clumped or adherent to suspended solids in water.
- The likelihood of a successful challenge by a pathogen, resulting in infection, depends upon the invasiveness and virulence of the pathogen, as well as upon the immunity of the individual.
- If infection is established, pathogens multiply in their host. Certain pathogenic bacteria are also able to multiply in food or beverages, thereby perpetuating or even increasing the chances of infection.
- Unlike many chemical agents, the dose–response of pathogens is not cumulative.

Faecal indicator bacteria, including *E. coli*, are important parameters for verification of microbial quality (see also section 2.2.1). Such water quality verification complements operational monitoring and assessments of contamination risks – for instance, through auditing of treatment works, evaluation of process control and sanitary inspection.

Faecal indicator bacteria should fulfil certain criteria to give meaningful results. They should be universally present in high numbers in the faeces of humans and other warm-blooded animals, should be readily detectable by simple methods and should not grow in natural water.

The indicator organism of choice for faecal pollution is *E. coli*. Thermotolerant coliforms can be used as an alternative to the test for *E. coli* in many circumstances.

Water intended for human consumption should contain no indicator organisms. In the majority of cases, monitoring for indicator bacteria provides a high degree of safety because of their large numbers in polluted waters.

Pathogens more resistant to conventional environmental conditions or treatment technologies may be present in treated drinking-water in the absence of *E. coli*. Retrospective studies of waterborne disease outbreaks and advances in the understanding of the behaviour of pathogens in water have shown that continued reliance on assumptions surrounding the absence or presence of *E. coli* does not ensure that optimal decisions are made regarding water safety.

Protozoa and some enteroviruses are more resistant to many disinfectants, including chlorine, and may remain viable (and pathogenic) in drinking-water following disinfection. Other organisms may be more appropriate indicators of persistent microbial hazards, and their selection as additional indicators should be evaluated in relation to local circumstances and scientific understanding. Therefore, verification may require analysis of a range of organisms, such as intestinal enterococci, (spores of) *Clostridium perfringens* and bacteriophages.

Table 7.7 presents guideline values for verification of microbial quality of drinking-water. Individual values should not be used directly from the tables. The

7. MICROBIAL ASPECTS

Table 7.7 Guideline values for verification of microbial quality[a] **(see also table 5.2)**

Organisms	Guideline value
All water directly intended for drinking	
E. coli or thermotolerant coliform bacteria[b,c]	Must not be detectable in any 100-ml sample
Treated water entering the distribution system	
E. coli or thermotolerant coliform bacteria[b]	Must not be detectable in any 100-ml sample
Treated water in the distribution system	
E. coli or thermotolerant coliform bacteria[b]	Must not be detectable in any 100-ml sample

[a] Immediate investigative action must be taken if E. coli are detected.
[b] Although E. coli is the more precise indicator of faecal pollution, the count of thermotolerant coliform bacteria is an acceptable alternative. If necessary, proper confirmatory tests must be carried out. Total coliform bacteria are not acceptable indicators of the sanitary quality of water supplies, particularly in tropical areas, where many bacteria of no sanitary significance occur in almost all untreated supplies.
[c] It is recognized that in the great majority of rural water supplies, especially in developing countries, faecal contamination is widespread. Especially under these conditions, medium-term targets for the progressive improvement of water supplies should be set.

guidelines values should be used and interpreted in conjunction with the information contained in these Guidelines and other supporting documentation.

A consequence of variable susceptibility to pathogens is that exposure to drinking-water of a particular quality may lead to different health effects in different populations. For guideline derivation, it is necessary to define reference populations or, in some cases, to focus on specific sensitive subgroups. National or local authorities may wish to apply specific characteristics of their populations in deriving national standards.

7.5 Methods of detection of faecal indicator bacteria

Analysis for faecal indicator bacteria provides a sensitive, although not the most rapid, indication of pollution of drinking-water supplies. Because the growth medium and the conditions of incubation, as well as the nature and age of the water sample, can influence the species isolated and the count, microbiological examinations may have variable accuracy. This means that the standardization of methods and of laboratory procedures is of great importance if criteria for the microbial quality of water are to be uniform in different laboratories and internationally.

International standard methods should be evaluated under local circumstances before being adopted. Established standard methods are available, such as those of the ISO (Table 7.8) or methods of equivalent efficacy and reliability. It is desirable that established standard methods be used for routine examinations. Whatever method is chosen for detection of *E. coli* or thermotolerant coliforms, the importance of "resuscitating" or recovering environmentally damaged or disinfectant-damaged strains must be considered.

Table 7.8 International Organization for Standardization (ISO) standards for detection and enumeration of faecal indicator bacteria in water

ISO standard	Title (water quality)
6461-1:1986	Detection and enumeration of the spores of sulfite-reducing anaerobes (clostridia) — Part 1: Method by enrichment in a liquid medium
6461-2:1986	Detection and enumeration of the spores of sulfite-reducing anaerobes (clostridia) — Part 2: Method by membrane filtration
7704:1985	Evaluation of membrane filters used for microbiological analyses
7899-1:1984	Detection and enumeration of faecal streptococci – Part 1: Method by enrichment in a liquid medium
7899-2:1984	Detection and enumeration of faecal streptococci – Part 2: Method by membrane filtration
9308-1:1990	Detection and enumeration of coliform organisms, thermotolerant coliform organisms and presumptive *Escherichia coli* – Part 1: Membrane filtration method
9308-2:1990	Detection and enumeration of coliform organisms, thermotolerant coliform organisms and presumptive *Escherichia coli* – Part 2: Multiple tube (most probable number) method

8
Chemical aspects

Most chemicals arising in drinking-water are of health concern only after extended exposure of years, rather than months. The principal exception is nitrate. Typically, changes in water quality occur progressively, except for those substances that are discharged or leach intermittently to flowing surface waters or groundwater supplies from, for example, contaminated landfill sites.

In some cases, there are groups of chemicals that arise from related sources – for example, the DBPs – and it may not be necessary to set standards for all of the substances for which there are guideline values. If chlorination is practised, the THMs, of which chloroform is the major component, are likely to be the main DBPs, together with the chlorinated acetic acids in some instances. In some cases, control of chloroform levels and, where appropriate, trichloroacetic acid levels will also provide an adequate measure of control over other chlorination by-products.

Several of the inorganic elements for which guideline values have been recommended are recognized to be essential elements in human nutrition. No attempt has been made here at this time to define a minimum desirable concentration of such substances in drinking-water.

Fact sheets for individual chemical contaminants are provided in chapter 12. For those contaminants for which a guideline value has been established, the fact sheets include a brief toxicological overview of the chemical, the basis for guideline derivation, treatment achievability and analytical limit of detection. More detailed chemical reviews are available (http://www.who.int/water_sanitation_health/dwq/guidelines/en/).

8.1 Chemical hazards in drinking-water

A number of chemical contaminants have been shown to cause adverse health effects in humans as a consequence of prolonged exposure through drinking-water. However, this is only a very small proportion of the chemicals that may reach drinking-water from various sources.

The substances considered here have been assessed for possible health effects, and guideline values have been proposed only on the basis of health concerns. Additional consideration of the potential effects of chemical contaminants on the acceptability of drinking-water to consumers is included in chapter 10. Some substances of health concern have effects on the acceptability of drinking-water that would normally lead to rejection of the water at concentrations significantly lower than those of health concern. For such substances, health-based guideline values are needed, for instance, for use in interpreting data collected in response to consumer complaints.

> The lists of chemicals addressed in these Guidelines do not imply that all of these chemicals will always be present or that other chemicals not addressed will be absent.

In section 2.3.2, it is indicated that "In developing national drinking-water standards based on these Guidelines, it will be necessary to take account of a variety of environmental, social, cultural, economic, dietary and other conditions affecting potential exposure. This may lead to national standards that differ appreciably from these Guidelines." This is particularly applicable to chemical contaminants, for which there is a long list, and setting standards for, or including, all of them in monitoring programmes is neither feasible nor desirable.

> It is important that chemical contaminants be prioritized so that the most important are considered for inclusion in national standards and monitoring programmes.

The probability that any particular chemical may occur in significant concentrations in any particular setting must be assessed on a case-by-case basis. The presence of certain chemicals may already be known within a particular country, but others may be more difficult to assess.

In most countries, whether developing or industrialized, water sector professionals are likely to be aware of a number of chemicals that are present in significant concentrations in drinking-water supplies. A body of local knowledge that has been built up by practical experience over a period of time is invaluable. Hence, the presence of a limited number of chemical contaminants in drinking-water is usually already known in many countries and in many local systems. Significant problems, even crises, can occur, however, when chemicals posing high health risk are widespread but their presence is unknown because their long-term health effect is caused by chronic exposure as opposed to acute exposure. Such has been the case of arsenic in groundwater in Bangladesh and West Bengal, for example.

For some contaminants, there will be exposure from sources other than drinking-water, and this may need to be taken into account when setting standards and considering the need for standards. It may also be important when considering the need for monitoring. In some cases, drinking-water will be a minor source of exposure, and controlling levels in water will have little impact on overall exposure. In other cases, controlling a contaminant in water may be the most cost-effective way of reducing exposure. Drinking-water monitoring strategies, therefore, should not be considered in isolation from other potential routes of exposure to chemicals in the environment.

8. CHEMICAL ASPECTS

Table 8.1 Categorization of source of chemical constituents

Source of chemical constituents	Examples of sources
Naturally occurring	Rocks, soils and the effects of the geological setting and climate
Industrial sources and human dwellings	Mining (extractive industries) and manufacturing and processing industries, sewage, solid wastes, urban runoff, fuel leakages
Agricultural activities	Manures, fertilizers, intensive animal practices and pesticides
Water treatment or materials in contact with drinking-water	Coagulants, DBPs, piping materials
Pesticides used in water for public health	Larvicides used in the control of insect vectors of disease
Cyanobacteria	Eutrophic lakes

The scientific basis for each of the guideline values is summarized in chapter 12. This information is important in helping to modify guideline values to suit national requirements or in assessing the significance for health of concentrations of a contaminant that are greater than the guideline value.

Chemical contaminants in drinking-water may be categorized in various ways; however, the most appropriate is to consider the primary source of the contaminant – i.e., to group chemicals according to where control may be effectively exercised. This aids in the development of approaches that are designed to prevent or minimize contamination, rather than those that rely primarily on the measurement of contaminant levels in final waters.

In general, approaches to the management of chemical hazards in drinking-water vary between those where the source water is a significant contributor (with control effected, for example, through source water selection, pollution control, treatment or blending) and those from materials and chemicals used in the production and distribution of drinking-water (controlled by process optimization or product specification). In these Guidelines, chemicals are therefore divided into six major source groups, as shown in Table 8.1.

Categories may not always be clear-cut. The group of naturally occurring contaminants, for example, includes many inorganic chemicals that are found in drinking-water as a consequence of release from rocks and soils by rainfall, some of which may become problematical where there is environmental disturbance, such as in mining areas.

8.2 Derivation of chemical guideline values

The criteria used to decide whether a guideline value is established for a particular chemical constituent are as follows:

—there is credible evidence of occurrence of the chemical in drinking-water, combined with evidence of actual or potential toxicity; or

— the chemical is of significant international concern; or
— the chemical is being considered for inclusion or is included in the WHO Pesticide Evaluation Scheme (WHOPES) programme (approval programme for direct application of pesticides to drinking-water for control of insect vectors of disease).

Guideline values are derived for many chemical constituents of drinking-water. A guideline value normally represents the concentration of a constituent that does not result in any significant risk to health over a lifetime of consumption. A number of provisional guideline values have been established at concentrations that are reasonably achievable through practical treatment approaches or in analytical laboratories; in these cases, the guideline value is above the concentration that would normally represent the calculated health-based value. Guideline values are also designated as provisional when there is a high degree of uncertainty in the toxicology and health data (see also section 8.2.6).

There are two principal sources of information on health effects resulting from exposure to chemicals that can be used in deriving guideline values. The first and preferred source is studies on human populations. However, the value of such studies for many substances is limited, owing to lack of quantitative information on the concentration to which people have been exposed or on simultaneous exposure to other agents. However, for some substances, such studies are the primary basis on which guideline values are developed. The second and most frequently used source of information is toxicity studies using laboratory animals. The limitations of toxicology studies include the relatively small number of animals used and the relatively high doses administered, which create uncertainty as to the relevance of particular findings to human health. This is because there is a need to extrapolate the results from animals to humans and to the low doses to which human populations are usually exposed. In most cases, the study used to derive the guideline value is supported by a range of other studies, including human data, and these are also considered in carrying out a health risk assessment.

In order to derive a guideline value to protect human health, it is necessary to select the most suitable study or studies. Data from well conducted studies, where a clear dose–response relationship has been demonstrated, are preferred. Expert judgement was exercised in the selection of the most appropriate study from the range of information available.

8.2.1 Approaches taken

Two approaches to the derivation of guideline values are used: one for "threshold chemicals" and the other for "non-threshold chemicals" (mostly genotoxic carcinogens).

It is generally considered that the initiating event in the process of genotoxic chemical carcinogenesis is the induction of a mutation in the genetic material (DNA) of somatic cells (i.e., cells other than ova or sperm) and that there is a theoretical risk at any level of exposure (i.e., no threshold). On the other hand, there are carcinogens that are capable of producing tumours in animals or humans without exerting a genotoxic activity, but acting through an indirect mechanism. It is generally believed that a demonstrable threshold dose exists for non-genotoxic carcinogens.

In deriving guideline values for carcinogens, consideration was given to the potential mechanism(s) by which the substance may cause cancer, in order to decide whether a threshold or non-threshold approach should be used (see sections 8.2.2 and 8.2.4).

The evaluation of the potential carcinogenicity of chemical substances is usually based on long-term animal studies. Sometimes data are available on carcinogenicity in humans, mostly from occupational exposure.

On the basis of the available evidence, the International Agency for Research on Cancer (IARC) categorizes chemical substances with respect to their potential carcinogenic risk into the following groups:

Group 1: the agent is carcinogenic to humans
Group 2A: the agent is probably carcinogenic to humans
Group 2B: the agent is possibly carcinogenic to humans
Group 3: the agent is not classifiable as to its carcinogenicity to humans
Group 4: the agent is probably not carcinogenic to humans

According to IARC, these classifications represent a first step in carcinogenic risk assessment, which leads to a second step of quantitative risk assessment where possible. In establishing guideline values for drinking-water, the IARC evaluation of carcinogenic compounds, where available, is taken into consideration.

8.2.2 Threshold chemicals

For most kinds of toxicity, it is believed that there is a dose below which no adverse effect will occur. For chemicals that give rise to such toxic effects, a tolerable daily intake (TDI) should be derived as follows, using the most sensitive end-point in the most relevant study, preferably involving administration in drinking-water:

$$TDI = (NOAEL \text{ or } LOAEL)/UF$$

where:

- NOAEL = no-observed-adverse-effect level
- LOAEL = lowest-observed-adverse-effect level
- UF = uncertainty factor

The guideline value (GV) is then derived from the TDI as follows:

$$GV = (TDI \times bw \times P)/C$$

where:

- bw = body weight (see Annex 3)
- P = fraction of the TDI allocated to drinking-water
- C = daily drinking-water consumption (see Annex 3)

Tolerable daily intake

The TDI is an estimate of the amount of a substance in food and drinking-water, expressed on a body weight basis (mg/kg or µg/kg of body weight), that can be ingested over a lifetime without appreciable health risk.

Acceptable daily intakes (ADIs) are established for food additives and pesticide residues that occur in food for necessary technological purposes or plant protection reasons. For chemical contaminants, which usually have no intended function in drinking-water, the term "tolerable daily intake" is more appropriate than "acceptable daily intake," as it signifies permissibility rather than acceptability.

Over many years, JECFA and JMPR have developed certain principles in the derivation of ADIs. These principles have been adopted where appropriate in the derivation of TDIs used in developing guideline values for drinking-water quality.

As TDIs are regarded as representing a tolerable intake for a lifetime, they are not so precise that they cannot be exceeded for short periods of time. Short-term exposure to levels exceeding the TDI is not a cause for concern, provided the individual's intake averaged over longer periods of time does not appreciably exceed the level set. The large uncertainty factors generally involved in establishing a TDI (see below) serve to provide assurance that exposure exceeding the TDI for short periods is unlikely to have any deleterious effects upon health. However, consideration should be given to any potential acute effects that may occur if the TDI is substantially exceeded for short periods of time.

No-observed-adverse-effect level and lowest-observed-adverse-effect level

The NOAEL is defined as the highest dose or concentration of a chemical in a single study, found by experiment or observation, that causes no detectable adverse health effect. Wherever possible, the NOAEL is based on long-term studies, preferably of ingestion in drinking-water. However, NOAELs obtained from short-term studies and studies using other sources of exposure (e.g., food, air) may also be used.

If a NOAEL is not available, a LOAEL may be used, which is the lowest observed dose or concentration of a substance at which there is a detectable adverse health effect. When a LOAEL is used instead of a NOAEL, an additional uncertainty factor is normally applied (see below).

Uncertainty factors

The application of uncertainty (or safety) factors has been widely used in the derivation of ADIs and TDIs for food additives, pesticides and environmental contaminants. The derivation of these factors requires expert judgement and careful consideration of the available scientific evidence.

8. CHEMICAL ASPECTS

Table 8.2 Source of uncertainty in derivation of guideline values

Source of uncertainty	Factor
Interspecies variation (animals to humans)	1–10
Intraspecies variation (individual variations within species)	1–10
Adequacy of studies or database	1–10
Nature and severity of effect	1–10

In the derivation of guideline values, uncertainty factors are applied to the NOAEL or LOAEL for the response considered to be the most biologically significant.

In relation to exposure of the general population, the NOAEL for the critical effect in animals is normally divided by an uncertainty factor of 100. This comprises two 10-fold factors, one for interspecies differences and one for interindividual variability in humans (see Table 8.2). Extra uncertainty factors may be incorporated to allow for database deficiencies and for the severity and irreversibility of effects.

Factors lower than 10 were used, for example, for interspecies variation when humans are known to be less sensitive than the animal species studied. Inadequate studies or databases include those where a LOAEL was used instead of a NOAEL and studies considered to be shorter in duration than desirable. Situations in which the nature or severity of effect might warrant an additional uncertainty factor include studies in which the end-point was malformation of a fetus or in which the end-point determining the NOAEL was directly related to possible carcinogenicity. In the latter case, an additional uncertainty factor was usually applied for carcinogenic compounds for which the guideline value was derived using a TDI approach rather than a theoretical risk extrapolation approach.

For substances for which the uncertainty factors were greater than 1000, guideline values are designated as provisional in order to emphasize the higher level of uncertainty inherent in these values. A high uncertainty factor indicates that the guideline value may be considerably lower than the concentration at which health effects would actually occur in a real human population. Guideline values with high uncertainty are more likely to be modified as new information becomes available.

The selection and application of uncertainty factors are important in the derivation of guideline values for chemicals, as they can make a considerable difference in the values set. For contaminants for which there is sufficient confidence in the database, the guideline value was derived using a smaller uncertainty factor. For most contaminants, however, there is greater scientific uncertainty, and a relatively large uncertainty factor was used. The use of uncertainty factors enables the particular attributes of the chemical and the data available to be considered in the derivation of guideline values.

Allocation of intake

Drinking-water is not usually the sole source of human exposure to the substances for which guideline values have been set. In many cases, the intake of chemical contaminants from drinking-water is small in comparison with that from other sources, such as food and air. Guideline values derived using the TDI approach take into account exposures from all sources by apportioning a percentage of the TDI to drinking-water. This approach ensures that total daily intake from all sources (including drinking-water containing concentrations of the substance at or near the guideline value) does not exceed the TDI.

Wherever possible, data concerning the proportion of total intake normally ingested in drinking-water (based on mean levels in food, air and drinking-water) or intakes estimated on the basis of consideration of physical and chemical properties were used in the derivation of the guideline values. Where such information was not available, an arbitrary (default) value of 10% for drinking-water was used. This default value is, in most cases, sufficient to account for additional routes of intake (i.e., inhalation and dermal absorption) of contaminants in water. In some cases, a specific discussion is made of the potential for exposure from intake through inhalation and dermal uptake in bathing and showering where there is evidence that this is likely to be significant, usually in circumstances where the allocation of the TDI to drinking-water is greater than 10%.

It is recognized that exposure from various media may vary with local circumstances. It should be emphasized, therefore, that the derived guideline values apply to a typical exposure scenario or are based on default values that may or may not be applicable for all areas. In those areas where relevant exposure data are available, authorities are encouraged to develop context-specific guideline values that are tailored to local circumstances and conditions. For example, in areas where the intake of a particular contaminant in drinking-water is known to be much greater than that from other sources (i.e., air and food), it may be appropriate to allocate a greater proportion of the TDI to drinking-water to derive a guideline value more suited to the local conditions. In addition, in cases in which guideline values are exceeded, efforts should be made to assess the contribution of other sources to total intake in order to interpret the health significance of the exceedance and to orient remedial measures to sources of exposure that are most relevant.

Significant figures

The calculated TDI is used to derive the guideline value, which is then rounded to one significant figure. In some instances, ADI values with only one significant figure set by JECFA or JMPR were used to calculate the guideline value. The guideline value was generally rounded to one significant figure to reflect the uncertainty in animal toxicity data and exposure assumptions made.

8.2.3 Alternative approaches

Alternative approaches being considered in the derivation of TDIs for threshold effects include the benchmark dose (BMD) (IPCS, 1994), categorical regression (IPCS, 1994) and chemical-specific adjustment factors (CSAF) (IPCS, 2001).

Benchmark dose[1]

The benchmark dose (BMD) is the lower confidence limit of the dose that produces a small increase in the level of adverse effects (e.g., 5% or 10%; Crump, 1984) to which uncertainty factors can be applied to develop a tolerable intake.

The BMD has a number of advantages over the NOAEL:

- It is derived on the basis of data from the entire dose–response curve for the critical effect rather than from the single dose group at the NOAEL (i.e., one of the few preselected dose levels).
- Use of the BMD facilitates the use and comparison of studies on the same agent or the potencies of different agents.
- The BMD can be calculated from data sets in which a NOAEL was not determined, eliminating the need for an additional uncertainty factor to be applied to the LOAEL.
- Definition of the BMD as a lower confidence limit accounts for the statistical power and quality of the data. That is, the confidence intervals around the dose–response curve for studies with small numbers of animals and, therefore, lower statistical power would be wide; similarly, confidence intervals in studies of poor quality with highly variable responses would also be wide. In either case, the wider confidence interval would lead to a lower BMD, reflecting the greater uncertainty of the database. On the other hand, narrow confidence limits (reflecting better studies) would result in higher BMDs.

Categorical regression[2]

The theory and application of categorical regression have been addressed by Hertzberg & Miller (1985), Hertzberg (1989), Guth et al. (1991) (inhalation exposure to methylisocyanate) and Farland & Dourson (1992) (oral exposure to arsenic). Data on toxicity are classified into one of several categories, such as no-observed-effect level (NOEL) or NOAEL, or others, as appropriate. These categories are then regressed on the basis of dose and, if required, duration of exposure. The result is a graph of probability of a given category of effect with dose or concentration, which is useful in the analysis of potential risks above the tolerable intake, especially for comparisons among chemicals.

Depending on the extent of the available data on toxicity, additional estimations regarding the percentage of individuals with specific adverse effects are possible. Such estimations require, however, an understanding of the mechanisms of toxicity of the critical effect, knowledge of the extrapolation between the experimental animal and humans and/or knowledge of the incidence of specific effects in humans.

Similar to the BMD, categorical regression utilizes information from the entire dose–response curve, resulting in more precise estimates of risk when compared with the current approach (NOAEL-based tolerable intakes). However, categorical

---

[1] This section has been taken from IPCS (1994).
[2] This section has been taken from IPCS (1994).

regression requires more information than the current tolerable intake method, and the interpretation of the probability scale can be problematic.

Chemical-specific adjustment factors

Approaches to the derivation of TDIs are increasingly being based on understanding of a chemical's mode of action in order to reduce reliance on empirical mathematical modelling and to eliminate the need to determine whether a threshold or non-threshold approach is more appropriate. This approach provides a departure from the use of default uncertainty factors and relies on the use of quantitative toxicokinetic and toxicodynamic data and "chemical-specific adjustment factors" (CSAFs) (IPCS, 2001) to assess interspecies and interindividual extrapolations in dose/concentration–response assessment. Previously, CSAFs were called "data-derived uncertainty factors" (Renwick, 1993; IPCS, 1994). The part of the CSAF approach that is at present best developed is the use of physiologically based pharmacokinetic models to replace the default values for extrapolation between species and between routes of exposure.

8.2.4 Non-threshold chemicals

In the case of compounds considered to be genotoxic carcinogens, guideline values were normally determined using a mathematical model. Although several models exist, the linearized multistage model was generally adopted. Other models were considered more appropriate in a few cases. Guideline values presented are the concentrations in drinking-water associated with an estimated upper-bound excess lifetime cancer risk of 10^{-5} (or one additional cancer per 100 000 of the population ingesting drinking-water containing the substance at the guideline value for 70 years).

The guideline values for carcinogenic substances have been computed from hypothetical mathematical models that cannot be verified experimentally. These models do not usually take into account a number of biologically important considerations, such as pharmacokinetics, DNA repair or protection by the immune system. They also assume the validity of a linear extrapolation of very high dose exposures in test animals to very low dose exposures in humans. As a consequence, the models used are conservative (i.e., err on the side of caution). The guideline values derived using these models should be interpreted differently from TDI-based values because of the lack of precision of the models. At best, these values must be regarded as rough estimates of cancer risk. Moderate short-term exposure to levels exceeding the guideline value for carcinogens does not significantly affect the risk.

8.2.5 Data quality

The following factors were taken into account in assessing the quality and reliability of available information:

- Oral studies are preferred (in particular, drinking-water studies), using the pure substance with appropriate dosing regime and a good-quality pathology.
- The database should be sufficiently broad that all potential toxicological end-points of concern have been identified.
- The quality of the studies is such that they are considered reliable; for example, there has been adequate consideration of confounding factors in epidemiological studies.
- There is reasonable consistency between studies; the end-point and study used to derive a guideline value do not contradict the overall weight of evidence.
- For inorganic substances, there is some consideration of speciation in drinking-water.
- There is appropriate consideration of multimedia exposure in the case of epidemiological studies.

In the development of guideline values, existing international approaches were carefully considered. In particular, previous risk assessments developed by the International Programme on Chemical Safety (IPCS) in EHC monographs and CICADs, IARC, JMPR and JECFA were reviewed. These assessments were relied upon except where new information justified a reassessment, but the quality of new data was critically evaluated before it was used in any risk assessment. Where international reviews were not available, other sources of data were used in the derivation of guideline values, including published reports from peer-reviewed open literature, national reviews recognized to be of high quality, information submitted by governments and other interested parties and, to a limited extent, unpublished proprietary data (primarily for the evaluation of pesticides). Future revisions and assessments of pesticides will take place primarily through WHO/IPCS/JMPR/JECFA processes.

8.2.6 Provisional guideline values

The use and designation of provisional guideline values are outlined in Table 8.3.

For non-threshold substances, in cases in which the concentration associated with an upper-bound excess lifetime cancer risk of 10^{-5} is not feasible as a result of inade-

Table 8.3 Use and designation of provisional guideline values

Situations where a provisional guideline applies	Designation	
Significant scientific uncertainties regarding derivation of health-based guideline value	P	
Calculated guideline value is below the practical quantification level	A	*(Guideline value is set at the achievable quantification level)*
Calculated guideline value is below the level that can be achieved through practical treatment methods	T	*(Guideline value is set at the practical treatment limit)*
Calculated guideline value is likely to be exceeded as a result of disinfection procedures	D	*(Guideline value is set on the basis of health, but disinfection of drinking-water remains paramount)*

quate analytical or treatment technology, a provisional guideline value (designated A or T, respectively) is recommended at a practicable level.

8.2.7 Chemicals with effects on acceptability

Some substances of health concern have effects on the taste, odour or appearance of drinking-water that would normally lead to rejection of water at concentrations significantly lower than those of concern for health. Such substances are not normally appropriate for routine monitoring. Nevertheless, health-based guideline values may be needed – for instance, for use in interpreting data collected in response to consumer complaints. In these circumstances, a health-based summary statement and guideline value are presented in the usual way. In the summary statement, the relationship between concentrations relevant to health and those relevant to the acceptability of the drinking-water is explained. In tables of guideline values, the health-based guideline values are designated with a "C."

8.2.8 Non-guideline chemicals

Additional information on many chemicals not included in these Guidelines is available from several credible sources, including WHO EHCs and CICADs (www.who.int/pcs/index), chemical risk assessment reports from JMPR, JECFA and IARC, and published documents from a number of national sources, such as the US EPA. Although these information sources may not have been reviewed for these Guidelines, they have been peer reviewed and provide readily accessible information on the toxicology of many additional chemicals. They can help drinking-water suppliers and health officials decide upon the significance (if any) of a detected chemical and on the response that might be appropriate.

8.2.9 Mixtures

Chemical contaminants of drinking-water supplies are present with numerous other inorganic and/or organic constituents. The guideline values are calculated separately for individual substances, without specific consideration of the potential for interaction of each substance with other compounds present. The large margin of uncertainty incorporated in the majority of the guideline values is considered to be sufficient to account for potential interactions. In addition, the majority of contaminants will not be continuously present at concentrations at or near their guideline value.

For many chemical contaminants, mechanisms of toxicity are different; consequently, there is no reason to assume that there are interactions. There may, however, be occasions when a number of contaminants with similar toxicological mechanisms are present at levels near their respective guideline values. In such cases, decisions concerning appropriate action should be made, taking into consideration local circumstances. Unless there is evidence to the contrary, it is appropriate to assume that the toxic effects of these compounds are additive.

8.3 Analytical aspects

As noted above, guideline values are not set at concentrations of substances that cannot reasonably be measured. In such circumstances, provisional guideline values are set at the reasonable analytical limits.

Guidance provided in this section is intended to assist readers to select appropriate analytical methods for specific circumstances.

8.3.1 Analytical achievability

Various collections of "standard" or "recommended" methods for water analysis are published by a number of national and international agencies. It is often thought that adequate analytical accuracy can be achieved provided that all laboratories use the same standard method. Experience shows that this is not always the case, as a variety of factors may affect the accuracy of the results. Examples include reagent purity, apparatus type and performance, degree of modification of the method in a particular laboratory and the skill and care of the analyst. These factors are likely to vary both between the laboratories and over time in an individual laboratory. Moreover, the precision and accuracy that can be achieved with a particular method frequently depend upon the adequacy of sampling and nature of the sample ("matrix"). While it is not essential to use standard methods, it is important that the methods used are properly validated and precision and accuracy determined before significant decisions are made based on the results. In the case of "non-specific" variables such as taste and odour, colour and turbidity, the result is method specific, and this needs to be considered when using the data to make comparisons.

A number of considerations are important in selecting methods:

- The overriding consideration is that the method chosen is demonstrated to have the required accuracy. Other factors, such as speed and convenience, should be considered only in selecting among methods that meet this primary criterion.
- There are a number of markedly different procedures for measuring and reporting the errors to which all methods are subject. This complicates and prejudices the effectiveness of method selection, and suggestions for standardizing such procedures have been made. It is therefore desirable that details of all analytical methods are published together with performance characteristics that can be interpreted unambiguously.
- If the analytical results from one laboratory are to be compared with those from others and/or with a numerical standard, it is obviously preferable for them not to have any associated systematic error. In practice, this is not possible, but each laboratory should select methods whose systematic errors have been thoroughly evaluated and shown to be acceptably small.

A qualitative ranking of analytical methods based on their degree of technical complexity is given in Table 8.4 for inorganic chemicals and in Table 8.5 for organic chemicals. These groups of chemicals are separated, as the analytical methods used differ

Table 8.4 Ranking of complexity of analytical methods for inorganic chemicals

Ranking	Example of analytical methods
1	Volumetric method, colorimetric method
2	Electrode method
3	Ion chromatography
4	High-performance liquid chromatography (HPLC)
5	Flame atomic absorption spectrometry (FAAS)
6	Electrothermal atomic absorption spectrometry (EAAS)
7	Inductively coupled plasma (ICP)/atomic emission spectrometry (AES)
8	ICP/mass spectrometry (MS)

greatly. The higher the ranking, the more complex the process in terms of equipment and/or operation. In general, higher rankings are also associated with higher total costs. Analytical achievabilities of the chemical guideline values based on detection limits are given in Tables 8.6–8.10.

There are many kinds of field test kits that are used for compliance examinations as well as operational monitoring of drinking-water quality. Although the field test kits are generally available at relatively low prices, their analytical accuracy is generally less than that of the methods shown in Tables 8.4 and 8.5. It is therefore necessary to check the validity of the field test kit before applying it.

Table 8.5 Ranking of complexity of analytical methods for organic chemicals

Ranking	Example of analytical methods
1	HPLC
2	Gas chromatography (GC)
3	GC/MS
4	Headspace GC/MS
5	Purge-and-trap GC
	Purge-and-trap GC/MS

8.3.2 *Analytical methods*

In *volumetric titration*, chemicals are analysed by titration with a standardized titrant. The titration end-point is identified by the development of colour resulting from the reaction with an indicator, by the change of electrical potential or by the change of pH value.

Colorimetric methods are based on measuring the intensity of colour of a coloured target chemical or reaction product. The optical absorbance is measured using light of a suitable wavelength. The concentration is determined by means of a calibration curve obtained using known concentrations of the determinant. The UV method is similar to this method except that UV light is used.

For ionic materials, the ion concentration can be measured using an *ion-selective electrode*. The measured potential is proportional to the logarithm of the ion concentration.

Table 8.6 Analytical achievability for inorganic chemicals for which guideline values have been established, by source category[a]

	Field methods		Laboratory methods				
	Col	Absor	IC	FAAS	EAAS	ICP	ICP/MS
Naturally occurring chemicals							
Arsenic		#		+(H)	++□+++(H)	++(H)	+++
Barium				+	+++	+++	+++
Boron		++				++	+++
Chromium		#		+	+++	+++	+++
Fluoride	#	+	++				
Manganese	+	++		++	+++	+++	+++
Molybdenum					+	+++	+++
Selenium		#		#	+++(H)	++(H)	+
Uranium						+	+++
Chemicals from industrial sources and human dwellings							
Cadmium		#			++	++	+++
Cyanide		#	+	+			
Mercury					+		
Chemicals from agricultural activities							
Nitrate/nitrite	+++	+++	#				
Chemicals used in water treatment or materials in contact with drinking-water							
Antimony				#	++(H)	++(H)	+++
Copper	#	+++		+++	+++	+++	+++
Lead		#			+	+	++
Nickel		+		#	+	+++	++

[a] For definitions and notes to Table 8.6, see below Table 8.10.

Some organic compounds absorb UV light (wavelength 190–380 nm) in proportion to their concentration. *UV absorption* is useful for qualitative estimation of organic substances, because a strong correlation may exist between UV absorption and organic carbon content.

Atomic absorption spectrometry (AAS) is used for determination of metals. It is based on the phenomenon that the atom in the ground state absorbs the light of wavelengths that are characteristic to each element when light is passed through the atoms in the vapour state. Because this absorption of light depends on the concentration of atoms in the vapour, the concentration of the target element in the water sample is determined from the measured absorbance. The Beer-Lambert law describes the relationship between concentration and absorbance.

In *flame atomic absorption spectrometry (FAAS)*, a sample is aspirated into a flame and atomized. A light beam from a hollow cathode lamp of the same element as the target metal is radiated through the flame, and the amount of absorbed light is measured by the detector. This method is much more sensitive than other methods and free from spectral or radiation interference by co-existing elements. Pretreatment is either unnecessary or straightforward. However, it is not suitable for simultaneous analysis of many elements, because the light source is different for each target element.

GUIDELINES FOR DRINKING-WATER QUALITY

Table 8.7 Analytical achievability for organic chemicals from industrial sources and human dwellings for which guideline values have been established[a]

	Col	GC	GC/PD	GC/EC	GC/FID	GC/FPD	GC/TID	GC/MS	PT-GC/MS	HPLC	HPLC/FD	HPLC/UVPAD	EAAS	IC/FD
Benzene				++	+				+++					
Carbon tetrachloride				+					+					
Di(2-ethylhexyl)phthalate								++						
1,2-Dichlorobenzene			+++	+++					+++					
1,4-Dichlorobenzene			+++	+++					+++					
1,2-Dichloroethane				+++					++					
1,1-Dichloroethene				+++	+				+++					
1,2-Dichloroethene				++	++				+++					
Dichloromethane				#	+									
Edetic acid (EDTA)								+++						
Ethylbenzene				+++	+++				+++					
Hexachlorobutadiene		+++							+					
Nitrilotriacetic acid (NTA)														
Pentachlorophenol				++					+++		+++			
Styrene				++	+				+++					
Tetrachloroethene				+++	++				+++					
Toluene					+++				+++					
Trichloroethene				+++	+				+++					
Xylenes					+++				+++					+

[a] For definitions and notes to Table 8.7, see below Table 8.10.

8. CHEMICAL ASPECTS

Table 8.8 Analytical achievability for organic chemicals from agricultural activities for which guideline values have been established[a]

	Col	GC	GC/PD	GC/EC	GC/FID	GC/FPD	GC/TID	GC/MS	PT-GC/MS	HPLC	HPLC/FD	HPLC/UVPAD	EAAS	IC/FD
Alachlor				□				+++						
Aldicarb												+		
Aldrin and dieldrin				+										
Atrazine								+++□				+		
Carbofuran							++				++			
Chlordane				+ □										
Chlorotoluron				□										
Cyanazine												+		
2,4-D				+++				+++						
2,4-DB				+++ □				++						
1,2-Dibromo-3-chloropropane								+++	++					
1,2-Dibromoethane								+	+					
1,2-Dichloropropane				+++				+++	+++					
1,3-Dichloropropene				+++				+++	+++					
Dichlorprop (2,4-DP)								+++						
Dimethoate								++						
Endrin				+				#						
Fenoprop				+										
Isoproturon				+								+++		
Lindane				+										
MCPA				+++				+++				+++		
Mecoprop				++				++						
Methoxychlor		+++						+++						
Metolachlor				+++				+++						
Molinate								+++						
Pendimethalin				+++			++	+++						
Simazine				+++			+	+++						
2,4,5-T			+					+++						
Terbuthylazine (TBA)												++		
Trifluralin		+++						+++				+		

[a] For definitions and notes to Table 8.8, see below Table 8.10.

GUIDELINES FOR DRINKING-WATER QUALITY

Table 8.9 Analytical achievability for chemicals used in water treatment or from materials in contact with water for which guideline values have been established[a]

	Col	GC	GC/PD	GC/EC	GC/FID	GC/FPD	GC/TID	GC/MS	PT-GC/MS	HPLC	HPLC/FD	HPLC/UVPAD	EAAS	IC
Disinfectants														
Monochloramine	+++													
Chlorine	+++													+++
Disinfection by-products														
Bromate														+
Bromodichloromethane				+++					+++					
Bromoform				+++					+++					
Chloral hydrate (trichloroacetaldehyde)				+				+						
Chlorate												□		□
Chlorite	□											□		
Chloroform				+++					+++					
Cyanogen chloride								□						
Dibromoacetonitrile				□				□						
Dibromochloromethane				+++				+	+++					
Dichloroacetate				+++				□						
Dichloroacetonitrile				+++				‡ □						
Formaldehyde										+++				
Monochloroacetate		‡						‡ □						
Trichloroacetate				□				+++						
2,4,6-Trichlorophenol				+++										
Trihalomethanes[b]				+++					+++					
Organic contaminants from treatment chemicals														
Acrylamide				+							+			
Epichlorohydrin				+	+			+						
Organic contaminants from pipes and fittings														
Benzo[a]pyrene								‡			‡			
Vinyl chloride				+					+					

[a] For definitions and notes to Table 8.9, see below Table 8.10.
[b] See also individual THMs.

8. CHEMICAL ASPECTS

Table 8.10 Analytical achievability for pesticides used in water for public health purposes for which guideline values have been established[a]

	Col	GC	GC/PD	GC/EC	GC/FID	GC/FPD	GC/TID	GC/MS	PT-GC/MS	HPLC	HPLC/FD	HPLC/UVPAD	EAAS	IC/FD
Chlorpyrifos				+++	+++		+++	+++						
DDT (and metabolites)				+++		+		+						
Pyriproxyfen								+++						

[a] For definitions and notes to Table 8.10, see below.

Definitions to Tables 8.6–8.10

Col	Colorimetry
Absor	Absorptiometry
GC	Gas chromatography
GC/PD	Gas chromatography photoionization detector
GC/EC	Gas chromatography electron capture
GC/FID	Gas chromatography flame ionization detector
GC/FPD	Gas chromatography flame photodiode detector
GC/TID	Gas chromatography thermal ionization detector
GC/MS	Gas chromatography mass spectrometry
PT-GC/MS	Purge-and-trap gas chromatography mass spectrometry
HPLC	High-performance liquid chromatography
HPLC/FD	High-performance liquid chromatography fluorescence detector
HPLC/UVPAD	High-performance liquid chromatography ultraviolet photodiode array detector
EAAS	Electrothermal atomic absorption spectrometry
IC	Ion chromatography
ICP	Inductively coupled plasma
ICP/MS	Inductively coupled plasma mass spectrometry
FAAS	Flame atomic absorption spectrometry
IC/FAAS	Ion chromatography flame atomic absorption spectrometry
IC/FD	Ion chromatography fluorescence detector

Notes to Tables 8.6–8.10

+	The detection limit is between the guideline value and 1/10 of its value.
++	The detection limit is between 1/10 and 1/50 of the guideline value.
+++	The detection limit is under 1/100 of the guideline value.
#	The analytical method is available for detection of the concentration of the guideline value, but it is difficult to detect the concentration of 1/10 of the guideline value.
□	The detection method(s) is/are available for the item.
(H)	This method is applicable to the determination by conversion to their hydrides by hydride generator.

Electrothermal atomic absorption spectrometry (EAAS) is based on the same principle as FAAS, but an electrically heated atomizer or graphite furnace replaces the standard burner head for determination of metals. In comparison with FAAS, EAAS gives higher sensitivities and lower detection limits, and a smaller sample volume is required. EAAS suffers from more interference through light scattering by co-existing elements and requires a longer analysis time than FAAS.

The principle of *inductively coupled plasma/atomic emission spectrometry (ICP/AES)* for determination of metals is as follows. An ICP source consists of a flowing stream of argon gas ionized by an applied radio frequency. A sample aerosol is generated in a nebulizer and spray chamber and then carried into the plasma through an injector tube. A sample is heated and excited in the high-temperature plasma. The high temperature of the plasma causes the atoms to become excited. On returning to the ground state, the excited atoms produce ionic emission spectra. A monochromator is used to separate specific wavelengths corresponding to different elements, and a detector measures the intensity of radiation of each wavelength. A significant reduction in chemical interference is achieved. In the case of water with low pollution, simultaneous or sequential analysis is possible without special pretreatment to achieve low detection limits for many elements. This, coupled with the extended dynamic range from three digits to five digits, means that multi-element determination of metals can be achieved. ICP/AES has similar sensitivity to FAAS or EAAS.

In *inductively coupled plasma/mass spectrometry (ICP/MS)*, elements are atomized and excited as in ICP/AES, then passed to a mass spectrometer. Once inside the mass spectrometer, the ions are accelerated by high voltage and passed through a series of ion optics, an electrostatic analyser and, finally, a magnet. By varying the strength of the magnet, ions are separated according to mass/charge ratio and passed through a slit into the detector, which records only a very small atomic mass range at a given time. By varying the magnet and electrostatic analyser settings, the entire mass range can be scanned within a relatively short period of time. In the case of water with low pollution, simultaneous or sequential analysis is possible without special pretreatment to achieve low detection limits for many elements. This, coupled with the extended dynamic range from three digits to five digits, means that multi-element determination of metals can be achieved.

Chromatography is a separation method based on the affinity difference between two phases, the stationary and mobile phases. A sample is injected into a column, either packed or coated with the stationary phase, and separated by the mobile phase based on the difference in interaction (distribution or adsorption) between compounds and the stationary phase. Compounds with a low affinity for the stationary phase move more quickly through the column and elute earlier. The compounds that elute from the end of the column are determined by a suitable detector.

In *ion chromatography*, an ion exchanger is used as the stationary phase, and the eluant for determination of anions is typically a dilute solution of sodium hydrogen carbonate and sodium carbonate. Colorimetric, electrometric or titrimetric detectors can be used for determining individual anions. In suppressed ion chromatography, anions are converted to their highly conductive acid forms; in the carbonate–bicarbonate eluant, anions are converted to weakly conductive carbonic acid. The separated acid forms are measured by conductivity and identified on the basis of retention time as compared with their standards.

High-performance liquid chromatography (HPLC) is an analytical technique using a liquid mobile phase and a column containing a liquid stationary phase. Detection of the separated compounds is achieved through the use of absorbance detectors for organic compounds and through conductivity or electrochemical detectors for metallic and inorganic compounds.

Gas chromatography (GC) permits the identification and quantification of trace organic compounds. In GC, gas is used as the mobile phase, and the stationary phase is a liquid that is coated either on an inert granular solid or on the walls of a capillary column. When the sample is injected into the column, the organic compounds are vaporized and moved through the column by the carrier gas at different rates depending on differences in partition coefficients between the mobile and stationary phases. The gas exiting the column is passed to a suitable detector. A variety of detectors can be used, including flame ionization (FID), electron capture (ECD) and nitrogen–phosphorus. Since separation ability is good in this method, mixtures of substances with similar structure are systematically separated, identified and determined quantitatively in a single operation.

The *gas chromatography/mass spectrometry (GC/MS)* method is based on the same principle as the GC method, using a mass spectrometer as the detector. As the gas emerges from the end of the GC column opening, it flows through a capillary column interface into the MS. The sample then enters the ionization chamber, where a collimated beam of electrons impacts the sample molecules, causing ionization and fragmentation. The next component is a mass analyser, which uses a magnetic field to separate the positively charged particles according to their mass. Several types of separating techniques exist; the most common are quadrupoles and ion traps. After the ions are separated according to their masses, they enter a detector.

The *purge-and-trap packed-column GC/MS method* or *purge-and-trap packed-column GC* method is applicable to the determination of various purgeable organic compounds that are transferred from the aqueous to the vapour phase by bubbling purge gas through a water sample at ambient temperature. The vapour is trapped with a cooled trap. The trap is heated and backflushed with the same purge gas to desorb the compounds onto a GC column. The principles of GC or GC/MS are as referred to above.

The principle of *enzyme-linked immunosorbent assay (ELISA)* is as follows. The protein (antibody) against the chemical of interest (antigen) is coated onto the solid material. The target chemical in the water sample binds to the antibody, and a second antibody with an enzyme attached is also added that will attach to the chemical of interest. After washing to remove any of the free reagents, a chromogen is added that will give a colour reaction due to cleavage by the enzyme that is proportional to the quantity of the chemical of interest. The ELISA method can be used to determine microcystin and synthetic surfactants.

8.4 Treatment

As noted above, where a health-based guideline value cannot be achieved by reasonably practicable treatment, then the guideline value is designated as provisional and set at the concentration that can be reasonably achieved through treatment.

8.4.1 Treatment achievability

The ability to achieve a guideline value within a drinking-water supply depends on a number of factors, including:

— the concentration of the chemical in the raw water;
— control measures employed throughout the drinking-water system;
— nature of the raw water (groundwater or surface water, presence of natural background and other components); and
— treatment processes already installed.

If a guideline value cannot be met with the existing system, then additional treatment may need to be considered, or water should be obtained from alternative sources.

The cost of achieving a guideline value will depend on the complexity of any additional treatment or other control measures required. It is not possible to provide general quantitative information on the cost of achieving individual guideline values. Treatment costs (capital and operating) will depend not only on the factors identified above, but also on issues such as plant throughput; local costs for labour, civil and mechanical works, chemicals and electricity; life expectancy of the plant; and so on.

A qualitative ranking of treatment processes based on their degree of technical complexity is given in Table 8.11. The higher the ranking, the more complex the

Table 8.11 Ranking of technical complexity and cost of water treatment processes

Ranking	Examples of treatment processes
1	Simple chlorination
	Plain filtration (rapid sand, slow sand)
2	Pre-chlorination plus filtration
	Aeration
3	Chemical coagulation
	Process optimization for control of DBPs
4	Granular activated carbon (GAC) treatment
	Ion exchange
5	Ozonation
6	Advanced oxidation processes
	Membrane treatment

8. CHEMICAL ASPECTS

Table 8.12 Treatment achievability for naturally occurring chemicals for which guideline values have been established[a,b]

	Chlorination	Coagulation	Ion exchange	Precipitation softening	Activated alumina	Activated carbon	Ozonation	Membranes
Arsenic		+++ <0.005	+++ <0.005	+++ <0.005	+++ <0.005			+++ <0.005
Fluoride		++			+++ <1			+++ <1
Manganese	+++ <0.05	++					+++ <0.05	+++ <0.05
Selenium		++	+++ <0.01		+++ <0.01			+++ <0.01
Uranium		++	+++ <0.001	++	+++ <0.001			

[a] Symbols are as follows:
  ++ 50% or more removal
  +++ 80% or more removal
[b] The table includes only those chemicals for which some treatment data are available. A blank entry in the table indicates either that the process is completely ineffective or that there are no data on the effectiveness of the process. For the most effective process(es), the table indicates the concentration of the chemical, in mg/litre, that should be achievable.

process in terms of plant and/or operation. In general, higher rankings are also associated with higher costs.

Tables 8.12–8.16 summarize the treatment processes that are capable of removing chemical contaminants of health significance. The tables include only those chemicals for which some treatment data are available.

These tables are provided to help inform decisions regarding the ability of existing treatment to meet guidelines and what additional treatment might need to be installed. They have been compiled on the basis of published literature, which includes mainly laboratory experiments, some pilot plant investigations and relatively few full-scale studies of water treatment processes. Consequently:

- Many of the treatments outlined are designed for larger treatment plants and may not necessarily be appropriate for smaller treatment plants or individual type treatment. In these cases, the choice of technology must be made on a case-by-case basis.
- The information is probably "best case," since the data would have been obtained under laboratory conditions or with a carefully controlled plant for the purposes of experimentation.
- Actual process performance will depend on the concentration of the chemical in the raw water and on general raw water quality. For example, chlorination and removal of organic chemicals and pesticides using activated carbon or ozonation will be impaired if there is a high concentration of natural organic matter.

Table 8.13 **Treatment achievability for chemicals from industrial sources and human dwellings for which guideline values have been established**[a,b]

	Air stripping	Coagulation	Ion exchange	Precipitation softening	Activated carbon	Ozonation	Advanced oxidation	Membranes
Cadmium		+++ <0.002	+++ <0.002	+++ <0.002				+++ <0.002
Mercury		+++ <0.0001		+++ <0.0001	+++ <0.0001			+++ <0.0001
Benzene	+++ <0.01				+++ <0.01	+++ <0.01		
Carbon tetrachloride	+++ <0.001	+			+++ <0.001			+++ <0.001
1,2-Dichlorobenzene	+++ <0.01				+++ <0.01	+++ <0.01		
1,4-Dichlorobenzene	+++ <0.01				+++ <0.01	+++ <0.01		
1,2-Dichloroethane	+				+++ <0.01	+	++	
1,2-Dichloroethene	+++ <0.01				+++ <0.01	+++ <0.01		
1,4-Dioxane					+++ no data			
Edetic acid (EDTA)					+++ <0.01			
Ethylbenzene	+++ <0.001	+			+++ <0.001	+++ <0.001		
Hexachlorobutadiene					+++ <0.001			
Nitrilotriacetic acid (NTA)					+++ no data			
Pentachlorophenol					+++ <0.0004			
Styrene	+++ <0.02				+++ <0.002			
Tetrachloroethene	+++ <0.001				+++ <0.001			
Toluene	+++ <0.001				+++ <0.001	+++ <0.001	+++ <0.001	
Trichloroethene	+++ <0.02				+++ <0.02	+++ <0.02	+++ <0.02	
Xylenes	+++ <0.005				+++ <0.005		+++ <0.005	

[a] Symbols are as follows:
  + Limited removal
  ++ 50% or more removal
  +++ 80% or more removal
[b] The table includes only those chemicals for which some treatment data are available. A blank entry in the table indicates either that the process is completely ineffective or that there are no data on the effectiveness of the process. For the most effective process(es), the table indicates the concentration of the chemical, in mg/litre, that should be achievable.

8. CHEMICAL ASPECTS

Table 8.14 Treatment achievability for chemicals from agricultural activities for which guideline values have been established[a,b]

	Chlorination	Air stripping	Coagulation	Ion exchange	Activated carbon	Ozonation	Advanced oxidation	Membranes	Biological treatment
Nitrate				+++ <5				+++ <5	+++ <5
Nitrite	+++ <0.1					+++ <0.1	+++ <0.1		
Alachlor					+++ <0.001	++	+++ <0.001	+++ <0.001	
Aldicarb	+++ <0.001				+++ <0.001	+++ <0.001		+++ <0.001	
Aldrin/dieldrin			++		+++ <0.00002	+++ <0.00002		+++ <0.00002	
Atrazine			+		+++ <0.0001	++	+++ <0.0001	+++ <0.0001	
Carbofuran			+		+++ <0.001			+++ <0.001	
Chlordane					+++ <0.0001	+++ <0.0001			
Chlorotoluron					+++ <0.0001	+++ <0.0001			
Cyanazine					+++ <0.0001	+		+++ <0.0001	
2,4-Dichlorophenoxyacetic acid (2,4-D)			+		+++ <0.001	+++ <0.001			
1,2-Dibromo-3-chloropropane		++ <0.001			+++ <0.0001				
1,2-Dibromoethane		+++ <0.0001			+++ <0.0001				
1,2-Dichloropropane (1,2-DCP)					+++ <0.001	+		+++ <0.001	
Dimethoate	+++ <0.001				++	++			
Endrin			+		+++ <0.0002				
Isoproturon		++			+++ <0.0001	+++ <0.0001	+++ <0.0001	+++ <0.0001	
Lindane					+++ <0.0001	++			
MCPA					+++ <0.0001	+++ <0.0001			
Mecoprop					+++ <0.0001	+++ <0.0001			
Methoxychlor			++		+++ <0.0001	+++ <0.0001			
Metalochlor					+++ <0.0001	++			

continued

Table 8.14 *Continued*

	Chlorination	Air stripping	Coagulation	Ion exchange	Activated carbon	Ozonation	Advanced oxidation	Membranes	Biological treatment
Simazine	+				+++ <0.0001	++	+++ <0.0001	+++ <0.0001	
2,4,5-T				++	+++ <0.001	+			
Terbuthylazine (TBA)			+		+++ <0.0001	++			
Trifluralin					+++ <0.0001			+++ <0.0001	

[a] Symbols are as follows:
  + Limited removal
  ++ 50% or more removal
  +++ 80% or more removal
[b] The table includes only those chemicals for which some treatment data are available. A blank entry in the table indicates either that the process is completely ineffective or that there are no data on the effectiveness of the process. For the most effective process(es), the table indicates the concentration of the chemical, in mg/litre, that should be achievable.

- For many contaminants, potentially several different processes could be appropriate, and the choice between processes should be made on the basis of technical complexity and cost, taking into account local circumstances. For example, membrane processes can remove a broad spectrum of chemicals, but simpler and cheaper alternatives are effective for the removal of most chemicals.
- It is normal practice to use a series of unit processes to achieve desired water quality objectives (e.g., coagulation, sedimentation, filtration, GAC, chlorination). Each of these may contribute to the removal of chemicals. It may be technically and

Table 8.15 Treatment achievability for pesticides used in water for public health for which guideline values have been established[a,b]

	Chlorination	Coagulation	Activated carbon	Ozonation	Advanced oxidation	Membranes
DDT and metabolites	+	+++ <0.0001	+++ <0.0001	+	+++ <0.0001	+++ <0.0001
Pyriproxyfen			+++ <0.001			

[a] Symbols are as follows:
  + Limited removal
  +++ 80% or more removal
[b] The table includes only those chemicals for which some treatment data are available. A blank entry in the table indicates either that the process is completely ineffective or that there are no data on the effectiveness of the process. For the most effective process(es), the table indicates the concentration of the chemical, in mg/litre, that should be achievable.

8. CHEMICAL ASPECTS

Table 8.16 Treatment achievability for cyanobacterial cells and cyanotoxins for which guideline values have been established[a,b,c]

	Chlorination	Coagulation	Activated carbon	Ozonation	Advanced oxidation	Membranes
Cyanobacterial cells		+++				+++
Cyanotoxins	+++		+++	+++	+++	

[a] Chlorination or ozonation may release cyanotoxins.
[b] +++ = 80% or more removal.
[c] The table includes only those chemicals for which some treatment data are available. A blank entry in the table indicates either that the process is completely ineffective or that there are no data on the effectiveness of the process.

economically advantageous to use a combination of processes (e.g., ozonation plus GAC) to remove particular chemicals.

- The effectiveness of potential processes should be assessed using laboratory or pilot plant tests on the actual raw water concerned. These tests should be of sufficient duration to identify potential seasonal or other temporal variations in contaminant concentrations and process performance.

8.4.2 Chlorination

Chlorination can be achieved by using liquefied chlorine gas, sodium hypochlorite solution or calcium hypochlorite granules and on-site chlorine generators. Liquefied chlorine gas is supplied in pressurized containers. The gas is withdrawn from the cylinder and is dosed into water by a chlorinator, which both controls and measures the gas flow rate. Sodium hypochlorite solution is dosed using a positive-displacement electric dosing pump or gravity feed system. Calcium hypochlorite has to be dissolved in water, then mixed with the main supply. Chlorine, whether in the form of chlorine gas from a cylinder, sodium hypochlorite or calcium hypochlorite, dissolves in water to form hypochlorous acid (HOCl) and hypochlorite ion (OCl$^-$).

Different techniques of chlorination can be used, including breakpoint chlorination, marginal chlorination and superchlorination/dechlorination. Breakpoint chlorination is a method in which the chlorine dose is sufficient to rapidly oxidize all the ammonia nitrogen in the water and to leave a suitable free residual chlorine available to protect the water against reinfection from the point of chlorination to the point of use. Superchlorination/dechlorination is the addition of a large dose of chlorine to effect rapid disinfection and chemical reaction, followed by reduction of excess free chlorine residual. Removing excess chlorine is important to prevent taste problems. It is used mainly when the bacterial load is variable or the detention time in a tank is not enough. Marginal chlorination is used where water supplies are of high quality and is the simple dosing of chlorine to produce a desired level of free residual chlorine. The chlorine demand in these supplies is very low, and a breakpoint might not even occur.

Chlorination is employed primarily for microbial disinfection. However, chlorine also acts as an oxidant and can remove or assist in the removal of some chemicals – for example, decomposition of easily oxidized pesticides such as aldicarb; oxidation of dissolved species (e.g., manganese(II)) to form insoluble products that can be removed by subsequent filtration; and oxidation of dissolved species to more easily removable forms (e.g., arsenite to arsenate).

A disadvantage of chlorine is its ability to react with natural organic matter to produce THMs and other halogenated DBPs. However, by-product formation may be controlled by optimization of the treatment system.

8.4.3 Ozonation

Ozone is a powerful oxidant and has many uses in water treatment, including oxidation of organic chemicals. Ozone can be used as a primary disinfectant. Ozone gas (O_3) is formed by passing dry air or oxygen through a high-voltage electric field. The resultant ozone-enriched air is dosed directly into the water by means of porous diffusers at the base of baffled contactor tanks. The contactor tanks, typically about 5 m deep, provide 10–20 min of contact time. Dissolution of at least 80% of the applied ozone should be possible, with the remainder contained in the off-gas, which is passed through an ozone destructor and vented to the atmosphere.

The performance of ozonation relies on achieving the desired concentration after a given contact period. For oxidation of organic chemicals, such as a few oxidizable pesticides, a residual of about 0.5 mg/litre after a contact time of up to 20 min is typically used. The doses required to achieve this vary with the type of water but are typically in the range 2–5 mg/litre. Higher doses are needed for untreated waters, because of the ozone demand of the natural background organics.

Ozone reacts with natural organics to increase their biodegradability, measured as assimilable organic carbon. To avoid undesirable bacterial growth in distribution, ozonation is normally used with subsequent treatment, such as filtration or GAC, to remove biodegradable organics, followed by a chlorine residual, since it does not provide a disinfectant residual. Ozone is effective for the degradation of a wide range of pesticides and other organic chemicals.

8.4.4 Other disinfection processes

Other disinfection methods include chloramination, the use of chlorine dioxide, UV radiation and advanced oxidation processes.

Chloramines (monochloramine, dichloramine and "trichloramine," or nitrogen trichloride) are produced by the reaction of aqueous chlorine with ammonia. Monochloramine is the only useful chloramine disinfectant, and conditions employed for chloramination are designed to produce only monochloramine. Monochloramine is a less effective disinfectant than free chlorine, but it is persistent, and it is therefore an attractive secondary disinfectant for the maintenance of a stable distribution system residual.

Although historically chlorine dioxide was not widely used for drinking-water disinfection, it has been used in recent years because of concerns about THM production associated with chlorine disinfection. Typically, chlorine dioxide is generated immediately prior to application by the addition of chlorine gas or an aqueous chlorine solution to aqueous sodium chlorite. Chlorine dioxide decomposes in water to form chlorite and chlorate. As chlorine dioxide does not oxidize bromide (in the absence of sunlight), water treatment with chlorine dioxide will not form bromoform or bromate.

Use of UV radiation in potable water treatment has typically been restricted to small facilities. UV radiation, emitted by a low-pressure mercury arc lamp, is biocidal between wavelengths of 180 and 320 nm. It can be used to inactivate protozoa, bacteria, bacteriophage, yeast, viruses, fungi and algae. Turbidity can inhibit UV disinfection. UV radiation can act as a strong catalyst in oxidation reactions when used in conjunction with ozone.

Processes aimed at generating hydroxyl radicals are known collectively as advanced oxidation processes and can be effective for the destruction of chemicals that are difficult to treat using other methods, such as ozone alone. Chemicals can react either directly with molecular ozone or with the hydroxyl radical ($HO\cdot$), which is a product of the decomposition of ozone in water and is an exceedingly powerful indiscriminate oxidant that reacts readily with a wide range of organic chemicals. The formation of hydroxyl radicals can be encouraged by using ozone at high pH. One advanced oxidation process using ozone plus hydrogen peroxide involves dosing hydrogen peroxide simultaneously with ozone at a rate of approximately 0.4 mg of hydrogen peroxide per litre per mg of ozone dosed per litre (the theoretical optimum ratio for hydroxyl radical production) and bicarbonate.

8.4.5 Filtration

Particulate matter can be removed from raw waters by rapid gravity, horizontal, pressure or slow sand filters. Slow sand filtration is essentially a biological process, whereas the others are physical treatment processes.

Rapid gravity, horizontal and pressure filters can be used for direct filtration of raw water, without pretreatment. Rapid gravity and pressure filters are commonly used to filter water that has been pretreated by coagulation and sedimentation. An alternative process is direct filtration, in which coagulation is added to the water, which then passes directly onto the filter where the precipitated floc (with contaminants) is removed; the application of direct filtration is limited by the available storage within the filter to accommodate solids.

Rapid gravity filters

Rapid gravity sand filters usually consist of open rectangular tanks (usually <100 m^2) containing silica sand (size range 0.5–1.0 mm) to a depth of between 0.6 and 2.0 m. The water flows downwards, and solids become concentrated in the upper layers of the bed. The flow rate is generally in the range 4–20 m^3/m^2·h. Treated water is collected via nozzles in the floor of the filter. The accumulated solids are removed periodically by backwashing with treated water, sometimes preceded by scouring of the sand with air. A dilute sludge that requires disposal is produced.

In addition to single-medium sand filters, dual-media or multimedia filters are used. Such filters incorporate different materials, such that the structure is from coarse to fine as the water passes through the filter. Materials of suitable density are used in order to maintain the segregation of the different layers following backwashing. A common example of a dual-media filter is the anthracite–sand filter, which typically consists of a 0.2-m-deep layer of 1.5-mm anthracite over a 0.6-m-deep layer of silica sand. Anthracite, sand and garnet can be used in multimedia filters. The advantage of dual- and multimedia filters is that there is more efficient use of the whole bed depth for particle retention – the rate of headloss development can be half that of single-medium filters, which can allow higher flow rates without increasing headloss development.

Rapid gravity filters are most commonly used to remove floc from coagulated waters (see section 8.4.7). They may also be used to reduce turbidity (including adsorbed chemicals) and oxidized iron and manganese from raw waters.

Roughing filters

Roughing filters can be applied as pre-filters prior to other processes such as slow sand filters. Roughing filters with coarse gravel or crushed stones as the filter medium can successfully treat water of high turbidity (>50 NTU). The main advantage of roughing filtration is that as the water passes through the filter, particles are removed by both filtration and gravity settling. Horizontal filters can be up to 10 m long and are operated at filtration rates of 0.3–1.0 m^3/m^2·h.

Pressure filters

Pressure filters are sometimes used where it is necessary to maintain head in order to eliminate the need for pumping into supply. The filter bed is enclosed in a cylindrical shell. Small pressure filters, capable of treating up to about 15 m^3/h, can be manufactured in glass-reinforced plastics. Larger pressure filters, up to 4 m in diameter, are manufactured in specially coated steel. Operation and performance are generally as described for the rapid gravity filter, and similar facilities are required for backwashing and disposal of the dilute sludge.

Slow sand filters

Slow sand filters usually consist of tanks containing sand (effective size range 0.15–0.3 mm) to a depth of between 0.5 and 1.5 m. The raw water flows downwards, and turbidity and microorganisms are removed primarily in the top few centimetres of the sand. A biological layer, known as the "schmutzdecke," develops on the surface of the filter and can be effective in removing microorganisms. Treated water is collected in underdrains or pipework at the bottom of the filter. The top few centimetres of sand containing the accumulated solids are removed and replaced periodically. Slow sand filters are operated at a water flow rate of between 0.1 and 0.3 m^3/m$^2 \cdot$h.

Slow sand filters are suitable only for low-turbidity water or water that has been pre-filtered. They are used to remove algae and microorganisms, including protozoa, and, if preceded by microstraining or coarse filtration, to reduce turbidity (including adsorbed chemicals). Slow sand filtration is effective for the removal of organics, including certain pesticides and ammonia.

8.4.6 Aeration

Aeration processes are designed to achieve removal of gases and volatile compounds by air stripping. Oxygen transfer can usually be achieved using a simple cascade or diffusion of air into water, without the need for elaborate equipment. Stripping of gases or volatile compounds, however, may require a specialized plant that provides a high degree of mass transfer from the liquid phase to the gas phase.

For oxygen transfer, cascade or step aerators are designed so that water flows in a thin film to achieve efficient mass transfer. Cascade aeration may introduce a significant headloss; design requirements are between 1 and 3 m to provide a loading of 10–30 m^3/m$^2 \cdot$h. Alternatively, compressed air can be diffused through a system of submerged perforated pipes. These types of aerator are used for oxidation and precipitation of iron and manganese.

Air stripping can be used for removal of volatile organics (e.g., solvents), some taste- and odour-causing compounds and radon. Aeration processes to achieve air stripping need to be much more elaborate to provide the necessary contact between the air and water. The most common technique is cascade aeration, usually in packed towers in which water is allowed to flow in thin films over plastic media with air blown counter-current. The required tower height and diameter are functions of the volatility and concentration of the compounds to be removed and the flow rate.

8.4.7 Chemical coagulation

Chemical coagulation-based treatment is the most common approach for treatment of surface waters and is almost always based on the following unit processes.

Chemical coagulants, usually salts of aluminium or iron, are dosed to the raw water under controlled conditions to form a solid flocculent metal hydroxide. Typical coagulant doses are 2–5 mg/litre as aluminium or 4–10 mg/litre as iron. The precipitated floc removes suspended and dissolved contaminants by mechanisms of charge neutralization, adsorption and entrapment. The efficiency of the coagulation process depends on raw water quality, the coagulant or coagulant aids used and operational factors, including mixing conditions, coagulation dose and pH. The floc is removed from the treated water by subsequent solid–liquid separation processes such as sedimentation or flotation and/or rapid or pressure gravity filtration.

Effective operation of the coagulation process depends on selection of the optimum coagulant dose and also the pH value. The required dose and pH can be determined by using small-scale batch coagulation tests, often termed "jar tests." Increasing doses of coagulant are applied to raw water samples that are stirred, then allowed to settle. The optimum dose is selected as that which achieves adequate removal of colour and turbidity; the optimum pH can be selected in a similar manner. These tests have to be conducted at a sufficient frequency to keep pace with changes in raw water quality and hence coagulant demand.

Powdered activated carbon (PAC) may be dosed during coagulation to adsorb organic chemicals such as some hydrophobic pesticides. The PAC will be removed as an integral fraction of the floc and disposed of with the waterworks sludge.

The floc may be removed by sedimentation to reduce the solids loading to the subsequent rapid gravity filters. Sedimentation is most commonly achieved in horizontal flow or floc blanket clarifiers. Alternatively, floc may be removed by dissolved air flotation, in which solids are contacted with fine bubbles of air that attach to the floc, causing them to float to the surface of the tank, where they are removed periodically as a layer of sludge. The treated water from either process is passed to rapid gravity filters (see section 8.4.5), where remaining solids are removed. Filtered water may be passed to a further stage of treatment, such as additional oxidation and filtration (for removal of manganese), ozonation and/or GAC adsorption (for removal of pesticides and other trace organics), prior to final disinfection before the treated water enters supply.

Coagulation is suitable for removal of certain heavy metals and low-solubility organic chemicals, such as certain organochlorine pesticides. For other organic chemicals, coagulation is generally ineffective, except where the chemical is bound to humic material or adsorbed onto particulates.

8.4.8 Activated carbon adsorption

Activated carbon is produced by the controlled thermalization of carbonaceous material, normally wood, coal, coconut shells or peat. This activation produces a porous material with a large surface area (500–1500 m^2/g) and a high affinity for organic compounds. It is normally used either in powdered (PAC) or in granular (GAC) form. When the adsorption capacity of the carbon is exhausted, it can be reactivated by burning off the organics in a controlled manner. However, PAC (and some GAC) is normally used only once before disposal. Different types of activated carbon have different affinities for types of contaminants.

The choice between PAC and GAC will depend upon the frequency and dose required. PAC would generally be preferred in the case of seasonal or intermittent contamination or where low dosage rates are required.

PAC is dosed as a slurry into the water and is removed by subsequent treatment processes together with the waterworks sludge. Its use is therefore restricted to surface water treatment works with existing filters. GAC in fixed-bed adsorbers is used much more efficiently than PAC dosed into the water, and the effective carbon use per water volume treated would be much lower than the dose of PAC required to achieve the same removal.

GAC is used for taste and odour control. It is normally used in fixed beds, either in purpose-built adsorbers for chemicals or in existing filter shells by replacement of sand with GAC of a similar particle size. Although at most treatment works it would be cheaper to convert existing filters rather than build separate adsorbers, use of existing filters usually allows only short contact times. It is therefore common practice to install additional GAC adsorbers (in some cases preceded by ozonation) between the rapid gravity filters and final disinfection. Most groundwater sources do not have existing filters, and separate adsorbers would need to be installed.

The service life of a GAC bed is dependent on the capacity of the carbon used and the contact time between the water and the carbon, the empty bed contact time (EBCT), controlled by the flow rate of the water. EBCTs are usually in the range 5–30 min. GACs vary considerably in their capacity for specific organic compounds, which can have a considerable effect upon their service life. A guide to capacity can be obtained from published isotherm data. Carbon capacity is strongly dependent on the water source and is greatly reduced by the presence of background organic compounds. The properties of a chemical that influence its adsorption onto activated carbon include the water solubility and octanol/water partition coefficient ($\log K_{ow}$). As a general rule, chemicals with low solubility and high $\log K_{ow}$ are well adsorbed.

Activated carbon is used for the removal of pesticides and other organic chemicals, taste and odour compounds, cyanobacterial toxins and total organic carbon.

8.4.9 Ion exchange

Ion exchange is a process in which ions of like charge are exchanged between the water phase and the solid resin phase. Water softening is achieved by cation exchange. Water is passed through a bed of cationic resin, and the calcium ions and magnesium ions in the water are replaced by sodium ions. When the ion exchange resin is exhausted (i.e., the sodium ions are depleted), it is regenerated using a solution of sodium chloride. The process of "dealkalization" can also soften water. Water is passed through a bed of weakly acidic resin, and the calcium and magnesium ions are replaced by hydrogen ions. The hydrogen ions react with the carbonate and bicarbonate ions to produce carbon dioxide. The hardness of the water is thus reduced without any increase in sodium levels. Anion exchange can be used to remove contaminants such as nitrate, which is exchanged for chloride. Nitrate-specific resins are available for this purpose.

An ion exchange plant normally consists of two or more resin beds contained in pressure shells with appropriate pumps, pipework and ancillary equipment for regeneration. The pressure shells are typically up to 4 m in diameter, containing 0.6–1.5 m depth of resin.

Cation exchange can be used for removal of certain heavy metals. Potential applications of anionic resins, in addition to nitrate removal, are for removal of arsenic and selenium species.

8.4.10 Membrane processes

The membrane processes of most significance in water treatment are reverse osmosis, ultrafiltration, microfiltration and nanofiltration. These processes have traditionally been applied to the production of water for industrial or pharmaceutical applications but are now being applied to the treatment of drinking-water.

High-pressure processes

If two solutions are separated by a semi-permeable membrane (i.e., a membrane that allows the passage of the solvent but not of the solute), the solvent will naturally pass from the lower-concentration solution to the higher-concentration solution. This process is known as osmosis. It is possible, however, to force the flow of solvent in the opposite direction, from the higher to the lower concentration, by increasing the pressure on the higher-concentration solution. The required pressure differential is known as the osmotic pressure, and the process is known as reverse osmosis.

Reverse osmosis results in the production of a treated water stream and a relatively concentrated waste stream. Typical operating pressures are in the range 15–50 bar, depending on the application. Reverse osmosis rejects monovalent ions and organics of molecular weight greater than about 50 (membrane pore sizes are less than 0.002 µm). The most common application of reverse osmosis is desalination of brackish water and seawater.

Nanofiltration uses a membrane with properties between those of reverse osmosis and ultrafiltration membranes; pore sizes are typically 0.001–0.01 µm. Nanofiltration membranes allow monovalent ions such as sodium or potassium to pass but reject a high proportion of divalent ions such as calcium and magnesium and organic molecules of molecular weight greater than 200. Operating pressures are typically about 5 bar. Nanofiltration may be effective for the removal of colour and organic compounds.

Lower-pressure processes

Ultrafiltration is similar in principle to reverse osmosis, but the membranes have much larger pore sizes (typically 0.002–0.03 µm) and operate at lower pressures. Ultrafiltration membranes reject organic molecules of molecular weight above about 800 and usually operate at pressures less than 5 bar.

Microfiltration is a direct extension of conventional filtration into the submicrometre range. Microfiltration membranes have pore sizes typically in the range 0.01–12 µm and do not separate molecules but reject colloidal and suspended material at operating pressures of 1–2 bar. Microfiltration is capable of sieving out particles greater than 0.05 µm. It has been used for water treatment in combination with coagulation or PAC to remove dissolved organic carbon and to improve permeate flux.

8.4.11 Other treatment processes

Other treatment processes that can be used in certain applications include:

- precipitation softening (addition of lime, lime plus sodium carbonate or sodium hydroxide to precipitate hardness at high pH);
- biological denitrification for removal of nitrate from surface waters;
- biological nitrification for removal of ammonia from surface waters; and
- activated alumina (or other adsorbents) for specialized applications, such as removal of fluoride and arsenic.

8.4.12 Disinfection by-products – process control measures

All chemical disinfectants produce inorganic and/or organic DBPs that may be of concern.

> In attempting to control DBP concentrations, it is of paramount importance that the efficiency of disinfection is not compromised and that a suitable residual level of disinfectant is maintained throughout the distribution system.

The principal DBPs formed during chlorination are THMs, chlorinated acetic acids, chlorinated ketones and haloacetonitriles, as a result of chlorination of naturally occurring organic precursors such as humic substances. Monochloramine produces lower THM concentrations than chlorine but produces other DBPs, including cyanogen chloride.

Ozone oxidizes bromide to produce hypohalous acids, which react with precursors to form brominated THMs. A range of other DBPs, including aldehydes and carboxylic acids, may also be formed. Of particular concern is bromate, formed by oxidation of bromide. Bromate may also be present in some sources of hypochlorite, but usually at concentrations that will give rise to levels in final water that are below the guideline value.

The main by-products from the use of chlorine dioxide are chlorite ion, which is an inevitable decomposition product, and chlorate ion. Chlorate is also produced in hypochlorate as it ages.

The basic strategies that can be adopted for reducing the concentrations of DBPs are:

- changing process conditions (including removal of precursor compounds prior to application);
- using a different chemical disinfectant with a lower propensity to produce by-products with the source water;
- using non-chemical disinfection; and/or
- removing DBPs prior to distribution.

Changes to process conditions

The formation of THMs during chlorination can be reduced by removing precursors prior to contact with chlorine – for example, by installing or enhancing coagulation (this may involve using higher coagulant doses and/or lower coagulation pH than are applied conventionally). DBP formation can also be reduced by lowering the applied chlorine dose; if this is done, it must be ensured that disinfection is still effective.

The pH value during chlorination affects the distribution of chlorinated by-products. Reducing the pH lowers the THM concentration, but at the expense of increased formation of haloacetic acids. Conversely, increasing the pH reduces haloacetic acid production but leads to increased THM formation.

The formation of bromate during ozonation depends on several factors, including concentrations of bromide and ozone and the pH. It is not practicable to remove bromide from raw water, and it is difficult to remove bromate once formed, although GAC filtration has been reported to be effective under certain circumstances. Bromate formation can be minimized by using lower ozone dose, shorter contact time and a lower residual ozone concentration. Operating at lower pH (e.g., pH 6.5) followed by raising the pH after ozonation also reduces bromate formation, and addition of ammonia can also be effective. Addition of hydrogen peroxide can increase or decrease bromate formation.

Changing disinfectants

It may be feasible to change disinfectant in order to achieve guideline values for DBPs. The extent to which this is possible will be dependent on raw water quality and installed treatment (e.g., for precursor removal).

It may be effective to change from chlorine to monochloramine, at least to provide a residual disinfectant within distribution, in order to reduce THM formation and subsequent development within the distribution system. While monochloramine provides a more stable residual within distribution, it is a less powerful disinfectant and should not be used as a primary disinfectant.

Chlorine dioxide can be considered as a potential alternative to both chlorine and ozone disinfection, although it does not provide a residual effect. The main concerns with chlorine dioxide are with the residual concentrations of chlorine dioxide and the by-products chlorite and chlorate. These can be addressed by controlling the dose of chlorine dioxide at the treatment plant.

Non-chemical disinfection

UV irradiation or membrane processes can be considered as alternatives to chemical disinfection. Neither of these provides any residual disinfection, and it may be considered appropriate to add a small dose of a persistent disinfectant such as chlorine or monochloramine to act as a preservative during distribution.

8.4.13 Treatment for corrosion control
General

Corrosion is the partial dissolution of the materials constituting the treatment and supply systems, tanks, pipes, valves and pumps. It may lead to structural failure, leaks, loss of capacity and deterioration of chemical and microbial water quality. The internal corrosion of pipes and fittings can have a direct impact on the concentration of some water constituents, including lead and copper. Corrosion control is therefore an important aspect of the management of a drinking-water system for safety.

Corrosion control involves many parameters, including the concentrations of calcium, bicarbonate, carbonate and dissolved oxygen, as well as pH. The detailed requirements differ depending on water quality and the materials used in the distribution system. The pH controls the solubility and rate of reaction of most of the metal species involved in corrosion reactions. It is particularly important in relation to the formation of a protective film at the metal surface. For some metals, alkalinity (carbonate and bicarbonate) and calcium (hardness) also affect corrosion rates.

Iron

Iron is frequently used in water distribution systems, and its corrosion is of concern. While structural failure as a result of iron corrosion is rare, water quality problems (e.g., "red water") can arise as a result of excessive corrosion of iron pipes. The corrosion of iron is a complex process that involves the oxidation of the metal, normally by dissolved oxygen, ultimately to form a precipitate of iron(III). This leads to the formation of tubercules on the pipe surface. The major water quality factors that determine whether the precipitate forms a protective scale are pH and alkalinity. The concentrations of calcium, chloride and sulfate also influence iron corrosion. Successful control of iron corrosion has been achieved by adjusting the pH to the range 6.8–7.3, hardness and alkalinity to at least 40 mg/litre (as calcium carbonate), oversaturation with calcium carbonate of 4–10 mg/litre and a ratio of alkalinity to Cl^- + SO_4^{2-} of at least 5 (when both are expressed as calcium carbonate).

Silicates and polyphosphates are often described as "corrosion inhibitors," but there is no guarantee that they will inhibit corrosion in water distribution systems. However, they can complex dissolved iron (in the iron(II) state) and prevent its precipitation as visibly obvious red "rust." These compounds may act by masking the effects of corrosion rather than by preventing it. Orthophosphate is a possible corrosion inhibitor and, like polyphosphates, is used to prevent "red water."

Lead

Lead corrosion (plumbosolvency) is of particular concern. Lead piping is still common in old houses in some countries, and lead solders have been used widely for jointing copper tube. The solubility of lead is governed by the formation of lead carbonates as pipe deposits. Wherever practicable, lead pipework should be replaced.

The solubility of lead increases markedly as the pH is reduced below 8 because of the substantial decrease in the equilibrium carbonate concentration. Thus, plumbosolvency tends to be at a maximum in waters with a low pH and low alkalinity, and a useful interim control procedure pending pipe replacement is to increase the pH to 8.0–8.5 after chlorination, and possibly to dose orthophosphate.

Lead can corrode more rapidly when it is coupled to copper. The rate of such galvanic corrosion is faster than that of simple oxidative corrosion, and lead concentrations are not limited by the solubility of the corrosion products. The rate of galvanic corrosion is affected principally by chloride concentration. Galvanic corrosion is less easily controlled but can be reduced by dosing zinc in conjunction with orthophosphate and by adjustment of pH.

Treatment to reduce plumbosolvency usually involves pH adjustment. When the water is very soft (less than 50 mg of calcium carbonate per litre), the optimum pH is about 8.0–8.5. Alternatively, dosing with orthophosphoric acid or sodium orthophosphate might be more effective, particularly when plumbosolvency occurs in non-acidic waters.

Copper

The corrosion of copper pipework and hot water cylinders can cause blue water, blue or green staining of bathroom fittings and, occasionally, taste problems. Copper tubing may be subject to general corrosion, impingement attack and pitting corrosion.

General corrosion is most often associated with soft, acidic waters; waters with pH below 6.5 and hardness of less than 60 mg of calcium carbonate per litre are very aggressive to copper. Copper, like lead, can enter water by dissolution of the corrosion product, basic copper carbonate. The solubility is mainly a function of pH and total inorganic carbon. Solubility decreases with increase in pH, but increases with increase in concentrations of carbonate species. Raising the pH to between 8 and 8.5 is the usual procedure to overcome these difficulties.

Impingement attack is the result of excessive flow velocities and is aggravated in soft water at high temperature and low pH.

The pitting of copper is commonly associated with hard groundwaters having a carbon dioxide concentration above 5 mg/litre and high dissolved oxygen. Surface waters with organic colour may also be associated with pitting corrosion. Copper pipes can fail by pitting corrosion, which involves highly localized attacks leading to perforations with negligible loss of metal. Two main types of attack are recognized. Type I pitting affects cold water systems (below 40 °C) and is associated, particularly, with hard borehole waters and the presence of a carbon film in the bore of the pipe, derived from the manufacturing process. Tubes that have had the carbon removed by cleaning are immune from Type I pitting. Type II pitting occurs in hot water systems (above 60 °C) and is associated with soft waters. A high proportion of general and pitting corrosion problems are associated with new pipe in which a protective oxide layer has not yet formed.

Brass

The main corrosion problem with brasses is dezincification, which is the selective dissolution of zinc from duplex brass, leaving behind copper as a porous mass of low mechanical strength. Meringue dezincification, in which a voluminous corrosion product of basic zinc carbonate forms on the brass surface, largely depends on the ratio of chloride to alkalinity. Meringue dezincification can be controlled by maintaining a low zinc to copper ratio (1:3 or lower) and by keeping pH below 8.3.

General dissolution of brass can also occur, releasing metals, including lead, into the water. Impingement attack can occur under conditions of high water velocity with waters that form poorly protective corrosion product layers and that contain large amounts of dissolved or entrained air.

Zinc

The solubility of zinc in water is a function of pH and total inorganic carbon concentrations; the solubility of basic zinc carbonate decreases with increase in pH and concentrations of carbonate species. For low-alkalinity waters, an increase of pH to 8.5 should be sufficient to control the dissolution of zinc.

With galvanized iron, the zinc layer initially protects the steel by corroding preferentially. In the long term, a protective deposit of basic zinc carbonate forms. Protective deposits do not form in soft waters where the alkalinity is less than 50 mg/litre as calcium carbonate or waters containing high carbon dioxide concentrations (>25 mg/litre as carbon dioxide), and galvanized steel is unsuitable for these waters. The corrosion of galvanized steel increases when it is coupled with copper tubing.

Nickel

Nickel may arise due to the leaching of nickel from new nickel/chromium-plated taps. Low concentrations may also arise from stainless steel pipes and fittings. Nickel leaching falls off over time. An increase of pH to control corrosion of other materials should also reduce leaching of nickel.

Concrete and cement

Concrete is a composite material consisting of a cement binder in which an inert aggregate is embedded. Cement is primarily a mixture of calcium silicates and aluminates together with some free lime. Cement mortar, in which the aggregate is fine sand, is used as a protective lining in iron and steel water pipes. In asbestos–cement pipe, the aggregate is asbestos fibres. Cement is subject to deterioration on prolonged exposure to aggressive water, due either to the dissolution of lime and other soluble compounds or to chemical attack by aggressive ions such as chloride or sulfate, and this may result in structural failure. Cement contains a variety of metals that can be leached into the water. Aggressiveness to cement is related to the "aggressivity index," which has been used specifically to assess the potential for the dissolution of concrete. A pH of 8.5 or higher may be necessary to control cement corrosion.

Characterizing corrosivity

Most of the indices that have been developed to characterize the corrosion potential of waters are based on the assumption that water with a tendency to deposit a calcium carbonate scale on metal surfaces will be less corrosive. The Langelier Index (LI) is the difference between the actual pH of a water and its "saturation pH," this being the pH at which a water of the same alkalinity and calcium hardness would be at equilibrium with solid calcium carbonate. Waters with positive LI are capable of depositing calcium carbonate scale from solution.

There is no corrosion index that applies to all materials, and corrosion indices, particularly those related to calcium carbonate saturation, have given mixed results. The parameters related to calcium carbonate saturation status are, strictly speaking, indicators of the tendency to deposit or dissolve calcium carbonate (calcite) scale, not indicators of the "corrosivity" of a water. For example, there are many waters with negative LI that are non-corrosive and many with positive LI that are corrosive. Nevertheless, there are many documented instances of the use of saturation indices for corrosion control based on the concept of laying down a protective "eggshell" scale of calcite in iron pipes. In general, waters with high pH, calcium and alkalinity are less corrosive, and this tends to be correlated with a positive LI.

The ratio of the chloride and sulfate concentrations to the bicarbonate concentration (Larson ratio) has been shown to be helpful in assessing the corrosiveness of water to cast iron and steel. A similar approach has been used in studying zinc dissolution from brass fittings – the Turner diagram.

Water treatment for corrosion control

To control corrosion in water distribution networks, the methods most commonly applied are adjusting pH, increasing the alkalinity and/or hardness or adding corrosion inhibitors, such as polyphosphates, silicates and orthophosphates. The quality and maximum dose to be used should be in line with specifications for such water treatment chemicals. Although pH adjustment is an important approach, its possible impact on other aspects of water supply technology, including disinfection, must always be taken into account.

It is not always possible to achieve the desired values for all parameters. For example, the pH of hard waters cannot be increased too much, or softening will occur. The application of lime and carbon dioxide to soft waters can be used to increase both the calcium concentration and the alkalinity to at least 40 mg/litre as calcium carbonate.

8.5 Guideline values for individual chemicals, by source category
8.5.1 Naturally occurring chemicals

There are a number of sources of naturally occurring chemicals in drinking-water. All natural water contains a range of inorganic and organic chemicals. The former derive from the rocks and soil through which water percolates or over which it flows. The latter derive from the breakdown of plant material or from algae and other microorganisms that grow in the water or on sediments. Most of the naturally occurring chemicals for which guideline values have been derived or that have been considered for guideline value derivation are inorganic. Only one, microcystin-LR, a toxin produced by cyanobacteria or blue-green algae, is organic; it is discussed in section 8.5.6.

The approach to dealing with naturally occurring chemicals will vary according to the nature of the chemical and the source. For inorganic contaminants that arise from rocks and sediments, it is important to screen possible water sources to determine whether the source is suitable for use or whether it will be necessary to treat the water to remove the contaminants of concern along with microbial contaminants. In some cases, where a number of sources may be available, dilution or blending of the water containing high levels of a contaminant with a water containing much lower levels may achieve the desired result.

A number of the most important chemical contaminants (i.e., those that have been shown to cause adverse health effects as a consequence of exposure through drinking-water) fall into the category of naturally occurring chemicals. Some naturally occurring chemicals have other primary sources and are therefore discussed in other sections of this chapter.

Guideline values have not been established for the chemicals listed in Table 8.17 for the reasons indicated in the table. Summary statements are included in chapter 12.

Guideline values have been established for the chemicals listed in Table 8.18, which meet the criteria for inclusion. Summary statements are included for each in chapter 12.

8.5.2 Chemicals from industrial sources and human dwellings

Chemicals from industrial sources can reach drinking-water directly from discharges or indirectly from diffuse sources arising from the use and disposal of materials and products containing the chemical. In some cases, inappropriate handling and disposal may lead to contamination, e.g., degreasing agents that are allowed to reach ground-

Table 8.17 Naturally occurring chemicals for which guideline values have not been established

Chemical	Reason for not establishing a guideline value	Remarks
Chloride	Occurs in drinking-water at concentrations well below those at which toxic effects may occur	May affect acceptability of drinking-water (see chapter 10)
Hardness	Occurs in drinking-water at concentrations well below those at which toxic effects may occur	May affect acceptability of drinking-water (see chapter 10)
Hydrogen sulfide	Occurs in drinking-water at concentrations well belowthose at which toxic effects may occur	May affect acceptability of drinking-water (see chapter 10)
pH	Values in drinking-water are well below those at which toxic effects may occur	An important operational water quality parameter
Sodium	Occurs in drinking-water at concentrations well below those at which toxic effects may occur	May affect acceptability of drinking-water (see chapter 10)
Sulfate	Occurs in drinking-water at concentrations well below those at which toxic effects may occur	May affect acceptability of drinking-water (see chapter 10)
Total dissolved solids (TDS)	Occurs in drinking-water at concentrations well below those at which toxic effects may occur	May affect acceptability of drinking-water (see chapter 10)

Table 8.18 **Guideline values for naturally occurring chemicals that are of health significance in drinking-water**

Chemical	Guideline value[a] (mg/litre)	Remarks
Arsenic	0.01 (P)	
Barium	0.7	
Boron	0.5 (T)	
Chromium	0.05 (P)	For total chromium
Fluoride	1.5	Volume of water consumed and intake from other sources should be considered when setting national standards
Manganese	0.4 (C)	
Molybdenum	0.07	
Selenium	0.01	
Uranium	0.015 (P, T)	Only chemical aspects of uranium addressed

[a] P = provisional guideline value, as there is evidence of a hazard, but the available information on health effects is limited; T = provisional guideline value because calculated guideline value is below the level that can be achieved through practical treatment methods, source protection, etc.; C = concentrations of the substance at or below the health-based guideline value may affect the appearance, taste or odour of the water, resulting in consumer complaints.

water. Some of these chemicals, particularly inorganic substances, may also be encountered as a consequence of natural contamination, but this may also be a by-product of industrial activity, such as mining, that changes drainage patterns. Many of these chemicals are used in small industrial units within human settlements, and, particularly where such units are found in groups of similar enterprises, they may be a significant source of pollution. Petroleum oils are widely used in human settlements, and improper handling or disposal can lead to significant pollution of surface water and groundwater. Where plastic pipes are used, the smaller aromatic molecules in petroleum oils can sometimes penetrate the pipes where they are surrounded by earth soaked in the oil, with subsequent pollution of the local water supply.

A number of chemicals can reach water as a consequence of disposal of general household chemicals; in particular, a number of heavy metals may be found in domestic wastewater. Where wastewater is treated, these will usually partition out into the sludge. Some chemicals that are widely used both in industry and in materials used in a domestic setting are found widely in the environment, e.g., di(2-ethylhexyl)phthalate, and these may be found in water sources, although usually at low concentrations.

Some chemicals that reach drinking-water from industrial sources or human settlements have other primary sources and are therefore discussed in other sections of this chapter. Where latrines and septic tanks are poorly sited, these can lead to contamination of drinking-water sources with nitrate (see section 8.5.3).

Identification of the potential for contamination by chemicals from industrial activities and human dwellings requires assessment of activities in the catchment and of the risk that particular contaminants may reach water sources. The primary approach to addressing these contaminants is prevention of contamination by encouraging good practices. However, if contamination has occurred, then it may be necessary to consider the introduction of treatment.

8. CHEMICAL ASPECTS

Table 8.19 Chemicals from industrial sources and human dwellings excluded from guideline value derivation

Chemical	Reason for exclusion
Beryllium	Unlikely to occur in drinking-water

The chemical listed in Table 8.19 has been excluded from guideline value derivation, as a review of the literature on occurrence and/or credibility of occurrence in drinking-water has shown evidence that it does not occur in drinking-water.

Guideline values have not been established for the chemicals listed in Table 8.20 for the reasons indicated in the table. Summary statements for each are included in chapter 12.

Guideline values have been established for the chemicals listed in Table 8.21, which meet all of the criteria for inclusion. Summary statements are included in chapter 12.

8.5.3 Chemicals from agricultural activities

Chemicals are used in agriculture on crops and in animal husbandry. Nitrate may be present as a consequence of tillage when there is no growth to take up nitrate released from decomposing plants, from the application of excess inorganic or organic fertilizer and in slurry from animal production. Most chemicals that may arise from agriculture are pesticides, although their presence will depend on many factors, and not all pesticides are used in all circumstances or climates. Contamination can result from application and subsequent movement following rainfall or from inappropriate disposal methods.

Some pesticides are also used in non-agricultural circumstances, such as the control of weeds on roads and railway lines. These pesticides are also included in this section.

Table 8.20 Chemicals from industrial sources and human dwellings for which guideline values have not been established

Chemical	Reason for not establishing a guideline value
Dichlorobenzene, 1,3-	Toxicological data are insufficient to permit derivation of health-based guideline value
Dichloroethane, 1,1-	Very limited database on toxicity and carcinogenicity
Di(2-ethylhexyl)adipate	Occurs in drinking-water at concentrations well below those at which toxic effects may occur
Hexachlorobenzene	Occurs in drinking-water at concentrations well below those at which toxic effects may occur
Monochlorobenzene	Occurs in drinking-water at concentrations well below those at which toxic effects may occur, and health-based value would far exceed lowest reported taste and odour threshold
Trichlorobenzenes (total)	Occur in drinking-water at concentrations well below those at which toxic effects may occur, and health-based value would exceed lowest reported odour threshold
Trichloroethane, 1,1,1-	Occurs in drinking-water at concentrations well below those at which toxic effects may occur

Table 8.21 **Guideline values for chemicals from industrial sources and human dwellings that are of health significance in drinking-water**

Inorganics	Guideline value (mg/litre)	Remarks
Cadmium	0.003	
Cyanide	0.07	
Mercury	0.001	For total mercury (inorganic plus organic)

Organics	Guideline value[a] (µg/litre)	Remarks
Benzene	10[b]	
Carbon tetrachloride	4	
Di(2-ethylhexyl)phthalate	8	
Dichlorobenzene, 1,2-	1000 (C)	
Dichlorobenzene, 1,4-	300 (C)	
Dichloroethane, 1,2-	30[b]	
Dichloroethene, 1,1-	30	
Dichloroethene, 1,2-	50	
Dichloromethane	20	
Edetic acid (EDTA)	600	Applies to the free acid
Ethylbenzene	300 (C)	
Hexachlorobutadiene	0.6	
Nitrilotriacetic acid (NTA)	200	
Pentachlorophenol	9[b] (P)	
Styrene	20 (C)	
Tetrachloroethene	40	
Toluene	700 (C)	
Trichloroethene	70 (P)	
Xylenes	500 (C)	

[a] P = provisional guideline value, as there is evidence of a hazard, but the available information on health effects is limited; C = concentrations of the substance at or below the health-based guideline value may affect the appearance, taste or odour of the water, leading to consumer complaints.

[b] For non-threshold substances, the guideline value is the concentration in drinking-water associated with an upper-bound excess lifetime cancer risk of 10^{-5} (one additional cancer per 100 000 of the population ingesting drinking-water containing the substance at the guideline value for 70 years). Concentrations associated with estimated upper-bound excess lifetime cancer risks of 10^{-4} and 10^{-6} can be calculated by multiplying and dividing, respectively, the guideline value by 10.

Guideline values have not been established for the chemicals listed in Table 8.22, as a review of the literature on occurrence and/or credibility of occurrence in drinking-water has shown evidence that the chemicals do not occur in drinking-water.

Guideline values have not been established for the chemicals listed in Table 8.23 for the reasons indicated in the table. Summary statements are included in chapter 12.

Guideline values have been established for the chemicals listed in Table 8.24, which meet the criteria for inclusion. Summary statements are included in chapter 12.

8.5.4 Chemicals used in water treatment or from materials in contact with drinking-water

Chemicals are used in water treatment and may give rise to residuals in the final water. In some cases, such as monochloramine and chlorine, this is intentional, and their

Table 8.22 Chemicals from agricultural activities excluded from guideline value derivation

Chemical	Reason for exclusion
Amitraz	Degrades rapidly in the environment and is not expected to occur at measurable concentrations in drinking-water supplies
Chlorobenzilate	Unlikely to occur in drinking-water
Chlorothalonil	Unlikely to occur in drinking-water
Cypermethrin	Unlikely to occur in drinking-water
Diazinon	Unlikely to occur in drinking-water
Dinoseb	Unlikely to occur in drinking-water
Ethylene thiourea	Unlikely to occur in drinking-water
Fenamiphos	Unlikely to occur in drinking-water
Formothion	Unlikely to occur in drinking-water
Hexachlorocyclohexanes (mixed isomers)	Unlikely to occur in drinking-water
MCPB	Unlikely to occur in drinking-water
Methamidophos	Unlikely to occur in drinking-water
Methomyl	Unlikely to occur in drinking-water
Mirex	Unlikely to occur in drinking-water
Monocrotophos	Has been withdrawn from use in many countries and is unlikely to occur in drinking-water
Oxamyl	Unlikely to occur in drinking-water
Phorate	Unlikely to occur in drinking-water
Propoxur	Unlikely to occur in drinking-water
Pyridate	Not persistent and only rarely found in drinking-water
Quintozene	Unlikely to occur in drinking-water
Toxaphene	Unlikely to occur in drinking-water
Triazophos	Unlikely to occur in drinking-water
Tributyltin oxide	Unlikely to occur in drinking-water
Trichlorfon	Unlikely to occur in drinking-water

presence confers a direct benefit to the consumer. Some arise as unwanted by-products of the disinfection process (see Table 8.25) and some as residuals from other parts of the treatment process, such as coagulation. Some may arise as contaminants in treatment chemicals, and others may arise as contaminants in, or as corrosion products from, materials used as pipes or in other parts of the drinking-water system. Some chemicals used in water treatment (e.g., fluoride) or in materials in contact with drinking-water (e.g., styrene) have other primary sources and are therefore discussed in detail in other sections of this chapter.

The approach to monitoring and management is preferably through control of the material or chemical, and this is covered in more detail in section 4.2. It is also important to optimize treatment processes and to ensure that such processes remain optimized in order to control residuals of chemicals used in treatment and to control the formation of DBPs.

Guideline values have not been established for the chemicals listed in Table 8.26 for the reasons indicated in the table. Summary statements are included in chapter 12.

Table 8.23 **Chemicals from agricultural activities for which guideline values have not been established**

Chemical	Reason for not establishing a guideline value
Ammonia	Occurs in drinking-water at concentrations well below those at which toxic effects may occur
Bentazone	Occurs in drinking-water at concentrations well below those at which toxic effects may occur
Dichloropropane, 1,3-	Data insufficient to permit derivation of health-based guideline value
Diquat	Rarely found in drinking-water, but may be used as an aquatic herbicide for the control of free-floating and submerged aquatic weeds in ponds, lakes and irrigation ditches
Endosulfan	Occurs in drinking-water at concentrations well below those at which toxic effects may occur
Fenitrothion	Occurs in drinking-water at concentrations well below those at which toxic effects may occur
Glyphosate and AMPA	Occurs in drinking-water at concentrations well below those at which toxic effects may occur
Heptachlor and heptachlor epoxide	Occurs in drinking-water at concentrations well below those at which toxic effects may occur
Malathion	Occurs in drinking-water at concentrations well below those at which toxic effects may occur
Methyl parathion	Occurs in drinking-water at concentrations well below those at which toxic effects may occur
Parathion	Occurs in drinking-water at concentrations well below those at which toxic effects may occur
Permethrin	Occurs in drinking-water at concentrations well below those at which toxic effects may occur
Phenylphenol, 2- and its sodium salt	Occurs in drinking-water at concentrations well below those at which toxic effects may occur
Propanil	Readily transformed into metabolites that are more toxic; a guideline value for the parent compound is considered inappropriate, and there are inadequate data to enable the derivation of guideline values for the metabolites

Guideline values have been established for the chemicals listed in Table 8.27, which meet the criteria for inclusion. Summary statements are included in chapter 12.

8.5.5 Pesticides used in water for public health purposes

Some pesticides are used for public health purposes, including the addition to water to control the aquatic larval stages of insects of public health significance (e.g., mosquitos for the control of malaria and typhus). There are currently four insecticide compounds and a bacterial larvicide recommended by WHO (under WHOPES) for addition to drinking-water as larvicides: temephos, methoprene, pyriproxyfen, permethrin and *Bacillus thuringiensis israelensis*. Of these, only pyriproxyfen has been reviewed to date. Other insecticides that are not recommended for addition to water for public health purposes by WHOPES but may be used in some countries as aquatic larvicides, or have been used as such in the past, include chlorpyrifos and DDT.

Table 8.24 Guideline values for chemicals from agricultural activities that are of health significance in drinking-water

Non-pesticides	Guideline value[a] (mg/litre)	Remarks
Nitrate (as NO_3^-)	50	Short-term exposure
Nitrite (as NO_2^-)	3	Short-term exposure
	0.2 (P)	Long-term exposure

Pesticides used in agriculture	Guideline value[a] (µg/litre)	Remarks
Alachlor	20[b]	
Aldicarb	10	Applies to aldicarb sulfoxide and aldicarb sulfone
Aldrin and dieldrin	0.03	For combined aldrin plus dieldrin
Atrazine	2	
Carbofuran	7	
Chlordane	0.2	
Chlorotoluron	30	
Cyanazine	0.6	
2,4-D (2,4-dichlorophenoxyacetic acid)	30	Applies to free acid
2,4-DB	90	
1,2-Dibromo-3-chloropropane	1[b]	
1,2-Dibromoethane	0.4[b] (P)	
1,2-Dichloropropane (1,2-DCP)	40 (P)	
1,3-Dichloropropene	20[b]	
Dichlorprop	100	
Dimethoate	6	
Endrin	0.6	
Fenoprop	9	
Isoproturon	9	
Lindane	2	
MCPA	2	
Mecoprop	10	
Methoxychlor	20	
Metolachlor	10	
Molinate	6	
Pendimethalin	20	
Simazine	2	
2,4,5-T	9	
Terbuthylazine	7	
Trifluralin	20	

[a] P = provisional guideline value, as there is evidence of a hazard, but the available information on health effects is limited.
[b] For substances that are considered to be carcinogenic, the guideline value is the concentration in drinking-water associated with an upper-bound excess lifetime cancer risk of 10^{-5} (one additional cancer per 100 000 of the population ingesting drinking-water containing the substance at the guideline value for 70 years). Concentrations associated with estimated upper-bound excess lifetime cancer risks of 10^{-4} and 10^{-6} can be calculated by multiplying and dividing, respectively, the guideline value by 10.

In considering those pesticides that may be added to water used for drinking-water for purposes of protection of public health, every effort should be made not to develop guidelines that are unnecessarily stringent as to impede their use. This approach enables a suitable balance to be achieved between the protection of drinking-water

Table 8.25 Disinfection by-products present in disinfected waters (from IPCS, 2000)

Disinfectant	Significant organohalogen products	Significant inorganic products	Significant non-halogenated products
Chlorine/ hypochlorous acid	THMs, haloacetic acids, haloacetonitriles, chloral hydrate, chloropicrin, chlorophenols, N-chloramines, halofuranones, bromohydrins	chlorate (mostly from hypochlorite use)	aldehydes, cyanoalkanoic acids, alkanoic acids, benzene, carboxylic acids
Chlorine dioxide		chlorite, chlorate	unknown
Chloramine	haloacetonitriles, cyanogen chloride, organic chloramines, chloramino acids, chloral hydrate, haloketones	nitrate, nitrite, chlorate, hydrazine	aldehydes, ketones
Ozone	bromoform, monobromoacetic acid, dibromoacetic acid, dibromoacetone, cyanogen bromide	chlorate, iodate, bromate, hydrogen peroxide, hypobromous acid, epoxides, ozonates	aldehydes, ketoacids, ketones, carboxylic acids

quality and the control of insects of public health significance. However, it is stressed that every effort should be made to keep overall exposure and the concentration of any larvicide as low as possible.

As for the other groups of chemicals discussed in this chapter, this category is not clear-cut. It includes pesticides that are extensively used for purposes other than public health protection – for example, agricultural purposes, in the case of chlorpyrifos.

Guideline values that have been derived for these larvicides are provided in Table 8.28. Summary statements are included in chapter 12.

8.5.6 Cyanobacterial toxins

Cyanobacteria (see also section 11.5) occur widely in lakes, reservoirs, ponds and slow-flowing rivers. Many species are known to produce toxins, i.e., "cyanotoxins," a number of which are of concern for health. Cyanotoxins vary in structure and may be found within cells or released into water. There is wide variation in the toxicity of recognized cyanotoxins (including different structural variants within a group, e.g., microcystins), and it is likely that further toxins remain unrecognized.

The toxins are classified, according to their mode of action, as hepatotoxins (microcystins and cylindrospermopsins), neurotoxins (anatoxin-a, saxitoxins and anatoxin-a(S)) and irritants or inflammatory agents (lipopolysaccharides). The hepatotoxins are produced by various species within the genera *Microcystis, Planktothrix, Anabaena, Aphanizomenon, Nodularia, Nostoc, Cylindrospermopsis* and *Umezakia*. The cyanotoxins occurring most frequently in elevated concentrations (i.e., >1 µg/litre) seem to be microcystins (oligopeptides) and cylindrospermopsin (an alkaloid), whereas the cyanobacterial neurotoxins appear to occur in high concentrations only occasionally.

8. CHEMICAL ASPECTS

Table 8.26 Chemicals used in water treatment or materials in contact with drinking-water for which guideline values have not been established

Chemical	Reason for not establishing a guideline value
Disinfectants	
Chlorine dioxide	Rapid breakdown of chlorine dioxide; also, the chlorite provisional guideline value is protective for potential toxicity from chlorine dioxide
Dichloramine	Available data inadequate to permit derivation of health-based guideline value
Iodine	Available data inadequate to permit derivation of health-based guideline value, and lifetime exposure to iodine through water disinfection is unlikely
Silver	Available data inadequate to permit derivation of health-based guideline value
Trichloramine	Available data inadequate to permit derivation of health-based guideline value
Disinfection by-products	
Bromochloroacetate	Available data inadequate to permit derivation of health-based guideline value
Bromochloroacetonitrile	Available data inadequate to permit derivation of health-based guideline value
Chloroacetones	Available data inadequate to permit derivation of health-based guideline values for any of the chloroacetones
Chlorophenol, 2-	Available data inadequate to permit derivation of health-based guideline value
Chloropicrin	Available data inadequate to permit derivation of health-based guideline value
Dibromoacetate	Available data inadequate to permit derivation of health-based guideline value
Dichlorophenol, 2,4-	Available data inadequate to permit derivation of health-based guideline value
Monobromoacetate	Available data inadequate to permit derivation of health-based guideline value
MX	Occurs in drinking-water at concentrations well below those at which toxic effects may occur
Trichloroacetonitrile	Available data inadequate to permit derivation of health-based guideline value
Contaminants from treatment chemicals	
Aluminium	Owing to limitations in the animal data as a model for humans and the uncertainty surrounding the human data, a health-based guideline value cannot be derived; however, practicable levels based on optimization of the coagulation process in drinking-water plants using aluminium-based coagulants are derived: 0.1 mg/litre or less in large water treatment facilities, and 0.2 mg/litre or less in small facilities
Iron	Not of health concern at concentrations normally observed in drinking-water, and taste and appearance of water are affected at concentrations below the health-based value
Contaminants from pipes and fittings	
Asbestos	No consistent evidence that ingested asbestos is hazardous to health
Dialkyltins	Available data inadequate to permit derivation of health-based guideline values for any of the dialkyltins
Fluoranthene	Occurs in drinking-water at concentrations well below those at which toxic effects may occur
Inorganic tin	Occurs in drinking-water at concentrations well below those at which toxic effects may occur
Zinc	Not of health concern at concentrations normally observed in drinking-water, but may affect the acceptability of water

Table 8.27 Guideline values for chemicals used in water treatment or materials in contact with drinking-water that are of health significance in drinking-water

Disinfectants	Guideline value[a] (mg/litre)	Remarks
Chlorine	5 (C)	For effective disinfection, there should be a residual concentration of free chlorine of ≥0.5 mg/litre after at least 30 min contact time at pH <8.0
Monochloramine	3	

Disinfection by-products	Guideline value[a] (µg/litre)	Remarks
Bromate	10[b] (A, T)	
Bromodichloromethane	60[b]	
Bromoform	100	
Chloral hydrate (trichloroacetaldehyde)	10 (P)	
Chlorate	700 (D)	
Chlorite	700 (D)	
Chloroform	200	
Cyanogen chloride	70	For cyanide as total cyanogenic compounds
Dibromoacetonitrile	70	
Dibromochloromethane	100	
Dichloroacetate	50 (T, D)	
Dichloroacetonitrile	20 (P)	
Formaldehyde	900	
Monochloroacetate	20	
Trichloroacetate	200	
Trichlorophenol, 2,4,6-	200[b] (C)	
Trihalomethanes		The sum of the ratio of the concentration of each to its respective guideline value should not exceed 1

Contaminants from treatment chemicals	Guideline value[a] (µg/litre)	Remarks
Acrylamide	0.5[b]	
Epichlorohydrin	0.4 (P)	

Contaminants from pipes and fittings	Guideline value[a] (µg/litre)	Remarks
Antimony	20	
Benzo[a]pyrene	0.7[b]	
Copper	2000	Staining of laundry and sanitary ware may occur below guideline value
Lead	10	
Nickel	20 (P)	
Vinyl chloride	0.3[b]	

[a] P = provisional guideline value, as there is evidence of a hazard, but the available information on health effects is limited; A = provisional guideline value because calculated guideline value is below the practical quantification level; T = provisional guideline value because calculated guideline value is below the level that can be achieved through practical treatment methods, source control, etc.; D = provisional guideline value because disinfection is likely to result in the guideline value being exceeded; C = concentrations of the substance at or below the health-based guideline value may affect the appearance, taste or odour of the water, causing consumer complaints.

[b] For substances that are considered to be carcinogenic, the guideline value is the concentration in drinking-water associated with an upper-bound excess lifetime cancer risk of 10^{-5} (one additional cancer per 100 000 of the population ingesting drinking-water containing the substance at the guideline value for 70 years). Concentrations associated with estimated upper-bound excess lifetime cancer risks of 10^{-4} and 10^{-6} can be calculated by multiplying and dividing, respectively, the guideline value by 10.

8. CHEMICAL ASPECTS

Table 8.28 Guideline values for pesticides used in water for public health purposes that are of health significance in drinking-water

Pesticides used in water for public health purposes[a]	Guideline value (µg/litre)
Chlorpyrifos	30
DDT and metabolites	1
Pyriproxyfen	300

[a] Only pyriproxyfen is recommended by WHOPES for addition to water for public health purposes.

Table 8.29. Guideline values for cyanotoxins that are of health significance in drinking-water

	Guideline value[a] (µg/litre)	Remarks
Microcystin-LR	1 (P)	For total microcystin-LR (free plus cell-bound)

[a] P = provisional guideline value, as there is evidence of a hazard, but the available information on health effects is limited.

Cyanotoxins can reach concentrations potentially hazardous to human health primarily in situations of high cell density through excessive growth, sometimes termed "bloom" events. These occur in response to elevated concentrations of nutrients (phosphorus and sometimes nitrogen) and may be triggered by conditions such as water body stratification and sufficiently high temperature. Blooms tend to recur in the same water bodies. Cells of some cyanobacterial species may accumulate at the surface as scums or at the theromocline of thermally stratified reservoirs. Such accumulations may develop rapidly, and they may be of short duration. In many circumstances, blooms and accumulations are seasonal.

A variety of resource protection and source management actions are available to decrease the probability of bloom occurrence, and some treatment methods, including filtration and chlorination, are available for removal of cyanobacteria and cyanotoxins. Filtration can effectively remove cyanobacterial cells and, with that, often a high share of the toxins. Oxidation through ozone or chlorine at sufficient concentrations and contact times can effectively remove most cyanotoxins dissolved in water.

Chemical analysis of cyanotoxins is not the preferred focus of routine monitoring. The preferred approach is monitoring of source water for evidence of blooms, or bloom-forming potential, and increased vigilance where such events occur. Analysis of cyanotoxins requires time, equipment and expertise, and quantitative analysis of some cyanotoxins is hampered by the lack of analytical standards. However, rapid methods, such as ELISA and enzyme assays, are becoming available for a small number, e.g., microcystins.

Chemical analysis of cyanotoxins is useful for assessing the efficacy of treatment and preventive strategies, i.e., as validation of control measures in a WSP (see chapter 4). While guideline values are derived where sufficient data exist, they are primarily intended to inform setting targets for control measures.

A provisional guideline value has been established for microcystin-LR, which meets the criteria for inclusion (see Table 8.29). Microcystin-LR is one of the most toxic of more than 70 structural variants of microcystin. Although, on a global scale, it appears to be one of the most widespread microcystins, in many regions it is not the most commonly occurring variant, and others may well be less toxic. If the provisional guideline value for microcystin-LR is used as a surrogate for their assessment and for setting targets, this serves as a worst-case estimate. A more detailed discussion of using "concentration equivalents" or "toxicity equivalents" for relating microcystins to microcystin-LR is given in Chorus & Bartram (1999).

9
Radiological aspects

The objective of this chapter is to provide criteria with which to assess the safety of drinking-water with respect to its radionuclide content. The Guidelines do not differentiate between naturally occurring and artificial or human-made radionuclides.

The guidance values for radioactivity in drinking-water recommended in the first edition of the Guidelines were based on the risks of exposure to radiation sources and the health consequences of exposure to radiation. The second edition of the Guidelines incorporated the 1990 recommendations of the International Commission on Radiological Protection (ICRP, 1991). The third edition incorporates recent developments, including the ICRP publications on prolonged exposures and on dose coefficients.

Radiological hazards may derive from ionizing radiation emitted by a number of radioactive substances (chemicals) in drinking-water. Such hazards from drinking-water are rarely of public health significance, and radiation exposure from drinking-water must be assessed alongside exposure from other sources.

The approach taken in the Guidelines for controlling radiological hazards has two stages:

— initial screening for gross alpha and/or beta activity to determine whether the activity concentrations (in Bq/litre) are below levels at which no further action is required; and
— if these screening levels are exceeded, investigation of the concentrations of individual radionuclides and comparison with specific guidance levels.

The risk due to radon in drinking-water derived from groundwater is typically low compared with that due to total inhaled radon but is distinct, as exposure occurs through both consumption of dissolved gas and inhalation of released radon and its daughter radionuclides. Greatest exposure is general ambient inhalation and inhalation from terrestrial sources, where the gas is infiltrating into dwellings, especially into basements. Radon of groundwater origin would usually be a small increment of the total, but may indicate deposits in the region that are emitting into basements.

The screening and guidance levels apply to routine ("normal") operational conditions of existing or new drinking-water supplies. They do not apply to a water supply contaminated during an emergency involving the release of radionuclides into the environment. Guidance and generic action levels covering emergency situations are available elsewhere (IAEA, 1996, 1997, 1999, 2002).

The current Guidelines are based on:

— a recommended reference dose level (RDL) of the committed effective dose, equal to 0.1 mSv from 1 year's consumption of drinking-water (from the possible total radioactive contamination of the annual drinking-water consumption). This comprises 10% of the intervention exemption level recommended by the ICRP for dominant commodities (e.g., food and drinking-water) for prolonged exposure situations, which is most relevant to long-term consumption of drinking-water by the public (ICRP, 2000). The RDL of 0.1 mSv is also equal to 10% of the dose limit for members of the population, recommended by both the ICRP (1991) and the International Basic Safety Standards (IAEA, 1996). These are accepted by most WHO Member States, the European Commission, FAO and WHO.
— dose coefficients for adults, provided by the ICRP.

The additional risk to health from exposure to an annual dose of 0.1 mSv associated with the intake of radionuclides from drinking-water is considered to be low for the following reasons:

- The nominal probability coefficient for radiation-induced stochastic health effects, which include fatal cancer, non-fatal cancer and severe hereditary effects for the whole population, is $7.3 \times 10^{-2}/Sv$ (ICRP, 1991). Multiplying this by an RDL equal to 0.1 mSv annual exposure via drinking-water gives an estimated lifetime risk of stochastic health effects of 10^{-5}, which can be considered small in comparison with other health risks. This risk level is comparable to the reference level of risk used elsewhere in these Guidelines.
- Background radiation exposures vary widely across the Earth, but the average is about 2.4 mSv/year, with the highest local levels being up to 10 times higher without any apparent health consequences; 0.1 mSv therefore represents a small addition to background levels.
- Despite the uncertainties in the determination of risk from radiation exposure at low levels, radiation risks are probably well below those due to microbes and some chemicals in drinking-water.

9.1 Sources and health effects of radiation exposure

Environmental radiation originates from a number of naturally occurring and human-made sources. Radioactive materials occur naturally everywhere in the environment (e.g., uranium, thorium and potassium-40). By far the largest proportion of human exposure to radiation comes from natural sources – from external sources of radiation, including cosmic and terrestrial radiation, and from inhalation or ingestion of radioactive materials (Figure 9.1). The United Nations Scientific Committee

9. RADIOLOGICAL ASPECTS

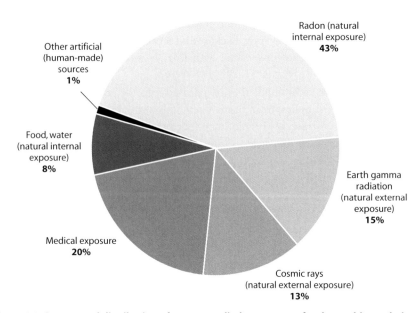

Figure 9.1 Sources and distribution of average radiation exposure for the world population

on the Effects of Atomic Radiation (UNSCEAR, 2000) has estimated that the global average annual human exposure from natural sources is 2.4 mSv/year (Table 9.1). Some sources (e.g., uranium) can be concentrated during extraction by mining and other industrial activities.

There are large local variations in human exposure to radiation, depending on a number of factors, such as height above sea level, the amount and type of radionuclides in the soil (terrestrial exposure), the composition of radionuclides in the air, food and drinking-water and the amount taken into the body via inhalation or ingestion. There are certain areas of the world, such as parts of the Kerala state in India and the Pocos del Caldas plateau in Brazil, where levels of background radiation are

Table 9.1 Average radiation dose from natural sources

Source	Worldwide average annual effective dose (mSv)	Typical range (mSv)
External exposure		
Cosmic rays	0.4	0.3–1.0
Terrestrial gamma rays[a]	0.5	0.3–0.6
Internal exposure		
Inhalation (mainly radon)	1.2	0.2–10[b]
Ingestion (food and drinking-water)	0.3	0.2–0.8
Total	**2.4**	**1–10**

[a] Terrestrial exposure is due to radionuclides in the soil and building materials.
[b] Dose from inhalation of radon may exceed 10 mSv/year in certain residential areas.
Source: UNSCEAR (2000).

relatively high. Levels of exposure for the general population in such areas may be up to 10 times higher than the average background level of 2.4 mSv given in Table 9.1. No deleterious health effects associated with this elevated radiation exposure have been detected.

Several radioactive compounds may be released into the environment, and hence into drinking-water supplies, from human activities and human-made sources (e.g., from medical or industrial use of radioactive sources). The worldwide per capita effective dose from diagnostic medical examination in 2000 was 0.4 mSv/year (typical range is 0.04–1.0 mSv/year, depending on level of health care). There is only a very small worldwide contribution from nuclear power production and nuclear weapons testing. The worldwide annual per capita effective dose from nuclear weapons testing in 2000 was estimated at 0.005 mSv; from the Chernobyl accident, 0.002 mSv; and from nuclear power production, 0.0002 mSv (UNSCEAR, 2000).

9.1.1 Radiation exposure through drinking-water

Radioactive constituents of drinking-water can result from:

— naturally occurring radioactive species (e.g., radionuclides of the thorium and uranium decay series in drinking-water sources), in particular radium-226/228 and a few others;
— technological processes involving naturally occurring radioactive materials (e.g., the mining and processing of mineral sands or phosphate fertilizer production);
— radionuclides discharged from nuclear fuel cycle facilities;
— manufactured radionuclides (produced and used in unsealed form), which might enter drinking-water supplies as a result of regular discharges and, in particular, in case of improper medical or industrial use and disposal of radioactive materials; such incidents are different from emergencies, which are outside the scope of these Guidelines; and
— past releases of radionuclides into the environment, including water sources.

The contribution of drinking-water to total exposure is typically very small and is due largely to naturally occurring radionuclides in the uranium and thorium decay series. Radionuclides from the nuclear fuel cycle and from medical and other uses of radioactive materials may, however, enter drinking-water supplies. The contributions from these sources are normally limited by regulatory control of the source or practice, and it is normally through this regulatory mechanism that remedial action should be taken in the event that such sources cause concern by contaminating drinking-water.

9.1.2 Radiation-induced health effects through drinking-water

There is evidence from both human and animal studies that radiation exposure at low to moderate doses may increase the long-term incidence of cancer. Animal studies in particular suggest that the rate of genetic malformations may be increased by radiation exposure.

No deleterious radiological health effects are expected from consumption of drinking-water if the concentrations of radionuclides are below the guidance levels (equivalent to a committed effective dose below 0.1 mSv/year).

Acute health effects of radiation, leading to reduced blood cell counts and, in very severe cases, death, occur at very high doses of exposure of the whole body or large part of the body (IAEA, 1998). Due to the low levels of radionuclides typically found in drinking-water supplies, acute health effects of radiation are not a concern for drinking-water supplies.

9.2 Units of radioactivity and radiation dose

The SI unit of radioactivity is the becquerel (Bq), where 1 Bq = 1 disintegration per second. Guidance levels for drinking-water are given as the activity of the radionuclide per litre, called the activity concentration (Bq/litre). The radiation dose resulting from ingestion of a radionuclide depends on a number of chemical and biological factors. These include the fraction of the intake that is absorbed from the gut, the organs or tissues to which the radionuclide is transported and the time during which the radionuclide remains in the organ or tissue before excretion. The nature of the radiation emitted on decay and the sensitivity of the irradiated organs or tissues to radiation must also be considered.

The absorbed dose refers to how much energy is deposited in material by the radiation. The SI unit for absorbed dose is the gray (Gy), where 1 Gy = 1 J/kg (joule per kilogram).

The equivalent dose is the product of the absorbed dose and a factor related to the particular type of radiation (depending on the ionizing capacity and density).

The effective dose of radiation received by a person is, in simple terms, the sum of the equivalent doses received by all tissues or organs, weighted for "tissue weighting factors." These reflect different sensitivities to radiation of different organs and tissues in the human body. The SI unit for the equivalent and effective dose is the sievert (Sv), where 1 Sv = 1 J/kg.

To reflect the persistence of radionuclides in the body once ingested, the committed effective dose is a measure of the total effective dose received over a lifetime (70 years) following intake of a radionuclide (internal exposure).

The term "dose" may be used as a general term to mean either absorbed dose (Gy) or effective dose (Sv), depending on the situation. For monitoring purposes, doses are determined from the activity concentration of the radionuclide in a given material. In the case of water, activity concentration is given in becquerels per litre (Bq/litre). This value can be related to an effective dose per year (mSv/year) using a dose coefficient (mSv/Bq) and the average annual consumption of water (litres/year).

The effective dose arising from the ingestion of a radioisotope in a particular chemical form can be estimated using a dose coefficient. Data for age-related dose coefficients for ingestion of radionuclides have been published by the ICRP and the International Atomic Energy Agency (IAEA). Table 9.2 shows the dose coefficients for

Table 9.2 Dose coefficients for ingestion of radionuclides by adult members of the public

Category	Radionuclide	Dose coefficient (mSv/Bq)
Natural uranium series	Uranium-238	4.5×10^{-5}
	Uranium-234	4.9×10^{-5}
	Thorium-230	2.1×10^{-4}
	Radium-226	2.8×10^{-4}
	Lead-210	6.9×10^{-4}
	Polonium-210	1.2×10^{-3}
Natural thorium series	Thorium-232	2.3×10^{-4}
	Radium-228	6.9×10^{-4}
	Thorium-228	7.2×10^{-5}
Fission products	Caesium-134	1.9×10^{-5}
	Caesium-137	1.3×10^{-5}
	Strontium-90	2.8×10^{-5}
	Iodine-131	2.2×10^{-5}
Other radionuclides	Tritium	1.8×10^{-8}
	Carbon-14	5.8×10^{-7}
	Plutonium-239	2.5×10^{-4}
	Americium-241	2.0×10^{-4}

naturally occurring radionuclides or those arising from human activities that might be found in drinking-water supplies (IAEA, 1996; ICRP, 1996).

9.3 Guidance levels for radionuclides in drinking-water

The guidance levels for radionuclides in drinking-water are presented in Table 9.3 for radionuclides originating from natural sources or discharged into the environment as the result of current or past activities. These levels also apply to radionuclides released due to nuclear accidents that occurred more than 1 year previously. The activity concentration values in Table 9.3 correspond to an RDL of 0.1 mSv/year from each radionuclide listed if their concentration in the drinking-water consumed during the year does not exceed these values. The associated risk estimate was given at the beginning of this chapter. However, for the first year immediately after an accident, generic action levels for foodstuffs apply as described in the International Basic Safety Standards (IAEA, 1996) and other relevant WHO and IAEA publications (WHO, 1988; IAEA, 1997, 1999).

The guidance levels for radionuclides in drinking-water were calculated by the following equation:

$$GL = IDC/(h_{ing} \cdot q)$$

where:

GL = guidance level of radionuclide in drinking-water (Bq/litre),
IDC = individual dose criterion, equal to 0.1 mSv/year for this calculation,
h_{ing} = dose coefficient for ingestion by adults (mSv/Bq),
q   = annual ingested volume of drinking-water, assumed to be 730 litres/year.

9. RADIOLOGICAL ASPECTS

Table 9.3 Guidance levels for radionuclides in drinking-water

Radionuclides	Guidance level (Bq/litre)[a]	Radionuclides	Guidance level (Bq/litre)[a]	Radionuclides	Guidance level (Bq/litre)[a]
^{3}H	10 000	^{93}Mo	100	^{140}La	100
^{7}Be	10 000	^{99}Mo	100	^{139}Ce	1000
^{14}C	100	^{96}Tc	100	^{141}Ce	100
^{22}Na	100	^{97}Tc	1000	^{143}Ce	100
^{32}P	100	^{97m}Tc	100	^{144}Ce	10
^{33}P	1 000	^{99}Tc	100	^{143}Pr	100
^{35}S	100	^{97}Ru	1000	^{147}Nd	100
^{36}Cl	100	^{103}Ru	100	^{147}Pm	1000
^{45}Ca	100	^{106}Ru	10	^{149}Pm	100
^{47}Ca	100	^{105}Rh	1000	^{151}Sm	1000
^{46}Sc	100	^{103}Pd	1000	^{153}Sm	100
^{47}Sc	100	^{105}Ag	100	^{152}Eu	100
^{48}Sc	100	^{110m}Ag	100	^{154}Eu	100
^{48}V	100	^{111}Ag	100	^{155}Eu	1000
^{51}Cr	10 000	^{109}Cd	100	^{153}Gd	1000
^{52}Mn	100	^{115}Cd	100	^{160}Tb	100
^{53}Mn	10 000	^{115m}Cd	100	^{169}Er	1000
^{54}Mn	100	^{111}In	1000	^{171}Tm	1000
^{55}Fe	1 000	^{114m}In	100	^{175}Yb	1000
^{59}Fe	100	^{113}Sn	100	^{182}Ta	100
^{56}Co	100	^{125}Sn	100	^{181}W	1000
^{57}Co	1 000	^{122}Sb	100	^{185}W	1000
^{58}Co	100	^{124}Sb	100	^{186}Re	100
^{60}Co	100	^{125}Sb	100	^{185}Os	100
^{59}Ni	1 000	^{123m}Te	100	^{191}Os	100
^{63}Ni	1 000	^{127}Te	1000	^{193}Os	100
^{65}Zn	100	^{127m}Te	100	^{190}Ir	100
^{71}Ge	10 000	^{129}Te	1000	^{192}Ir	100
^{73}As	1 000	^{129m}Te	100	^{191}Pt	1000
^{74}As	100	^{131}Te	1000	^{193m}Pt	1000
^{76}As	100	^{131m}Te	100	^{198}Au	100
^{77}As	1 000	^{132}Te	100	^{199}Au	1000
^{75}Se	100	^{125}I	10	^{197}Hg	1000
^{82}Br	100	^{126}I	10	^{203}Hg	100
^{86}Rb	100	^{129}I	1000	^{200}Tl	1000
^{85}Sr	100	^{131}I	10	^{201}Tl	1000
^{89}Sr	100	^{129}Cs	1000	^{202}Tl	1000
^{90}Sr	10	^{131}Cs	1000	^{204}Tl	100
^{90}Y	100	^{132}Cs	100	^{203}Pb	1000
^{91}Y	100	^{134}Cs	10	^{206}Bi	100
^{93}Zr	100	^{135}Cs	100	^{207}Bi	100
^{95}Zr	100	^{136}Cs	100	^{210}Bi[b]	100
^{93m}Nb	1 000	^{137}Cs	10	^{210}Pb[b]	0.1
^{94}Nb	100	^{131}Ba	1000	^{210}Po[b]	0.1
^{95}Nb	100	^{140}Ba	100	^{223}Ra[b]	1
^{224}Ra[b]	1	^{235}U[b]	1	^{242}Cm	10
^{225}Ra	1	^{236}U[b]	1	^{243}Cm	1
^{226}Ra[b]	1	^{237}U	100	^{244}Cm	1
^{228}Ra[b]	0.1	^{238}U[b,c]	10	^{245}Cm	1

continued

Table 9.3 Continued

Radionuclides	Guidance level (Bq/litre)	Radionuclides	Guidance level (Bq/litre)	Radionuclides	Guidance level (Bq/litre)
^{227}Th[b]	10	^{237}Np	1	^{246}Cm	1
^{228}Th[b]	1	^{239}Np	100	^{247}Cm	1
^{229}Th	0.1	^{236}Pu	1	^{248}Cm	0.1
^{230}Th[b]	1	^{237}Pu	1000	^{249}Bk	100
^{231}Th[b]	1 000	^{238}Pu	1	^{246}Cf	100
^{232}Th[b]	1	^{239}Pu	1	^{248}Cf	10
^{234}Th[b]	100	^{240}Pu	1	^{249}Cf	1
^{230}Pa	100	^{241}Pu	10	^{250}Cf	1
^{231}Pa[b]	0.1	^{242}Pu	1	^{251}Cf	1
^{233}Pa	100	^{244}Pu	1	^{252}Cf	1
^{230}U	1	^{241}Am	1	^{253}Cf	100
^{231}U	1 000	^{242}Am	1000	^{254}Cf	1
^{232}U	1	^{242m}Am	1	^{253}Es	10
^{233}U	1	^{243}Am	1	^{254}Es	10
^{234}U[b]	10			^{254m}Es	100

[a] Guidance levels are rounded according to averaging the log scale values (to 10^n if the calculated value was below 3×10^n and above $3 \times 10^{n-1}$).
[b] Natural radionuclides.
[c] The provisional guideline value for uranium in drinking-water is 15 µg/litre based on its chemical toxicity for the kidney (see section 8.5).

The higher age-dependent dose coefficients calculated for children (accounting for the higher uptake and/or metabolic rates) do not lead to significantly higher doses due to the lower mean volume of drinking-water consumed by infants and children. Consequently, the recommended RDL of committed effective dose of 0.1 mSv/year from 1 year's consumption of drinking-water applies independently of age.

9.4 Monitoring and assessment for dissolved radionuclides

9.4.1 Screening of drinking-water supplies

The process of identifying individual radioactive species and determining their concentration requires sophisticated and expensive analysis, which is normally not justified, because the concentrations of radionuclides in most circumstances are very low. A more practical approach is to use a screening procedure, where the total radioactivity present in the form of alpha and beta radiation is first determined, without regard to the identity of specific radionuclides.

Screening levels for drinking-water below which no further action is required are 0.5 Bq/litre for gross alpha activity and 1 Bq/litre for gross beta activity. The gross beta activity screening level was published in the second edition of the Guidelines and, in the worse case (radium-222), would lead to a dose close to the guidance RDL of 0.1 mSv/year. The screening level for gross alpha activity is 0.5 Bq/litre (instead of the former 0.1 Bq/litre), as this activity concentration reflects values nearer the radionuclide-specific guidance RDL.

9.4.2 Strategy for assessing drinking-water

If either of the screening levels is exceeded, then the specific radionuclides producing this activity should be identified and their individual activity concentrations measured. From these data, an estimate of committed effective dose for each radionuclide should be made and the sum of these doses determined. If the following additive formula is satisfied, no further action is required:

$$\sum_i \frac{C_i}{GL_i} \leq 1$$

where:

C_i = the measured activity concentration of radionuclide i, and
GL_i = the guidance level value (see Table 9.3) of radionuclide i that, at an intake of 2 litres/day for 1 year, will result in a committed effective dose of 0.1 mSv/year.

Where the sum exceeds unity for a single sample, the RDL of 0.1 mSv would be exceeded only if the exposure to the same measured concentrations were to continue for a full year. *Hence, such a sample does not in itself imply that the water is unsuitable for consumption* but should be regarded as an indication that further investigation, including additional sampling, is needed. Gross beta and gross alpha activity screening has to be repeated first, then radionuclide-specific analysis conducted only if subsequently measured gross values exceed the recommended practical screening values (1 Bq/litre and 0.5 Bq/litre, respectively).

The application of these recommendations is summarized in Figure 9.2.

The gross beta measurement includes a contribution from potassium-40, a beta emitter that occurs naturally in a fixed ratio to stable potassium. Potassium is an essential element for humans and is absorbed mainly from ingested food. Potassium-40 does not accumulate in the body but is maintained at a constant level independent of intake. The contribution of potassium-40 to beta activity should therefore be subtracted following a separate determination of total potassium. The specific activity of potassium-40 is 30.7 Bq/g of potassium. However, not all the radiation from potassium-40 appears as beta activity. The beta activity of potassium-40 is 27.6 Bq/g of stable potassium, which is the factor that should be used to calculate the beta activity due to potassium-40.

9.4.3 Remedial measures

If the RDL of 0.1 mSv/year is being exceeded on aggregate, then the options available to the competent authority to reduce the dose should be examined. Where remedial measures are contemplated, any strategy considered should first be justified (in the sense that it achieves a net benefit) and then optimized in accordance with the recommendations of ICRP (1989, 1991) in order to produce the maximum net benefit.

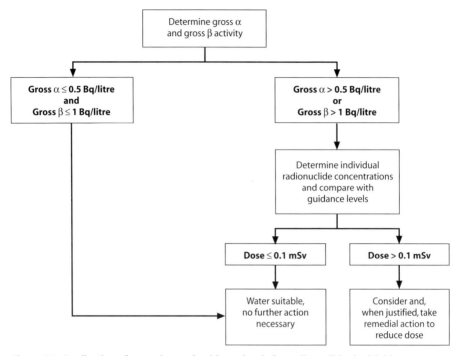

Figure 9.2 Application of screening and guidance levels for radionuclides in drinking-water

9.5 Radon
9.5.1 Radon in air and water
The largest fraction of natural radiation exposure comes from radon, a radioactive gas (see Table 9.1 and Figure 9.1), due to decay of radium contained in rocks and soil as part of the uranium radionuclide chain. The term radon in general refers mostly to radon-222. Radon is present virtually everywhere on Earth, but particularly in the air over land and in buildings.

Underground rock containing natural uranium continuously releases radon into water in contact with it (groundwater). Radon is readily released from surface water; consequently, groundwater normally has much higher concentrations of radon than surface water. The average concentration of radon is usually less than 0.4 Bq/litre in public water supplies derived from surface waters and about 20 Bq/litre from groundwater sources. However, some wells have been identified with higher concentrations, up to 400 times the average, and in rare cases exceeding 10 kBq/litre.

For assessing the dose from radon ingestion, it is important that water processing technology before consumption is taken into account. Moreover, the use of some groundwater supplies for general domestic purposes will increase the levels of radon in the air, thus increasing the dose from inhalation. This dose depends markedly on the form of domestic usage and housing construction (NCRP, 1989). The amount and form of water intake, the other domestic uses of water and the construction of houses vary widely throughout the world.

UNSCEAR (2000) refers to a US NAS (1999) report and calculates the "average doses from radon in drinking water to be as low as 0.025 mSv/year via inhalation and 0.002 mSv/year from ingestion" compared with the inhalation dose of 1.1 mSv/year from radon and its decay products in air.

9.5.2 Risk
One report estimates that 12% of lung cancer deaths in the USA are linked to radon (radon-222 and its short-lived decay products) in indoor air (US NAS, 1999). Thus, radon causes about 19 000 deaths (in the range of 15 000–22 000) due to lung cancer annually out of a total of about 160 000 deaths from lung cancer, which are mainly as a result of smoking tobacco (US NRC, 1999).

US NAS (1999) reports an approximately 100-fold smaller risk from exposure to radon in drinking-water (i.e., 183 deaths each year). In addition to the 19 000 deaths from lung cancer caused by radon in indoor air, a further 160 were estimated to result from inhaling radon that was emitted from water used in the home. For comparison, about 700 lung cancer deaths each year were attributed to exposure to natural levels of radon while people are outdoors.

The US NAS (1999) also assessed that the risk of stomach cancer caused by drinking-water that contains dissolved radon is extremely small, with the probability of about 20 deaths annually compared with the 13 000 deaths from stomach cancer that arise each year from other causes in the USA.

9.5.3 Guidance on radon in drinking-water supplies
Controls should be implemented if the radon concentration of drinking-water for public water supplies exceeds 100 Bq/litre. Any new drinking-water supply should be tested prior to being used for general consumption. If the radon concentration exceeds 100 Bq/litre, treatment of the water source should be undertaken to reduce the radon levels to well below 100 Bq/litre. If there are significant amounts of radon-producing minerals around the water source, then it may be appropriate for larger drinking-water supplies to test for radon concentration periodically – for example, every 5 years.

9.6 Sampling, analysis and reporting
9.6.1 Measuring gross alpha and gross beta activity concentrations
To analyse drinking-water for gross alpha and gross beta activities (excluding radon), the most common approach is to evaporate a known volume of the sample to dryness and measure the activity of the residue. As alpha radiation is easily absorbed within a thin layer of solid material, the reliability and sensitivity of the method for alpha determination may be reduced in samples with a high TDS content.

Table 9.4 **Methods for the analysis of gross alpha and gross beta activities in drinking-water**

Method, reference	Technique	Detection limit	Application
International Organization for Standardization: ISO-9695 (for gross beta) ISO-9696 (gross alpha) (ISO, 1991a, 1991b)	Evaporation	0.02–0.1 Bq/litre	Groundwater with TDS greater than 0.1 g/litre
American Public Health Association (APHA, 1998)	Co-precipitation	0.02 Bq/litre	Surface water and groundwater (TDS is not a factor)

Where possible, standardized methods should be used to determine concentrations of gross alpha and gross beta activities. Three procedures for this analysis are listed in Table 9.4.

The determination of gross beta activity using the evaporation method includes the contribution from potassium-40. An additional analysis of total potassium is therefore required if the gross beta screening value is exceeded.

The co-precipitation technique (APHA, 1998) excludes the contribution due to potassium-40; therefore, determination of total potassium is not necessary. This method is not applicable to assessment of water samples containing certain fission products, such as caesium-137. However, under normal circumstances, concentrations of fission products in drinking-water supplies are extremely low.

9.6.2 *Measuring potassium-40*

It is impractical to use a radioactive measurement technique to determine the concentration of potassium-40 in a water sample due to the lack of sensitivity in gamma-ray analysis and the difficulty of chemically isolating the radionuclide from solution. Because of the fixed ratio between potassium-40 and stable potassium, chemical analysis for potassium is recommended. A measurement sensitivity of 1 mg/litre for potassium is adequate, and suitable techniques that can readily achieve this are atomic absorption spectrophotometry and specific ion analysis. The activity due to potassium-40 can then be calculated using a factor of 27.6 Bq of beta activity per gram of total potassium.

9.6.3 *Measuring radon*

There are difficulties in deriving activity concentrations of radon-222 in drinking-water arising from the ease with which radon is released from water during handling. Stirring and transferring water from one container to another will liberate dissolved radon. According to the widely used Pylon technique (Pylon, 1989, 2003), detection of radon in drinking-water is performed using a water degassing unit and Lucas scintillation chambers. Water that has been left to stand will have reduced radon activity, and boiling will remove radon completely.

9. RADIOLOGICAL ASPECTS

9.6.4 Sampling

New water sources should be sampled (e.g., every 3 months for the first 12 months) to determine their suitability for drinking-water supply before design and construction to characterize the radiological quality of the water supply and to assess any seasonal variation in radionuclide concentrations. This should include analysis for radon and radon daughters.

Once measurements indicate the normal range of the supply, then the sampling frequency can be reduced to, for example, annually or every 5 years. However, if sources of potential radionuclide contamination exist nearby (e.g., mining activity or nuclear reactors), then sampling should be more frequent. Less significant surface and underground water supplies can be sampled less frequently.

Levels of radon and radon daughters in groundwater supplies are usually stable over time. Monitoring of water for radon and its daughters can therefore be relatively infrequent. Knowledge of the geology of the area should be considered in determining whether the source is likely to contain significant concentrations of radon and radon daughters. An additional risk factor would be the presence of mining in the vicinity; in such circumstances, more frequent monitoring may be appropriate.

Guidance on assessing water quality, sampling techniques and programmes and the preservation and handling of samples is given in the Australian and New Zealand Standard (AS, 1998).

9.6.5 Reporting of results

The analytical results for each sample should contain the following information:

- sample identifying code or information;
- reference date and time for the reported results (e.g., sample collection date);
- identification of the standard analytical method used or a brief description of any non-standard method used;
- identification of the radionuclide(s) or type and total radioactivity determined;
- measurement-based concentration or activity value calculated using the appropriate blank for each radionuclide;
- estimates of the counting uncertainty and total projected uncertainty; and
- minimum detectable concentration for each radionuclide or parameter analysed.

The estimate of total projected uncertainty of the reported result should include the contributions from all the parameters within the analytical method (i.e., counting and other random and systematic uncertainties or errors).

10
Acceptability aspects

The most undesirable constituents of drinking-water are those capable of having a direct adverse impact on public health. Many of these are described in other chapters of these Guidelines.

To a large extent, consumers have no means of judging the safety of their drinking-water themselves, but their attitude towards their drinking-water supply and their drinking-water suppliers will be affected to a considerable extent by the aspects of water quality that they are able to perceive with their own senses. It is natural for consumers to regard with suspicion water that appears dirty or discoloured or that has an unpleasant taste or smell, even though these characteristics may not in themselves be of direct consequence to health.

The provision of drinking-water that is not only safe but also acceptable in appearance, taste and odour is of high priority. Water that is aesthetically unacceptable will undermine the confidence of consumers, lead to complaints and, more importantly, possibly lead to the use of water from sources that are less safe.

> The appearance, taste and odour of drinking-water should be acceptable to the consumer.

It is important to consider whether existing or proposed water treatment and distribution practices can affect the acceptability of drinking-water. For example, a change in disinfection practice may generate an odorous compound such as trichloramine in the treated water. Other effects may be indirect, such as the disturbance of internal pipe deposits and biofilms when changing between or blending waters from different sources in distribution systems.

The acceptability of drinking-water to consumers is subjective and can be influenced by many different constituents. The concentration at which constituents are objectionable to consumers is variable and dependent on individual and local factors, including the quality of the water to which the community is accustomed and a variety of social, environmental and cultural considerations. Guideline values have not been established for constituents influencing water quality that have no direct link to adverse health impacts.

10. ACCEPTABILITY ASPECTS

In the summaries in this chapter and chapter 12, reference is made to levels likely to give rise to complaints from consumers. These are not precise numbers, and problems may occur at lower or higher levels, depending on individual and local circumstances.

It is not normally appropriate to directly regulate or monitor substances of health concern whose effects on the acceptability of water would normally lead to rejection of the water at concentrations significantly lower than those of concern for health; rather, these substances may be addressed through a general requirement that water be acceptable to the majority of consumers. For such substances, a health-based summary statement and guideline value are derived in these Guidelines in the usual way. In the summary statement, this is explained, and information on acceptability is described. In the tables of guideline values (see chapter 8 and Annex 4), the health-based guideline value is designated with a "C," with a footnote explaining that while the substance is of health significance, water would normally be rejected by consumers at concentrations well below the health-based guideline value. Monitoring of such substances should be undertaken in response to consumer complaints.

There are other water constituents that are of no direct consequence to health at the concentrations at which they normally occur in water but which nevertheless may be objectionable to consumers for various reasons.

10.1 Taste, odour and appearance

Taste and odour can originate from natural inorganic and organic chemical contaminants and biological sources or processes (e.g., aquatic microorganisms), from contamination by synthetic chemicals, from corrosion or as a result of water treatment (e.g., chlorination). Taste and odour may also develop during storage and distribution due to microbial activity.

Taste and odour in drinking-water may be indicative of some form of pollution or of a malfunction during water treatment or distribution. It may therefore be an indication of the presence of potentially harmful substances. The cause should be investigated and the appropriate health authorities should be consulted, particularly if there is a sudden or substantial change.

Colour, cloudiness, particulate matter and visible organisms may also be noticed by consumers and may create concerns about the quality and acceptability of a drinking-water supply.

10.1.1 Biologically derived contaminants

There are a number of diverse organisms that may have no public health significance but which are undesirable because they produce taste and odour. As well as affecting the acceptability of the water, they indicate that water treatment and/or the state of maintenance and repair of the distribution system are insufficient.

Actinomycetes and fungi

Actinomycetes and fungi can be abundant in surface water sources, including reservoirs, and they also can grow on unsuitable materials in the water supply distribution systems, such as rubber. They can give rise to geosmin, 2-methyl isoborneol and other substances, resulting in objectionable tastes and odours in the drinking-water.

Animal life[1]

Invertebrate animals are naturally present in many water resources used as sources for the supply of drinking-water and often infest shallow, open wells. Small numbers of invertebrates may also pass through water treatment works where the barriers to particulate matter are not completely effective and colonize the distribution system. Their motility may enable them and their larvae to penetrate filters at the treatment works and vents on storage reservoirs.

The types of animal concerned can be considered, for control purposes, as belonging to two groups. First, there are free-swimming organisms in the water itself or on water surfaces, such as the crustaceans *Gammarus pulex* (freshwater shrimp), *Crangonyx pseudogracilis*, *Cyclops* spp. and *Chydorus sphaericus*. Second, there are other animals that either move along surfaces or are anchored to them (e.g., water louse *Asellus aquaticus*, snails, zebra mussel *Dreissena polymorpha*, other bivalve molluscs and the bryozoan *Plumatella* sp.) or inhabit slimes (e.g., *Nais* spp., nematodes and the larvae of chironomids). In warm weather, slow sand filters can sometimes discharge the larvae of gnats (*Chironomus* and *Culex* spp.) into the water.

Many of these animals can survive, deriving food from bacteria, algae and protozoa in the water or present on slimes on pipe and tank surfaces. Few, if any, water distribution systems are completely free of animals. However, the density and composition of animal populations vary widely, from heavy infestations, including readily visible species that are objectionable to consumers, to sparse occurrences of microscopic species.

The presence of animals has largely been regarded by piped drinking-water suppliers in temperate regions as an acceptability problem, either directly or through their association with discoloured water. In tropical and subtropical countries, on the other hand, there are species of aquatic animal that act as secondary hosts for parasites. For example, the small crustacean *Cyclops* is the intermediate host of the guinea worm *Dracunculus medinensis* (see sections 7.1.1 and 11.4). However, there is no evidence that guinea worm transmission occurs from piped drinking-water supplies. The presence of animals in drinking-water, especially if visible, raises consumer concern about the quality of the drinking-water supply and should be controlled.

Penetration of waterworks and mains is more likely to be a problem when low-quality raw waters are abstracted and high-rate filtration processes are used. Pre-chlorination assists in destroying animal life and in its removal by filtration.

---

[1] The section was drawn largely from Evins (2004).

10. ACCEPTABILITY ASPECTS

Production of high-quality water, maintenance of chlorine residuals in the distribution system and the regular cleaning of water mains by flushing or swabbing will usually control infestation.

Treatment of invertebrate infestations in piped distribution systems is discussed in detail in chapter 6 of the supporting document *Safe, Piped Water* (section 1.3).

Cyanobacteria and algae

Blooms of cyanobacteria and other algae in reservoirs and in river waters may impede coagulation and filtration, causing coloration and turbidity of water after filtration. They can also give rise to geosmin, 2-methyl isoborneol and other chemicals, which have taste thresholds in drinking-water of a few nanograms per litre. Some cyanobacterial products – cyanotoxins – are also of direct health significance (see section 8.5.6).

Iron bacteria

In waters containing ferrous and manganous salts, oxidation by iron bacteria (or by exposure to air) may cause rust-coloured deposits on the walls of tanks, pipes and channels and carry-over of deposits into the water.

10.1.2 Chemically derived contaminants

Aluminium

Naturally occurring aluminium as well as aluminium salts used as coagulants in drinking-water treatment are the most common sources of aluminium in drinking-water. The presence of aluminium at concentrations in excess of 0.1–0.2 mg/litre often leads to consumer complaints as a result of deposition of aluminium hydroxide floc in distribution systems and the exacerbation of discoloration of water by iron. It is therefore important to optimize treatment processes in order to minimize any residual aluminium entering the supply. Under good operating conditions, aluminium concentrations of less than 0.1 mg/litre are achievable in many circumstances. Available evidence does not support the derivation of a health-based guideline value for aluminium in drinking-water (see sections 8.5.4 and 12.5).

Ammonia

The threshold odour concentration of ammonia at alkaline pH is approximately 1.5 mg/litre, and a taste threshold of 35 mg/litre has been proposed for the ammonium cation. Ammonia is not of direct relevance to health at these levels, and no health-based guideline value has been proposed (see sections 8.5.3 and 12.6).

Chloride

High concentrations of chloride give a salty taste to water and beverages. Taste thresholds for the chloride anion depend on the associated cation and are in the range of 200–300 mg/litre for sodium, potassium and calcium chloride. Concentrations in excess of 250 mg/litre are increasingly likely to be detected by taste, but some consumers may become accustomed to low levels of chloride-induced taste. No health-based guideline value is proposed for chloride in drinking-water (see sections 8.5.4 and 12.22).

Chlorine

Most individuals are able to taste or smell chlorine in drinking-water at concentrations well below 5 mg/litre, and some at levels as low as 0.3 mg/litre. At a residual free chlorine concentration of between 0.6 and 1.0 mg/litre, there is an increasing likelihood that some consumers may object to the taste. The taste threshold for chlorine is below the health-based guideline value (see sections 8.5.4 and 12.23).

Chlorophenols

Chlorophenols generally have very low taste and odour thresholds. The taste thresholds in water for 2-chlorophenol, 2,4-dichlorophenol and 2,4,6-trichlorophenol are 0.1, 0.3 and 2 µg/litre, respectively. Odour thresholds are 10, 40 and 300 µg/litre, respectively. If water containing 2,4,6-trichlorophenol is free from taste, it is unlikely to present a significant risk to health (see section 12.26). Microorganisms in distribution systems may sometimes methylate chlorophenols to produce chlorinated anisoles, for which the odour threshold is considerably lower.

Colour

Drinking-water should ideally have no visible colour. Colour in drinking-water is usually due to the presence of coloured organic matter (primarily humic and fulvic acids) associated with the humus fraction of soil. Colour is also strongly influenced by the presence of iron and other metals, either as natural impurities or as corrosion products. It may also result from the contamination of the water source with industrial effluents and may be the first indication of a hazardous situation. The source of colour in a drinking-water supply should be investigated, particularly if a substantial change has taken place.

Most people can detect colours above 15 true colour units (TCU) in a glass of water. Levels of colour below 15 TCU are usually acceptable to consumers, but acceptability may vary. High colour could also indicate a high propensity to produce by-products from disinfection processes. No health-based guideline value is proposed for colour in drinking-water.

Copper

Copper in a drinking-water supply usually arises from the corrosive action of water leaching copper from copper pipes. Concentrations can vary significantly with the period of time the water has been standing in contact with the pipes; for example, first-draw water would be expected to have a higher copper concentration than a fully flushed sample. High concentrations can interfere with the intended domestic uses of the water. Copper in drinking-water may increase the corrosion of galvanized iron and steel fittings. Staining of laundry and sanitary ware occurs at copper concentrations above 1 mg/litre. At levels above 5 mg/litre, copper also imparts a colour and an undesirable bitter taste to water. Although copper can give rise to taste, it should be acceptable at the health-based guideline value (see sections 8.5.4 and 12.31).

Dichlorobenzenes

Odour thresholds of 2–10 and 0.3–30 µg/litre have been reported for 1,2- and 1,4-dichlorobenzene, respectively. Taste thresholds of 1 and 6 µg/litre have been reported for 1,2- and 1,4-dichlorobenzene, respectively. The health-based guideline values derived for 1,2- and 1,4-dichlorobenzene (see sections 8.5.4 and 12.42) far exceed the lowest reported taste and odour thresholds for these compounds.

Dissolved oxygen

The dissolved oxygen content of water is influenced by the source, raw water temperature, treatment and chemical or biological processes taking place in the distribution system. Depletion of dissolved oxygen in water supplies can encourage the microbial reduction of nitrate to nitrite and sulfate to sulfide. It can also cause an increase in the concentration of ferrous iron in solution, with subsequent discoloration at the tap when the water is aerated. No health-based guideline value is recommended.

Ethylbenzene

Ethylbenzene has an aromatic odour; the reported odour threshold in water ranges from 2 to 130 µg/litre. The lowest reported odour threshold is 100-fold lower than the health-based guideline value (see sections 8.5.4 and 12.60). The taste threshold ranges from 72 to 200 µg/litre.

Hardness

Hardness caused by calcium and magnesium is usually indicated by precipitation of soap scum and the need for excess use of soap to achieve cleaning. Public acceptability of the degree of hardness of water may vary considerably from one community to another, depending on local conditions. In particular, consumers are likely to notice changes in hardness.

The taste threshold for the calcium ion is in the range of 100–300 mg/litre, depending on the associated anion, and the taste threshold for magnesium is probably lower than that for calcium. In some instances, consumers tolerate water hardness in excess of 500 mg/litre.

Depending on the interaction of other factors, such as pH and alkalinity, water with a hardness above approximately 200 mg/litre may cause scale deposition in the treatment works, distribution system and pipework and tanks within buildings. It will also result in excessive soap consumption and subsequent "scum" formation. On heating, hard waters form deposits of calcium carbonate scale. Soft water, with a hardness of less than 100 mg/litre, may, on the other hand, have a low buffering capacity and so be more corrosive for water pipes.

No health-based guideline value is proposed for hardness in drinking-water.

Hydrogen sulfide

The taste and odour thresholds of hydrogen sulfide in water are estimated to be between 0.05 and 0.1 mg/litre. The "rotten eggs" odour of hydrogen sulfide is particularly noticeable in some groundwaters and in stagnant drinking-water in the distribution system, as a result of oxygen depletion and the subsequent reduction of sulfate by bacterial activity.

Sulfide is oxidized rapidly to sulfate in well aerated or chlorinated water, and hydrogen sulfide levels in oxygenated water supplies are normally very low. The presence of hydrogen sulfide in drinking-water can be easily detected by the consumer and requires immediate corrective action. It is unlikely that a person could consume a harmful dose of hydrogen sulfide from drinking-water, and hence a health-based guideline value has not been derived for this compound (see sections 8.5.1 and 12.71).

Iron

Anaerobic groundwater may contain ferrous iron at concentrations of up to several milligrams per litre without discoloration or turbidity in the water when directly pumped from a well. On exposure to the atmosphere, however, the ferrous iron oxidizes to ferric iron, giving an objectionable reddish-brown colour to the water.

Iron also promotes the growth of "iron bacteria," which derive their energy from the oxidation of ferrous iron to ferric iron and in the process deposit a slimy coating on the piping. At levels above 0.3 mg/litre, iron stains laundry and plumbing fixtures. There is usually no noticeable taste at iron concentrations below 0.3 mg/litre, although turbidity and colour may develop. No health-based guideline value is proposed for iron (see sections 8.5.4 and 12.74).

Manganese

At levels exceeding 0.1 mg/litre, manganese in water supplies causes an undesirable taste in beverages and stains sanitary ware and laundry. The presence of manganese in drinking-water, like that of iron, may lead to the accumulation of deposits in the distribution system. Concentrations below 0.1 mg/litre are usually acceptable to consumers. Even at a concentration of 0.2 mg/litre, manganese will often form a coating on pipes, which may slough off as a black precipitate. The health-based guideline value for manganese is 4 times higher than this acceptability threshold of 0.1 mg/litre (see sections 8.5.1 and 12.79).

Monochloramine

Most individuals are able to taste or smell monochloramine, generated from the reaction of chlorine with ammonia, in drinking-water at concentrations well below

5 mg/litre, and some at levels as low as 0.3 mg/litre. The taste threshold for monochloramine is below the health-based guideline value (see sections 8.5.4 and 12.89).

Monochlorobenzene

Taste and odour thresholds of 10–20 µg/litre and odour thresholds ranging from 40 to 120 µg/litre have been reported for monochlorobenzene. A health-based guideline value has not been derived for monochlorobenzene (see sections 8.5.4 and 12.91), although the health-based value that could be derived far exceeds the lowest reported taste and odour threshold in water.

Petroleum oils

Petroleum oils can give rise to the presence of a number of low molecular weight hydrocarbons that have low odour thresholds in drinking-water. Although there are no formal data, experience indicates that these may have lower odour thresholds when several are present as a mixture. Benzene, toluene, ethylbenzene and xylenes are considered individually in this section, as health-based guideline values have been derived for these chemicals. However, a number of other hydrocarbons, particularly alkylbenzenes such as trimethylbenzene, may give rise to a very unpleasant "diesel-like" odour at concentrations of a few micrograms per litre.

pH and corrosion

Although pH usually has no direct impact on consumers, it is one of the most important operational water quality parameters. Careful attention to pH control is necessary at all stages of water treatment to ensure satisfactory water clarification and disinfection (see the supporting document *Safe, Piped Water*; section 1.3). For effective disinfection with chlorine, the pH should preferably be less than 8; however, lower-pH water is likely to be corrosive. The pH of the water entering the distribution system must be controlled to minimize the corrosion of water mains and pipes in household water systems. Alkalinity and calcium management also contribute to the stability of water and control its aggressiveness to pipe and appliance. Failure to minimize corrosion can result in the contamination of drinking-water and in adverse effects on its taste and appearance. The optimum pH required will vary in different supplies according to the composition of the water and the nature of the construction materials used in the distribution system, but it is usually in the range 6.5–8. Extreme values of pH can result from accidental spills, treatment breakdowns and insufficiently cured cement mortar pipe linings or cement mortar linings applied when the alkalinity of the water is low. No health-based guideline value has been proposed for pH (see sections 8.5.1 and 12.100).

Sodium

The taste threshold concentration of sodium in water depends on the associated anion and the temperature of the solution. At room temperature, the average taste threshold for sodium is about 200 mg/litre. No health-based guideline value has been derived (see sections 8.5.1 and 12.108).

Styrene

Styrene has a sweet odour, and reported odour thresholds for styrene in water range from 4 to 2600 µg/litre, depending on temperature. Styrene may therefore be detected in water at concentrations below its health-based guideline value (see sections 8.5.2 and 12.109).

Sulfate

The presence of sulfate in drinking-water can cause noticeable taste, and very high levels might cause a laxative effect in unaccustomed consumers. Taste impairment varies with the nature of the associated cation; taste thresholds have been found to range from 250 mg/litre for sodium sulfate to 1000 mg/litre for calcium sulfate. It is generally considered that taste impairment is minimal at levels below 250 mg/litre. No health-based guideline value has been derived for sulfate (see sections 8.5.1 and 12.110).

Synthetic detergents

In many countries, persistent types of anionic detergent have been replaced by others that are more easily biodegraded, and hence the levels found in water sources have decreased substantially. The concentration of detergents in drinking-water should not be allowed to reach levels giving rise to either foaming or taste problems. The presence of any detergent may indicate sanitary contamination of source water.

Toluene

Toluene has a sweet, pungent, benzene-like odour. The reported taste threshold ranges from 40 to 120 µg/litre. The reported odour threshold for toluene in water ranges from 24 to 170 µg/litre. Toluene may therefore affect the acceptability of water at concentrations below its health-based guideline value (see sections 8.5.2 and 12.114).

Total dissolved solids

The palatability of water with a TDS level of less than 600 mg/litre is generally considered to be good; drinking-water becomes significantly and increasingly unpalatable at TDS levels greater than about 1000 mg/litre. The presence of high levels of TDS may also be objectionable to consumers, owing to excessive scaling in water pipes, heaters, boilers and household appliances. No health-based guideline value for TDS has been proposed (see sections 8.5.1 and 12.115).

Trichlorobenzenes

Odour thresholds of 10, 5–30 and 50 µg/litre have been reported for 1,2,3-, 1,2,4- and 1,3,5-trichlorobenzene, respectively. A taste and odour threshold concentration of

30 µg/litre has been reported for 1,2,4-trichlorobenzene. A health-based guideline value was not derived for trichlorobenzenes, although the health-based value that could be derived (see sections 8.5.2 and 12.117) exceeds the lowest reported odour threshold in water of 5 µg/litre.

Turbidity

Turbidity in drinking-water is caused by particulate matter that may be present from source water as a consequence of inadequate filtration or from resuspension of sediment in the distribution system. It may also be due to the presence of inorganic particulate matter in some groundwaters or sloughing of biofilm within the distribution system. The appearance of water with a turbidity of less than 5 NTU is usually acceptable to consumers, although this may vary with local circumstances.

Particulates can protect microorganisms from the effects of disinfection and can stimulate bacterial growth. In all cases where water is disinfected, the turbidity must be low so that disinfection can be effective. The impact of turbidity on disinfection efficiency is discussed in more detail in section 4.1.

Turbidity is also an important operational parameter in process control and can indicate problems with treatment processes, particularly coagulation/sedimentation and filtration.

No health-based guideline value for turbidity has been proposed; ideally, however, median turbidity should be below 0.1 NTU for effective disinfection, and changes in turbidity are an important process control parameter.

Xylenes

Xylene concentrations in the range of 300 µg/litre produce a detectable taste and odour. The odour threshold for xylene isomers in water has been reported to range from 20 to 1800 µg/litre. The lowest odour threshold is well below the health-based guideline value derived for the compound (see sections 8.5.2 and 12.124).

Zinc

Zinc imparts an undesirable astringent taste to water at a taste threshold concentration of about 4 mg/litre (as zinc sulfate). Water containing zinc at concentrations in excess of 3–5 mg/litre may appear opalescent and develop a greasy film on boiling. Although drinking-water seldom contains zinc at concentrations above 0.1 mg/litre, levels in tap water can be considerably higher because of the zinc used in older galvanized plumbing materials. No health-based guideline value has been proposed for zinc in drinking-water (see sections 8.5.4 and 12.125).

10.1.3 Treatment of taste, odour and appearance problems

The following water treatment techniques are generally effective in removing organic chemicals that cause tastes and odours:

— aeration (see section 8.4.6);
— activated carbon (GAC or PAC) (see section 8.4.8); and
— ozonation (see section 8.4.3).

Tastes and odours caused by disinfectants and DBPs are best controlled through careful operation of the disinfection process. In principle, they can be removed by activated carbon.

Manganese can be removed by chlorination followed by filtration. Techniques for removing hydrogen sulfide include aeration, GAC, filtration and oxidation. Ammonia can be removed by biological nitrification. Precipitation softening or cation exchange can reduce hardness. Other taste- and odour-causing inorganic chemicals (e.g., chloride and sulfate) are generally not amenable to treatment (see the supporting document *Chemical Safety of Drinking-water*; section 1.3).

10.2 Temperature

Cool water is generally more palatable than warm water, and temperature will impact on the acceptability of a number of other inorganic constituents and chemical contaminants that may affect taste. High water temperature enhances the growth of microorganisms and may increase taste, odour, colour and corrosion problems.

11
Microbial fact sheets

Fact sheets are provided on potential waterborne pathogens as well as on indicator and index microorganisms.

The potential waterborne pathogens include:

— bacteria, viruses, protozoa and helminths identified in Table 7.1 and Figure 7.1, with the exception of *Schistosoma*, which is primarily spread by contact with contaminated surface water during bathing and washing;
— potentially emerging pathogens, including *Helicobacter pylori*, *Tsukamurella*, *Isospora belli* and microsporidia, for which waterborne transmission is plausible but unconfirmed;
— *Bacillus*, which includes the foodborne pathogenic species *Bacillus cereus* but for which there is no evidence at this time of waterborne transmission; and
— hazardous cyanobacteria.

The human health effects caused by waterborne transmission vary in severity from mild gastroenteritis to severe and sometimes fatal diarrhoea, dysentery, hepatitis and typhoid fever. Contaminated water can be the source of large outbreaks of disease, including cholera, dysentery and cryptosporidiosis; for the majority of waterborne pathogens, however, there are other important sources of infection, such as person-to-person contact and food.

Most waterborne pathogens are introduced into drinking-water supplies in human or animal faeces, do not grow in water and initiate infection in the gastrointestinal tract following ingestion. However, *Legionella*, atypical mycobacteria, *Burkholderia pseudomallei* and *Naegleria fowleri* are environmental organisms that can grow in water and soil. Besides ingestion, other routes of transmission can include inhalation, leading to infections of the respiratory tract (e.g., *Legionella*, atypical mycobacteria), and contact, leading to infections at sites as diverse as the skin and brain (e.g., *Naegleria fowleri*, *Burkholderia pseudomallei*).

Of all the waterborne pathogens, the helminth *Dracunculus medinensis* is unique in that it is the only pathogen that is solely transmitted through drinking-water.

The fact sheets on potential pathogens include information on human health effects, sources and occurrence, routes of transmission and the significance of drinking-water as a source of infection. The fact sheets on microorganisms that can be used as indicators of the effectiveness of control measures or as indices for the potential presence of pathogenic microorganisms provide information on indicator value, source and occurrence, application and significance of detection.

11.1 Bacterial pathogens

Most bacterial pathogens potentially transmitted by water infect the gastrointestinal tract and are excreted in the faeces of infected humans and other animals. However, there are also some waterborne bacterial pathogens, such as *Legionella*, *Burkholderia pseudomallei* and atypical mycobacteria, that can grow in water and soil. The routes of transmission of these bacteria include inhalation and contact (bathing), with infections occurring in the respiratory tract, in skin lesions or in the brain.

11.1.1 *Acinetobacter*

General description

Acinetobacter spp. are Gram-negative, oxidase-negative, non-motile coccobacilli (short plump rods). Owing to difficulties in naming individual species and biovars, the term *Acinetobacter calcoaceticus baumannii* complex is used in some classification schemes to cover all subgroups of this species, such as *A. baumannii*, *A. iwoffii* and *A. junii*.

Human health effects

Acinetobacter spp. are usually commensal organisms, but they occasionally cause infections, predominantly in susceptible patients in hospitals. They are opportunistic pathogens that may cause urinary tract infections, pneumonia, bacteraemia, secondary meningitis and wound infections. These diseases are predisposed by factors such as malignancy, burns, major surgery and weakened immune systems, such as in neonates and elderly individuals. The emergence and rapid spread of multidrug-resistant *A. calcoaceticus baumannii* complex, causing nosocomial infections, are of concern in health care facilities.

Source and occurrence

Acinetobacter spp. are ubiquitous inhabitants of soil, water and sewage environments. *Acinetobacter* has been isolated from 97% of natural surface water samples in numbers of up to 100/ml. The organisms have been found to represent 1.0–5.5% of the HPC flora in drinking-water samples and have been isolated from 5–92% of distribution water samples. In a survey of untreated groundwater supplies in the USA, *Acinetobacter* spp. were detected in 38% of the groundwater supplies at an arithmetic mean density of 8/100 ml. The study also revealed that slime production, a virulence factor for *A. calcoaceticus*, was not significantly different between well water isolates and clinical strains, suggesting some degree of pathogenic potential for strains isolated from groundwater. *Acinetobacter* spp. are part of the natural microbial flora of the skin and occasionally the respiratory tract of healthy individuals.

Routes of exposure

Environmental sources within hospitals and person-to-person transmission are the likely sources for most outbreaks of hospital infections. Infection is most commonly associated with contact with wounds and burns or inhalation by susceptible individuals. In patients with *Acinetobacter* bacteraemia, intravenous catheters have also been identified as a source of infection. Outbreaks of infection have been associated with water baths and room humidifiers. Ingestion is not a usual source of infection.

Significance in drinking-water

While *Acinetobacter* spp. are often detected in treated drinking-water supplies, an association between the presence of *Acinetobacter* spp. in drinking-water and clinical disease has not been confirmed. There is no evidence of gastrointestinal infection through ingestion of *Acinetobacter* spp. in drinking-water among the general population. However, transmission of non-gastrointestinal infections by drinking-water may be possible in susceptible individuals, particularly in settings such as health care facilities and hospitals. As discussed in chapter 6, specific WSPs should be developed for buildings, including hospitals and other health care facilities. These plans need to take account of particular sensitivities of occupants. *Acinetobacter* spp. are sensitive to disinfectants such as chlorine, and numbers will be low in the presence of a disinfectant residual. Control measures that can limit growth of the bacteria in distribution systems include treatment to optimize organic carbon removal, restriction of the residence time of water in distribution systems and maintenance of disinfectant residuals. *Acinetobacter* spp. are detected by HPC, which can be used together with parameters such as disinfectant residuals to indicate conditions that could support growth of these organisms. However, *E. coli* (or, alternatively, thermotolerant coliforms) cannot be used as an index for the presence/absence of *Acinetobacter* spp.

Selected bibliography

Bartram J et al., eds. (2003) *Heterotrophic plate counts and drinking-water safety: the significance of HPCs for water quality and human health.* WHO Emerging Issues in Water and Infectious Disease Series. London, IWA Publishing.

Bergogne-Berezin E, Towner KJ (1996) *Acinetobacter* as nosocomial pathogens: microbiological, clinical and epidemiological features. *Clinical Microbiology Reviews*, 9:148–165.

Bifulco JM, Shirey JJ, Bissonnette GK (1989) Detection of *Acinetobacter* spp. in rural drinking water supplies. *Applied and Environmental Microbiology*, 55:2214–2219.

Jellison TK, McKinnon PS, Rybak MJ (2001) Epidemiology, resistance and outcomes of *Acinetobacter baumannii* bacteremia treated with imipenem-cilastatin or ampicillin-sulbactam. *Pharmacotherapy*, 21:142–148.

Rusin PA et al. (1997) Risk assessment of opportunistic bacterial pathogens in drinking-water. *Reviews of Environmental Contamination and Toxicology*, 152:57–83.

11.1.2 Aeromonas

General description

Aeromonas spp. are Gram-negative, non-spore-forming, facultative anaerobic bacilli belonging to the family Vibrionaceae. They bear many similarities to the Enterobacteriaceae. The genus is divided into two groups. The group of psychrophilic non-motile aeromonads consists of only one species, *A. salmonicida*, an obligate fish pathogen that is not considered further here. The group of mesophilic motile (single polar flagellum) aeromonads is considered of potential human health significance and consists of the species *A. hydrophila*, *A. caviae*, *A. veronii* subsp. *sobria*, *A. jandaei*, *A. veronii* subsp. *veronii* and *A. schubertii*. The bacteria are normal inhabitants of fresh water and occur in water, soil and many foods, particularly meat and milk.

Human health effects

Aeromonas spp. can cause infections in humans, including septicaemia, particularly in immunocompromised patients, wound infections and respiratory tract infections. There have been some claims that *Aeromonas* spp. can cause gastrointestinal illness, but epidemiological evidence is not consistent. Despite marked toxin production by *Aeromonas* spp. *in vitro*, diarrhoea has not yet been introduced in test animals or human volunteers.

Source and occurrence

Aeromonas spp. occur in water, soil and food, particularly meat, fish and milk. *Aeromonas* spp. are generally readily found in most fresh waters, and they have been detected in many treated drinking-water supplies, mainly as a result of growth in distribution systems. The factors that affect the occurrence of *Aeromonas* spp. in water distribution systems are not fully understood, but organic content, temperature, the residence time of water in the distribution network and the presence of residual chlorine have been shown to influence population sizes.

Routes of exposure

Wound infections have been associated with contaminated soil and water-related activities, such as swimming, diving, boating and fishing. Septicaemia can follow from such wound infections. In immunocompromised individuals, septicaemia may arise from aeromonads present in their own gastrointestinal tract.

Significance in drinking-water

Despite frequent isolation of *Aeromonas* spp. from drinking-water, the body of evidence does not provide significant support for waterborne transmission. Aeromonads typically found in drinking-water do not belong to the same DNA homology groups as those associated with cases of gastroenteritis. The presence of *Aeromonas* spp. in drinking-water supplies is generally considered a nuisance. Entry of aeromonads into distribution systems can be minimized by adequate disinfection. Control measures that can limit growth of the bacteria in distribution systems include treatment to optimize organic carbon removal, restriction of the residence time of water in distribution systems and maintenance of disinfectant residuals. *Aeromonas* spp. are detected by HPC, which can be used together with parameters such as disinfectant residuals to indicate conditions that could support growth of these organisms. However, *E. coli* (or, alternatively, thermotolerant coliforms) cannot be used as an index for the presence/absence of *Aeromonas* spp.

Selected bibliography

Bartram J et al., eds. (2003) *Heterotrophic plate counts and drinking-water safety: the significance of HPCs for water quality and human health*. WHO Emerging Issues in Water and Infectious Disease Series. London, IWA Publishing.

Borchardt MA, Stemper ME, Standridge JH (2003) *Aeromonas* isolates from human diarrheic stool and groundwater compared by pulsed-field gel electrophoresis. *Emerging Infectious Diseases*, 9:224–228.

WHO (2002) *Aeromonas*. In: *Guidelines for drinking-water quality*, 2nd ed. *Addendum: Microbiological agents in drinking water*. Geneva, World Health Organization.

11.1.3 Bacillus

General description

Bacillus spp. are large (4–10 µm), Gram-positive, strictly aerobic or facultatively anaerobic encapsulated bacilli. They have the important feature of producing spores that are exceptionally resistant to unfavourable conditions. *Bacillus* spp. are classified into the subgroups *B. polymyxa*, *B. subtilis* (which includes *B. cereus* and *B. licheniformis*), *B. brevis* and *B. anthracis*.

Human health effects

Although most *Bacillus* spp. are harmless, a few are pathogenic to humans and animals. *Bacillus cereus* causes food poisoning similar to staphylococcal food poisoning. Some strains produce heat-stable toxin in food that is associated with spore germination and gives rise to a syndrome of vomiting within 1–5 h of ingestion. Other strains produce a heat-labile enterotoxin after ingestion that causes diarrhoea within 10–15 h. *Bacillus cereus* is known to cause bacteraemia in immunocompromised patients as well as symptoms such as vomiting and diarrhoea. *Bacillus anthracis* causes anthrax in humans and animals.

Source and occurrence
Bacillus spp. commonly occur in a wide range of natural environments, such as soil and water. They form part of the HPC bacteria, which are readily detected in most drinking-water supplies.

Routes of exposure
Infection with *Bacillus* spp. is associated with the consumption of a variety of foods, especially rice, pastas and vegetables, as well as raw milk and meat products. Disease may result from the ingestion of the organisms or toxins produced by the organisms. Drinking-water has not been identified as a source of infection of pathogenic *Bacillus* spp., including *Bacillus cereus*. Waterborne transmission of *Bacillus* gastroenteritis has not been confirmed.

Significance in drinking-water
Bacillus spp. are often detected in drinking-water supplies, even supplies treated and disinfected by acceptable procedures. This is largely due to the resistance of spores to disinfection processes. Owing to a lack of evidence that waterborne *Bacillus* spp. are clinically significant, specific management strategies are not required.

Selected bibliography
Bartram J et al., eds. (2003) *Heterotrophic plate counts and drinking-water safety: the significance of HPCs for water quality and human health*. WHO Emerging Issues in Water and Infectious Disease Series. London, IWA Publishing.

11.1.4 Burkholderia pseudomallei
General description
Burkholderia pseudomallei is a Gram-negative bacillus commonly found in soil and muddy water, predominantly in tropical regions such as northern Australia and southeast Asia. The organism is acid tolerant and survives in water for prolonged periods in the absence of nutrients.

Human health effects
Burkholderia pseudomallei can cause the disease melioidosis, which is endemic in northern Australia and other tropical regions. The most common clinical manifestation is pneumonia, which may be fatal. In some of these areas, melioidosis is the most common cause of community-acquired pneumonia. Cases appear throughout the year but peak during the rainy season. Many patients present with milder forms of pneumonia, which respond well to appropriate antibiotics, but some may present with a severe septicaemic pneumonia. Other symptoms include skin abscesses or ulcers, abscesses in internal organs and unusual neurological illnesses, such as brainstem encephalitis and acute paraplegia. Although melioidosis can occur in healthy children and adults, it occurs mainly in people whose defence mechanisms against infection

are impaired by underlying conditions or poor general health associated with poor nutrition or living conditions.

Source and occurrence
The organism occurs predominantly in tropical regions, typically in soil or surface-accumulated muddy water, from where it may reach raw water sources and also drinking-water supplies. The number of organisms in drinking-water that would constitute a significant risk of infection is not known.

Routes of exposure
Most infections appear to be through contact of skin cuts or abrasions with contaminated water. In south-east Asia, rice paddies represent a significant source of infection. Infection may also occur via other routes, particularly through inhalation or ingestion. The relative importance of these routes of infection is not known.

Significance in drinking-water
In two Australian outbreaks of melioidosis, indistinguishable isolates of *B. pseudomallei* were cultured from cases and the drinking-water supply. The detection of the organisms in one drinking-water supply followed replacement of water pipes and chlorination failure, while the second supply was unchlorinated. Within a WSP, control measures that should provide effective protection against this organism include application of established treatment and disinfection processes for drinking-water coupled with protection of the distribution system from contamination, including during repairs and maintenance. HPC and disinfectant residual as measures of water treatment effectiveness and application of appropriate mains repair procedures could be used to indicate protection against *B. pseudomallei*. Because of the environmental occurrence of *B. pseudomallei*, *E. coli* (or, alternatively, thermotolerant coliforms) is not a suitable index for the presence/absence of this organism.

Selected bibliography
Ainsworth R, ed. (2004) *Safe, piped water: Managing microbial water quality in piped distribution systems*. IWA Publishing, London, for the World Health Organization, Geneva.

Currie BJ (2000) The epidemiology of melioidosis in Australia and Papua New Guinea. *Acta Tropica*, 74:121–127.

Currie BJ et al. (2001) A cluster of melioidosis cases from an endemic region is clonal and is linked to the water supply using molecular typing of *Burkholderia pseudomallei* isolates. *American Journal of Tropical Medicine and Hygiene*, 65:177–179.

Inglis TJJ et al. (2000) Outbreak strain of *Burkholderia pseudomallei* traced to water treatment plant. *Emerging Infectious Diseases*, 6:56–59.

11.1.5 Campylobacter

General description

Campylobacter spp. are microaerophilic (require decreased oxygen) and capnophilic (require increased carbon dioxide), Gram-negative, curved spiral rods with a single unsheathed polar flagellum. *Campylobacter* spp. are one of the most important causes of acute gastroenteritis worldwide. *Campylobacter jejuni* is the most frequently isolated species from patients with acute diarrhoeal disease, whereas *C. coli*, *C. laridis* and *C. fetus* have also been isolated in a small proportion of cases. Two closely related genera, *Helicobacter* and *Archobacter*, include species previously classified as *Campylobacter* spp.

Human health effects

An important feature of *C. jejuni* is relatively high infectivity compared with other bacterial pathogens. As few as 1000 organisms can cause infection. Most symptomatic infections occur in infancy and early childhood. The incubation period is usually 2–4 days. Clinical symptoms of *C. jejuni* infection are characterized by abdominal pain, diarrhoea (with or without blood or faecal leukocytes), vomiting, chills and fever. The infection is self-limited and resolves in 3–7 days. Relapses may occur in 5–10% of untreated patients. Other clinical manifestations of *C. jejuni* infections in humans include reactive arthritis and meningitis. Several reports have associated *C. jejuni* infection with Guillain-Barré syndrome, an acute demyelinating disease of the peripheral nerves.

Source and occurrence

Campylobacter spp. occur in a variety of environments. Wild and domestic animals, especially poultry, wild birds and cattle, are important reservoirs. Pets and other animals may also be reservoirs. Food, including meat and unpasteurized milk, are important sources of *Campylobacter* infections. Water is also a significant source. The occurrence of the organisms in surface waters has proved to be strongly dependent on rainfall, water temperature and the presence of waterfowl.

Routes of exposure

Most *Campylobacter* infections are reported as sporadic in nature, with food considered a common source of infection. Transmission to humans typically occurs by the consumption of animal products. Meat, particularly poultry products, and unpasteurized milk are important sources of infection. Contaminated drinking-water supplies have been identified as a source of outbreaks. The number of cases in these outbreaks ranged from a few to several thousand, with sources including unchlorinated or inadequately chlorinated surface water supplies and faecal contamination of water storage reservoirs by wild birds.

Significance in drinking-water

Contaminated drinking-water supplies have been identified as a significant source of outbreaks of campylobacteriosis. The detection of waterborne outbreaks and cases appears to be increasing. Waterborne transmission has been confirmed by the isolation of the same strains from patients and drinking-water they had consumed. Within a WSP, control measures that can be applied to manage potential risk from *Campylobacter* spp. include protection of raw water supplies from animal and human waste, adequate treatment and protection of water during distribution. Storages of treated and disinfected water should be protected from bird faeces. *Campylobacter* spp. are faecally borne pathogens and are not particularly resistant to disinfection. Hence, *E. coli* (or thermotolerant coliforms) is an appropriate indicator for the presence/absence of *Campylobacter* spp. in drinking-water supplies.

Selected bibliography

Frost JA (2001) Current epidemiological issues in human campylobacteriosis. *Journal of Applied Microbiology*, 90:85S–95S.

Koenraad PMFJ, Rombouts FM, Notermans SHW (1997) Epidemiological aspects of thermophilic *Campylobacter* in water-related environments: A review. *Water Environment Research*, 69:52–63.

Kuroki S et al. (1991) Guillain-Barré syndrome associated with *Campylobacter* infection. *Pediatric Infectious Diseases Journal*, 10:149–151.

11.1.6 *Escherichia coli* pathogenic strains

General description

Escherichia coli is present in large numbers in the normal intestinal flora of humans and animals, where it generally causes no harm. However, in other parts of the body, *E. coli* can cause serious disease, such as urinary tract infections, bacteraemia and meningitis. A limited number of enteropathogenic strains can cause acute diarrhoea. Several classes of enteropathogenic *E. coli* have been identified on the basis of different virulence factors, including enterohaemorrhagic *E. coli* (EHEC), enterotoxigenic *E. coli* (ETEC), enteropathogenic *E. coli* (EPEC), enteroinvasive *E. coli* (EIEC), enteroaggregative *E. coli* (EAEC) and diffusely adherent *E. coli* (DAEC). More is known about the first four classes named; the pathogenicity and prevalence of EAEC and DAEC strains are less well established.

Human health effects

EHEC serotypes, such as *E. coli* O157:H7 and *E. coli* O111, cause diarrhoea that ranges from mild and non-bloody to highly bloody, which is indistinguishable from haemorrhagic colitis. Between 2% and 7% of cases can develop the potentially fatal haemolytic uraemic syndrome (HUS), which is characterized by acute renal failure and haemolytic anaemia. Children under 5 years of age are at most risk of developing HUS. The infectivity of EHEC strains is substantially higher than that of the other strains. As few as 100 EHEC organisms can cause infection. ETEC produces heat-labile or heat-stable *E. coli* enterotoxin, or both toxins simultaneously, and is an important cause of diarrhoea in developing countries, especially in young children. Symptoms of ETEC infection include mild watery diarrhoea, abdominal cramps, nausea and headache. Infection with EPEC has been associated with severe, chronic, non-bloody diarrhoea, vomiting and fever in infants. EPEC infections are rare in developed countries, but occur commonly in developing countries, with infants presenting with malnutrition, weight loss and growth retardation. EIEC causes watery and occasionally bloody diarrhoea where strains invade colon cells by a pathogenic mechanism similar to that of *Shigella*.

Source and occurrence
Enteropathogenic *E. coli* are enteric organisms, and humans are the major reservoir, particularly of EPEC, ETEC and EIEC strains. Livestock, such as cattle and sheep and, to a lesser extent, goats, pigs and chickens, are a major source of EHEC strains. The latter have also been associated with raw vegetables, such as bean sprouts. The pathogens have been detected in a variety of water environments.

Routes of exposure
Infection is associated with person-to-person transmission, contact with animals, food and consumption of contaminated water. Person-to-person transmissions are particularly prevalent in communities where there is close contact between individuals, such as nursing homes and day care centres.

Significance in drinking-water
Waterborne transmission of pathogenic *E. coli* has been well documented for recreational waters and contaminated drinking-water. A well publicized waterborne outbreak of illness caused by *E. coli* O157:H7 (and *Campylobacter jejuni*) occurred in the farming community of Walkerton in Ontario, Canada. The outbreak took place in May 2000 and led to 7 deaths and more than 2300 illnesses. The drinking-water supply was contaminated by rainwater runoff containing cattle excreta. Within a WSP, control measures that can be applied to manage potential risk from enteropathogenic *E. coli* include protection of raw water supplies from animal and human waste, adequate treatment and protection of water during distribution. There is no indication that the response of enteropathogenic strains of *E. coli* to water treatment and disinfection procedures differs from that of other *E. coli*. Hence, conventional testing for *E. coli* (or, alternatively, thermotolerant coliform bacteria) provides an appropriate index for the enteropathogenic serotypes in drinking-water. This applies even though standard tests will generally not detect EHEC strains.

Selected bibliography

Nataro JP, Kaper JB (1998) Diarrheagenic *Escherichia coli. Clinical Microbiology Reviews*, 11:142–201.

O'Connor DR (2002) *Report of the Walkerton Inquiry: The events of May 2000 and related issues. Part 1: A summary*. Toronto, Ontario, Ontario Ministry of the Attorney General, Queen's Printer for Ontario.

11.1.7 *Helicobacter pylori*

General description

Helicobacter pylori, originally classified as *Campylobacter pylori*, is a Gram-negative, microaerophilic, spiral-shaped, motile bacterium. There are at least 14 species of *Helicobacter*, but only *H. pylori* has been identified as a human pathogen.

Human health effects

Helicobacter pylori is found in the stomach; although most infections are asymptomatic, the organism is associated with chronic gastritis, which may lead to complications such as peptic and duodenal ulcer disease and gastric cancer. Whether the organism is truly the cause of these conditions remains unclear. The majority of *H. pylori* infections are initiated in childhood and without treatment are chronic. The infections are more prevalent in developing countries and are associated with overcrowded living conditions. Interfamilial clustering is common.

Source and occurrence

Humans appear to be the primary host of *H. pylori*. Other hosts may include domestic cats. There is evidence that *H. pylori* is sensitive to bile salts, which would reduce the likelihood of faecal excretion, although it has been isolated from faeces of young children. *Helicobacter pylori* has been detected in water. Although *H. pylori* is unlikely to grow in the environment, it has been found to survive for 3 weeks in biofilms and up to 20–30 days in surface waters. In a study conducted in the USA, *H. pylori* was found in the majority of surface water and shallow groundwater samples. The presence of *H. pylori* was not correlated with the presence of *E. coli*. Possible contamination of the environment can be through children with diarrhoea or through vomiting by children as well as adults.

Routes of exposure

Person-to-person contact within families has been identified as the most likely source of infection through oral–oral transmission. *Helicobacter pylori* can survive well in mucus or vomit. However, it is difficult to detect in mouth or faecal samples. Faecal–oral transmission is also considered possible.

Significance in drinking-water
Consumption of contaminated drinking-water has been suggested as a potential source of infection, but further investigation is required to establish any link with waterborne transmission. Humans are the principal source of *H. pylori*, and the organism is sensitive to oxidizing disinfectants. Hence, control measures that can be applied to protect drinking-water supplies from *H. pylori* include preventing contamination by human waste and adequate disinfection. *Escherichia coli* (or, alternatively, thermotolerant coliforms) is not a reliable index for the presence/absence of this organism.

Selected bibliography

Dunn BE, Cohen H, Blaser MJ (1997) Helicobacter pylori. *Clinical Microbiology Reviews*, 10:720–741.

Hegarty JP, Dowd MT, Baker KH (1999) Occurrence of Helicobacter pylori in surface water in the United States. *Journal of Applied Microbiology*, 87:697–701.

Hulten K et al. (1996) Helicobacter pylori in drinking-water in Peru. *Gastroenterology*, 110:1031–1035.

Mazari-Hiriart M, López-Vidal Y, Calva JJ (2001) Helicobacter pylori in water systems for human use in Mexico City. *Water Science and Technology*, 43:93–98.

11.1.8 Klebsiella

General description

Klebsiella spp. are Gram-negative, non-motile bacilli that belong to the family Enterobacteriaceae. The genus *Klebsiella* consists of a number of species, including *K. pneumoniae*, *K. oxytoca*, *K. planticola* and *K. terrigena*. The outermost layer of *Klebsiella* spp. consists of a large polysaccharide capsule that distinguishes the organisms from other members of the family. Approximately 60–80% of all *Klebsiella* spp. isolated from faeces and clinical specimens are *K. pneumoniae* and are positive in the thermotolerant coliform test. *Klebsiella oxytoca* has also been identified as a pathogen.

Human health effects

Klebsiella spp. have been identified as colonizing hospital patients, where spread is associated with the frequent handling of patients (e.g., in intensive care units). Patients at highest risk are those with impaired immune systems, such as the elderly or very young, patients with burns or excessive wounds, those undergoing immunosuppressive therapy or those with HIV/AIDS infection. Colonization may lead to invasive infections. On rare occasions, *Klebsiella* spp., notably *K. pneumoniae* and *K. oxytoca*, may cause serious infections, such as destructive pneumonia.

Source and occurrence

Klebsiella spp. are natural inhabitants of many water environments, and they may multiply to high numbers in waters rich in nutrients, such as pulp mill wastes, textile finishing plants and sugar-cane processing operations. In drinking-water distribution systems, they are known to colonize washers in taps. The organisms can grow in water distribution systems. *Klebsiella* spp. are also excreted in the faeces of many healthy humans and animals, and they are readily detected in sewage-polluted water.

Routes of exposure
Klebsiella can cause nosocomial infections, and contaminated water and aerosols may be a potential source of the organisms in hospital environments and other health care facilities.

Significance in drinking-water
Klebsiella spp. are not considered to represent a source of gastrointestinal illness in the general population through ingestion of drinking-water. *Klebsiella* spp. detected in drinking-water are generally biofilm organisms and are unlikely to represent a health risk. The organisms are reasonably sensitive to disinfectants, and entry into distribution systems can be prevented by adequate treatment. Growth within distribution systems can be minimized by strategies that are designed to minimize biofilm growth, including treatment to optimize organic carbon removal, restriction of the residence time of water in distribution systems and maintenance of disinfectant residuals. *Klebsiella* is a coliform and can be detected by traditional tests for total coliforms.

Selected bibliography
Ainsworth R, ed. (2004) *Safe, piped water: Managing microbial water quality in piped distribution systems*. IWA Publishing, London, for the World Health Organization, Geneva.

Bartram J et al., eds. (2003) *Heterotrophic plate counts and drinking-water safety: the significance of HPCs for water quality and human health*. WHO Emerging Issues in Water and Infectious Disease Series. London, IWA Publishing.

11.1.9 Legionella
General description
The genus *Legionella*, a member of the family Legionellaceae, has at least 42 species. Legionellae are Gram-negative, rod-shaped, non-spore-forming bacteria that require L-cysteine for growth and primary isolation. *Legionella* spp. are heterotrophic bacteria found in a wide range of water environments and can proliferate at temperatures above 25 °C.

Human health effects
Although all *Legionella* spp. are considered potentially pathogenic for humans, *L. pneumophila* is the major waterborne pathogen responsible for legionellosis, of which two clinical forms are known: Legionnaires' disease and Pontiac fever. The former is a pneumonic illness with an incubation period of 3–6 days. Host factors influence the likelihood of illness: males are more frequently affected than females, and most cases

occur in the 40- to 70-year age group. Risk factors include smoking, alcohol abuse, cancer, diabetes, chronic respiratory or kidney disease and immunosuppression, as in transplant recipients. Pontiac fever is a milder, self-limiting disease with a high attack rate and an onset (5h to 3 days) and symptoms similar to those of influenza: fever, headache, nausea, vomiting, aching muscles and coughing. Studies of seroprevalence of antibodies indicate that many infections are asymptomatic.

Source and occurrence
Legionella spp. are members of the natural flora of many freshwater environments, such as rivers, streams and impoundments, where they occur in relatively low numbers. However, they thrive in certain human-made water environments, such as water cooling devices (cooling towers and evaporative condensers) associated with air conditioning systems, hot water distribution systems and spas, which provide suitable temperatures (25–50 °C) and conditions for their multiplication. Devices that support multiplication of *Legionella* have been associated with outbreaks of Legionnaires' disease. *Legionella* survive and grow in biofilms and sediments and are more easily detected from swab samples than from flowing water. Legionellae can be ingested by trophozoites of certain amoebae such as *Acanthamoeba*, *Hartmanella* and *Naegleria*, which may play a role in their persistence in water environments.

Routes of exposure
The most common route of infection is the inhalation of aerosols containing the bacteria. Such aerosols can be generated by contaminated cooling towers, warm water showers, humidifiers and spas. Aspiration has also been identified as a route of infection in some cases associated with contaminated water, food and ice. There is no evidence of person-to-person transmission.

Significance in drinking-water
Legionella spp. are common waterborne organisms, and devices such as cooling towers, hot water systems and spas that utilize mains water have been associated with outbreaks of infection. Owing to the prevalence of *Legionella*, the potential for ingress into drinking-water systems should be considered as a possibility, and control measures should be employed to reduce the likelihood of survival and multiplication. Disinfection strategies designed to minimize biofilm growth and temperature control can minimize the potential risk from *Legionella* spp. The organisms are sensitive to disinfection. Monochloramine has been shown to be particularly effective, probably due to its stability and greater effectiveness against biofilms. Water temperature is an important element of control strategies. Wherever possible, water temperatures should be kept outside the range of 25–50 °C. In hot water systems, storages should be maintained above 55 °C, and similar temperatures throughout associated pipework will prevent growth of the organism. However, maintaining temperatures of hot water above 50 °C may represent a scalding risk in young children, the elderly and other vul-

nerable groups. Where temperatures in hot or cold water distribution systems cannot be maintained outside the range of 25–50 °C, greater attention to disinfection and strategies aimed at limiting development of biofilms are required. Accumulation of sludge, scale, rust, algae or slime deposits in water distribution systems supports the growth of *Legionella* spp., as does stagnant water. Systems that are kept clean and flowing are less likely to support excess growth of *Legionella* spp. Care should also be taken to select plumbing materials that do not support microbial growth and the development of biofilms.

Legionella spp. represent a particular concern in devices such as cooling towers and hot water systems in large buildings. As discussed in chapter 6, specific WSPs incorporating control measures for *Legionella* spp. should be developed for these buildings. *Legionella* are not detected by HPC techniques, and *E. coli* (or, alternatively, thermotolerant coliforms) is not a suitable index for the presence/absence of this organism.

Selected bibliography

Codony F et al. (2002) Factors promoting colonization by legionellae in residential water distribution systems: an environmental case–control survey. *European Journal of Clinical Microbiology and Infectious Diseases*, 21:717–721.

Emmerson AM (2001) Emerging waterborne infections in health-care settings. *Emerging Infectious Diseases*, 7:272–276.

Rusin PA et al. (1997) Risk assessment of opportunistic bacterial pathogens in drinking-water. *Reviews of Environmental Contamination and Toxicology*, 152:57–83.

WHO (in preparation) *Legionella and the prevention of legionellosis*. Geneva, World Health Organization.

11.1.10 *Mycobacterium*

General description

The tuberculous or "typical" species of *Mycobacterium*, such as *M. tuberculosis*, *M. bovis*, *M. africanum* and *M. leprae*, have only human or animal reservoirs and are not transmitted by water. In contrast, the non-tuberculous or "atypical" species of *Mycobacterium* are natural inhabitants of a variety of water environments. These aerobic, rod-shaped and acid-fast bacteria grow slowly in suitable water environments and on culture media. Typical examples include the species *M. gordonae*, *M. kansasii*, *M. marinum*, *M. scrofulaceum*, *M. xenopi*, *M. intracellulare* and *M. avium* and the more rapid growers *M. chelonae* and *M. fortuitum*. The term *M. avium* complex has been used to describe a group of pathogenic species including *M. avium* and *M. intracellulare*. However, other atypical mycobacteria are also pathogenic. A distinct feature of all *Mycobacterium* spp. is a cell wall with high lipid content, which is used in identification of the organisms using acid-fast staining.

Human health effects

Atypical *Mycobacterium* spp. can cause a range of diseases involving the skeleton, lymph nodes, skin and soft tissues, as well as the respiratory, gastrointestinal and genitourinary tracts. Manifestations include pulmonary disease, Buruli ulcer, osteomyelitis and septic arthritis in people with no known predisposing factors. These bacteria are a major cause of disseminated infections in immunocompromised patients and are a common cause of death in HIV-positive persons.

Source and occurrence

Atypical *Mycobacterium* spp. multiply in a variety of suitable water environments, notably biofilms. One of the most commonly occurring species is *M. gordonae*. Other species have also been isolated from water, including *M. avium*, *M. intracellulare*, *M. kansasii*, *M. fortuitum* and *M. chelonae*. High numbers of atypical *Mycobacterium* spp. may occur in distribution systems after events that dislodge biofilms, such as flushing or flow reversals. They are relatively resistant to treatment and disinfection and have been detected in well operated and maintained drinking-water supplies with HPC less than 500/ml and total chlorine residuals of up to 2.8 mg/litre. The growth of these organisms in biofilms reduces the effectiveness of disinfection. In one survey, the organisms were detected in 54% of ice and 35% of public drinking-water samples.

Routes of exposure

Principal routes of infection appear to be inhalation, contact and ingestion of contaminated water. Infections by various species have been associated with their presence in drinking-water supplies. In 1968, an endemic of *M. kansasii* infections was associated with the presence of the organisms in the drinking-water supply, and the spread of the organisms was associated with aerosols from showerheads. In Rotterdam, Netherlands, an investigation into the frequent isolation of *M. kansasii* from clinical specimens revealed the presence of the same strains, confirmed by phage type and weak nitrase activity, in tap water. An increase in numbers of infections by the *M. avium* complex in Massachusetts, USA, has also been attributed to their incidence in drinking-water. In all these cases, there is only circumstantial evidence of a causal relationship between the occurrence of the bacteria in drinking-water and human disease. Infections have been linked to contaminated water in spas.

Significance in drinking-water

Detections of atypical mycobacteria in drinking-water and the identified routes of transmission suggest that drinking-water supplies are a plausible source of infection. There are limited data on the effectiveness of control measures that could be applied to reduce the potential risk from these organisms. One study showed that a water treatment plant could achieve a 99% reduction in numbers of mycobacteria from raw water. Atypical mycobacteria are relatively resistant to disinfection. Persistent residual disinfectant should reduce numbers of mycobacteria in the water column but is unlikely to be effective against organisms present in biofilms. Control measures that are designed to minimize biofilm growth, including treatment to optimize organic carbon removal, restriction of the residence time of water in distribution systems and maintenance of disinfectant residuals, could result in less growth of these organisms. Mycobacteria are not detected by HPC techniques, and *E. coli* (or, alternatively, thermotolerant coliforms) is not a suitable index for the presence/absence of this organism.

Selected bibliography

Bartram J et al., eds. (2003) *Heterotrophic plate counts and drinking-water safety: the significance of HPCs for water quality and human health*. WHO Emerging Issues in Water and Infectious Disease Series. London, IWA Publishing.

Bartram J et al., eds. (2004) *Pathogenic mycobacteria in water: A guide to public health consequences, monitoring and management*. Geneva, World Health Organization.

Covert TC et al. (1999) Occurrence of nontuberculous mycobacteria in environmental samples. *Applied and Environmental Microbiology*, 65:2492–2496.

Falkinham JO, Norton CD, LeChevallier MW (2001) Factors influencing numbers of *Mycobacterium avium*, *Mycobacterium intracellulare* and other mycobacteria in drinking water distribution systems. *Applied and Environmental Microbiology*, 66:1225–1231.

Grabow WOK (1996) Waterborne diseases: Update on water quality assessment and control. *Water SA*, 22:193–202.

Rusin PA et al. (1997) Risk assessment of opportunistic bacterial pathogens in drinking-water. *Reviews of Environmental Contamination and Toxicology*, 152:57–83.

Singh N, Yu VL (1994) Potable water and *Mycobacterium avium* complex in HIV patients: is prevention possible? *Lancet*, 343:1110–1111.

Von Reyn CF et al. (1994) Persistent colonization of potable water as a source of *Mycobacterium avium* infection in AIDS. *Lancet*, 343:1137–1141.

11.1.11 Pseudomonas aeruginosa

General description

Pseudomonas aeruginosa is a member of the family Pseudomonadaceae and is a polarly flagellated, aerobic, Gram-negative rod. When grown in suitable media, it produces the non-fluorescent bluish pigment pyocyanin. Many strains also produce the fluorescent green pigment pyoverdin. *Pseudomonas aeruginosa*, like other fluorescent pseudomonads, produces catalase, oxidase and ammonia from arginine and can grow on citrate as the sole source of carbon.

Human health effects

Pseudomonas aeruginosa can cause a range of infections but rarely causes serious illness in healthy individuals without some predisposing factor. It predominantly colonizes damaged sites such as burn and surgical wounds, the respiratory tract of people with underlying disease and physically damaged eyes. From these sites, it may invade the body, causing destructive lesions or septicaemia and meningitis. Cystic fibrosis and immunocompromised patients are prone to colonization with *P. aeruginosa*, which may lead to serious progressive pulmonary infections. Water-related folliculitis and ear infections are associated with warm, moist environments such as swimming pools and spas. Many strains are resistant to a range of antimicrobial agents, which can increase the significance of the organism in hospital settings.

Source and occurrence
Pseudomonas aeruginosa is a common environmental organism and can be found in faeces, soil, water and sewage. It can multiply in water environments and also on the surface of suitable organic materials in contact with water. *Pseudomonas aeruginosa* is a recognized cause of hospital-acquired infections with potentially serious complications. It has been isolated from a range of moist environments such as sinks, water baths, hot water systems, showers and spa pools.

Routes of exposure
The main route of infection is by exposure of susceptible tissue, notably wounds and mucous membranes, to contaminated water or contamination of surgical instruments. Cleaning of contact lenses with contaminated water can cause a form of keratitis. Ingestion of drinking-water is not an important source of infection.

Significance in drinking-water
Although *P. aeruginosa* can be significant in certain settings such as health care facilities, there is no evidence that normal uses of drinking-water supplies are a source of infection in the general population. However, the presence of high numbers of *P. aeruginosa* in potable water, notably in packaged water, can be associated with complaints about taste, odour and turbidity. *Pseudomonas aeruginosa* is sensitive to disinfection, and entry into distribution systems can be minimized by adequate disinfection. Control measures that are designed to minimize biofilm growth, including treatment to optimize organic carbon removal, restriction of the residence time of water in distribution systems and maintenance of disinfectant residuals, should reduce the growth of these organisms. *Pseudomonas aeruginosa* is detected by HPC, which can be used together with parameters such as disinfectant residuals to indicate conditions that could support growth of these organisms. However, as *P. aeruginosa* is a common environmental organism, *E. coli* (or, alternatively, thermotolerant coliforms) cannot be used for this purpose.

Selected bibliography
Bartram J et al., eds. (2003) *Heterotrophic plate counts and drinking-water safety: the significance of HPCs for water quality and human health*. WHO Emerging Issues in Water and Infectious Disease Series. London, IWA Publishing.

de Victorica J, Galván M (2001) *Pseudomonas aeruginosa* as an indicator of health risk in water for human consumption. *Water Science and Technology*, 43:49–52.

Hardalo C, Edberg SC (1997) *Pseudomonas aeruginosa*: Assessment of risk from drinking-water. *Critical Reviews in Microbiology*, 23:47–75.

11.1.12 Salmonella

General description

Salmonella spp. belong to the family Enterobacteriaceae. They are motile, Gram-negative bacilli that do not ferment lactose, but most produce hydrogen sulfide or gas from carbohydrate fermentation. Originally, they were grouped into more than 2000 species (serotypes) according to their somatic (O) and flagellar (H) antigens (Kauffmann-White classification). It is now considered that this classification is below species level and that there are actually no more than 2–3 species (*Salmonella enterica* or *Salmonella choleraesuis*, *Salmonella bongori* and *Salmonella typhi*), with the serovars being subspecies. All of the enteric pathogens except *S. typhi* are members of the species *S. enterica*. Convention has dictated that subspecies are abbreviated, so that *S. enterica* serovar Paratyphi A becomes *S.* Paratyphi A.

Human health effects

Salmonella infections typically cause four clinical manifestations: gastroenteritis (ranging from mild to fulminant diarrhoea, nausea and vomiting), bacteraemia or septicaemia (high spiking fever with positive blood cultures), typhoid fever / enteric fever (sustained fever with or without diarrhoea) and a carrier state in persons with previous infections. In regard to enteric illness, *Salmonella* spp. can be divided into two fairly distinct groups: the typhoidal species/serovars (*Salmonella typhi* and *S.* Paratyphi) and the remaining non-typhoidal species/serovars. Symptoms of non-typhoidal gastroenteritis appear from 6 to 72 h after ingestion of contaminated food or water. Diarrhoea lasts 3–5 days and is accompanied by fever and abdominal pain. Usually the disease is self-limiting. The incubation period for typhoid fever can be 1–14 days but is usually 3–5 days. Typhoid fever is a more severe illness and can be fatal. Although typhoid is uncommon in areas with good sanitary systems, it is still prevalent elsewhere, and there are many millions of cases each year.

Source and occurrence

Salmonella spp. are widely distributed in the environment, but some species or serovars show host specificity. Notably, *S. typhi* and generally *S.* Paratyphi are restricted to humans, although livestock can occasionally be a source of *S.* Paratyphi. A large number of serovars, including *S.* Typhimurium and *S.* Enteritidis, infect humans and also a wide range of animals, including poultry, cows, pigs, sheep, birds and even reptiles. The pathogens typically gain entry into water systems through faecal contamination from sewage discharges, livestock and wild animals. Contamination has been detected in a wide variety of foods and milk.

Routes of exposure

Salmonella is spread by the faecal–oral route. Infections with non-typhoidal serovars are primarily associated with person-to-person contact, the consumption of a variety of contaminated foods and exposure to animals. Infection by typhoid species is associated with the consumption of contaminated water or food, with direct person-to-person spread being uncommon.

Significance in drinking-water

Waterborne typhoid fever outbreaks have devastating public health implications. However, despite their widespread occurrence, non-typhoidal *Salmonella* spp. rarely cause drinking-water-borne outbreaks. Transmission, most commonly involving *S.* Typhimurium, has been associated with the consumption of contaminated groundwater and surface water supplies. In an outbreak of illness associated with a communal rainwater supply, bird faeces were implicated as a source of contamination. *Salmonella* spp. are relatively sensitive to disinfection. Within a WSP, control measures that can be applied to manage risk include protection of raw water supplies from animal and human waste, adequate treatment and protection of water during distribution. *Escherichia coli* (or, alternatively, thermotolerant coliforms) is a generally reliable index for *Salmonella* spp. in drinking-water supplies.

Selected bibliography

Angulo FJ et al. (1997) A community waterborne outbreak of salmonellosis and the effectiveness of a boil water order. *American Journal of Public Health*, 87:580–584.

Escartin EF et al. (2002) Potential *Salmonella* transmission from ornamental fountains. *Journal of Environmental Health*, 65:9–12.

Koplan JP et al. (1978) Contaminated roof-collected rainwater as a possible cause of an outbreak of salmonellosis. *Journal of Hygiene*, 81:303–309.

11.1.13 Shigella

General description

Shigella spp. are Gram-negative, non-spore-forming, non-motile, rod-like members of the family Enterobacteriaceae, which grow in the presence or absence of oxygen. Members of the genus have a complex antigenic pattern, and classification is based on their somatic O antigens, many of which are shared with other enteric bacilli, including *E. coli*. There are four species: *S. dysenteriae*, *S. flexneri*, *S. boydii* and *S. sonnei*.

Human health effects

Shigella spp. can cause serious intestinal diseases, including bacillary dysentery. Over 2 million infections occur each year, resulting in about 600 000 deaths, predominantly in developing countries. Most cases of *Shigella* infection occur in children under 10 years of age. The incubation period for shigellosis is usually 24–72 h. Ingestion of as few as 10–100 organisms may lead to infection, which is substantially less than the infective dose of most other enteric bacteria. Abdominal cramps, fever and watery diarrhoea occur early in the disease. All species can produce severe disease, but illness due to *S. sonnei* is usually relatively mild and self-limiting. In the case of *S. dysenteriae*, clinical manifestations may proceed to an ulceration process, with bloody diarrhoea and high concentrations of neutrofils in the stool. The production of Shiga toxin by the pathogen plays an important role in this outcome. *Shigella* spp. seem to be better adapted to cause human disease than most other enteric bacterial pathogens.

Source and occurrence
Humans and other higher primates appear to be the only natural hosts for the shigellae. The bacteria remain localized in the intestinal epithelial cells of their hosts. Epidemics of shigellosis occur in crowded communities and where hygiene is poor. Many cases of shigellosis are associated with day care centres, prisons and psychiatric institutions. Military field groups and travellers to areas with poor sanitation are also prone to infection.

Routes of exposure
Shigella spp. are enteric pathogens predominantly transmitted by the faecal–oral route through person-to-person contact, contaminated food and water. Flies have also been identified as a transmission vector from contaminated faecal waste.

Significance in drinking-water
A number of large waterborne outbreaks of shigellosis have been recorded. As the organisms are not particularly stable in water environments, their presence in drinking-water indicates recent human faecal pollution. Available data on prevalence in water supplies may be an underestimate, because detection techniques generally used can have a relatively low sensitivity and reliability. The control of *Shigella* spp. in drinking-water supplies is of special public health importance in view of the severity of the disease caused. *Shigella* spp. are relatively sensitive to disinfection. Within a WSP, control measures that can be applied to manage potential risk include protection of raw water supplies from human waste, adequate treatment and protection of water during distribution. *Escherichia coli* (or, alternatively, thermotolerant coliforms) is a generally reliable index for *Shigella* spp. in drinking-water supplies.

Selected bibliography
Alamanos Y et al. (2000) A community waterborne outbreak of gastro-enteritis attributed to *Shigella sonnei*. *Epidemiology and Infection*, 125:499–503.

Pegram GC, Rollins N, Espay Q (1998) Estimating the cost of diarrhoea and epidemic dysentery in Kwa-Zulu-Natal and South Africa. *Water SA*, 24:11–20.

11.1.14 *Staphylococcus aureus*
General description

Staphylococcus aureus is an aerobic or anaerobic, non-motile, non-spore-forming, catalase- and coagulase-positive, Gram-positive coccus, usually arranged in grapelike irregular clusters. The genus *Staphylococcus* contains at least 15 different species. Apart from *S. aureus*, the species *S. epidermidis* and *S. saprophyticus* are also associated with disease in humans.

Human health effects

Although *Staphylococcus aureus* is a common member of the human microflora, it can produce disease through two different mechanisms. One is based on the ability of the organisms to multiply and spread widely in tissues, and the other is based on the ability of the organisms to produce extracellular enzymes and toxins. Infections based on the multiplication of the organisms are a significant problem in hospitals and other health care facilities. Multiplication in tissues can result in manifestations such as boils, skin sepsis, post-operative wound infections, enteric infections, septicaemia, endocarditis, osteomyelitis and pneumonia. The onset of clinical symptoms for these infections is relatively long, usually several days. Gastrointestinal disease (enterocolitis or food poisoning) is caused by a heat-stable staphylococcal enterotoxin and characterized by projectile vomiting, diarrhoea, fever, abdominal cramps, electrolyte imbalance and loss of fluids. Onset of disease in this case has a characteristic short incubation period of 1–8 h. The same applies to the toxic shock syndrome caused by toxic shock syndrome toxin-1.

Source and occurrence

Staphylococcus aureus is relatively widespread in the environment but is found mainly on the skin and mucous membranes of animals. The organism is a member of the normal microbial flora of the human skin and is found in the nasopharynx of 20–30% of adults at any one time. Staphylococci are occasionally detected in the gastrointestinal tract and can be detected in sewage. *Staphylococcus aureus* can be released by human contact into water environments such as swimming pools, spa pools and other recreational waters. It has also been detected in drinking-water supplies.

Routes of exposure

Hand contact is by far the most common route of transmission. Inadequate hygiene can lead to contamination of food. Foods such as ham, poultry and potato and egg salads kept at room or higher temperature offer an ideal environment for the multiplication of *S. aureus* and the release of toxins. The consumption of foods containing *S. aureus* toxins can lead to enterotoxin food poisoning within a few hours.

Significance in drinking-water

Although *S. aureus* can occur in drinking-water supplies, there is no evidence of transmission through the consumption of such water. Although staphylococci are slightly more resistant to chlorine residuals than *E. coli*, their presence in water is readily controlled by conventional treatment and disinfection processes. Since faecal material is not their usual source, *E. coli* (or, alternatively, thermotolerant coliforms) is not a suitable index for *S. aureus* in drinking-water supplies.

Selected bibliography

Antai SP (1987) Incidence of *Staphylococcus aureus*, coliforms and antibiotic-resistant strains of *Escherichia coli* in rural water supplies in Port Harcourt. *Journal of Applied Bacteriology*, 62:371–375.

LeChevallier MW, Seidler RJ (1980) *Staphylococcus aureus* in rural drinking-water. *Applied and Environmental Microbiology*, 39:739–742.

11.1.15 Tsukamurella

General description

The genus *Tsukamurella* belongs to the family Nocardiaceae. *Tsukamurella* spp. are Gram-positive, weakly or variably acid-fast, non-motile, obligate aerobic, irregular rod-shaped bacteria. They are actinomycetes related to *Rhodococcus*, *Nocardia* and *Mycobacterium*. The genus was created in 1988 to accommodate a group of chemically unique organisms characterized by a series of very long chain (68–76 carbons), highly unsaturated mycolic acids, *meso*-diaminopimelic acid and arabinogalactan, common to the genus *Corynebacterium*. The type species is *T. paurometabola*, and the following additional species were proposed in the 1990s: *T. wratislaviensis*, *T. inchonensis*, *T. pulmonis*, *T. tyrosinosolvens* and *T. strandjordae*.

Human health effects

Tsukamurella spp. cause disease mainly in immunocompromised individuals. Infections with these microorganisms have been associated with chronic lung diseases, immune suppression (leukaemia, tumours, HIV/AIDS infection) and post-operative wound infections. *Tsukamurella* were reported in four cases of catheter-related bacteraemia and in individual cases including chronic lung infection, necrotizing tenosynovitis with subcutaneous abscesses, cutaneous and bone infections, meningitis and peritonitis.

Source and occurrence

Tsukamurella spp. exist primarily as environmental saprophytes in soil, water and foam (thick stable scum on aeration vessels and sedimentation tanks) of activated sludge. *Tsukamurella* are represented in HPC populations in drinking-water.

Routes of exposure
Tsukamurella spp. appear to be transmitted through devices such as catheters or lesions. The original source of the contaminating organisms is unknown.

Significance in drinking-water
Tsukamurella organisms have been detected in drinking-water supplies, but the significance is unclear. There is no evidence of a link between organisms in water and illness. As *Tsukamurella* is an environmental organism, *E. coli* (or, alternatively, thermotolerant coliforms) is not a suitable index for this organism.

Selected bibliography
Bartram J et al., eds. (2003) *Heterotrophic plate counts and drinking-water safety: the significance of HPCs for water quality and human health*. WHO Emerging Issues in Water and Infectious Disease Series. London, IWA Publishing.
Kattar MM et al. (2001) *Tsukamurella strandjordae* sp. nov., a proposed new species causing sepsis. *Journal of Clinical Microbiology*, 39:1467–1476.
Larkin JA et al. (1999) Infection of a knee prosthesis with *Tsukamurella* species. *Southern Medical Journal*, 92:831–832.

11.1.16 Vibrio
General description
Vibrio spp. are small, curved (comma-shaped), Gram-negative bacteria with a single polar flagellum. Species are typed according to their O antigens. There are a number of pathogenic species, including *V. cholerae*, *V. parahaemolyticus* and *V. vulnificus*. *Vibrio cholerae* is the only pathogenic species of significance from freshwater environments. While a number of serotypes can cause diarrhoea, only O1 and O139 currently cause the classical cholera symptoms in which a proportion of cases suffer fulminating and severe watery diarrhoea. The O1 serovar has been further divided into "classical" and "El Tor" biotypes. The latter is distinguished by features such as the ability to produce a dialysable heat-labile haemolysin, active against sheep and goat red blood cells. The classical biotype is considered responsible for the first six cholera pandemics, while the El Tor biotype is responsible for the seventh pandemic that commenced in 1961. Strains of *V. cholerae* O1 and O139 that cause cholera produce an enterotoxin (cholera toxin) that alters the ionic fluxes across the intestinal mucosa, resulting in substantial loss of water and electrolytes in liquid stools. Other factors associated with infection are an adhesion factor and an attachment pilus. Not all strains of serotypes O1 or O139 possess the virulence factors, and they are rarely possessed by non-O1/O139 strains.

Human health effects
Cholera outbreaks continue to occur in many areas of the developing world. Symptoms are caused by heat-labile cholera enterotoxin carried by toxigenic strains of *V.*

cholerae O1/O139. A large percentage of infected persons do not develop illness; about 60% of the classical and 75% of the El Tor group infections are asymptomatic. Symptomatic illness ranges from mild or moderate to severe disease. The initial symptoms of cholera are an increase in peristalses followed by loose, watery and mucus-flecked "rice-water" stools that may cause a patient to lose as much as 10–15 litres of liquid per day. Decreasing gastric acidity by administration of sodium bicarbonate reduces the infective dose of *V. cholerae* O1 from more than 10^8 to about 10^4 organisms. Case fatality rates vary according to facilities and preparedness. As many as 60% of untreated patients may die as a result of severe dehydration and loss of electrolytes, but well established diarrhoeal disease control programmes can reduce fatalities to less than 1%. Non-toxigenic strains of *V. cholerae* can cause self-limiting gastroenteritis, wound infections and bacteraemia.

Source and occurrence
Non-toxigenic *V. cholerae* is widely distributed in water environments, but toxigenic strains are not distributed as widely. Humans are an established source of toxigenic *V. cholerae*; in the presence of disease, the organism can be detected in sewage. Although *V. cholerae* O1 can be isolated from water in areas without disease, the strains are not generally toxigenic. Toxigenic *V. cholerae* has also been found in association with live copepods as well as other aquatic organisms, including molluscs, crustaceans, plants, algae and cyanobacteria. Numbers associated with these aquatic organisms are often higher than in the water column. Non-toxigenic *V. cholerae* has been isolated from birds and herbivores in areas far away from marine and coastal waters. The prevalence of *V. cholerae* decreases as water temperatures fall below 20 °C.

Routes of exposure
Cholera is typically transmitted by the faecal–oral route, and the infection is predominantly contracted by the ingestion of faecally contaminated water and food. The high numbers required to cause infection make person-to-person contact an unlikely route of transmission.

Significance in drinking-water
Contamination of water due to poor sanitation is largely responsible for transmission, but this does not fully explain the seasonality of recurrence, and factors other than poor sanitation must play a role. The presence of the pathogenic *V. cholerae* O1 and O139 serotypes in drinking-water supplies is of major public health importance and can have serious health and economic implications in the affected communities. *Vibrio cholerae* is highly sensitive to disinfection processes. Within a WSP, control measures that can be applied to manage potential risk from toxigenic *V. cholerae* include protection of raw water supplies from human waste, adequate treatment and protection of water during distribution. *Vibrio cholerae* O1 and non-O1 have been

detected in the absence of *E. coli*, and this organism (or, alternatively, thermotolerant coliforms) is not a reliable index for *V. cholerae* in drinking-water.

Selected bibliography

Kaper JB, Morris JG, Levine MM (1995) Cholera. *Clinical Microbiology Reviews*, 8:48–86.

Ogg JE, Ryder RA, Smith HL (1989) Isolation of *Vibrio cholerae* from aquatic birds in Colorado and Utah. *Applied and Environmental Microbiology*, 55:95–99.

Rhodes JB, Schweitzer D, Ogg JE (1985) Isolation of non-O1 *Vibrio cholerae* associated with enteric disease of herbivores in western Colorado. *Journal of Clinical Microbiology*, 22:572–575.

WHO (2002) *Vibrio cholerae*. In: *Guidelines for drinking-water quality*, 2nd ed. *Addendum: Microbiological agents in drinking water*. Geneva, World Health Organization, pp. 119–142.

11.1.17 Yersinia

General description

The genus *Yersinia* is classified in the family Enterobacteriaceae and comprises seven species. The species *Y. pestis*, *Y. pseudotuberculosis* and certain serotypes of *Y. enterocolitica* are pathogens for humans. *Yersinia pestis* is the cause of bubonic plague through contact with rodents and their fleas. *Yersinia* spp. are Gram-negative rods that are motile at 25 °C but not at 37 °C.

Human health effects

Yersinia enterocolitica penetrates cells of the intestinal mucosa, causing ulcerations of the terminal ilium. Yersiniosis generally presents as an acute gastroenteritis with diarrhoea, fever and abdominal pain. Other clinical manifestations include greatly enlarged painful lymph nodes referred to as "buboes." The disease seems to be more acute in children than in adults.

Source and occurrence

Domestic and wild animals are the principal reservoir for *Yersinia* spp.; pigs are the major reservoir of pathogenic *Y. enterocolitica*, whereas rodents and small animals are the major reservoir of *Y. pseudotuberculosis*. Pathogenic *Y. enterocolitica* has been detected in sewage and polluted surface waters. However, *Y. enterocolitica* strains detected in drinking-water are more commonly non-pathogenic strains of probable environmental origin. At least some species and strains of *Yersinia* seem to be able to replicate in water environments if at least trace amounts of organic nitrogen are present, even at temperatures as low as 4 °C.

Routes of exposure
Yersinia spp. are transmitted by the faecal–oral route, with the major source of infection considered to be foods, particularly meat and meat products, milk and dairy products. Ingestion of contaminated water is also a potential source of infection. Direct transmission from person to person and from animals to humans is also known to occur.

Significance in drinking-water
Although most *Yersinia* spp. detected in water are probably non-pathogenic, circumstantial evidence has been presented to support transmission of *Y. enterocolitica* and *Y. pseudotuberculosis* to humans from untreated drinking-water. The most likely source of pathogenic *Yersinia* spp. is human or animal waste. The organisms are sensitive to disinfection processes. Within a WSP, control measures that can be used to minimize the presence of pathogenic *Yersinia* spp. in drinking-water supplies include protection of raw water supplies from human and animal waste, adequate disinfection and protection of water during distribution. Owing to the long survival and/or growth of some strains of *Yersinia* spp. in water, *E. coli* (or, alternatively, thermotolerant coliforms) is not a suitable index for the presence/absence of these organisms in drinking-water.

Selected bibliography
Aleksic S, Bockemuhl J (1988) Serological and biochemical characteristics of 416 *Yersinia* strains from well water and drinking water plants in the Federal Republic of Germany: lack of evidence that these strains are of public health significance. *Zentralblatt für Bakteriologie, Mikrobiologie und Hygiene B*, 185:527–533.

Inoue M et al. (1988) Three outbreaks of *Yersinia pseudotuberculosis* infection. *Zentralblatt für Bakteriologie, Mikrobiologie und Hygiene B*, 186:504–511.

Ostroff SM et al. (1994) Sources of sporadic *Yersinia enterocolitica* infections in Norway: a prospective case control study. *Epidemiology and Infection*, 112:133–141.

Waage AS et al. (1999) Detection of low numbers of pathogenic *Yersinia enterocolitica* in environmental water and sewage samples by nested polymerase chain reaction. *Journal of Applied Microbiology*, 87:814–821.

11.2 Viral pathogens
Viruses associated with waterborne transmission are predominantly those that can infect the gastrointestinal tract and are excreted in the faeces of infected humans (enteric viruses). With the exception of hepatitis E, humans are considered to be the only source of human infectious species. Enteric viruses typically cause acute disease with a short incubation period. Water may also play a role in the transmission of other viruses with different modes of action. As a group, viruses can cause a wide variety of infections and symptoms involving different routes of transmission, routes and sites

of infection and routes of excretion. The combination of these routes and sites of infection can vary and will not always follow expected patterns. For example, viruses that are considered to primarily cause respiratory infections and symptoms are usually transmitted by person-to-person spread of respiratory droplets. However, some of these respiratory viruses may be discharged in faeces, leading to potential contamination of water and subsequent transmission through aerosols and droplets. Another example is viruses excreted in urine, such as polyomaviruses, which could contaminate and then be potentially transmitted by water, with possible long-term health effects, such as cancer, that are not readily associated epidemiologically with waterborne transmission.

11.2.1 Adenoviruses
General description
The family Adenoviridae is classified into the two genera *Mastadenovirus* (mammal hosts) and *Aviadenovirus* (avian hosts). Adenoviruses are widespread in nature, infecting birds, mammals and amphibians. To date, 51 antigenic types of human adenoviruses (HAds) have been described. HAds have been classified into six groups (A–F) on the basis of their physical, chemical and biological properties. Adenoviruses consist of a double-stranded DNA genome in a non-enveloped icosahedral capsid with a diameter of about 80 nm and unique fibres. The subgroups A–E grow readily in cell culture, but serotypes 40 and 41 are fastidious and do not grow well. Identification of serotypes 40 and 41 in environmental samples is generally based on polymerase chain reaction (PCR) techniques with or without initial cell culture amplification.

Human health effects
HAds cause a wide range of infections with a spectrum of clinical manifestations. These include infections of the gastrointestinal tract (gastroenteritis), the respiratory tract (acute respiratory diseases, pneumonia, pharyngoconjunctival fever), the urinary tract (cervicitis, urethritis, haemorrhagic cystitis) and the eyes (epidemic keratoconjunctivitis, also known as "shipyard eye"; pharyngoconjunctival fever, also known as "swimming pool conjunctivitis"). Different serotypes are associated with specific illnesses; for example, types 40 and 41 are the main cause of enteric illness. Adenoviruses are an important source of childhood gastroenteritis. In general, infants and children are most susceptible to adenovirus infections, and many infections are asymptomatic. High attack rates in outbreaks imply that infecting doses are low.

Source and occurrence
Adenoviruses are excreted in large numbers in human faeces and are known to occur in sewage, raw water sources and treated drinking-water supplies worldwide. Although the subgroup of enteric adenoviruses (mainly types 40 and 41) is a major cause of gastroenteritis worldwide, notably in developing communities, little is known about the prevalence of these enteric adenoviruses in water sources. The limited availability

of information on enteric adenoviruses is largely due to the fact that they are not detectable by conventional cell culture isolation.

Routes of exposure

Owing to the diverse epidemiology of the wide spectrum of HAds, exposure and infection are possible by a variety of routes. Person-to-person contact plays a major role in the transmission of illness; depending on the nature of illness, this can include faecal–oral, oral–oral and hand–eye contact transmission, as well as indirect transfer through contaminated surfaces or shared utensils. There have been numerous outbreaks associated with hospitals, military establishments, child care centres and schools. Symptoms recorded in most outbreaks were acute respiratory disease, keratoconjunctivitis and conjunctivitis. Outbreaks of gastroenteritis have also been reported. The consumption of contaminated food or water may be an important source of enteric illness, although there is no substantial evidence supporting this route of transmission. Eye infections may be contracted by the exposure of eyes to contaminated water, the sharing of towels at swimming pools or the sharing of goggles, as in the case of "shipyard eye." Confirmed outbreaks of adenovirus infections associated with water have been limited to pharyngitis and/or conjunctivitis, with exposure arising from use of swimming pools.

Significance in drinking-water

HAds have been shown to occur in substantial numbers in raw water sources and treated drinking-water supplies. In one study, the incidence of HAds in such waters was exceeded only by the group of enteroviruses among viruses detectable by PCR-based techniques. In view of their prevalence as an enteric pathogen and detection in water, contaminated drinking-water represents a likely but unconfirmed source of HAd infections. HAds are also considered important because they are exceptionally resistant to some water treatment and disinfection processes, notably UV light irradiation. HAds have been detected in drinking-water supplies that met accepted specifications for treatment, disinfection and conventional indicator organisms. Within a WSP, control measures to reduce potential risk from HAds should focus on prevention of source water contamination by human waste, followed by adequate treatment and disinfection. The effectiveness of treatment processes used to remove HAds will require validation. Drinking-water supplies should also be protected from contamination during distribution. Because of the high resistance of the viruses to disinfection, *E. coli* (or, alternatively, thermotolerant coliforms) is not a reliable index of the presence/absence of HAds in drinking-water supplies.

Selected bibliography

Chapron CD et al. (2000) Detection of astroviruses, enteroviruses and adenoviruses types 40 and 41 in surface waters collected and evaluated by the information collection rule and integrated cell culture-nested PCR procedure. *Applied and Environmental Microbiology*, 66:2520–2525.

D'Angelo LJ et al. (1979) Pharyngoconjunctival fever caused by adenovirus type 4: Report of a swimming pool-related outbreak with recovery of virus from pool water. *Journal of Infectious Diseases*, 140:42–47.

Grabow WOK, Taylor MB, de Villiers JC (2001) New methods for the detection of viruses: call for review of drinking water quality guidelines. *Water Science and Technology*, 43:1–8.

Puig M et al. (1994) Detection of adenoviruses and enteroviruses in polluted water by nested PCR amplification. *Applied and Environmental Microbiology*, 60:2963–2970.

11.2.2 Astroviruses

General description

Human and animal strains of astroviruses are single-stranded RNA viruses classified in the family Astroviridae. Astroviruses consist of a single-stranded RNA genome in a non-enveloped icosahedral capsid with a diameter of about 28 nm. In a proportion of the particles, a distinct surface star-shaped structure can be seen by electron microscopy. Eight different serotypes of human astroviruses (HAstVs) have been described. The most commonly identified is HAstV serotype 1. HAstVs can be detected in environmental samples using PCR techniques with or without initial cell culture amplification.

Human health effects

HAstVs cause gastroenteritis, predominantly diarrhoea, mainly in children under 5 years of age, although it has also been reported in adults. Seroprevalence studies showed that more than 80% of children between 5 and 10 years of age have antibodies against HAstVs. Occasional outbreaks in schools, nurseries and families have been reported. The illness is self-limiting, is of short duration and has a peak incidence in the winter. HAstVs are the cause of only a small proportion of reported gastroenteritis infections. However, the number of infections may be underestimated, since the illness is usually mild, and many cases will go unreported.

Source and occurrence

Infected individuals generally excrete large numbers of HAstVs in faeces; hence, the viruses will be present in sewage. HAstVs have been detected in water sources and in drinking-water supplies.

Routes of exposure

HAstVs are transmitted by the faecal–oral route. Person-to-person spread is considered the most common route of transmission, and clusters of cases are seen in child care centres, paediatric wards, families, homes for the elderly and military establishments. Ingestion of contaminated food or water could also be important.

Significance in drinking-water
The presence of HAstVs in treated drinking-water supplies has been confirmed. Since the viruses are typically transmitted by the faecal–oral route, transmission by drinking-water seems likely, but has not been confirmed. HAstVs have been detected in drinking-water supplies that met accepted specifications for treatment, disinfection and conventional indicator organisms. Within a WSP, control measures to reduce potential risk from HAstVs should focus on prevention of source water contamination by human waste, followed by adequate treatment and disinfection. The effectiveness of treatment processes used to remove HAstVs will require validation. Drinking-water supplies should also be protected from contamination during distribution. Owing to the higher resistance of the viruses to disinfection, *E. coli* (or, alternatively, thermotolerant coliforms) is not a reliable index of the presence/absence of HAstVs in drinking-water supplies.

Selected bibliography
Grabow WOK, Taylor MB, de Villiers JC (2001) New methods for the detection of viruses: call for review of drinking water quality guidelines. *Water Science and Technology*, 43:1–8.
Nadan S et al. (2003) Molecular characterization of astroviruses by reverse transcriptase PCR and sequence analysis: comparison of clinical and environmental isolates from South Africa. *Applied and Environmental Microbiology*, 69:747–753.
Pintó RM et al. (2001) Astrovirus detection in wastewater. *Water Science and Technology*, 43:73–77.

11.2.3 Caliciviruses
General description
The family Caliciviridae consists of four genera of single-stranded RNA viruses with a non-enveloped capsid (diameter 35–40 nm), which generally displays a typical surface morphology resembling cup-like structures. Human caliciviruses (HuCVs) include the genera *Norovirus* (Norwalk-like viruses) and *Sapovirus* (Sapporo-like viruses). *Sapovirus* spp. demonstrate the typical calicivirus morphology and are called classical caliciviruses. Noroviruses generally fail to reveal the typical morphology and were in the past referred to as small round-structured viruses. The remaining two genera of the family contain viruses that infect animals other than humans. HuCVs cannot be propagated in available cell culture systems. The viruses were originally discovered by electron microscopy. Some *Norovirus* spp. can be detected by ELISA using antibodies raised against baculovirus-expressed *Norovirus* capsid proteins. Several reverse transcriptase PCR procedures have been described for the detection of HuCVs.

Human health effects

HuCVs are a major cause of acute viral gastroenteritis in all age groups. Symptoms include nausea, vomiting and abdominal cramps. Usually about 40% of infected individuals present with diarrhoea; some have fever, chills, headache and muscular pain. Since some cases present with vomiting only and no diarrhoea, the condition is also known as "winter vomiting disease." Infections by HuCVs induce a short-lived immunity. The symptoms are usually relatively mild and rarely last for more than 3 days. High attack rates in outbreaks indicate that the infecting dose is low.

Source and occurrence

HuCVs are excreted in faeces of infected individuals and will therefore be present in domestic wastewaters as well as faecally contaminated food and water, including drinking-water supplies.

Routes of exposure

The epidemiology of the disease indicates that person-to-person contact and the inhalation of contaminated aerosols and dust particles, as well as airborne particles of vomitus, are the most common routes of transmission. Drinking-water and a wide variety of foods contaminated with human faeces have been confirmed as major sources of exposure. Numerous outbreaks have been associated with contaminated drinking-water, ice, water on cruise ships and recreational waters. Shellfish harvested from sewage-contaminated waters have also been identified as a source of outbreaks.

Significance in drinking-water

Many HuCV outbreaks have been epidemiologically linked to contaminated drinking-water supplies. Within a WSP, control measures to reduce potential risk from HuCV should focus on prevention of source water contamination by human waste, followed by adequate treatment and disinfection. The effectiveness of treatment processes used to remove HuCV will require validation. Drinking-water supplies should also be protected from contamination during distribution. Owing to the higher resistance of the viruses to disinfection, *E. coli* (or, alternatively, thermotolerant coliforms) is not a reliable index of the presence/absence of HuCVs in drinking-water supplies.

Selected bibliography

Berke T et al. (1997) Phylogenetic analysis of the Caliciviridae. *Journal of Medical Virology*, 52:419–424.

Jiang X et al. (1999) Design and evaluation of a primer pair that detects both Norwalk- and Sapporo-like caliciviruses by RT-PCR. *Journal of Virological Methods*, 83:145–154.

Mauer AM, Sturchler DA (2000) A waterborne outbreak of small round-structured virus, *Campylobacter* and *Shigella* co-infections in La Neuveville, Switzerland, 1998. *Epidemiology and Infection*, 125:325–332.

Monroe SS, Ando T, Glass R (2000) Introduction: Human enteric caliciviruses – An emerging pathogen whose time has come. *Journal of Infectious Diseases*, 181(Suppl. 2):S249–251.

11.2.4 Enteroviruses

General description

The genus *Enterovirus* is a member of the family Picornaviridae. This genus consists of 69 serotypes (species) that infect humans: poliovirus types 1–3, coxsackievirus types A1–A24, coxsackievirus types B1–B6, echovirus types 1–33 and the numbered enterovirus types EV68–EV73. Members of the genus are collectively referred to as enteroviruses. Other species of the genus infect animals other than humans – for instance, the bovine group of enteroviruses. Enteroviruses are among the smallest known viruses and consist of a single-stranded RNA genome in a non-enveloped icosahedral capsid with a diameter of 20–30 nm. Some members of the genus are readily isolated by cytopathogenic effect in cell cultures, notably poliovirus, coxsackievirus B, echovirus and enterovirus.

Human health effects

Enteroviruses are one of the most common causes of human infections. They have been estimated to cause about 30 million infections in the USA each year. The spectrum of diseases caused by enteroviruses is broad and ranges from a mild febrile illness to myocarditis, meningoencephalitis, poliomyelitis, herpangina, hand-foot-and-mouth disease and neonatal multi-organ failure. The persistence of the viruses in chronic conditions such as polymyositis, dilated cardiomyopathy and chronic fatigue syndrome has been described. Most infections, particularly in children, are asymptomatic, but still lead to the excretion of large numbers of the viruses, which may cause clinical disease in other individuals.

Source and occurrence

Enteroviruses are excreted in the faeces of infected individuals. Among the types of viruses detectable by conventional cell culture isolation, enteroviruses are generally the most numerous in sewage, water resources and treated drinking-water supplies. The viruses are also readily detected in many foods.

Routes of exposure

Person-to-person contact and inhalation of airborne viruses or viruses in respiratory droplets are considered to be the predominant routes of transmission of enteroviruses in communities. Transmission from drinking-water could also be important, but this has not yet been confirmed. Waterborne transmission of enteroviruses (coxsackievirus

A16 and B5) has been epidemiologically confirmed for only two outbreaks, and these were associated with children bathing in lake water in the 1970s.

Significance in drinking-water
Enteroviruses have been shown to occur in substantial numbers in raw water sources and treated drinking-water supplies. In view of their prevalence, drinking-water represents a likely, although unconfirmed, source of enterovirus infection. The limited knowledge on the role of waterborne transmission could be related to a number of factors, including the wide range of clinical symptoms, frequent asymptomatic infection, the diversity of serotypes and the dominance of person-to-person spread. Enteroviruses have been detected in drinking-water supplies that met accepted specifications for treatment, disinfection and conventional indicator organisms. Within a WSP, control measures to reduce potential risk from enteroviruses should focus on prevention of source water contamination by human waste, followed by adequate treatment and disinfection. The effectiveness of treatment processes used to remove enteroviruses will require validation. Drinking-water supplies should also be protected from contamination during distribution. Owing to the higher resistance of the viruses to disinfection, *E. coli* (or, alternatively, thermotolerant coliforms) is not a reliable index of the presence/absence of enteroviruses in drinking-water supplies.

Selected bibliography
Grabow WOK, Taylor MB, de Villiers JC (2001) New methods for the detection of viruses: call for review of drinking water quality guidelines. *Water Science and Technology*, 43:1–8.
Hawley HB et al. (1973) Coxsackie B epidemic at a boys' summer camp. *Journal of the American Medical Association*, 226:33–36.

11.2.5 Hepatitis A virus
General description
HAV is the only species of the genus *Hepatovirus* in the family Picornaviridae. The virus shares basic structural and morphological features with other members of the family, as described for enteroviruses. Human and simian HAVs are genotypically distinguishable. HAV cannot be readily detected or cultivated in conventional cell culture systems, and identification in environmental samples is based on the use of PCR techniques.

Human health effects
HAV is highly infectious, and the infecting dose is considered to be low. The virus causes the disease hepatitis A, commonly known as "infectious hepatitis." Like other members of the group enteric viruses, HAV enters the gastrointestinal tract by ingestion, where it infects epithelial cells. From here, the virus enters the bloodstream and reaches the liver, where it may cause severe damage to liver cells. In as many as 90%

of cases, particularly in children, there is little, if any, liver damage, and the infection passes without clinical symptoms and elicits lifelong immunity. In general, the severity of illness increases with age. The damage to liver cells results in the release of liver-specific enzymes such as aspartate aminotransferase, which are detectable in the bloodstream and used as a diagnostic tool. The damage also results in the failure of the liver to remove bilirubin from the bloodstream; the accumulation of bilirubin causes the typical symptoms of jaundice and dark urine. After a relatively long incubation period of 28–30 days on average, there is a characteristic sudden onset of illness, including symptoms such as fever, malaise, nausea, anorexia, abdominal discomfort and eventually jaundice. Although mortality is generally less than 1%, repair of the liver damage is a slow process that may keep patients incapacitated for 6 weeks or longer. This has substantial burden of disease implications. Mortality is higher in those over 50 years of age.

Source and occurrence
HAV occurs worldwide, but the prevalence of clinical disease has typical geographically based characteristics. HAV is excreted in faecal material of infected people, and there is strong epidemiological evidence that faecally contaminated food and water are common sources of the virus. In areas with poor sanitation, children are often infected at a very early age and become immune for life without clinical symptoms of disease. In areas with good sanitation, infection tends to occur later in life.

Routes of exposure
Person-to-person spread is probably the most common route of transmission, but contaminated food and water are important sources of infection. There is stronger epidemiological evidence for waterborne transmission of HAV than for any other virus. Foodborne outbreaks are also relatively common, with sources of infection including infected food handlers, shellfish harvested from contaminated water and contaminated produce. Travel of people from areas with good sanitation to those with poor sanitation provides a high risk of infection. Infection can also be spread in association with injecting and non-injecting drug use.

Significance in drinking-water
The transmission of HAV by drinking-water supplies is well established, and the presence of HAV in drinking-water constitutes a substantial health risk. Within a WSP, control measures to reduce potential risk from HAV should focus on prevention of source water contamination by human waste, followed by adequate treatment and disinfection. The effectiveness of treatment processes used to remove HAV will require validation. Drinking-water supplies should also be protected from contamination during distribution. Owing to the higher resistance of the viruses to disinfection, *E. coli* (or, alternatively, thermotolerant coliforms) is not a reliable index of the presence/absence of HAV in drinking-water supplies.

Selected bibliography
Cuthbert JA (2001) Hepatitis A: Old and new. *Clinical Microbiology Reviews*, 14:38–58.
WHO (2002) Enteric hepatitis viruses. In: *Guidelines for drinking-water quality*, 2nd ed. *Addendum: Microbiological agents in drinking water*. Geneva, World Health Organization, pp. 18–39.

11.2.6 Hepatitis E virus
General description
HEV consists of a single-stranded RNA genome in a non-enveloped icosahedral capsid with a diameter of 27–34 nm. HEV shares properties with a number of viruses, and classification is a challenge. At one stage, HEV was classified as a member of the family Caliciviridae, but most recently it has been placed in a separate family called hepatitis E-like viruses. There are indications of antigenic variation, and possibly even differences in serotypes of the virus, whereas human HAV consists of only one clearly defined serotype. HEV cannot be readily detected or cultivated in conventional cell culture systems, and identification in environmental samples is based on the use of PCR techniques.

Human health effects
HEV causes hepatitis that is in many respects similar to that caused by HAV. However, the incubation period tends to be longer (average 40 days), and infections typically have a mortality rate of up to 25% in pregnant women. In endemic regions, first infections are typically seen in young adults rather than young children. Despite evidence of antigenic variation, single infection appears to provide lifelong immunity to HEV. Global prevalence has a characteristic geographic distribution. HEV is endemic and causes clinical diseases in certain developing parts of the world, such as India, Nepal, central Asia, Mexico and parts of Africa. In many of these areas, HEV is the most important cause of viral hepatitis. Although seroprevalence can be high, clinical cases and outbreaks are rare in certain parts of the world, such as Japan, South Africa, the United Kingdom, North and South America, Australasia and central Europe. The reason for the lack of clinical cases in the presence of the virus is unknown.

Source and occurrence
HEV is excreted in faeces of infected people, and the virus has been detected in raw and treated sewage. Contaminated water has been associated with very large outbreaks. HEV is distinctive, in that it is the only enteric virus with a meaningful animal reservoir, including domestic animals, particularly pigs, as well as cattle, goats and even rodents.

Routes of exposure
Secondary transmission of HEV from cases to contacts and particularly nursing staff has been reported, but appears to be much less common than for HAV. The lower

level of person-to-person spread suggests that faecally polluted water could play a much more important role in the spread of HEV than of HAV. Waterborne outbreaks involving thousands of cases are on record. These include one outbreak in 1954 with approximately 40 000 cases in Delhi, India; one with more than 100 000 cases in 1986–1988 in the Xinjiang Uighar region of China; and one in 1991 with some 79 000 cases in Kanpur, India. Animal reservoirs may also serve as a route of exposure, but the extent to which humans contract HEV infection from animals remains to be elucidated.

Significance in drinking-water
The role of contaminated water as a source of HEV has been confirmed, and the presence of the virus in drinking-water constitutes a major health risk. There is no laboratory information on the resistance of the virus to disinfection processes, but data on waterborne outbreaks suggest that HEV may be as resistant as other enteric viruses. Within a WSP, control measures to reduce potential risk from HEV should focus on prevention of source water contamination by human and animal waste, followed by adequate treatment and disinfection. The effectiveness of treatment processes used to remove HEV will require validation. Drinking-water supplies should also be protected from contamination during distribution. Due to the likelihood that the virus has a higher resistance to disinfection, *E. coli* (or, alternatively, thermotolerant coliforms) is not a reliable index of the presence/absence of HEV in drinking-water supplies.

Selected bibliography
Pina S et al. (1998) Characterization of a strain of infectious hepatitis E virus isolated from sewage in an area where hepatitis E is not endemic. *Applied and Environmental Microbiology*, 64:4485–4488.
Van der Poel WHM et al. (2001) Hepatitis E virus sequence in swine related to sequences in humans, the Netherlands. *Emerging Infectious Diseases*, 7:970–976.
WHO (2002) Enteric hepatitis viruses. In: *Guidelines for drinking-water quality*, 2nd ed. *Addendum: Microbiological agents in drinking water*. Geneva, World Health Organization, pp. 18–39.

11.2.7 Rotaviruses and orthoreoviruses
General description
Members of the genus *Rotavirus* consist of a segmented double-stranded RNA genome in a non-enveloped icosahedral capsid with a diameter of 50–65 nm. This capsid is surrounded by a double-layered shell, giving the virus the appearance of a wheel – hence the name rotavirus. The diameter of the entire virus is about 80 nm. *Rotavirus* and *Orthoreovirus* are the two genera of the family Reoviridae typically associated with human infection. Orthoreoviruses are readily isolated by cytopathogenic effect on cell cultures. The genus *Rotavirus* is serologically divided into seven groups, A–G, each of which consists of a number of subgroups; some of these subgroups specifically infect

humans, whereas others infect a wide spectrum of animals. Groups A–C are found in humans, with group A being the most important human pathogens. Wild-type strains of rotavirus group A are not readily grown in cell culture, but there are a number of PCR-based detection methods available for testing environmental samples.

Human health effects

Human rotaviruses (HRVs) are the most important single cause of infant death in the world. Typically, 50–60% of cases of acute gastroenteritis of hospitalized children throughout the world are caused by HRVs. The viruses infect cells in the villi of the small intestine, with disruption of sodium and glucose transport. Acute infection has an abrupt onset of severe watery diarrhoea with fever, abdominal pain and vomiting; dehydration and metabolic acidosis may develop, and the outcome may be fatal if the infection is not appropriately treated. The burden of disease of rotavirus infections is extremely high. Members of the genus *Orthoreovirus* infect many humans, but they are typical "orphan viruses" and not associated with any meaningful disease.

Source and occurrence

HRVs are excreted by patients in numbers up to 10^{11} per gram of faeces for periods of about 8 days. This implies that domestic sewage and any environments polluted with the human faeces are likely to contain large numbers of HRVs. The viruses have been detected in sewage, rivers, lakes and treated drinking-water. Orthoreoviruses generally occur in wastewater in substantial numbers.

Routes of exposure

HRVs are transmitted by the faecal–oral route. Person-to-person transmission and the inhalation of airborne HRVs or aerosols containing the viruses would appear to play a much more important role than ingestion of contaminated food or water. This is confirmed by the spread of infections in children's wards in hospitals, which takes place much faster than can be accounted for by the ingestion of food or water contaminated by the faeces of infected patients. The role of contaminated water in transmission is lower than expected, given the prevalence of HRV infections and presence in contaminated water. However, occasional waterborne and foodborne outbreaks have been described. Two large outbreaks in China in 1982–1983 were linked to contaminated water supplies.

Significance in drinking-water

Although ingestion of drinking-water is not the most common route of transmission, the presence of HRVs in drinking-water constitutes a public health risk. There is some evidence that the rotaviruses are more resistant to disinfection than other enteric viruses. Within a WSP, control measures to reduce potential risk from HRVs should focus on prevention of source water contamination by human waste, followed by adequate treatment and disinfection. The effectiveness of treatment processes used to remove HRVs will require validation. Drinking-water supplies should also be protected from contamination during distribution. Due to a higher resistance of the viruses to disinfection, *E. coli* (or, alternatively, thermotolerant coliforms) is not a reliable index of the presence/absence of HRVs in drinking-water supplies.

Selected bibliography

Baggi F, Peduzzi R (2000) Genotyping of rotaviruses in environmental water and stool samples in southern Switzerland by nucleotide sequence analysis of 189 base pairs at the 5' end of the VP7 gene. *Journal of Clinical Microbiology*, 38:3681–3685.

Gerba CP et al. (1996) Waterborne rotavirus: a risk assessment. *Water Research*, 30:2929–2940.

Hopkins RS et al. (1984) A community waterborne gastroenteritis outbreak: evidence for rotavirus as the agent. *American Journal of Public Health*, 74:263–265.

Hung T et al. (1984) Waterborne outbreak of rotavirus diarrhoea in adults in China caused by a novel rotavirus. *Lancet*, i:1139–1142.

Sattar SA, Raphael RA, Springthorpe VS (1984) Rotavirus survival in conventionally treated drinking water. *Canadian Journal of Microbiology*, 30:653–656.

11.3 Protozoan pathogens

Protozoa and helminths are among the most common causes of infection and disease in humans and other animals. The diseases have a major public health and socioeconomic impact. Water plays an important role in the transmission of some of these pathogens. The control of waterborne transmission presents real challenges, because most of the pathogens produce cysts, oocysts or eggs that are extremely resistant to processes generally used for the disinfection of water and in some cases can be difficult to remove by filtration processes. Some of these organisms cause "emerging diseases." In the last 25 years, the most notable example of an emerging disease caused by a protozoan pathogen is cryptosporidiosis. Other examples are diseases caused by microsporidia and *Cyclospora*. As evidence for waterborne transmission of "emerging diseases" has been reported relatively recently, some questions about their epidemiology and behaviour in water treatment and disinfection processes remain to be elucidated. It would appear that the role of water in the transmission of this group of pathogens may increase substantially in importance and complexity as human and animal populations grow and the demands for potable drinking-water escalate.

Further information on emerging diseases is provided in *Emerging Issues in Water and Infectious Disease* (WHO, 2003) and associated texts.

11.3.1 Acanthamoeba

General description

Acanthamoeba spp. are free-living amoebae (10–50 μm in diameter) common in aquatic environments and one of the prominent protozoa in soil. The genus contains some 20 species, of which *A. castellanii*, *A. polyphaga* and *A. culbertsoni* are known to be human pathogens. However, the taxonomy of the genus may change substantially when evolving molecular biological knowledge is taken into consideration. *Acanthamoeba* has a feeding, replicative trophozoite, which, under unfavourable conditions, such as an anaerobic environment, will develop into a dormant cyst that can withstand extremes of temperature (−20 to 56 °C), disinfection and desiccation.

Human health effects

Acanthamoeba culbertsoni causes granulomatous amoebic encephalitis (GAE), whereas *A. castellanii* and *A. polyphaga* are associated with acanthamoebic keratitis and acanthamoebic uveitis.

GAE is a multifocal, haemorrhagic and necrotizing encephalitis that is generally seen only in debilitated or immunodeficient persons. It is a rare but usually fatal disease. Early symptoms include drowsiness, personality changes, intense headaches, stiff neck, nausea, vomiting, sporadic low fevers, focal neurological changes, hemiparesis and seizures. This is followed by an altered mental status, diplopia, paresis, lethargy, cerebellar ataxia and coma. Death follows within a week to a year after the appearance of the first symptoms, usually as a result of bronchopneumonia. Associated disorders of GAE include skin ulcers, liver disease, pneumonitis, renal failure and pharyngitis.

Acanthamoebic keratitis is a painful infection of the cornea and can occur in healthy individuals, especially among contact lens wearers. It is a rare disease that may lead to impaired vision, permanent blindness and loss of the eye. The prevalence of antibodies to *Acanthamoeba* and the detection of the organism in the upper airways of healthy persons suggest that infection may be common with few apparent symptoms in the vast majority of cases.

Source and occurrence

The wide distribution of *Acanthamoeba* in the natural environment makes soil, airborne dust and water all potential sources. *Acanthamoeba* can be found in many types of aquatic environments, including surface water, tap water, swimming pools and contact lens solutions. Depending on the species, *Acanthamoeba* can grow over a wide temperature range in water, with the optimum temperature for pathogenic species being 30 °C. Trophozoites can exist and replicate in water while feeding on bacteria, yeasts and other organisms. Infections occur in most temperate and tropical regions of the world.

Routes of exposure

Acanthamoebic keratitis has been associated with soft contact lenses being washed with contaminated home-made saline solutions or contamination of the contact lens containers. Although the source of the contaminating organisms has not been established, tap water is one possibility. Warnings have been issued by a number of health agencies that only sterile water should be used to prepare wash solutions for contact lenses. The mode of transmission of GAE has not been established, but water is not considered to be a source of infection. The more likely routes of transmission are via the blood from other sites of colonization, such as skin lesions or lungs.

Significance in drinking-water

Cases of acanthamoebic keratitis have been associated with drinking-water due to use of tap water in preparing solutions for washing contact lenses. Cleaning of contact lenses is not considered to be a normal use for tap water, and a higher-quality water may be required. Compared with *Cryptosporidium* and *Giardia*, *Acanthamoeba* is relatively large and is amenable to removal from raw water by filtration. Reducing the presence of biofilm organisms is likely to reduce food sources and growth of the organism in distribution systems, but the organism is highly resistant to disinfection. However, as normal uses of drinking-water lack significance as a source of infection, setting a health-based target for *Acanthamoeba* spp. is not warranted.

Selected bibliography

Marshall MM et al. (1997) Waterborne protozoan pathogens. *Clinical Microbiology Reviews*, 10:67–85.

Yagita K, Endo T, De Jonckheere JF (1999) Clustering of *Acanthamoeba* isolates from human eye infections by means of mitochondrial DNA digestion patterns. *Parasitology Research*, 85:284–289.

11.3.2 Balantidium coli

General description

Balantidium coli is a unicellular protozoan parasite with a length up to 200 µm, making it the largest of the human intestinal protozoa. The trophozoites are oval in shape and covered with cilia for motility. The cysts are 60–70 µm in length and resistant to unfavourable environmental conditions, such as pH and temperature extremes. *Balantidium coli* belongs to the largest protozoan group, the ciliates, with about 7200 species, of which only *B. coli* is known to infect humans.

Human health effects

Infections in humans are relatively rare, and most are asymptomatic. The trophozoites invade the mucosa and submucosa of the large intestine and destroy the host cells when multiplying. The multiplying parasites form nests and small abscesses that break down into oval, irregular ulcers. Clinical symptoms may include dysentery similar to amoebiasis, colitis, diarrhoea, nausea, vomiting, headache and anorexia. The infections are generally self-limiting, with complete recovery.

Source and occurrence

Humans seem to be the most important host of *B. coli*, and the organism can be detected in domestic sewage. Animal reservoirs, particularly swine, also contribute to the prevalence of the cysts in the environment. The cysts have been detected in water sources, but the prevalence in tap water is unknown.

Routes of exposure
Transmission of *B. coli* is by the faecal–oral route, from person to person, from contact with infected swine or by consumption of contaminated water or food. One waterborne outbreak of balantidiasis has been reported. This outbreak occurred in 1971 when a drinking-water supply was contaminated with stormwater runoff containing swine faeces after a typhoon.

Significance in drinking-water
Although water does not appear to play an important role in the spread of this organism, one waterborne outbreak is on record. *Balantidium coli* is large and amenable to removal by filtration, but cysts are highly resistant to disinfection. Within a WSP, control measures to reduce potential risk from *B. coli* should focus on prevention of source water contamination by human and swine waste, followed by adequate treatment. Due to resistance to disinfection, *E. coli* (or, alternatively, thermotolerant coliforms) is not a reliable index for the presence/absence of *B. coli* in drinking-water supplies.

Selected bibliography
Garcia LS (1999) Flagellates and ciliates. *Clinics in Laboratory Medicine*, 19:621–638.
Walzer PD et al. (1973) Balantidiasis outbreak in Truk. *American Journal of Tropical Medicine and Hygiene*, 22:33–41.

11.3.3 Cryptosporidium
General description
Cryptosporidium is an obligate, intracellular, coccidian parasite with a complex life cycle including sexual and asexual replication. Thick-walled oocysts with a diameter of 4–6 µm are shed in faeces. The genus *Cryptosporidium* has about eight species, of which *C. parvum* is responsible for most human infections, although other species can cause illness. *Cryptosporidium* is one of the best examples of an "emerging disease"-causing organism. It was discovered to infect humans only in 1976, and waterborne transmission was confirmed for the first time in 1984.

Human health effects
Cryptosporidium generally causes a self-limiting diarrhoea, sometimes including nausea, vomiting and fever, which usually resolves within a week in normally healthy people, but can last for a month or more. Severity of cryptosporidiosis varies according to age and immune status, and infections in severely immunocompromised people can be life-threatening. The impact of cryptosporidiosis outbreaks is relatively high due to the large numbers of people that may be involved and the associated socioe-

conomic implications. The total cost of illness associated with the 1993 outbreak in Milwaukee, USA, has been estimated at US$96.2 million.

Source and occurrence

A large range of animals are reservoirs of *C. parvum*, but humans and livestock, particularly young animals, are the most significant source of human infectious organisms. Calves can excrete 10^{10} oocysts per day. Concentrations of oocysts as high as 14 000 per litre for raw sewage and 5800 per litre for surface water have been reported. Oocysts can survive for weeks to months in fresh water. *Cryptosporidium* oocysts have been detected in many drinking-water supplies. However, in most cases, there is little information about whether human infectious species were present. The currently available standard analytical techniques provide an indirect measure of viability and no indication of human infectivity. Oocysts also occur in recreational waters.

Routes of exposure

Cryptosporidium is transmitted by the faecal–oral route. The major route of infection is person-to-person contact. Other sources of infection include the consumption of contaminated food and water and direct contact with infected farm animals and possibly domestic pets. Contaminated drinking-water, recreational water and, to a lesser extent, food have been associated with outbreaks. In 1993, *Cryptosporidium* caused the largest waterborne outbreak of disease on record, when more than 400 000 people were infected by the drinking-water supply of Milwaukee, USA. The infectivity of *Cryptosporidium* oocysts is relatively high. Studies on healthy human volunteers revealed that ingestion of fewer than 10 oocysts can lead to infection.

Significance in drinking-water

The role of drinking-water in the transmission of *Cryptosporidium*, including in large outbreaks, is well established. Attention to these organisms is therefore important. The oocysts are extremely resistant to oxidizing disinfectants such as chlorine, but investigations based on assays for infectivity have shown that UV light irradiation inactivates oocysts. Within a WSP, control measures to reduce potential risk from *Cryptosporidium* should focus on prevention of source water contamination by human and livestock waste, adequate treatment and protection of water during distribution. Because of their relatively small size, the oocysts represent a challenge for removal by conventional granular media-based filtration processes. Acceptable removal requires well designed and operated systems. Membrane filtration processes that provide a direct physical barrier may represent a viable alternative for the effective removal of *Cryptosporidium* oocysts. Owing to the exceptional resistance of the oocysts to disinfectants, *E. coli* (or, alternatively, thermotolerant coliforms) cannot be relied upon as an index for the presence/absence of *Cryptosporidium* oocysts in drinking-water supplies.

Selected bibliography

Corso PS et al. (2003) Cost of illness in the 1993 waterborne *Cryptosporidium* outbreak, Milwaukee, Wisconsin. *Emerging Infectious Diseases*, 9:426–431.

Haas CN et al. (1996) Risk assessment of *Cryptosporidium parvum* oocysts in drinking water. *Journal of the American Water Works Association*, 88:131–136.

Leav BA, Mackay M, Ward HD (2003) *Cryptosporidium* species: new insight and old challenges. *Clinical Infectious Diseases*, 36:903–908.

Linden KG, Shin G, Sobsey MD (2001) Comparative effectiveness of UV wavelengths for the inactivation of *Cryptosporidium parvum* oocysts in water. *Water Science and Technology*, 43:171–174.

Okhuysen PC et al. (1999) Virulence of three distinct *Cryptosporidium parvum* isolates for healthy adults. *Journal of Infectious Diseases*, 180:1275–1281.

WHO (2002) Protozoan parasites (*Cryptosporidium, Giardia, Cyclospora*). In: *Guidelines for drinking-water quality*, 2nd ed. *Addendum: Microbiological agents in drinking water*. Geneva, World Health Organization, pp. 70–118.

11.3.4 Cyclospora cayetanensis

General description

Cyclospora cayetanensis is a single-cell, obligate, intracellular, coccidian protozoan parasite, which belongs to the family Eimeriidae. It produces thick-walled oocysts of 8–10 μm in diameter that are excreted in the faeces of infected individuals. *Cyclospora cayetanensis* is considered an emerging waterborne pathogen.

Human health effects

Sporozoites are released from the oocysts when ingested and penetrate epithelial cells in the small intestine of susceptible individuals. Clinical symptoms of cyclosporiasis include watery diarrhoea, abdominal cramping, weight loss, anorexia, myalgia and occasionally vomiting and/or fever. Relapsing illness often occurs.

Source and occurrence

Humans are the only host identified for this parasite. The unsporulated oocysts pass into the external environment with faeces and undergo sporulation, which is complete in 7–12 days, depending on environmental conditions. Only the sporulated oocysts are infectious. Due to the lack of a quantification technique, there is limited information on the prevalence of *Cyclospora* in water environments. However, *Cyclospora* has been detected in sewage and water sources.

Routes of exposure

Cyclospora cayetanensis is transmitted by the faecal–oral route. Person-to-person transmission is virtually impossible, because the oocysts must sporulate outside the host to become infectious. The primary routes of exposure are contaminated water and food. The initial source of organisms in foodborne outbreaks has generally not been established, but contaminated water has been implicated in several cases. Drinking-water has also been implicated as a cause of outbreaks. The first report was among staff of a hospital in Chicago, USA, in 1990. The infections were associated with drinking tap water that had possibly been contaminated with stagnant water from a rooftop storage reservoir. Another outbreak was reported from Nepal, where drinking-water consisting of a mixture of river and municipal water was associated with infections in 12 of 14 soldiers.

Significance in drinking-water
Transmission of the pathogens by drinking-water has been confirmed. The oocysts are resistant to disinfection and are not inactivated by chlorination practices generally applied in the production of drinking-water. Within a WSP, control measures that can be applied to manage potential risk from *Cyclospora* include prevention of source water contamination by human waste, followed by adequate treatment and protection of water during distribution. Owing to the resistance of the oocysts to disinfectants, *E. coli* (or, alternatively, thermotolerant coliforms) cannot be relied upon as an index of the presence/absence of *Cyclospora* in drinking-water supplies.

Selected bibliography
Curry A, Smith HV (1998) Emerging pathogens: *Isospora*, *Cyclospora* and microsporidia. *Parasitology*, 117:S143–159.
Dowd SE et al. (2003) Confirmed detection of *Cyclospora cayetanensis*, *Encephalitozoon intestinalis* and *Cryptosporidium parvum* in water used for drinking. *Journal of Water and Health*, 1:117–123.
Goodgame R (2003) Emerging causes of traveller's diarrhea: *Cryptosporidium*, *Cyclospora*, *Isospora* and microsporidia. *Current Infectious Disease Reports*, 5:66–73.
Herwaldt BL (2000) *Cyclospora cayetanensis*: A review, focusing on the outbreaks of cyclosporiasis in the 1990s. *Clinical Infectious Diseases*, 31:1040–1057.
Rabold JG et al. (1994) *Cyclospora* outbreak associated with chlorinated drinking-water [letter]. *Lancet*, 344:1360–1361.
WHO (2002) Protozoan parasites (*Cryptosporidium*, *Giardia*, *Cyclospora*). In: *Guidelines for drinking-water quality*, 2nd ed. Addendum: *Microbiological agents in drinking water*. Geneva, World Health Organization, pp. 70–118.

11.3.5 Entamoeba histolytica
General description
Entamoeba histolytica is the most prevalent intestinal protozoan pathogen worldwide and belongs to the superclass Rhizopoda in the subphylum Sarcodina. *Entamoeba* has a feeding, replicative trophozoite (diameter 10–60 μm), which, under unfavourable conditions, will develop into a dormant cyst (diameter 10–20 μm). Infection is contracted by the ingestion of cysts. Recent studies with RNA and DNA probes demon-

strated genetic differences between pathogenic and non-pathogenic *E. histolytica*; the latter has been separated and reclassified as *E. dispar*.

Human health effects

About 85–95% of human infections with *E. histolytica* are asymptomatic. Acute intestinal amoebiasis has an incubation period of 1–14 weeks. Clinical disease results from the penetration of the epithelial cells in the gastrointestinal tract by the amoebic trophozoites. Approximately 10% of infected individuals present with dysentery or colitis. Symptoms of amoebic dysentery include diarrhoea with cramping, lower abdominal pain, low-grade fever and the presence of blood and mucus in the stool. The ulcers produced by the invasion of the trophozoites may deepen into the classic flask-shaped ulcers of amoebic colitis. *Entamoeba histolytica* may invade other parts of the body, such as the liver, lungs and brain, sometimes with fatal outcome.

Source and occurrence

Humans are the reservoir of infection, and there would not appear to be other meaningful animal reservoirs of *E. histolytica*. In the acute phase of infection, patients excrete only trophozoites that are not infectious. Chronic cases and asymptomatic carriers who excrete cysts are more important sources of infection and can discharge up to 1.5×10^7 cysts daily. *Entamoeba histolytica* can be present in sewage and contaminated water. Cysts may remain viable in suitable aquatic environments for several months at low temperature. The potential for waterborne transmission is greater in the tropics, where the carrier rate sometimes exceeds 50%, compared with more temperate regions, where the prevalence in the general population may be less than 10%.

Routes of exposure

Person-to-person contact and contamination of food by infected food handlers appear to be the most significant means of transmission, although contaminated water also plays a substantial role. Ingestion of faecally contaminated water and consumption of food crops irrigated with contaminated water can both lead to transmission of amoebiasis. Sexual transmission, particularly among male homosexuals, has also been documented.

Significance in drinking-water

The transmission of *E. histolytica* by contaminated drinking-water has been confirmed. The cysts are relatively resistant to disinfection and may not be inactivated by chlorination practices generally applied in the production of drinking-water. Within a WSP, control measures that can be applied to manage potential risk from *E. histolytica* include prevention of source water contamination by human waste, followed by adequate treatment and protection of water during distribution. Owing to the resistance of the oocysts to disinfectants, *E. coli* (or, alternatively, thermotolerant coliforms) cannot be relied upon as an index of the presence/absence of *E. histolytica* in drinking-water supplies.

Selected bibliography
Marshall MM et al. (1997) Waterborne protozoan pathogens. *Clinical Microbiology Reviews*, 10:67–85.

11.3.6 *Giardia intestinalis*
General description
Giardia spp. are flagellated protozoa that parasitize the gastrointestinal tract of humans and certain animals. The genus *Giardia* consists of a number of species, but human infection (giardiasis) is usually assigned to *G. intestinalis*, also known as *G. lamblia* or *G. duodenalis*. *Giardia* has a relatively simple life cycle consisting of a flagellate trophozoite that multiplies in the gastrointestinal tract and an infective thick-walled cyst that is shed intermittently but in large numbers in faeces. The trophozoites are bilaterally symmetrical and ellipsoidal in shape. The cysts are ovoid in shape and 8–12 µm in diameter.

Human health effects
Giardia has been known as a human parasite for 200 years. After ingestion and excystation of cysts, the trophozoites attach to surfaces of the gastrointestinal tract. Infections in both children and adults may be asymptomatic. In day care centres, as many as 20% of children may carry *Giardia* and excrete cysts without clinical symptoms. The symptoms of giardiasis may result from damage caused by the trophozoites, although the mechanisms by which *Giardia* causes diarrhoea and intestinal malabsorption remain controversial. Symptoms generally include diarrhoea and abdominal cramps; in severe cases, however, malabsorption deficiencies in the small intestine may be present, mostly among young children. Giardiasis is self-limiting in most cases, but it may be chronic in some patients, lasting more than 1 year, even in otherwise healthy people. Studies on human volunteers revealed that fewer than 10 cysts constitute a meaningful risk of infection.

Source and occurrence
Giardia can multiply in a wide range of animal species, including humans, which excrete cysts into the environment. Numbers of cysts as high as 88 000 per litre in raw sewage and 240 per litre in surface water resources have been reported. These cysts are robust and can survive for weeks to months in fresh water. The presence of cysts in raw water sources and drinking-water supplies has been confirmed. However, there is no information on whether human infectious species were present. The currently available standard analytical techniques provide an indirect measure of viability and no indication of human infectivity. Cysts also occur in recreational waters and contaminated food.

Routes of exposure
By far the most common route of transmission of *Giardia* is person-to-person contact, particularly between children. Contaminated drinking-water, recreational water and, to a lesser extent, food have been associated with outbreaks. Animals have been implicated as a source of human infectious *G. intestinalis*, but further investigations are required to determine their role.

Significance in drinking-water
Waterborne outbreaks of giardiasis have been associated with drinking-water supplies for over 30 years; at one stage, *Giardia* was the most commonly identified cause of waterborne outbreaks in the USA. *Giardia* cysts are more resistant than enteric bacteria to oxidative disinfectants such as chlorine, but they are not as resistant as *Cryptosporidium* oocysts. The time required for 90% inactivation at a free chlorine residual of 1 mg/litre is about 25–30 min. Within a WSP, control measures that can be applied to manage potential risk from *Giardia* include prevention of source water contamination by human and animal waste, followed by adequate treatment and disinfection and protection of water during distribution. Owing to the resistance of the cysts to disinfectants, *E. coli* (or, alternatively, thermotolerant coliforms) cannot be relied upon as an index of the presence/absence of *Giardia* in drinking-water supplies.

Selected bibliography
LeChevallier MW, Norton WD, Lee RG (1991) Occurrence of *Giardia* and *Cryptosporidium* species in surface water supplies. *Applied and Environmental Microbiology*, 57:2610–2616.
Ong C et al. (1996) Studies of *Giardia* spp. and *Cryptosporidium* spp. in two adjacent watersheds. *Applied and Environmental Microbiology*, 62:2798–2805.
Rimhanen-Finne R et al. (2002) An IC-PCR method for detection of *Cryptosporidium* and *Giardia* in natural surface waters in Finland. *Journal of Microbiological Methods*, 50:299–303.
Slifko TR, Smith HV, Rose JB (2000) Emerging parasite zoonoses associated with water and food. *International Journal for Parasitology*, 30:1379–1393.
Stuart JM et al. (2003) Risk factors for sporadic giardiasis: a case–control study in southwestern England. *Emerging Infectious Diseases*, 9:229–233.
WHO (2002) Protozoan parasites (*Cryptosporidium, Giardia, Cyclospora*). In: *Guidelines for drinking-water quality*, 2nd ed. *Addendum: Microbiological agents in drinking water*. Geneva, World Health Organization, pp. 70–118.

11.3.7 Isospora belli
General description
Isospora is a coccidian, single-celled, obligate parasite related to *Cryptosporidium* and *Cyclospora*. There are many species of *Isospora* that infect animals, but only *I. belli* is known to infect humans, the only known host for this species. *Isospora belli* is one of

the few coccidia that undergo sexual reproduction in the human intestine. Sporulated oocysts are ingested, and, after complete asexual and sexual life cycles in the mucosal epithelium of the upper small intestine, unsporulated oocysts are released in faeces.

Human health effects

Illness caused by *I. belli* is similar to that caused by *Cryptosporidium* and *Giardia*. About 1 week after ingestion of viable cysts, a low-grade fever, lassitude and malaise may appear, followed soon by mild diarrhoea and vague abdominal pain. The infection is usually self-limited after 1–2 weeks, but occasionally diarrhoea, weight loss and fever may last for 6 weeks to 6 months. Symptomatic isosporiasis is more common in children than in adults. Infection is often associated with immunocompromised patients, in whom symptoms are more severe and likely to be recurrent or chronic, leading to malabsorption and weight loss. Infections are usually sporadic and most common in the tropics and subtropics, although they also occur elsewhere, including industrialized countries. They have been reported from Central and South America, Africa and south-east Asia.

Source and occurrence

Unsporulated oocysts are excreted in the faeces of infected individuals. The oocysts sporulate within 1–2 days in the environment to produce the potentially infectious form of the organism. Few data are available on numbers of oocysts in sewage and raw and treated water sources. This is largely because sensitive and reliable techniques for the quantitative enumeration of oocysts in water environments are not available. Little is known about the survival of oocysts in water and related environments.

Routes of exposure

Poor sanitation and faecally contaminated food and water are the most likely sources of infection, but waterborne transmission has not been confirmed. The oocysts are less likely than *Cryptosporidium* oocysts or *Giardia* cysts to be transmitted directly from person to person, because freshly shed *I. belli* oocysts require 1–2 days in the environment to sporulate before they are capable of infecting humans.

Significance in drinking-water

The characteristics of *I. belli* suggest that illness could be transmitted by contaminated drinking-water supplies, but this has not been confirmed. No information is available on the effectiveness of water treatment processes for removal of *I. belli*, but it is likely that the organism is relatively resistant to disinfectants. It is considerably larger than *Cryptosporidium* and should be easier to remove by filtration. Within a WSP, control measures that can be applied to manage potential risk from *I. belli* include prevention of source water contamination by human waste, followed by adequate treatment and disinfection and protection of water during distribution. Owing to the likely resistance of the oocysts to disinfectants, *E. coli* (or, alternatively, thermotolerant coliforms) cannot be relied upon as an index of the presence/absence of *I. belli* in drinking-water supplies.

Selected bibliography

Ballal M et al. (1999) *Cryptosporidium* and *Isospora belli* diarrhoea in immunocompromised hosts. *Indian Journal of Cancer*, 36:38–42.

Bialek R et al. (2002) Comparison of autofluorescence and iodine staining for detection of *Isospora belli* in feces. *American Journal of Tropical Medicine and Hygiene*, 67:304–305.

Curry A, Smith HV (1998) Emerging pathogens: *Isospora*, *Cyclospora* and microsporidia. *Parasitology*, 117:S143–159.

Goodgame R (2003) Emerging causes of traveller's diarrhea: *Cryptosporidium*, *Cyclospora*, *Isospora* and microsporidia. *Current Infectious Disease Reports*, 5:66–73.

11.3.8 Microsporidia

General description

The term "microsporidia" is a non-taxonomic designation commonly used to describe a group of obligate intracellular protozoa belonging to the phylum Microspora. More than 100 microsporidial genera and almost 1000 species have been identified. Infections occur in every major animal group, including vertebrates and invertebrates. A number of genera have been implicated in human infections, including *Enterocytozoon*, *Encephalitozoon* (including *Septata*), *Nosema*, *Pleistophora*, *Vittaforma* and *Trachipleistophora*, as well as a collective group of unclassified microsporidia referred to as microsporidium. Microsporidia are among the smallest eukaryotes. They produce unicellular spores with a diameter of 1.0–4.5 µm and a characteristic coiled polar filament for injecting the sporoplasm into a host cell to initiate infection. Within an infected cell, a complex process of multiplication takes place, and new spores are produced and released in faeces, urine, respiratory secretions or other body fluids, depending on the type of species and the site of infection.

Human health effects

Microsporidia are emerging human pathogens identified predominantly in persons with AIDS, but their ability to cause disease in immunologically normal hosts has been recognized. Reported human infections are globally dispersed and have been documented in persons from all continents. The most common clinical manifestation in AIDS patients is a severe enteritis involving chronic diarrhoea, dehydration and weight loss. Prolonged illness for up to 48 months has been reported. Infections in the general population are less pronounced. *Enterocytozoon* infection generally appears to be limited to intestinal enterocytes and biliary epithelium. *Encephalitozoon* spp. infect a variety of cells, including epithelial and endothelial cells, fibroblasts, kidney tubule cells, macrophages and possibly other cell types. Unusual complications include keratoconjunctivitis, myositis and hepatitis.

Source and occurrence
The sources of microsporidia infecting humans are uncertain. Spores are likely to be excreted in faeces and are also excreted in urine and respiratory secretions. Due to the lack of a quantification technique, there is limited information on the prevalence of microsporidia spores in water environments. However, microsporidia have been detected in sewage and water sources. Indications are that their numbers in raw sewage may be similar to those of *Cryptosporidium* and *Giardia*, and they may survive in certain water environments for many months. Certain animals, notably swine, may serve as a host for human infectious species.

Routes of exposure
Little is known about transmission of microsporidia. Person-to-person contact and ingestion of spores in water or food contaminated with human faeces or urine are probably important routes of exposure. A waterborne outbreak of microsporidiosis has been reported involving about 200 cases in Lyon, France, during the summer of 1995. However, the source of the organism and faecal contamination of the drinking-water supply were not demonstrated. Transmission by the inhalation of airborne spores or aerosols containing spores seems possible. The role of animals in transmission to humans remains unclear. Epidemiological and experimental studies in mammals suggest that *Encephalitozoon* spp. can be transmitted transplacentally from mother to offspring. No information is available on the infectivity of the spores. However, in view of the infectivity of spores of closely related species, the infectivity of microsporidia may be high.

Significance in drinking-water
Waterborne transmission has been reported, and infection arising from contaminated drinking-water is plausible but unconfirmed. Little is known about the response of microsporidia to water treatment processes. One study has suggested that the spores may be susceptible to chlorine. The small size of the organism is likely to make them difficult to remove by filtration processes. Within a WSP, control measures that can be applied to manage potential risk from microsporidia include prevention of source water contamination by human and animal waste, followed by adequate treatment and disinfection and protection of water during distribution. Owing to the lack of information on sensitivity of infectious species of microsporidia to disinfection, the reliability of *E. coli* (or, alternatively, thermotolerant coliforms) as an index for the presence/absence of these organisms from drinking-water supplies is unknown.

Selected bibliography
Coote L et al. (2000) Waterborne outbreak of intestinal microsporidiosis in persons with and without human immunodeficiency virus infection. *Journal of Infectious Diseases*, 180:2003–2008.

Dowd SE et al. (2003) Confirmed detection of *Cyclospora cayetanensis, Encephalitozoon intestinalis* and *Cryptosporidium parvum* in water used for drinking. *Journal of Water and Health*, 1:117–123.

Goodgame R (2003) Emerging causes of traveller's diarrhea: *Cryptosporidium, Cyclospora, Isospora* and microsporidia. *Current Infectious Disease Reports*, 5:66–73.

Joynson DHM (1999) Emerging parasitic infections in man. *The Infectious Disease Review*, 1:131–134.

Slifko TR, Smith HV, Rose JB (2000) Emerging parasite zoonoses associated with water and food. *International Journal for Parasitology*, 30:1379–1393.

11.3.9 Naegleria fowleri

General description

Naegleria are free-living amoeboflagellates distributed widely in the environment. There are several species of *Naegleria*, of which *N. fowleri* is the primary infectious species. *Naegleria* spp. exist as a trophozoite, a flagellate and a cyst stage. The trophozoite (10–20 µm) moves by eruptive pseudopod formation feeding on bacteria and reproduces by binary fission. The trophozoite can transform into a flagellate stage with two anterior flagella. The flagellate does not divide but reverts to the trophozoite stage. Under adverse conditions, the trophozoite transforms into a circular cyst (7–15 µm), which is resistant to unfavourable conditions.

Human health effects

Naegleria fowleri causes primary amoebic meningoencephalitis (PAM) in healthy individuals. The amoeba enters the brain by penetrating the olfactory mucosa and cribiform plate. The disease is acute, and patients often die within 5–10 days and before the infectious agent can be diagnosed. Treatment is difficult. Although the infection is rare, new cases are reported every year.

Source and occurrence

Naegleria fowleri is thermophilic and grows well at temperatures up to 45 °C. It occurs naturally in fresh water of suitable temperature, and prevalence is only indirectly related to human activity, inasmuch as such activity may modify temperature or promote bacterial (food source) production. The pathogen has been reported from many countries, usually associated with thermally polluted water environments such as geothermal water or heated swimming pools. However, the organism has been detected in drinking-water supplies, particularly where water temperature can exceed 25–30 °C. Water is the only known source of infection. The first cases of amoebic meningitis were diagnosed in 1965 in Australia and Florida. Since that time, about 100 cases of PAM have been reported throughout the world.

Routes of exposure

Infection with *N. fowleri* is almost exclusively contracted by exposure of the nasal passages to contaminated water. Infection is predominantly associated with recreational use of water, including swimming pools and spas, as well as surface waters naturally heated by the sun, industrial cooling waters and geothermal springs. In a limited number of cases, a link to recreational water exposure is lacking. The occurrence of PAM is highest during hot summer months, when many people engage in water recreation and when the temperature of water is conducive to growth of the organism. Consumption of contaminated water or food and person-to-person spread have not been reported as routes of transmission.

Significance in drinking-water

Naegleria fowleri has been detected in drinking-water supplies. Although unproven, a direct or indirect role of drinking-water-derived organisms – for example, through use of drinking-water in swimming pools – is possible. Any water supply that seasonally exceeds 30 °C or that continually exceeds 25 °C can potentially support the growth of *N. fowleri*. In such cases, a periodic prospective study would be valuable. Free chlorine or monochloramine residuals in excess of 0.5 mg/litre have been shown to control *N. fowleri*, providing the disinfectant persists through the water distribution system. In addition to maintaining persistent disinfectant residuals, other control measures aimed at limiting the presence of biofilm organisms will reduce food sources and hence growth of the organism in distribution systems. Owing to the environmental nature of this amoeba, *E. coli* (or, alternatively, thermotolerant coliforms) cannot be relied upon as an index for the presence/absence of *N. fowleri* in drinking-water supplies.

Selected bibliography

Behets J et al. (2003) Detection of *Naegleria* spp. and *Naegleria fowleri*: a comparison of flagellation tests, ELISA and PCR. *Water Science and Technology*, 47:117–122.

Cabanes P-A et al. (2001) Assessing the risk of primary amoebic meningoencephalitis from swimming in the presence of environmental *Naegleria fowleri*. *Applied and Environmental Microbiology*, 67:2927–2931.

Dorsch MM, Cameron AS, Robinson BS (1983) The epidemiology and control of primary amoebic meningoencephalitis with particular reference to South Australia. *Transactions of the Royal Society of Tropical Medicine and Hygiene*, 77:372–377.

Martinez AJ, Visvesvara GS (1997) Free-living amphizoic and opportunistic amebas. *Brain Pathology*, 7:583–598.

Parija SC, Jayakeerthee SR (1999) *Naegleria fowleri*: a free living amoeba of emerging medical importance. *Communicable Diseases*, 31:153–159.

11.3.10 Toxoplasma gondii

General description

Many species of *Toxoplasma* and *Toxoplasma*-like organisms have been described, but it would appear that *T. gondii* is the only human infectious species. *Toxoplasma gondii* is a coccidian parasite, and the cat is the definitive host. Only cats harbour the parasite in the intestinal tract, where sexual reproduction takes place. The actively multiplying asexual form in the human host is an obligate, intracellular parasite (diameter 3–6 µm) called a tachyzoite. A chronic phase of the disease develops as the tachyzoites transform into slowly replicating bradyzoites, which eventually become cysts in the host tissue. In the natural cycle, mice and rats containing infective cysts are eaten by cats, which host the sexual stage of the parasite. The cyst wall is digested, and bradyzoites penetrate epithelial cells of the small intestine. Several generations of intracellular multiplication lead to the development of micro- and macrogametes. Fertilization of the latter leads to the development of oocysts that are excreted in faeces as early as 5 days after a cat has ingested the cysts. Oocysts require 1–5 days to sporulate in the environment. Sporulated oocysts and tissue-borne cysts can both cause infections in susceptible hosts.

Human health effects

Toxoplasmosis is usually asymptomatic in humans. In a small percentage of cases, flu-like symptoms, lymphadenopathy and hepatosplenomegaly present 5–23 days after the ingestion of cysts or oocysts. Dormant cysts, formed in organ tissue after primary infection, can be reactivated when the immune system becomes suppressed, producing disseminated disease involving the central nervous system and lungs and leading to severe neurological disorders or pneumonia. When these infection sites are involved, the disease can be fatal in immunocompromised patients. Congenital toxoplasmosis is mostly asymptomatic, but can produce chorioretinitis, cerebral calcifications, hydrocephalus, severe thrombocytopenia and convulsions. Primary infection during early pregnancy can lead to spontaneous abortion, stillbirth or fetal abnormality.

Source and occurrence

Toxoplasmosis is found worldwide. Estimates indicate that in many parts of the world, 15–30% of lamb and pork meat is infected with cysts. The prevalence of oocyst-shedding cats may be 1%. By the third decade of life, about 50% of the European population is infected, and in France this proportion is close to 80%. *Toxoplasma gondii* oocysts may occur in water sources and supplies contaminated with the faeces of infected cats. Due to a lack of practical methods for the detection of *T. gondii* oocysts, there is little information on the prevalence of the oocysts in raw and treated water supplies. Details on the survival and behaviour of the oocysts in water environments are also not available. However, qualitative evidence of the presence of oocysts in faecally polluted water has been reported, and results suggest that *T. gondii* oocysts may be as resistant to unfavourable conditions in water environments as the oocysts of related parasites.

Routes of exposure

Both *T. gondii* oocysts that sporulate after excretion by cats and tissue-borne cysts are potentially infectious. Humans can become infected by ingestion of oocysts excreted by cats by direct contact or through contact with contaminated soil or water. Two outbreaks of toxoplasmosis have been associated with consumption of contaminated water. In Panama, creek water contaminated by oocysts from jungle cats was identified as the most likely source of infection, while in 1995, an outbreak in Canada was associated with a drinking-water reservoir being contaminated by excreta from domestic or wild cats. A study in Brazil during 1997–1999 identified the consumption of unfiltered drinking-water as a risk factor for *T. gondii* seropositivity. More commonly, humans contract toxoplasmosis through the consumption of undercooked or raw meat and meat products containing *T. gondii* cysts. Transplacental infection also occurs.

Significance in drinking-water

Contaminated drinking-water has been identified as a source of toxoplasmosis outbreaks. Little is known about the response of *T. gondii* to water treatment processes. The oocysts are larger than *Cryptosporidium* oocysts and should be amenable to removal by filtration. Within a WSP, control measures to manage potential risk from *T. gondii* should be focused on prevention of source water contamination by wild and domesticated cats. If necessary, the organisms can be removed by filtration. Owing to the lack of information on sensitivity of *T. gondii* to disinfection, the reliability of *E. coli* (or, alternatively, thermotolerant coliforms) as an indicator for the presence/absence of these organisms in drinking-water supplies is unknown.

Selected bibliography

Aramini JJ et al. (1999) Potential contamination of drinking water with *Toxoplasma gondii* oocysts. *Epidemiology and Infection*, 122:305–315.

Bahia-Oliveira LMG et al. (2003) Highly endemic, waterborne toxoplasmosis in North Rio de Janeiro State, Brazil. *Emerging Infectious Diseases*, 9:55–62.

Bowie WR et al. (1997) Outbreak of toxoplasmosis associated with municipal drinking water. The BC Toxoplasma Investigation Team. *Lancet*, 350:173–177.

Kourenti C et al. (2003) Development and application of different methods for the detection of *Toxoplasma gondii* in water. *Applied and Environmental Microbiology*, 69:102–106.

11.4 Helminth pathogens

The word "helminth" comes from the Greek word meaning "worm" and refers to all types of worms, both free-living and parasitic. The major parasitic worms are classified primarily in the phylum Nematoda (roundworms) and the phylum Platyhelminthes (flatworms including trematodes). Helminth parasites infect a large number of people and animals worldwide. For most helminths, drinking-water is not a significant route of transmission. There are two exceptions: *Dracunculus medinensis* (guinea worm) and *Fasciola* spp. (*F. hepatica* and *F. gigantica*) (liver flukes). Dracunculiasis and fascioliasis both require intermediate hosts to complete their life cycles but are transmitted through drinking-water by different mechanisms. Other helminthiases can be transmitted through water contact (schistosomiasis) or are associated with the use of untreated wastewater in agriculture (ascariasis, trichuriasis, hookworm infections and strongyloidiasis) but are not usually transmitted through drinking-water.

11.4.1 Dracunculus medinensis

Dracunculus medinensis, commonly known as "guinea worm," belongs to the phylum Nematoda and is the only nematode associated with significant transmission by drinking-water.

The eradication of guinea worm infection from the world by 1995 was a target of the International Drinking Water Supply and Sanitation Decade (1981–1990), and the World Health Assembly formally committed itself to this goal in 1991. The Dracunculus Eradication Programme has achieved a massive reduction in the number of cases. There were an estimated 3.3 million cases in 1986, 625 000 cases in 1990 and fewer than 60 000 cases in 2002, with the majority occurring in Sudan. Dracunculiasis is restricted to a central belt of countries in sub-Saharan Africa.

General description

The *D. medinensis* worms inhabit the cutaneous and subcutaneous tissues of infected individuals, the female reaching a length of up to 700 mm, and the male 25 mm. When the female is ready to discharge larvae (embryos), its anterior end emerges from a blister or ulcer, usually on the foot or lower limb, and releases large numbers of rhabditiform larvae when the affected part of the body is immersed in water. The larvae can move about in water for approximately 3 days and during that time can be ingested by many species of *Cyclops* (cyclopoid Copepoda, Crustacea). The larvae penetrate into the haemocoelom, moult twice and are infective to a new host in about 2 weeks. If the *Cyclops* (0.5–2.0 mm) are swallowed in drinking-water, the larvae are released in the stomach, penetrate the intestinal and peritoneal walls and inhabit the subcutaneous tissues.

Human health effects

The onset of symptoms occurs just prior to the local eruption of the worm. The early manifestations of urticaria, erythema, dyspnoea, vomiting, pruritus and giddiness are of an allergic nature. In about 50% of cases, the whole worm is extruded in a few weeks; the lesion then heals rapidly, and disability is of limited duration. In the remaining cases, however, complications ensue, and the track of the worm becomes secondarily infected, leading to a severe inflammatory reaction that may result in abscess formation with disabling pain that lasts for months. Mortality is extremely rare, but permanent disability can result from contractures of tendons and chronic arthritis. The economic impact can be substantial. One study reported an 11% annual reduction in rice production from an area of eastern Nigeria, at a cost of US$20 million.

Source and occurrence
Infection with guinea worm is geographically limited to a central belt of countries in sub-Saharan Africa. Drinking-water containing infected *Cyclops* is the only source of infection with *Dracunculus*. The disease typically occurs in rural areas where piped water supplies are not available. Transmission tends to be highly seasonal, depending on changes in water sources. For instance, transmission is highest in the early rainy season in a dry savannah zone of Mali with under 800 mm annual rainfall but in the dry season in the humid savannah area of southern Nigeria with over 1300 mm annual rainfall. The eradication strategy combines a variety of interventions, including integrated surveillance systems, intensified case containment measures, provision of safe water and health education.

Routes of exposure
The only route of exposure is the consumption of drinking-water containing *Cyclops* spp. carrying infectious *Dracunculus* larvae.

Significance in drinking-water
Dracunculus medinensis is the only human parasite that may be eradicated in the near future by the provision of safe drinking-water. Infection can be prevented by a number of relatively simple control measures. These include intervention strategies to prevent the release of *D. medinensis* larvae from female worms in infected patients into water and control of *Cyclops* spp. in water resources by means of fish. Prevention can also be achieved through the provision of boreholes and safe wells. Wells and springs should be surrounded by cement curbings, and bathing and washing in these waters should be avoided. Other control measures include filtration of water carrying infectious *Dracunculus* larvae through a fine mesh cloth to remove *Cyclops* spp. or inactivation of *Cyclops* spp. in drinking-water by treatment with chlorine.

Selected bibliography
Cairncross S, Muller R, Zagaria N (2002) Dracunculiasis (guinea worm disease) and the eradication initiative. *Clinical Microbiology Reviews*, 15:223–246.

Hopkins DR, Ruiz-Tiben E (1991) Strategies for dracunculiasis eradication. *Bulletin of the World Health Organization*, 69:533–540.

11.4.2 Fasciola spp.

Fascioliasis is caused by two trematode species of the genus *Fasciola*: *F. hepatica*, present in Europe, Africa, Asia, the Americas and Oceania, and *F. gigantica*, mainly distributed in Africa and Asia. Human fascioliasis was considered a secondary zoonotic disease until the mid-1990s. In most regions, fascioliasis is a foodborne disease. However, the discovery of floating metacercariae in hyperendemic regions (including the Andean Altiplano region in South America) indicates that drinking-water may be a significant transmission route for fascioliasis in certain locations.

General description

The life cycle of *F. hepatica* and *F. gigantica* takes about 14–23 weeks and requires two hosts. The life cycle comprises four phases. In the first phase, the definitive host ingests metacercariae. The metacercariae excyst in the intestinal tract and then migrate to the liver and bile ducts. After 3–4 months, the flukes attain sexual maturity and produce eggs, which are excreted into the bile and intestine. Adult flukes can live for 9–14 years in the host. In the second phase, the eggs are excreted by the human or animal. Once in fresh water, a miracidium develops inside. In the third phase, miracidia penetrate a snail host and develop into cercaria, which are released into the water. In the fourth and final phase, cercaria swim for a short period of time until they reach a suitable attachment site (aquatic plants), where they encyst to form metacercariae, which become infective within 24 h. Some metacercariae do not attach to plants but remain floating in the water.

Human health effects

The parasites inhabit the large biliary passages and the gall-bladder. Disease symptoms are different for the acute and chronic phases of the infection. The invasive or acute phase may last from 2 to 4 months and is characterized by symptoms such as dyspepsia, nausea and vomiting, abdominal pain and a high fever (up to 40 °C). Anaemia and allergic responses (e.g., pruritis, urticaria) may also occur. In children, the acute infection can be accompanied by severe symptoms and sometimes causes death. The obstructive or chronic phase (after months to years of infection) may be characterized by painful liver enlargement and in some cases obstructive jaundice, chest pains, loss of weight and cholelithiasis. The most important pathogenic sequelae are hepatic lesions and fibrosis and chronic inflammation of the bile ducts. Immature flukes may deviate during migration, enter other organs and cause ectopic fascioliasis in a range of subcutaneous tissues. Fascioliasis can be treated with triclabendazole.

Source and occurrence

Human cases have been increasing in 51 countries on five continents. Estimates of the numbers of humans with fascioliasis range from 2.4 to 17 million people or even higher, depending on unquantified prevalence in many African and Asian countries.

Analysis of the geographical distribution of human cases shows that the correlation between animal and human fascioliasis occurs only at a basic level. High prevalences in humans are not necessarily related to areas where fascioliasis is a great veterinary problem. Major health problems associated with fascioliasis occur in Andean countries (Bolivia, Peru, Chile, Ecuador), the Caribbean (Cuba), northern Africa (Egypt), Near East (Iran and neighbouring countries) and western Europe (Portugal, France and Spain).

Routes of exposure
Humans can contract fascioliasis when they ingest infective metacercariae by eating raw aquatic plants (and, in some cases, terrestrial plants, such as lettuce, irrigated with contaminated water), drinking contaminated water, using utensils washed in contaminated water or eating raw liver infected with immature flukes.

Significance in drinking-water
Water is often cited as a human infection source. In the Bolivian Altiplano, 13% of metacercariae isolates are floating. Untreated drinking-water in hyperendemic regions often contains floating metacercariae; for example, a small stream crossing in the Altiplano region of Bolivia contained up to 7 metacercariae per 500 ml. The importance of fascioliasis transmission through water is supported by indirect evidence. There are significant positive associations between liver fluke infection and infection by other waterborne protozoans and helminths in Andean countries and in Egypt. In many human hyperendemic areas of the Americas, people do not have a history of eating watercress or other water plants. In the Nile Delta region, people living in houses with piped water had a higher infection risk. Metacercariae are likely to be resistant to chlorine disinfection but should be removed by various filtration processes. For example, in Tiba, Egypt, human prevalence was markedly decreased after filtered water was supplied to specially constructed washing units.

Selected bibliography
Mas-Coma S (2004) Human fascioliasis. In: *Waterborne zoonoses: Identification, causes, and controls*. IWA Publishing, London, on behalf of the World Health Organization, Geneva.
Mas-Coma S, Esteban JG, Bargues MD (1999) Epidemiology of human fascioliasis: a review and proposed new classification. *Bulletin of the World Health Organization*, 77(4):340–346.
WHO (1995) *Control of foodborne trematode infections*. Geneva, World Health Organization (WHO Technical Report Series 849).

11.5 Toxic cyanobacteria
More detailed information on toxic cyanobacteria is available in the supporting document *Toxic Cyanobacteria in Water* (see section 1.3).

General description

Cyanobacteria are photosynthetic bacteria that share some properties with algae. Notably, they possess chlorophyll-a and liberate oxygen during photosynthesis. The first species to be recognized were blue-green in colour; hence, a common term for these organisms is blue-green algae. However, owing to the production of different pigments, there are a large number that are not blue-green, and they can range in colour from blue-green to yellow-brown and red. Most cyanobacteria are aerobic phototrophs, but some exhibit heterotrophic growth. They may grow as separate cells or in multicellular filaments or colonies. They can be identified by their morphology to genus level under a microscope. Some species form surface blooms or scums, while others stay mixed in the water column or are bottom dwelling (benthic). Some cyanobacteria possess the ability to regulate their buoyancy via intracellular gas vacuoles, and some species can fix elemental nitrogen dissolved in water. The most notable feature of cyanobacteria in terms of public health impact is that a range of species can produce toxins.

Human health effects

Many cyanobacteria produce potent toxins, as shown in Table 11.1. Cyanobacterial toxins are also discussed in section 8.5.6. Each toxin has specific properties, with distinct concerns including liver damage, neurotoxicity and tumour promotion. Acute symptoms reported after exposure include gastrointestinal disorders, fever and irritations of the skin, ears, eyes, throat and respiratory tract. Cyanobacteria do not multiply in the human body and hence are not infectious.

Source and occurrence

Cyanobacteria are widespread and found in a diverse range of environments, including soils, seawater and, most notably, freshwater environments. Some environmental conditions, including sunlight, warm weather, low turbulence and high nutrient levels, can promote growth. Depending on the species, this may result in greenish discol-

Table 11.1 Cyanotoxins produced by cyanobacteria

Toxic species	Cyanotoxin
Potentially *Anabaena* spp.	Anatoxin-a(S), anatoxin-a, microcystins, saxitoxins
Anabaenopsis millenii	Microcystins
Aphanizomenon spp.	Anatoxin-a, saxitoxins, cylindrospermopsin
Cylindrospermum spp.	Cylindrospermopsin, saxitoxins, anatoxin-a
Lyngbya spp.	Saxitoxins, lyngbyatoxins
Microcystis spp.	Microcystins, anatoxin-a (minor amounts)
Nodularia spp.	Nodularins
Nostoc spp.	Microcystins
Oscillatoria spp.	Anatoxin-a, microcystins
Planktothrix spp.	Anatoxin-a, homoanatoxin-a, microcystins
Raphidiopsis curvata	Cylindrospermopsin
Umezakia natans	Cylindrospermopsin

oration of water due to a high density of suspended cells and, in some cases, the formation of surface scums. Such cell accumulations may lead to high toxin concentrations.

Routes of exposure
Potential health concerns arise from exposure to the toxins through ingestion of drinking-water, during recreation, through showering and potentially through consumption of algal food supplement tablets. Repeated or chronic exposure is the primary concern for many of the cyanotoxins; in some cases, however, acute toxicity is more important (e.g., lyngbyatoxins and the neurotoxins saxitoxin and anatoxin). Human fatalities have occurred through use of inadequately treated water containing high cyanotoxin levels for renal dialysis. Dermal exposure may lead to irritation of the skin and mucous membranes and to allergic reactions.

Significance in drinking-water
Cyanobacteria occur in low cell density in most surface waters. However, in suitable environmental conditions, high-density "blooms" can occur. Eutrophication (increased biological growth associated with increased nutrients) can support the development of cyanobacterial blooms (see also section 8.5.6).

Selected bibliography
Backer LC (2002) Cyanobacterial harmful algal blooms (CyanoHABs): Developing a public health response. *Lake and Reservoir Management*, 18:20–31.

Chorus I, Bartram J, eds. (1999) *Toxic cyanobacteria in water: A guide to their public health consequences, monitoring and management.* Published by E & FN Spon, London, on behalf of the World Health Organization, Geneva.

Lahti K et al. (2001) Occurrence of microcystins in raw water sources and treated drinking water of Finnish waterworks. *Water Science and Technology*, 43:225–228.

11.6 Indicator and index organisms
Owing to issues relating to complexity, cost and timeliness of obtaining results, testing for specific pathogens is generally limited to validation, where monitoring is used to determine whether a treatment or other process is effective in removing target organisms. Very occasionally, pathogen testing may be performed to verify that a specific treatment or process has been effective. However, microbial testing included as part of operational and verification (including surveillance) monitoring is usually limited to that for indicator organisms, either to measure the effectiveness of control measures or as an index of faecal pollution.

The concept of using indicator organisms as signals of faecal pollution is a well established practice in the assessment of drinking-water quality. The criteria determined for such indicators were that they should not be pathogens themselves and should:

— be universally present in faeces of humans and animals in large numbers;
— not multiply in natural waters;
— persist in water in a similar manner to faecal pathogens;
— be present in higher numbers than faecal pathogens;
— respond to treatment processes in a similar fashion to faecal pathogens; and
— be readily detected by simple, inexpensive methods.

These criteria reflect an assumption that the same indicator organism could be used as both an index of faecal pollution and an indicator of treatment/process efficacy. However, it has become clear that one indicator cannot fulfil these two roles. Increased attention has focused on shortcomings of traditional indicators, such as *E. coli*, as surrogates for enteric viruses and protozoa, and alternative indicators of these pathogens, such as bacteriophages and bacterial spores, have been suggested. In addition, greater reliance is being placed on parameters that can be used as indicators for the effectiveness of treatments and processes designed to remove faecal pathogens, including bacteria, viruses, protozoa and helminths.

It is important to distinguish between microbial testing undertaken to signal the presence of faecal pathogens or alternatively to measure the effectiveness of treatments/processes. As a first step, the separate terms *index* and *indicator* have been proposed, whereby:

— an *index organism* is one that points to the presence of pathogenic organisms – for example, as an index of faecal pathogens; and
— an *indicator organism* is one that is used to measure the effectiveness of a process – for example, a process indicator or disinfection indicator.

These terms can also be applied to non-microbial parameters; hence, turbidity can be used a filtration indicator.

Further discussion on index and indicator organisms is contained in the supporting document *Assessing Microbial Safety of Drinking Water* (see section 1.3).

11.6.1 Total coliform bacteria
General description
Total coliform bacteria include a wide range of aerobic and facultatively anaerobic, Gram-negative, non-spore-forming bacilli capable of growing in the presence of relatively high concentrations of bile salts with the fermentation of lactose and production of acid or aldehyde within 24 h at 35–37 °C. *Escherichia coli* and thermotolerant coliforms are a subset of the total coliform group that can ferment lactose at higher temperatures (see section 11.6.2). As part of lactose fermentation, total coliforms produce the enzyme β-galactosidase. Traditionally, coliform bacteria were regarded as belonging to the genera *Escherichia*, *Citrobacter*, *Klebsiella* and *Enterobacter*, but the group is more heterogeneous and includes a wider range of genera, such as *Serratia* and *Hafnia*. The total coliform group includes both faecal and environmental species.

Indicator value
Total coliforms include organisms that can survive and grow in water. Hence, they are not useful as an index of faecal pathogens, but they can be used as an indicator of treatment effectiveness and to assess the cleanliness and integrity of distribution systems and the potential presence of biofilms. However, there are better indicators for these purposes. As a disinfection indicator, the test for total coliforms is far slower and less reliable than direct measurement of disinfectant residual. In addition, total coliforms are far more sensitive to disinfection than are enteric viruses and protozoa. HPC measurements detect a wider range of microorganisms and are generally considered a better indicator of distribution system integrity and cleanliness.

Source and occurrence
Total coliform bacteria (excluding *E. coli*) occur in both sewage and natural waters. Some of these bacteria are excreted in the faeces of humans and animals, but many coliforms are heterotrophic and able to multiply in water and soil environments. Total coliforms can also survive and grow in water distribution systems, particularly in the presence of biofilms.

Application in practice
Total coliforms are generally measured in 100-ml samples of water. A variety of relatively simple procedures are available based on the production of acid from lactose or the production of the enzyme β-galactosidase. The procedures include membrane filtration followed by incubation of the membranes on selective media at 35–37 °C and counting of colonies after 24 h. Alternative methods include most probable number procedures using tubes or micro-titre plates and P/A tests. Field test kits are available.

Significance in drinking-water
Total coliforms should be absent immediately after disinfection, and the presence of these organisms indicates inadequate treatment. The presence of total coliforms in distribution systems and stored water supplies can reveal regrowth and possible biofilm formation or contamination through ingress of foreign material, including soil or plants.

Selected bibliography
Ashbolt NJ, Grabow WOK, Snozzi M (2001) Indicators of microbial water quality. In: Fewtrell L, Bartram J, eds. *Water quality: Guidelines, standards and health – Assessment of risk and risk management for water-related infectious disease.* WHO Water Series. London, IWA Publishing, pp. 289–315.

Grabow WOK (1996) Waterborne diseases: Update on water quality assessment and control. *Water SA*, 22:193–202.

Sueiro RA et al. (2001) Evaluation of Coli-ID and MUG Plus media for recovering *Escherichia coli* and other coliform bacteria from groundwater samples. *Water Science and Technology*, 43:213–216.

11.6.2 *Escherichia coli* and thermotolerant coliform bacteria

General description
Total coliform bacteria that are able to ferment lactose at 44–45 °C are known as thermotolerant coliforms. In most waters, the predominant genus is *Escherichia*, but some types of *Citrobacter*, *Klebsiella* and *Enterobacter* are also thermotolerant. *Escherichia coli* can be differentiated from the other thermotolerant coliforms by the ability to produce indole from tryptophan or by the production of the enzyme β-glucuronidase. *Escherichia coli* is present in very high numbers in human and animal faeces and is rarely found in the absence of faecal pollution, although there is some evidence for growth in tropical soils. Thermotolerant coliform species other than *E. coli* can include environmental organisms.

Indicator value
Escherichia coli is considered the most suitable index of faecal contamination. In most circumstances, populations of thermotolerant coliforms are composed predominantly of *E. coli*; as a result, this group is regarded as a less reliable but acceptable index of faecal pollution. *Escherichia coli* (or, alternatively, thermotolerant coliforms) is the first organism of choice in monitoring programmes for verification, including surveillance of drinking-water quality. These organisms are also used as disinfection indicators, but testing is far slower and less reliable than direct measurement of disinfectant residual. In addition, *E. coli* is far more sensitive to disinfection than are enteric viruses and protozoa.

Source and occurrence
Escherichia coli occurs in high numbers in human and animal faeces, sewage and water subject to recent faecal pollution. Water temperatures and nutrient conditions present in drinking-water distribution systems are highly unlikely to support the growth of these organisms.

Application in practice
Escherichia coli (or, alternatively, thermotolerant coliforms) are generally measured in 100-ml samples of water. A variety of relatively simple procedures are available based on the production of acid and gas from lactose or the production of the enzyme β-glucuronidase. The procedures include membrane filtration followed by incubation of the membranes on selective media at 44–45 °C and counting of colonies after 24 h. Alternative methods include most probable number procedures using tubes or microtitre plates and P/A tests, some for volumes of water larger than 100 ml. Field test kits are available.

Significance in drinking-water

The presence of *E. coli* (or, alternatively, thermotolerant coliforms) provides evidence of recent faecal contamination, and detection should lead to consideration of further action, which could include further sampling and investigation of potential sources such as inadequate treatment or breaches in distribution system integrity.

Selected bibliography

Ashbolt NJ, Grabow WOK, Snozzi M (2001) Indicators of microbial water quality. In: Fewtrell L, Bartram J, eds. *Water quality: Guidelines, standards and health – Assessment of risk and risk management for water-related infectious disease*. WHO Water Series. London, IWA Publishing, pp. 289–315.

George I et al. (2001) Use of rapid enzymatic assays to study the distribution of faecal coliforms in the Seine river (France). *Water Science and Technology*, 43:77–80.

Grabow WOK (1996) Waterborne diseases: Update on water quality assessment and control. *Water SA*, 22:193–202.

Sueiro RA et al. (2001) Evaluation of Coli-ID and MUG Plus media for recovering *Escherichia coli* and other coliform bacteria from groundwater samples. *Water Science and Technology*, 43:213–216.

11.6.3 Heterotrophic plate counts

A substantial review of the use of HPC is available (Bartram et al., 2003).

General description

HPC measurement detects a wide spectrum of heterotrophic microorganisms, including bacteria and fungi, based on the ability of the organisms to grow on rich growth media, without inhibitory or selective agents, over a specified incubation period and at a defined temperature. The spectrum of organisms detected by HPC testing includes organisms sensitive to disinfection processes, such as coliform bacteria; organisms resistant to disinfection, such as spore formers; and organisms that rapidly proliferate in treated water in the absence of residual disinfectants. The tests detect only a small proportion of the microorganisms that are present in water. The population recovered will differ according to the method and conditions applied. Although standard methods have been developed, there is no single universal HPC measurement. A range of media is available, incubation temperatures used vary from 20 °C to 37 °C and incubation periods range from a few hours to 7 days or more.

Indicator value

The test has little value as an index of pathogen presence but can be useful in operational monitoring as a treatment and disinfectant indicator, where the objective is to keep numbers as low as possible. In addition, HPC measurement can be used in assessing the cleanliness and integrity of distribution systems and the presence of biofilms.

Source and occurrence

Heterotrophic microorganisms include both members of the natural (typically non-hazardous) microbial flora of water environments and organisms present in a range of pollution sources. They occur in large numbers in raw water sources. The actual organisms detected by HPC tests vary widely between locations and between consecutive samples. Some drinking-water treatment processes, such as coagulation and sedimentation, reduce the number of HPC organisms in water. However, the organisms proliferate in other treatment processes, such as biologically active carbon and sand filtration. Numbers of HPC organisms are reduced significantly by disinfection practices, such as chlorination, ozonation and UV light irradiation. However, in practice, none of the disinfection processes sterilizes water; under suitable conditions, such as the absence of disinfectant residuals, HPC organisms can grow rapidly. HPC organisms can grow in water and on surfaces in contact with water as biofilms. The principal determinants of growth or "regrowth" are temperature, availability of nutrients, including assimilable organic carbon, lack of disinfectant residual and stagnation.

Application in practice

No sophisticated laboratory facilities or highly trained staff are required. Results on simple aerobically incubated agar plates are available within hours to days, depending on the characteristics of the procedure used.

Significance in drinking-water

After disinfection, numbers would be expected to be low; for most uses of HPC test results, however, actual numbers are of less value than changes in numbers at particular locations. In distribution systems, increasing numbers can indicate a deterioration in cleanliness, possibly stagnation and the potential development of biofilms. HPC can include potentially "opportunistic" pathogens such as *Acinetobacter, Aeromonas, Flavobacterium, Klebsiella, Moraxella, Serratia, Pseudomonas* and *Xanthomonas*. However, there is no evidence of an association of any of these organisms with gastrointestinal infection through ingestion of drinking-water in the general population.

Selected bibliography

Ashbolt NJ, Grabow WOK, Snozzi M (2001) Indicators of microbial water quality. In: Fewtrell L, Bartram J, eds. *Water quality: Guidelines, standards and health – Assessment of risk and risk management for water-related infectious disease.* WHO Water Series. London, IWA Publishing, pp. 289–315.

Bartram J et al., eds. (2003) *Heterotrophic plate counts and drinking-water safety: the significance of HPCs for water quality and human health.* WHO Emerging Issues in Water and Infectious Disease Series. London, IWA Publishing.

11.6.4 Intestinal enterococci

General description

Intestinal enterococci are a subgroup of the larger group of organisms defined as faecal streptococci, comprising species of the genus *Streptococcus*. These bacteria are Gram-positive and relatively tolerant of sodium chloride and alkaline pH levels. They are facultatively anaerobic and occur singly, in pairs or as short chains. Faecal streptococci including intestinal enterococci all give a positive reaction with Lancefield's Group D antisera and have been isolated from the faeces of warm-blooded animals. The subgroup intestinal enterococci consists of the species *Enterococcus faecalis, E. faecium, E. durans* and *E. hirae*. This group was separated from the rest of the faecal streptococci because they are relatively specific for faecal pollution. However, some intestinal enterococci isolated from water may occasionally also originate from other habitats, including soil, in the absence of faecal pollution.

Indicator value

The intestinal enterococci group can be used as an index of faecal pollution. Most species do not multiply in water environments. The numbers of intestinal enterococci in human faeces are generally about an order of magnitude lower than those of *E. coli*. Important advantages of this group are that they tend to survive longer in water environments than *E. coli* (or thermotolerant coliforms), are more resistant to drying and are more resistant to chlorination. Intestinal enterococci have been used in testing of raw water as an index of faecal pathogens that survive longer than *E. coli* and in drinking-water to augment testing for *E. coli*. In addition, they have been used to test water quality after repairs to distribution systems or after new mains have been laid.

Source and occurrence

Intestinal enterococci are typically excreted in the faeces of humans and other warm-blooded animals. Some members of the group have also been detected in soil in the absence of faecal contamination. Intestinal enterococci are present in large numbers in sewage and water environments polluted by sewage or wastes from humans and animals.

Application in practice

Enterococci are detectable by simple, inexpensive cultural methods that require basic bacteriology laboratory facilities. Commonly used methods include membrane filtration with incubation of membranes on selective media and counting of colonies after incubation at 35–37 °C for 48 h. Other methods include a most probable number technique using micro-titre plates where detection is based on the ability of intestinal enterococci to hydrolyse 4-methyl-umbelliferyl-β-D-glucoside in the presence of thallium acetate and nalidixic acid within 36 h at 41 °C.

Significance in drinking-water
The presence of intestinal enterococci provides evidence of recent faecal contamination, and detection should lead to consideration of further action, which could include further sampling and investigation of potential sources such as inadequate treatment or breaches in distribution system integrity.

Selected bibliography
Ashbolt NJ, Grabow WOK, Snozzi M (2001) Indicators of microbial water quality. In: Fewtrell L, Bartram J, eds. *Water quality: Guidelines, standards and health – Assessment of risk and risk management for water-related infectious disease.* WHO Water Series. London, IWA Publishing, pp. 289–315.
Grabow WOK (1996) Waterborne diseases: Update on water quality assessment and control. *Water SA*, 22:193–202.
Junco TT et al. (2001) Identification and antibiotic resistance of faecal enterococci isolated from water samples. *International Journal of Hygiene and Environmental Health*, 203:363–368.
Pinto B et al. (1999) Characterization of "faecal streptococci" as indicators of faecal pollution and distribution in the environment. *Letters in Applied Microbiology*, 29:258–263.

11.6.5 *Clostridium perfringens*
General description
Clostridium spp. are Gram-positive, anaerobic, sulfite-reducing bacilli. They produce spores that are exceptionally resistant to unfavourable conditions in water environments, including UV irradiation, temperature and pH extremes, and disinfection processes, such as chlorination. The characteristic species of the genus, *C. perfringens*, is a member of the normal intestinal flora of 13–35% of humans and other warm-blooded animals. Other species are not exclusively of faecal origin. Like *E. coli*, *C. perfringens* does not multiply in most water environments and is a highly specific indicator of faecal pollution.

Indicator value
In view of the exceptional resistance of *C. perfringens* spores to disinfection processes and other unfavourable environmental conditions, *C. perfringens* has been proposed as an index of enteric viruses and protozoa in treated drinking-water supplies. In addition, *C. perfringens* can serve as an index of faecal pollution that took place previously and hence indicate sources liable to intermittent contamination. *Clostridium perfringens* is not recommended for routine monitoring, as the exceptionally long survival times of its spores are likely to far exceed those of enteric pathogens, including viruses and protozoa. *Clostridium perfringens* spores are smaller than protozoan (oo)cysts and may be useful indicators of the effectiveness of filtration processes. Low numbers in

some source waters suggest that use of *C. perfringens* spores for this purpose may be limited to validation of processes rather than routine monitoring.

Source and occurrence
Clostridium perfringens and its spores are virtually always present in sewage. The organism does not multiply in water environments. *Clostridium perfringens* is present more often and in higher numbers in the faeces of some animals, such as dogs, than in the faeces of humans and less often in the faeces of many other warm-blooded animals. The numbers excreted in faeces are normally substantially lower than those of *E. coli*.

Application in practice
Vegetative cells and spores of *C. perfringens* are usually detected by membrane filtration techniques in which membranes are incubated on selective media under strict anaerobic conditions. These detection techniques are not as simple and inexpensive as those for other indicators, such as *E. coli* and intestinal enterococci.

Significance in drinking-water
The presence of *C. perfringens* in drinking-water can be an index of intermittent faecal contamination. Potential sources of contamination should be investigated. Filtration processes designed to remove enteric viruses or protozoa should also remove *C. perfringens*. Detection in water immediately after treatment should lead to investigation of filtration plant performance.

Selected bibliography
Araujo M et al. (2001) Evaluation of fluorogenic TSC agar for recovering *Clostridium perfringens* in groundwater samples. *Water Science and Technology*, 43:201–204.

Ashbolt NJ, Grabow WOK, Snozzi M (2001) Indicators of microbial water quality. In: Fewtrell L, Bartram J, eds. *Water quality: Guidelines, standards and health – Assessment of risk and risk management for water-related infectious disease.* WHO Water Series. London, IWA Publishing, pp. 289–315.

Nieminski EC, Bellamy WD, Moss LR (2000) Using surrogates to improve plant performance. *Journal of the American Water Works Association*, 92(3):67–78.

Payment P, Franco E (1993) *Clostridium perfringens* and somatic coliphages as indicators of the efficiency of drinking-water treatment for viruses and protozoan cysts. *Applied and Environmental Microbiology*, 59:2418–2424.

11.6.6 Coliphages
General description
Bacteriophages (phages) are viruses that use only bacteria as hosts for replication. Coliphages use *E. coli* and closely related species as hosts and hence can be released

by these bacterial hosts into the faeces of humans and other warm-blooded animals. Coliphages used in water quality assessment are divided into the major groups of somatic coliphages and F-RNA coliphages. Differences between the two groups include the route of infection.

Somatic coliphages initiate infection by attaching to receptors permanently located on the cell wall of hosts. They replicate more frequently in the gastrointestinal tract of warm-blooded animals but can also replicate in water environments. Somatic coliphages consist of a wide range of phages (members of the phage families Myoviridae, Siphoviridae, Podoviridae and Microviridae) with a spectrum of morphological types.

F-RNA coliphages initiate infection by attaching to fertility (F-, sex) fimbriae on *E. coli* hosts. These F-fimbriae are produced only by bacteria carrying the fertility (F-) plasmid. Since F-fimbriae are produced only in the logarithmic growth phase at temperatures above 30 °C, F-RNA phages are not likely to replicate in environments other than the gastrointestinal tract of warm-blooded animals. F-RNA coliphages comprise a restricted group of closely related phages, which belong to the family Leviviridae, and consist of a single-stranded RNA genome and an icosahedral capsid that is morphologically similar to that of picornaviruses. F-RNA coliphages have been divided into serological types I–IV, which can be identified as genotypes by molecular techniques such as gene probe hybridization. Members of groups I and IV have to date been found exclusively in animal faeces, and group III in human faeces. Group II phages have been detected in human faeces and no animal faeces other than about 28% of porcine faeces. This specificity, which is not fully understood, offers a potential tool to distinguish between faecal pollution of human and animal origin under certain conditions and limitations.

Indicator value

Phages share many properties with human viruses, notably composition, morphology, structure and mode of replication. As a result, coliphages are useful models or surrogates to assess the behaviour of enteric viruses in water environments and the sensitivity to treatment and disinfection processes. In this regard, they are superior to faecal bacteria. However, there is no direct correlation between numbers of coliphages and numbers of enteric viruses. In addition, coliphages cannot be absolutely relied upon as an index for enteric viruses. This has been confirmed by the isolation of enteric viruses from treated and disinfected drinking-water supplies that yielded negative results in conventional tests for coliphages.

F-RNA coliphages provide a more specific index of faecal pollution than somatic phages. In addition, F-RNA coliphages are better indicators of the behaviour of enteric viruses in water environments and their response to treatment and disinfection processes than are somatic coliphages. This has been confirmed by studies in which the behaviour and survival of F-RNA coliphages, somatic phages, faecal bacteria and enteric viruses have been compared. Available data indicate that the specificity of F-

RNA serogroups (genotypes) for human and animal excreta may prove useful in the distinction between faecal pollution of human and animal origin. However, there are shortcomings and conflicting data that need to be resolved, and the extent to which this tool can be applied in practice remains to be elucidated. Due to the limitations of coliphages, they are best used in laboratory investigations, pilot trials and possibly validation testing. They are not suitable for operational or verification (including surveillance) monitoring.

Source and occurrence

Coliphages are excreted by humans and animals in relatively low numbers. As a result of their respective modes of replication and host specificity, somatic coliphages are generally excreted by most humans and animals, whereas F-RNA coliphages are excreted by a variable and generally lower percentage of humans and animals. Available data indicate that in some communities, F-RNA phages are detectable in 10% of human, 45% of bovine, 60% of porcine and 70% of poultry faecal specimens. Somatic coliphages have been found to generally outnumber F-RNA phages in water environments by a factor of about 5 and cytopathogenic human viruses by a factor of about 500, although these ratios vary considerably. Sewage contains somatic coliphages in numbers of the order of 10^6–10^8 per litre; in one study, slaughterhouse wastewater was found to contain somatic coliphages in numbers up to 10^{10} per litre. There are indications that they may multiply in sewage, and somatic coliphages may multiply in natural water environments using saprophytic hosts. Somatic phages and F-RNA phages have been detected in numbers up to 10^5 per litre in lake and river water.

Application in practice

Somatic coliphages are detectable by relatively simple and inexpensive plaque assays, which yield results within 24 h. Plaque assays for F-RNA coliphages are not quite as simple, because the culture of host bacteria has to be in the logarithmic growth phase at a temperature above 30 °C to ensure that F-fimbriae are present. Plaque assays using large petri dishes have been designed for the quantitative enumeration of plaques in 100-ml samples, and P/A tests have been developed for volumes of water of 500 ml or more.

Significance in drinking-water

Since coliphages typically replicate in the gastrointestinal tract of humans and warm-blooded animals, their presence in drinking-water provides an index of faecal pollution and hence the potential presence of enteric viruses and possibly also other pathogens. The presence of coliphages in drinking-water also indicates shortcomings in treatment and disinfection processes designed to remove enteric viruses. F-RNA coliphages provide a more specific index for faecal pollution. The absence of coliphages from treated drinking-water supplies does not confirm the absence of pathogens such as enteric viruses and protozoan parasites.

Selected bibliography

Ashbolt NJ, Grabow WOK, Snozzi M (2001) Indicators of microbial water quality. In: Fewtrell L, Bartram J, eds. *Water quality: Guidelines, standards and health – Assessment of risk and risk management for water-related infectious disease.* WHO Water Series. London, IWA Publishing, pp. 289–315.

Grabow WOK (2001) Bacteriophages: Update on application as models for viruses in water. *Water SA*, 27:251–268.

Mooijman KA et al. (2001) Optimisation of the ISO-method on enumeration of somatic coliphages (draft ISO 10705–2). *Water Science and Technology*, 43:205–208.

Schaper M et al. (2002) Distribution of genotypes of F-specific RNA bacteriophages in human and non-human sources of faecal pollution in South Africa and Spain. *Journal of Applied Microbiology*, 92:657–667.

Storey MV, Ashbolt NJ (2001) Persistence of two model enteric viruses (B40-8 and MS-2 bacteriophages) in water distribution pipe biofilms. *Water Science and Technology*, 43:133–138.

11.6.7 Bacteroides fragilis phages

General description

The bacterial genus *Bacteroides* inhabits the human gastrointestinal tract in greater numbers than *E. coli*. Faeces can contain 10^9–10^{10} *Bacteroides* per gram compared with 10^6–10^8 *E. coli* per gram. *Bacteroides* are rapidly inactivated by environmental oxygen levels, but *Bacteroides* bacteriophages are resistant to unfavourable conditions. Two groups of *B. fragilis* phages are used as indicators in water quality assessment. One is a restricted group of phages that specifically uses *B. fragilis* strain HSP40 as host. This group of phages appears unique, because it is found only in human faeces and not in faeces of other animals. The numbers of these phages in sewage appear to be relatively low, and they are almost absent in some geographical areas. The *B. fragilis* HSP40 phages belong to the family Siphoviridae, with flexible non-contractile tails, double-stranded DNA and capsids with a diameter of up to 60 nm. The second group of *Bacteroides* phages used as indicators is those that use *B. fragilis* strain RYC2056 as a host. This group includes a substantially wider spectrum of phages, occurring in the faeces of humans and many other animals. The numbers of these phages in sewage are generally substantially higher than those of *B. fragilis* HSP40 phages.

Indicator value

Bacteroides bacteriophages have been proposed as a possible index of faecal pollution due to their specific association with faecal material and exceptional resistance to environmental conditions. In particular, *B. fragilis* HSP40 phages are found only in human faeces. *Bacteroides fragilis* phage B40-8, a typical member of the group of *B. fragilis* HSP40 phages, has been found to be more resistant to inactivation by chlorine than poliovirus type 1, simian rotavirus SA11, coliphage f2, *E. coli* and *Streptococcus faecalis*. *Bacteroides fragilis* strain RYC2056 phages seem to be likewise relatively resistant to disinfection. Indicator shortcomings of *B. fragilis* phages include relatively low numbers in sewage and polluted water environments. This applies in particular to *B. fragilis* HSP40 phages. Human enteric viruses have been detected in drinking-water supplies that yielded negative results in conventional tests for *B. fragilis* HSP40 phages. Owing to the limitations of *Bacteroides* bacteriophages, they are best used in laboratory investigations, pilot trials and possibly validation testing. They are not suitable for operational or verification (including surveillance) monitoring.

Source and occurrence

Bacteroides fragilis HSP40 phages are excreted by about 10–20% of humans in certain parts of the world; consequently, their numbers in sewage are substantially lower than those of somatic and even F-RNA coliphages. A mean count of 67 *B. fragilis* HSP40 phages per litre in a sewage-polluted river has been reported. In some parts of the world, *B. fragilis* HSP40 phages would appear not to be detectable in sewage at all. Phages using *B. fragilis* RYC2056 as host are excreted in larger numbers and seem to occur more universally. On average, these phages are excreted by more than 25% of humans. In a survey of water environments, *B. fragilis* HSP40 phages have been found to outnumber cytopathogenic enteric viruses on average by only about 5-fold. Theoretically, wastewaters could be expected to contain higher levels of *B. fragilis* phages than those detected. The reason for the discrepancy may be due to failure in maintaining sufficiently anaerobic conditions during the performance of plaque assays. Improvement of detection methods may result in the recording of higher numbers of *B. fragilis* phages in sewage and polluted water environments.

Application in practice

Disadvantages of *B. fragilis* phages are that the detection methods are more complex and expensive than those for coliphages. Costs are increased by the need to use antibiotics for purposes of selection and to incubate cultures and plaque assays under absolute anaerobic conditions. Results of plaque assays are usually available after about 24 h compared with about 8 h for coliphages.

Significance in drinking-water

The presence of *B. fragilis* phages in drinking-water is sound evidence of faecal pollution as well as shortcomings in water treatment and disinfection processes. In addition, the presence of *B. fragilis* HSP40 phages strongly indicates faecal pollution of human origin. However, *B. fragilis* phages occur in relatively low numbers in sewage, polluted water environments and drinking-water supplies. This implies that the absence of *B. fragilis* phages from treated drinking-water supplies does not confirm the absence of pathogens such as enteric viruses and protozoan parasites.

Selected bibliography

Bradley G et al. (1999) Distribution of the human faecal bacterium *Bacteroides fragilis* and their relationship to current sewage pollution indicators in bathing waters. *Journal of Applied Microbiology*, 85(Suppl.):90S–100S.

Grabow WOK (2001) Bacteriophages: Update on application as models for viruses in water. *Water SA*, 27:251–268.

Puig A et al. (1999) Diversity of *Bacteroides fragilis* strains in their capacity to recover phages from human and animal wastes and from fecally polluted wastewater. *Applied and Environmental Microbiology*, 65:1772–1776.

Storey MV, Ashbolt NJ (2001) Persistence of two model enteric viruses (B40-8 and MS-2 bacteriophages) in water distribution pipe biofilms. *Water Science and Technology*, 43:133–138.

Tartera C, Lucena F, Jofre J (1989) Human origin of *Bacteroides fragilis* bacteriophages present in the environment. *Applied and Environmental Microbiology*, 10:2696–2701.

11.6.8 Enteric viruses

General description

The viruses referred to here are a combined group of those that infect the human gastrointestinal tract and are predominantly transmitted by the faecal–oral route. Well known members of this group include the enteroviruses, astroviruses, enteric adenoviruses, orthoreoviruses, rotaviruses, caliciviruses and hepatitis A and E viruses. The enteric viruses cover a wide spectrum of viruses, members of which are a major cause of morbidity and mortality worldwide. Members of the group of enteric viruses differ with regard to structure, composition, nucleic acid and morphology. There are also differences in the numbers and frequency of excretion, survival in the environment and resistance to water treatment processes. Enteric viruses have robust capsids that enable them to survive unfavourable conditions in the environment as well as allowing passage through the acidic and proteolytic conditions in the stomach on their way to the duodenum, where they infect susceptible epithelial cells.

Indicator value

The use of enteric viruses as indicator or index organisms is based on the shortcomings of the existing choices. The survival of faecal bacteria in water environments and the sensitivity to treatment and disinfection processes differ substantially from those of enteric viruses. Monitoring based on one or more representatives of the large group of enteric viruses themselves would, therefore, be more valuable for assessment of the presence of any of the enteric viruses in water and the response to control measures.

Source and occurrence

Enteric viruses are excreted by individuals worldwide at a frequency and in numbers that result in many of these viruses being universally present in substantial numbers in wastewater. However, the prevalence of individual members may vary to a large extent due to variations in rates of infection and excretion. Much higher numbers would be present during outbreaks.

Application in practice
Practical methods are not yet available for the routine monitoring of water supplies for a broad spectrum of enteric viruses. Viruses that are more readily detectable include members of the enterovirus, adenovirus and orthoreovirus groups. These viruses occur in polluted environments in relatively high numbers and can be detected by reasonably practical and moderate-cost techniques based on cytopathogenic effect in cell culture that yield results within 3–12 days (depending on the type of virus). In addition, progress in technology and expertise is decreasing costs. The cost for the recovery of enteric viruses from large volumes of drinking-water has been reduced extensively. Some techniques – for instance, those based on glass wool adsorption–elution – are inexpensive. The cost of cell culture procedures has also been reduced. Consequently, the cost of testing drinking-water supplies for cytopathogenic viruses has become acceptable for certain purposes. Testing could be used to validate effectiveness of treatment processes and, in certain circumstances, as part of specific investigations to verify performance of processes. The incubation times, cost and relative complexity of testing mean that enteric virus testing is not suitable for operational or verification (including surveillance) monitoring. Orthoreoviruses, and at least the vaccine strains of polioviruses detected in many water environments, also have the advantage of not constituting a health risk to laboratory workers.

Significance in drinking-water
The presence of any enteric viruses in drinking-water should be regarded as an index for the potential presence of other enteric viruses, is conclusive evidence of faecal pollution and also provides evidence of shortcomings in water treatment and disinfection processes.

Selected bibliography
Ashbolt NJ, Grabow WOK, Snozzi M (2001) Indicators of microbial water quality. In: Fewtrell L, Bartram J, eds. *Water quality: Guidelines, standards and health – Assessment of risk and risk management for water-related infectious disease*. WHO Water Series. London, IWA Publishing, pp. 289–315.

Grabow WOK, Taylor MB, de Villiers JC (2001) New methods for the detection of viruses: call for review of drinking-water quality guidelines. *Water Science and Technology*, 43:1–8.

12
Chemical fact sheets

The background documents referred to in this chapter may be found on the Water Sanitation and Health website at http://www.who.int/water_sanitation_health/dwq/guidelines/en/.

12.1 Acrylamide

Residual acrylamide monomer occurs in polyacrylamide coagulants used in the treatment of drinking-water. In general, the maximum authorized dose of polymer is 1 mg/litre. At a monomer content of 0.05%, this corresponds to a maximum theoretical concentration of 0.5 µg/litre of the monomer in water. Practical concentrations may be lower by a factor of 2–3. This applies to the anionic and non-ionic polyacrylamides, but residual levels from cationic polyacrylamides may be higher. Polyacrylamides are also used as grouting agents in the construction of drinking-water reservoirs and wells. Additional human exposure might result from food, owing to the use of polyacrylamide in food processing and the potential formation of acrylamide in foods cooked at high temperatures.

Guideline value	0.0005 mg/litre (0.5 µg/litre)
Occurrence	Concentrations of a few micrograms per litre have been detected in tap water.
Basis of guideline derivation	Combined mammary, thyroid and uterine tumours observed in female rats in a drinking-water study, and using the linearized multistage model
Limit of detection	0.032 µg/litre by GC; 0.2 µg/litre by HPLC; 10 µg/litre by HPLC with UV detection
Treatment achievability	Conventional treatment processes do not remove acrylamide. Acrylamide concentrations in drinking-water are controlled by limiting either the acrylamide content of polyacrylamide flocculants or the dose used, or both.
Additional comments	Although the practical quantification level for acrylamide in most laboratories is above the guideline value (generally in the order of 1 µg/litre), concentrations in drinking-water can be controlled by product and dose specification.

Toxicological review
Following ingestion, acrylamide is readily absorbed from the gastrointestinal tract and widely distributed in body fluids. Acrylamide can cross the placenta. It is neurotoxic, affects germ cells and impairs reproductive function. In mutagenicity assays, acrylamide was negative in the Ames test but induced gene mutations in mammalian cells and chromosomal aberrations *in vitro* and *in vivo*. In a long-term carcinogenicity study in rats exposed via drinking-water, acrylamide induced scrotal, thyroid and adrenal tumours in males and mammary, thyroid and uterine tumours in females. IARC has placed acrylamide in Group 2A. Recent data have shown that exposure to acrylamide from cooked food is much higher than previously thought. The significance of this new information for the risk assessment has not yet been determined.

History of guideline development
The 1958, 1963 and 1971 WHO *International Standards for Drinking-water* and the first edition of the *Guidelines for Drinking-water Quality*, published in 1984, did not refer to acrylamide. The 1993 Guidelines established a guideline value of 0.0005 mg/litre associated with an upper-bound excess lifetime cancer risk of 10^{-5}, noting that although the practical quantification level for acrylamide is generally in the order of 0.001 mg/litre, concentrations in drinking-water can be controlled by product and dose specification.

Assessment date
The risk assessment was conducted in 2003.

Principal reference
WHO (2003) *Acrylamide in drinking-water. Background document for preparation of WHO Guidelines for drinking-water quality*. Geneva, World Health Organization (WHO/SDE/WSH/03.04/71).

12.2 Alachlor

Alachlor (CAS No. 15972-60-8) is a pre- and post-emergence herbicide used to control annual grasses and many broad-leaved weeds in maize and a number of other crops. It is lost from soil mainly through volatilization, photodegradation and biodegradation. Many alachlor degradation products have been identified in soil.

Guideline value	0.02 mg/litre
Occurrence	Has been detected in groundwater and surface water; has also been detected in drinking-water at levels below 0.002 mg/litre
Basis of guideline derivation	Calculated by applying the linearized multistage model to data on the incidence of nasal tumours in rats
Limit of detection	0.1 µg/litre by gas–liquid chromatography with electrolytic conductivity detection in the nitrogen mode or by capillary column GC with a nitrogen–phosphorus detector
Treatment achievability	0.001 mg/litre should be achievable using GAC

Toxicological review

On the basis of available experimental data, evidence for the genotoxicity of alachlor is considered to be equivocal. However, a metabolite of alachlor, 2,6-diethylaniline, has been shown to be mutagenic. Available data from two studies in rats clearly indicate that alachlor is carcinogenic, causing benign and malignant tumours of the nasal turbinate, malignant stomach tumours and benign thyroid tumours.

History of guideline development

The 1958 and 1963 WHO *International Standards for Drinking-water* did not refer to alachlor, but the 1971 International Standards suggested that pesticide residues that may occur in community water supplies make only a minimal contribution to the total daily intake of pesticides for the population served. Alachlor was not evaluated in the first edition of the *Guidelines for Drinking-water Quality*, published in 1984, but the 1993 Guidelines calculated a guideline value of 0.02 mg/litre for alachlor in drinking-water, corresponding to an upper-bound excess lifetime cancer risk of 10^{-5}.

Assessment date

The risk assessment was originally conducted in 1993. The Final Task Force Meeting in 2003 agreed that this risk assessment be brought forward to this edition of the *Guidelines for Drinking-water Quality*.

Principal reference

WHO (2003) *Alachlor in drinking-water. Background document for preparation of WHO Guidelines for drinking-water quality*. Geneva, World Health Organization (WHO/SDE/WSH/03.04/31).

12.3 Aldicarb

Aldicarb (CAS No. 116-06-3) is a systemic pesticide used to control nematodes in soil and insects and mites on a variety of crops. It is very soluble in water and highly mobile in soil. It degrades mainly by biodegradation and hydrolysis, persisting for weeks to months.

Guideline value	0.01 mg/litre
Occurrence	Frequently found as a contaminant in groundwater, particularly when associated with sandy soil; concentrations in well water as high as 500 µg/litre have been measured. Aldicarb sulfoxide and aldicarb sulfone residues are found in an approximately 1:1 ratio in groundwater.
ADI	0.003 mg/kg of body weight based on cholinesterase depression in a single oral dose study in human volunteers
Limit of detection	0.001 mg/litre by reverse-phase HPLC with fluorescence detection
Treatment achievability	0.001 mg/litre should be achievable using GAC or ozonation
Guideline derivation • allocation to water • weight • consumption	10% of ADI 60-kg adult 2 litres/day
Additional comments	The guideline value derived from the 1992 JMPR assessment was very similar to the guideline value derived in the second edition, which was therefore retained.

Toxicological review

Aldicarb is one of the most acutely toxic pesticides in use, although the only consistently observed toxic effect with both long-term and single-dose administration is acetylcholinesterase inhibition. It is metabolized to the sulfoxide and sulfone. Aldicarb sulfoxide is a more potent inhibitor of acetylcholinesterase than aldicarb itself, while aldicarb sulfone is considerably less toxic than either aldicarb or the sulfoxide. The weight of evidence indicates that aldicarb, aldicarb sulfoxide and aldicarb sulfone are not genotoxic or carcinogenic. IARC has concluded that aldicarb is not classifiable as to its carcinogenicity (Group 3).

History of guideline development

The 1958 and 1963 WHO *International Standards for Drinking-water* did not refer to aldicarb, but the 1971 International Standards suggested that pesticide residues that may occur in community water supplies make only a minimal contribution to the total daily intake of pesticides for the population served. Aldicarb was not evaluated in the first edition of the *Guidelines for Drinking-water Quality*, published in 1984, but a health-based guideline value of 0.01 mg/litre was derived for aldicarb in the 1993 Guidelines.

Assessment date

The risk assessment was conducted in 2003.

Principal references
FAO/WHO (1993) *Pesticide residues in food – 1992*. Rome, Food and Agriculture Organization of the United Nations, Joint FAO/WHO Meeting on Pesticide Residues (Report No. 116).

WHO (2003) *Aldicarb in drinking-water. Background document for preparation of WHO Guidelines for drinking-water quality*. Geneva, World Health Organization (WHO/SDE/WSH/03.04/72).

12.4 Aldrin and dieldrin

Aldrin (CAS No. 309-00-2) and dieldrin (CAS No. 60-57-1) are chlorinated pesticides that are used against soil-dwelling pests, for wood protection and, in the case of dieldrin, against insects of public health importance. Since the early 1970s, a number of countries have either severely restricted or banned the use of both compounds, particularly in agriculture. The two compounds are closely related with respect to their toxicology and mode of action. Aldrin is rapidly converted to dieldrin under most environmental conditions and in the body. Dieldrin is a highly persistent organochlorine compound that has low mobility in soil, can be lost to the atmosphere and bioaccumulates. Dietary exposure to aldrin/dieldrin is very low and decreasing.

Guideline value	0.00003 mg/litre (0.03 µg/litre) combined aldrin and dieldrin
Occurrence	Concentrations of aldrin and dieldrin in drinking-water normally less than 0.01 µg/litre; rarely present in groundwater
PTDI	0.1 µg/kg of body weight (combined total for aldrin and dieldrin), based on NOAELs of 1 mg/kg of diet in the dog and 0.5 mg/kg of diet in the rat, which are equivalent to 0.025 mg/kg of body weight per day in both species, and applying an uncertainty factor of 250 based on concern about carcinogenicity observed in mice
Limit of detection	0.003 µg/litre for aldrin and 0.002 µg/litre for dieldrin by GC with ECD
Treatment achievability	0.02 µg/litre should be achievable using coagulation, GAC or ozonation
Guideline derivation • allocation to water • weight • consumption	1% of PTDI 60-kg adult 2 litres/day
Additional comments	Aldrin and dieldrin are listed under the Stockholm Convention on Persistent Organic Pollutants. Hence, monitoring may occur in addition to that required by drinking-water guidelines.

Toxicological review

Both compounds are highly toxic in experimental animals, and cases of poisoning in humans have occurred. Aldrin and dieldrin have more than one mechanism of toxicity. The target organs are the central nervous system and the liver. In long-term studies, dieldrin was shown to produce liver tumours in both sexes of two strains of mice. It did not produce an increase in tumours in rats and does not appear to be genotoxic. IARC has classified aldrin and dieldrin in Group 3. It is considered that all the available information on aldrin and dieldrin taken together, including studies on humans, supports the view that, for practical purposes, these chemicals make very little contribution, if any, to the incidence of cancer in humans.

History of guideline development
The 1958 and 1963 WHO *International Standards for Drinking-water* did not refer to aldrin and dieldrin, but the 1971 International Standards suggested that pesticide residues that may occur in community water supplies make only a minimal contribution to the total daily intake of pesticides for the population served. In the first edition of the *Guidelines for Drinking-water Quality*, published in 1984, a health-based guideline value of 0.03 µg/litre was recommended for aldrin and dieldrin, based on the ADI recommended by JMPR in 1970 for aldrin and dieldrin residues separately or together and reaffirmed by toxicological data available in 1977. The 1993 Guidelines confirmed the health-based guideline value of 0.03 µg/litre for aldrin and dieldrin, based on the reaffirmation of the ADI recommended in 1977 by JMPR.

Assessment date
The risk assessment was conducted in 2003.

Principal references
FAO/WHO (1995) *Pesticide residues in food – 1994. Report of the Joint Meeting of the FAO Panel of Experts on Pesticide Residues in Food and the Environment and WHO Toxicological and Environmental Core Assessment Groups*. Rome, Food and Agriculture Organization of the United Nations (FAO Plant Production and Protection Paper 127).

WHO (2003) *Aldrin and dieldrin in drinking-water. Background document for preparation of WHO Guidelines for drinking-water quality*. Geneva, World Health Organization (WHO/SDE/WSH/03.04/73).

12.5 Aluminium
Aluminium is the most abundant metallic element and constitutes about 8% of the Earth's crust. Aluminium salts are widely used in water treatment as coagulants to reduce organic matter, colour, turbidity and microorganism levels. Such use may lead to increased concentrations of aluminium in finished water. Where residual concentrations are high, undesirable colour and turbidity may ensue. Concentrations of aluminium at which such problems may occur are highly dependent on a number of water quality parameters and operational factors at the water treatment plant. Aluminium intake from foods, particularly those containing aluminium compounds used as food additives, represents the major route of aluminium exposure for the general

public. The contribution of drinking-water to the total oral exposure to aluminium is usually less than 5% of the total intake.

In humans, aluminium and its compounds appear to be poorly absorbed, although the rate and extent of absorption have not been adequately studied for all sectors of the population. The degree of aluminium absorption depends on a number of parameters, such as the aluminium salt administered, pH (for aluminium speciation and solubility), bioavailability and dietary factors. These parameters should be taken into consideration during tissue dosimetry and response assessment. The use of currently available animal studies to develop a guideline value for aluminium is not appropriate because of these specific toxicokinetic/toxicodynamic considerations.

There is little indication that orally ingested aluminium is acutely toxic to humans despite the widespread occurrence of the element in foods, drinking-water and many antacid preparations. It has been hypothesized that aluminium exposure is a risk factor for the development or acceleration of onset of Alzheimer disease (AD) in humans. The 1997 WHO EHC document for aluminium concludes that:

> On the whole, the positive relationship between aluminium in drinking-water and AD, which was demonstrated in several epidemiological studies, cannot be totally dismissed. However, strong reservations about inferring a causal relationship are warranted in view of the failure of these studies to account for demonstrated confounding factors and for total aluminium intake from all sources.
>
> Taken together, the relative risks for AD from exposure to aluminium in drinking-water above 100 µg/litre, as determined in these studies, are low (less than 2.0). But, because the risk estimates are imprecise for a variety of methodological reasons, a population-attributable risk cannot be calculated with precision. Such imprecise predictions may, however, be useful in making decisions about the need to control exposures to aluminium in the general population.

Owing to the limitations of the animal data as a model for humans and the uncertainty surrounding the human data, a health-based guideline value for aluminium cannot be derived at this time.

The beneficial effects of the use of aluminium as a coagulant in water treatment are recognized. Taking this into account, and considering the health concerns about aluminium (i.e., its potential neurotoxicity), a practicable level is derived, based on optimization of the coagulation process in drinking-water plants using aluminium-based coagulants, to minimize aluminium levels in finished water.

Several approaches are available for minimizing residual aluminium concentrations in treated water. These include use of optimum pH in the coagulation process, avoiding excessive aluminium dosage, good mixing at the point of application of the coagulant, optimum paddle speeds for flocculation and efficient filtration of the aluminium floc. Under good operating conditions, concentrations of aluminium of 0.1 mg/litre or less are achievable in large water treatment facilities. Small facilities (e.g., those serving fewer than 10 000 people) might experience some difficulties in attaining this level, because the small size of the plant provides little buffering for fluctuation in operation; moreover, such facilities often have limited resources and limited access to the expertise needed to solve specific operational problems. For these small facilities, 0.2 mg/litre or less is a practicable level for aluminium in finished water.

History of guideline development
The 1958, 1963 and 1971 WHO *International Standards for Drinking-water* did not refer to aluminium. In the first edition of the *Guidelines for Drinking-water Quality*, published in 1984, a guideline value of 0.2 mg/litre was established for aluminium, based on aesthetic considerations (as a compromise between the use of aluminium compounds in water treatment and discoloration that may be observed if levels above 0.1 mg/litre remain in the distributed water). No health-based guideline value was recommended in the 1993 Guidelines, but the Guidelines confirmed that a concentration of 0.2 mg/litre in drinking-water provides a compromise between the practical use of aluminium salts in water treatment and discoloration of distributed water. No health-based guideline value was derived for aluminium in the addendum to the Guidelines published in 1998, owing to the limitations of the animal data as a model for humans and the uncertainty surrounding the human data. However, taking the beneficial effects of the use of aluminium as a coagulant in water treatment into account and considering the health concerns about aluminium (i.e., its potential neurotoxicity), a practicable level was derived based on optimization of the coagulation process in drinking-water plants using aluminium-based coagulants, to minimize aluminium levels in finished water. Under good operating conditions, concentrations of aluminium of 0.1 mg/litre or less are achievable in large water treatment facilities. For small facilities, 0.2 mg/litre or less is a practicable level for aluminium in finished water.

Assessment date
The risk assessment was originally conducted in 1998. The Final Task Force Meeting in 2003 agreed that this risk assessment be brought forward to this edition of the *Guidelines for Drinking-water Quality*.

Principal reference
WHO (2003) *Aluminium in drinking-water. Background document for preparation of WHO Guidelines for drinking-water quality*. Geneva, World Health Organization (WHO/SDE/WSH/03.04/53).

12.6 Ammonia
The term ammonia includes the non-ionized (NH_3) and ionized (NH_4^+) species. Ammonia in the environment originates from metabolic, agricultural and industrial processes and from disinfection with chloramine. Natural levels in groundwater and surface water are usually below 0.2 mg/litre. Anaerobic groundwaters may contain up to 3 mg/litre. Intensive rearing of farm animals can give rise to much higher levels in surface water. Ammonia contamination can also arise from cement mortar pipe

linings. Ammonia in water is an indicator of possible bacterial, sewage and animal waste pollution.

Ammonia is a major component of the metabolism of mammals. Exposure from environmental sources is insignificant in comparison with endogenous synthesis of ammonia. Toxicological effects are observed only at exposures above about 200 mg/kg of body weight.

Ammonia in drinking-water is not of immediate health relevance, and therefore no health-based guideline value is proposed. However, ammonia can compromise disinfection efficiency, result in nitrite formation in distribution systems, cause the failure of filters for the removal of manganese and cause taste and odour problems (see also chapter 10).

History of guideline development
The 1958, 1963 and 1971 WHO *International Standards for Drinking-water* and the first edition of the *Guidelines for Drinking-water Quality*, published in 1984, did not refer to ammonia. In the 1993 Guidelines, no health-based guideline value was recommended, but the Guidelines stated that ammonia could cause taste and odour problems at concentrations above 35 and 1.5 mg/litre, respectively.

Assessment date
The risk assessment was originally conducted in 1993. The Final Task Force Meeting in 2003 agreed that this risk assessment be brought forward to this edition of the *Guidelines for Drinking-water Quality*.

Principal reference
WHO (2003) *Ammonia in drinking-water. Background document for preparation of WHO Guidelines for drinking-water quality.* Geneva, World Health Organization (WHO/SDE/WSH/03.04/1).

12.7 Antimony
Elemental antimony forms very hard alloys with copper, lead and tin. Antimony compounds have various therapeutic uses. Antimony was considered as a possible replacement for lead in solders, but there is no evidence of any significant contribution to drinking-water concentrations from this source. Daily oral uptake of antimony appears to be significantly higher than exposure by inhalation, although total exposure from environmental sources, food and drinking-water is very low compared with occupational exposure.

Guideline value	0.02 mg/litre
Occurrence	Concentrations in groundwater and surface water normally range from 0.1 to 0.2 µg/litre; concentrations in drinking-water appear to be less than 5 µg/litre.
TDI	6 µg/kg of body weight, based on a NOAEL of 6.0 mg/kg of body weight per day for decreased body weight gain and reduced food and water intake in a 90-day study in which rats were administered potassium antimony tartrate in drinking-water, using an uncertainty factor of 1000 (100 for inter- and intraspecies variation, 10 for the short duration of the study)
Limit of detection	0.01 µg/litre by EAAS; 0.1–1 µg/litre by ICP/MS; 0.8 µg/litre by graphite furnace atomic absorption spectrophotometry; 5 µg/litre by hydride generation AAS
Treatment achievability	Conventional treatment processes do not remove antimony. However, antimony is not normally a raw water contaminant. As the most common source of antimony in drinking-water appears to be dissolution from metal plumbing and fittings, control of antimony from such sources would be by product control.
Guideline derivation • allocation to water • weight • consumption	 10% of TDI 60-kg adult 2 litres/day

Toxicological review

There has been a significant increase in the toxicity data available since the previous review, although much of it pertains to the intraperitoneal route of exposure. The form of antimony in drinking-water is a key determinant of the toxicity, and it would appear that antimony leached from antimony-containing materials would be in the form of the antimony(V) oxo-anion, which is the less toxic form. The subchronic toxicity of antimony trioxide is lower than that of potassium antimony tartrate, which is the most soluble form. Antimony trioxide, due to its low bioavailability, is genotoxic only in some *in vitro* tests, but not *in vivo*, whereas soluble antimony(III) salts exert genotoxic effects *in vitro* and *in vivo*. Animal experiments from which the carcinogenic potential of soluble or insoluble antimony compounds may be quantified are not available. IARC has concluded that antimony trioxide is possibly carcinogenic to humans (Group 2B) on the basis of an inhalation study in rats, but that antimony trisulfide was not classifiable as to its carcinogenicity to humans (Group 3). However, chronic oral uptake of potassium antimony tartrate may not be associated with an additional carcinogenic risk, since antimony after inhalation exposure was carcinogenic only in the lung but not in other organs and is known to cause direct lung damage following chronic inhalation as a consequence of overload with insoluble particulates. Although there is some evidence for the carcinogenicity of certain antimony compounds by inhalation, there are no data to indicate carcinogenicity by the oral route.

History of guideline development
The 1958, 1963 and 1971 WHO *International Standards for Drinking-water* did not refer to antimony. In the first edition of the *Guidelines for Drinking-water Quality*, published in 1984, it was concluded that no action was required for antimony. A provisional guideline value for antimony was set at a practical quantification level of 0.005 mg/litre in the 1993 Guidelines, based on available toxicological data.

Assessment date
The risk assessment was conducted in 2003.

Principal reference
WHO (2003) *Antimony in drinking-water. Background document for preparation of WHO Guidelines for drinking-water quality*. Geneva, World Health Organization (WHO/SDE/WSH/03.04/74).

12.8 Arsenic
Arsenic is widely distributed throughout the Earth's crust, most often as arsenic sulfide or as metal arsenates and arsenides. Arsenicals are used commercially and industrially, primarily as alloying agents in the manufacture of transistors, lasers and semiconductors. Arsenic is introduced into drinking-water sources primarily through the dissolution of naturally occurring minerals and ores. Except for individuals who are occupationally exposed to arsenic, the most important route of exposure is through the oral intake of food and beverages. There are a number of regions where arsenic may be present in drinking-water sources, particularly groundwater, at elevated concentrations. Arsenic in drinking-water is a significant cause of health effects in some areas, and arsenic is considered to be a high-priority substance for screening in drinking-water sources. Concentrations are often highly dependent on the depth to which the well is sunk.

Provisional guideline value	0.01 mg/litre The guideline value is designated as provisional in view of the scientific uncertainties.
Occurrence	Levels in natural waters generally rangey between 1 and 2 µg/litre, although concentrations may be elevated (up to 12 mg/litre) in areas containing natural sources.
Basis of guideline derivation	There remains considerable uncertainty over the actual risks at low concentrations, and available data on mode of action do not provide a biological basis for using either linear or non-linear extrapolation. In view of the significant uncertainties surrounding the risk assessment for arsenic carcinogenicity, the practical quantification limit in the region of 1–10 µg/litre and the practical difficulties in removing arsenic from drinking-water, the guideline value of 10 µg/litre is retained. In view of the scientific uncertainties, the guideline value is designated as provisional.

Limit of detection	0.1 µg/litre by ICP/MS; 2 µg/litre by hydride generation AAS or FAAS
Treatment achievability	It is technically feasible to achieve arsenic concentrations of 5 µg/litre or lower using any of several possible treatment methods. However, this requires careful process optimization and control, and a more reasonable expectation is that 10 µg/litre should be achievable by conventional treatment, e.g., coagulation.
Additional comments	• A management guidance document on arsenic is available. • In many countries, this guideline value may not be attainable. Where this is the case, every effort should be made to keep concentrations as low as possible.

Toxicological review

Arsenic has not been demonstrated to be essential in humans. It is an important drinking-water contaminant, as it is one of the few substances shown to cause cancer in humans through consumption of drinking-water. There is overwhelming evidence from epidemiological studies that consumption of elevated levels of arsenic through drinking-water is causally related to the development of cancer at several sites, particularly skin, bladder and lung. In several parts of the world, arsenic-induced disease, including cancer, is a significant public health problem. Because trivalent inorganic arsenic has greater reactivity and toxicity than pentavalent inorganic arsenic, it is generally believed that the trivalent form is the carcinogen. However, there remain considerable uncertainty and controversy over both the mechanism of carcinogenicity and the shape of the dose–response curve at low intakes. Inorganic arsenic compounds are classified by IARC in Group 1 (carcinogenic to humans) on the basis of sufficient evidence for carcinogenicity in humans and limited evidence for carcinogenicity in animals.

History of guideline development

The 1958 WHO *International Standards for Drinking-water* recommended a maximum allowable concentration of 0.2 mg/litre for arsenic, based on health concerns. In the 1963 International Standards, this value was lowered to 0.05 mg/litre, which was retained as a tentative upper concentration limit in the 1971 International Standards. The guideline value of 0.05 mg/litre was also retained in the first edition of the *Guidelines for Drinking-water Quality*, published in 1984. A provisional guideline value for arsenic was set at the practical quantification limit of 0.01 mg/litre in the 1993 Guidelines, based on concern regarding its carcinogenicity in humans.

Assessment date

The risk assessment was conducted in 2003.

Principal references

IPCS (2001) *Arsenic and arsenic compounds*. Geneva, World Health Organization, International Programme on Chemical Safety (Environmental Health Criteria 224).

WHO (2003) *Arsenic in drinking-water. Background document for preparation of WHO Guidelines for drinking-water quality*. Geneva, World Health Organization (WHO/SDE/WSH/03.04/75).

12.9 Asbestos

Asbestos is introduced into water by the dissolution of asbestos-containing minerals and ores as well as from industrial effluents, atmospheric pollution and asbestos-cement pipes in the distribution system. Exfoliation of asbestos fibres from asbestos-cement pipes is related to the aggressiveness of the water supply. Limited data indicate that exposure to airborne asbestos released from tap water during showers or humidification is negligible.

Asbestos is a known human carcinogen by the inhalation route. Although well studied, there has been little convincing evidence of the carcinogenicity of ingested asbestos in epidemiological studies of populations with drinking-water supplies containing high concentrations of asbestos. Moreover, in extensive studies in animal species, asbestos has not consistently increased the incidence of tumours of the gastrointestinal tract. There is, therefore, no consistent evidence that ingested asbestos is hazardous to health, and thus it is concluded that there is no need to establish a health-based guideline value for asbestos in drinking-water.

History of guideline development

The 1958, 1963 and 1971 WHO *International Standards for Drinking-water* did not refer to asbestos. In the first edition of the *Guidelines for Drinking-water Quality*, published in 1984, it was noted that available data were insufficient to determine whether a guideline value was needed for asbestos. The 1993 Guidelines concluded that there was no consistent evidence that ingested asbestos was hazardous to health and that there was therefore no need to establish a health-based guideline value for asbestos in drinking-water.

Assessment date

The risk assessment was originally conducted in 1993. The Final Task Force Meeting in 2003 agreed that this risk assessment be brought forward to this edition of the *Guidelines for Drinking-water Quality*.

Principal reference

WHO (2003) *Asbestos in drinking-water. Background document for preparation of WHO Guidelines for drinking-water quality*. Geneva, World Health Organization (WHO/SDE/WSH/03.04/2).

12.10 Atrazine

Atrazine (CAS No. 1912-24-9) is a selective pre- and early post-emergence herbicide. It has been found in surface water and groundwater as a result of its mobility in soil.

It is relatively stable in soil and aquatic environments, with a half-life measured in months, but is degraded by photolysis and microbial action in soil.

Guideline value	0.002 mg/litre
Occurrence	Found in groundwater and drinking-water at levels below 10 µg/litre
TDI	0.5 µg/kg of body weight based on a NOAEL of 0.5 mg/kg of body weight per day in a carcinogenicity study in the rat and an uncertainty factor of 1000 (100 for inter- and intraspecies variation and 10 to reflect potential neoplasia)
Limit of detection	0.01 µg/litre by GC/MS
Treatment achievability	0.1 µg/litre should be achievable using GAC
Guideline derivation • allocation to water • weight • consumption	 10% of TDI 60-kg adult 2 litres/day

Toxicological review

The weight of evidence from a wide variety of genotoxicity assays indicates that atrazine is not genotoxic. There is evidence that atrazine can induce mammary tumours in rats. It is highly probable that the mechanism for this process is non-genotoxic. No significant increase in neoplasia has been observed in mice. IARC has concluded that there is inadequate evidence in humans and limited evidence in experimental animals for the carcinogenicity of atrazine (Group 2B).

History of guideline development

The 1958 and 1963 WHO *International Standards for Drinking-water* did not refer to atrazine, but the 1971 International Standards suggested that pesticide residues that may occur in community water supplies make only a minimal contribution to the total daily intake of pesticides for the population served. Atrazine was not evaluated in the first edition of the *Guidelines for Drinking-water Quality*, published in 1984, but the 1993 Guidelines established a health-based guideline value of 0.002 mg/litre for atrazine in drinking-water.

Assessment date

The risk assessment was originally conducted in 1993. The Final Task Force Meeting in 2003 agreed that this risk assessment be brought forward to this edition of the *Guidelines for Drinking-water Quality*.

Principal reference

WHO (2003) *Atrazine in drinking-water. Background document for preparation of WHO Guidelines for drinking-water quality*. Geneva, World Health Organization (WHO/SDE/WSH/03.04/32).

12.11 Barium

Barium is present as a trace element in both igneous and sedimentary rocks, and barium compounds are used in a variety of industrial applications; however, barium in water comes primarily from natural sources. Food is the primary source of intake for the non-occupationally exposed population. However, where barium levels in water are high, drinking-water may contribute significantly to total intake.

Guideline value	0.7 mg/litre
Occurrence	Concentrations in drinking-water are generally below 100 µg/litre, although concentrations above 1 mg/litre have been measured in drinking-water derived from groundwater.
NOAEL in humans	7.3 mg/litre in the most sensitive epidemiological study conducted to date, in which there were no significant differences in blood pressure or in the prevalence of cardiovascular disease between a population drinking water containing a mean barium concentration of 7.3 mg/litre and one whose water contained a barium concentration of 0.1 mg/litre
Guideline derivation	Uncertainty factor of 10 for intraspecies variation applied to NOAEL in humans
Limit of detection	0.1 µg/litre by ICP/MS; 2 µg/litre by AAS; 3 µg/litre by ICP/optical emission spectroscopy
Treatment achievability	0.1 mg/litre should be achievable using either ion exchange or precipitation softening; other conventional processes are ineffective

Toxicological review

There is no evidence that barium is carcinogenic or mutagenic. Barium has been shown to cause nephropathy in laboratory animals, but the toxicological end-point of greatest concern to humans appears to be its potential to cause hypertension.

History of guideline development

The 1958 WHO *International Standards for Drinking-water* did not refer to barium. The 1963 International Standards recommended a maximum allowable concentration of 1.0 mg/litre, based on health concerns. The 1971 International Standards stated that barium should be controlled in drinking-water, but that insufficient information was available to enable a tentative limit to be established. In the first edition of the *Guidelines for Drinking-water Quality*, published in 1984, it was concluded that it was not necessary to establish a guideline value for barium in drinking-water, as there was no firm evidence of any health effects associated with the normally low levels of barium in water. A health-based guideline value of 0.7 mg/litre was derived for barium in the 1993 Guidelines, based on concern regarding the potential of barium to cause hypertension.

Assessment date

The risk assessment was conducted in 2003.

Principal references

IPCS (2001) *Barium and barium compounds.* Geneva, World Health Organization, International Programme on Chemical Safety (Concise International Chemical Assessment Document 33).

WHO (2003) *Barium in drinking-water. Background document for preparation of WHO Guidelines for drinking-water quality.* Geneva, World Health Organization (WHO/SDE/WSH/03.04/76).

12.12 Bentazone

Bentazone (CAS No. 25057-89-0) is a broad-spectrum herbicide used for a variety of crops. Photodegradation occurs in soil and water; however, bentazone is very mobile in soil and moderately persistent in the environment. Bentazone has been reported to occur in surface water, groundwater and drinking-water at concentrations of a few micrograms per litre or less. Although it has been found in groundwater and has a high affinity for the water compartment, it does not seem to accumulate in the environment. Exposure from food is unlikely to be high.

Long-term studies conducted in rats and mice have not indicated a carcinogenic potential, and a variety of *in vitro* and *in vivo* assays have indicated that bentazone is not genotoxic. A health-based value of 300 µg/litre can be calculated on the basis of an ADI of 0.1 mg/kg of body weight established by JMPR, based on haematological effects observed in a 2-year dietary study in rats. However, because bentazone occurs at concentrations well below those at which toxic effects are observed, it is not considered necessary to derive a health-based guideline value.

History of guideline development

The 1958 and 1963 WHO *International Standards for Drinking-water* did not refer to bentazone, but the 1971 International Standards suggested that pesticide residues that may occur in community water supplies make only a minimal contribution to the total daily intake of pesticides for the population served. Bentazone was not evaluated in the first edition of the *Guidelines for Drinking-water Quality*, published in 1984, but the 1993 Guidelines established a health-based guideline value of 0.03 mg/litre for bentazone, based on an ADI established by JMPR in 1991. This guideline value was amended to 0.3 mg/litre in the addendum to the Guidelines, published in 1998, based on new information on the environmental behaviour of bentazone and exposure from food.

Assessment date

The risk assessment was conducted in 2003.

Principal references

FAO/WHO (1999) *Pesticide residues in food – 1998. Evaluations – 1998. Part II – Toxicology.* Geneva, World Health Organization, Joint FAO/WHO Meeting on Pesticide Residues (WHO/PCS/01.12).

WHO (2003) *Bentazone in drinking-water. Background document for preparation of WHO Guidelines for drinking-water quality*. Geneva, World Health Organization (WHO/SDE/WSH/03.04/77).

12.13 Benzene

Benzene is used principally in the production of other organic chemicals. It is present in petrol, and vehicular emissions constitute the main source of benzene in the environment. Benzene may be introduced into water by industrial effluents and atmospheric pollution.

Guideline value	0.01 mg/litre
Occurrence	Concentrations in drinking-water generally less than 5 µg/litre
Basis of guideline derivation	Robust linear extrapolation model (because of statistical lack of fit of some of the data with the linearized multistage model) applied to leukaemia and lymphomas in female mice and oral cavity squamous cell carcinomas in male rats in a 2-year gavage study in rats and mice
Limit of detection	0.2 µg/litre by GC with photoionization detection and confirmation by MS
Treatment achievability	0.01 mg/litre should be achievable using GAC or air stripping
Additional comments	Lower end of estimated range of concentrations in drinking-water corresponding to an upper-bound excess lifetime cancer risk of 10^{-5} (10–80 µg/litre) corresponds to the estimate derived from data on leukaemia from epidemiological studies involving inhalation exposure, which formed the basis for the previous guideline value. The previous guideline value is therefore retained.

Toxicological review

Acute exposure of humans to high concentrations of benzene primarily affects the central nervous system. At lower concentrations, benzene is toxic to the haematopoietic system, causing a continuum of haematological changes, including leukaemia. Because benzene is carcinogenic to humans, IARC has classified it in Group 1. Haematological abnormalities similar to those observed in humans have been observed in animal species exposed to benzene. In animal studies, benzene was shown to be carcinogenic following both inhalation and ingestion. It induced several types of tumours in both rats and mice in a 2-year carcinogenesis bioassay by gavage in corn oil. Benzene has not been found to be mutagenic in bacterial assays, but it has been shown to cause chromosomal aberrations *in vivo* in a number of species, including humans, and to be positive in the mouse micronucleus test.

History of guideline development

The 1958, 1963 and 1971 WHO *International Standards for Drinking-water* did not refer to benzene. In the first edition of the *Guidelines for Drinking-water Quality*, published in 1984, a health-based guideline value of 0.01 mg/litre was recommended for benzene based on human leukaemia data from inhalation exposure applied to a linear multistage extrapolation model. The 1993 Guidelines estimated the range of benzene concentrations in drinking-water corresponding to an upper-bound excess lifetime cancer risk of 10^{-5} to be 0.01–0.08 mg/litre based on carcinogenicity in female mice and male rats. As the lower end of this estimate corresponds to the estimate derived from epidemiological data, which formed the basis for the previous guideline value of 0.01 mg/litre associated with a 10^{-5} upper-bound excess lifetime cancer risk, the guideline value of 0.01 mg/litre was retained.

Assessment date
The risk assessment was originally conducted in 1993. The Final Task Force Meeting in 2003 agreed that this risk assessment be brought forward to this edition of the *Guidelines for Drinking-water Quality*.

Principal reference
WHO (2003) *Benzene in drinking-water. Background document for preparation of WHO Guidelines for drinking-water quality*. Geneva, World Health Organization (WHO/SDE/WSH/03.04/24).

12.14 Boron
Boron compounds are used in the manufacture of glass, soaps and detergents and as flame retardants. The general population obtains the greatest amount of boron through food intake, as it is naturally found in many edible plants. Boron is found naturally in groundwater, but its presence in surface water is frequently a consequence of the discharge of treated sewage effluent, in which it arises from use in some detergents, to surface waters.

Provisional guideline value	0.5 mg/litre
	The guideline is designated as provisional because it will be difficult to achieve in areas with high natural boron levels with the treatment technology available.
Occurrence	Concentrations vary widely and depend on the surrounding geology and wastewater discharges. For most of the world, the concentration range of boron in drinking-water is judged to be between 0.1 and 0.3 mg/litre.
TDI	0.16 mg/kg of body weight, based on a NOAEL of 9.6 mg/kg of body weight per day for developmental toxicity (decreased fetal body weight in rats) and an uncertainty factor of 60 (10 for interspecies variation and 6 for intraspecies variation)
Limit of detection	0.2 µg/litre by ICP/MS; 6–10 µg/litre by ICP/AES

Treatment achievability	Conventional water treatment (coagulation, sedimentation, filtration) does not significantly remove boron, and special methods need to be installed in order to remove boron from waters with high boron concentrations. Ion exchange and reverse osmosis processes may enable substantial reduction but are likely to be prohibitively expensive. Blending with low-boron supplies may be the only economical method to reduce boron concentrations in waters where these concentrations are high.
Guideline derivation • allocation to water • weight • consumption	 10% of TDI 60-kg adult 2 litres/day

Toxicological review

Short- and long-term oral exposures to boric acid or borax in laboratory animals have demonstrated that the male reproductive tract is a consistent target of toxicity. Testicular lesions have been observed in rats, mice and dogs given boric acid or borax in food or drinking-water. Developmental toxicity has been demonstrated experimentally in rats, mice and rabbits. Negative results in a large number of mutagenicity assays indicate that boric acid and borax are not genotoxic. In long-term studies in mice and rats, boric acid and borax caused no increase in tumour incidence.

History of guideline development

The 1958, 1963 and 1971 WHO *International Standards for Drinking-water* did not refer to boron. In the first edition of the *Guidelines for Drinking-water Quality*, published in 1984, it was concluded that no action was required for boron. A health-based guideline value of 0.3 mg/litre for boron was established in the 1993 Guidelines, while noting that boron's removal by drinking-water treatment appears to be poor. This guideline value was increased to 0.5 mg/litre in the addendum to the Guidelines published in 1998 and was designated as provisional because, with the treatment technology available, the guideline value will be difficult to achieve in areas with high natural boron levels.

Assessment date

The risk assessment was originally conducted in 1998. The Final Task Force Meeting in 2003 agreed that this risk assessment be brought forward to this edition of the *Guidelines for Drinking-water Quality*.

Principal reference

WHO (2003) *Boron in drinking-water. Background document for preparation of WHO Guidelines for drinking-water quality*. Geneva, World Health Organization (WHO/SDE/WSH/03.04/54).

12.15 Bromate

Sodium and potassium bromate are powerful oxidizers used mainly in permanent wave neutralizing solutions and the dyeing of textiles using sulfur dyes. Potassium bromate is also used as an oxidizer to mature flour during milling, in treating barley in beer making and in fish paste products, although JECFA has concluded that the use of potassium bromate in food processing is not appropriate. Bromate is not normally found in water, but may be formed during ozonation when the bromide ion is present in water. Under certain conditions, bromate may also be formed in concentrated hypochlorite solutions used to disinfect drinking-water.

Provisional guideline value	0.01 mg/litre The guideline value is provisional because of limitations in available analytical and treatment methods and uncertainties in the toxicological data.
Occurrence	Has been reported in drinking-water with a variety of source water characteristics after ozonation at concentrations ranging from <2 to 293 µg/litre, depending on bromide ion concentration, ozone dosage, pH, alkalinity and dissolved organic carbon
Basis of guideline derivation	Upper-bound estimate of cancer potency for bromate is 0.19 per mg/kg of body weight per day, based on low-dose linear extrapolation (a one-stage Weibull time-to-tumour model was applied to the incidence of mesotheliomas, renal tubule tumours and thyroid follicular tumours in male rats given potassium bromate in drinking-water, using the 12-, 26-, 52- and 77-week interim kill data). A health-based value of 2 µg/litre is associated with the upper-bound excess cancer risk of 10^{-5}. A similar conclusion may be reached through several other methods of extrapolation, leading to values in the range 2–6 µg/litre.
Limit of detection	1.5 µg/litre by ion chromatography with suppressed conductivity detection; 0.2 µg/litre by ion chromatography with UV/visible absorbance detection; 0.3 µg/litre by ion chromatography with detection by ICP/MS
Treatment achievability	Bromate is difficult to remove once formed. By appropriate control of disinfection conditions, it is possible to achieve bromate concentrations below 0.01 mg/litre.

Toxicological review

IARC has concluded that although there is inadequate evidence of carcinogenicity in humans, there is sufficient evidence for the carcinogenicity of potassium bromate in experimental animals and has classified it in Group 2B (possibly carcinogenic to humans). Bromate is mutagenic both *in vitro* and *in vivo*. At this time, there is not sufficient evidence to conclude the mode of carcinogenic action for potassium bromate. Observation of tumours at a relatively early time and the positive response of bromate in a variety of genotoxicity assays suggest that the predominant mode of action at low doses is due to DNA reactivity. Although there is limited evidence to suggest that the DNA reactivity in kidney tumours may have a non-linear dose–response relationship, there is no evidence to suggest that this same dose–response relationship operates in the development of mesotheliomas or thyroid tumours. Oxidative stress may play a role in the formation of kidney tumours, but the evidence is insufficient to establish lipid peroxidation and free radical production as key events responsible for induction of kidney tumours. Also, there are no data currently available to suggest that any single mechanism, including oxidative stress, is responsible for the production of thyroid and peritoneal tumours by bromate.

History of guideline development
The 1958, 1963 and 1971 WHO *International Standards for Drinking-water* and the first edition of the *Guidelines for Drinking-water Quality*, published in 1984, did not refer to bromate. The 1993 Guidelines calculated the concentration of bromate in drinking-water associated with an upper-bound excess lifetime cancer risk of 10^{-5} to be 0.003 mg/litre. However, because of limitations in available analytical and treatment methods, a provisional guideline value of 0.025 mg/litre, associated with an upper-bound excess lifetime cancer risk of 7×10^{-5}, was recommended.

Assessment date
The risk assessment was conducted in 2003.

Principal reference
WHO (2003) *Bromate in drinking-water. Background document for preparation of WHO Guidelines for drinking-water quality*. Geneva, World Health Organization (WHO/SDE/WSH/03.04/78).

12.16 Brominated acetic acids
Brominated acetic acids are formed during disinfection of water that contains bromide ions and organic matter. Bromide ions occur naturally in surface water and groundwater and exhibit seasonal fluctuations in levels. Bromide ion levels can increase due to saltwater intrusion resulting from drought conditions or due to pollution. Brominated acetates are generally present in surface water and groundwater distribution systems at mean concentrations below 5 µg/litre.

The database for dibromoacetic acid is considered inadequate for the derivation of a guideline value. There are no systemic toxicity studies of subchronic duration or longer. The database also lacks suitable toxicokinetic studies, a carcinogenicity study, a developmental study in a second species and a multigeneration reproductive toxicity study (one has been conducted but is currently being evaluated by the US EPA). Available mutagenicity data suggest that dibromoacetate is genotoxic.

Data are also limited on the oral toxicity of monobromoacetic acid and bromochloroacetic acid. Limited mutagenicity and genotoxicity data give mixed results for monobromoacetic acid and generally positive results for bromochloroacetic acid.

Data gaps include subchronic or chronic toxicity studies, multigeneration reproductive toxicity studies, standard developmental toxicity studies and carcinogenicity studies. The available data are considered inadequate to establish guideline values for these chemicals.

History of guideline development
The 1958, 1963 and 1971 WHO *International Standards for Drinking-water* did not refer to brominated acetic acids. Brominated acetic acids were not evaluated in the first edition of the *Guidelines for Drinking-water Quality*, published in 1984, in the second edition, published in 1993, or in the addendum to the second edition, published in 1998.

Assessment date
The risk assessment was conducted in 2003.

Principal references
IPCS (2000) *Disinfectants and disinfectant by-products*. Geneva, World Health Organization, International Programme on Chemical Safety (Environmental Health Criteria 216).

WHO (2003) *Brominated acetic acids in drinking-water. Background document for preparation of WHO Guidelines for drinking-water quality*. Geneva, World Health Organization (WHO/SDE/WSH/03.04/79).

12.17 Cadmium

Cadmium metal is used in the steel industry and in plastics. Cadmium compounds are widely used in batteries. Cadmium is released to the environment in wastewater, and diffuse pollution is caused by contamination from fertilizers and local air pollution. Contamination in drinking-water may also be caused by impurities in the zinc of galvanized pipes and solders and some metal fittings. Food is the main source of daily exposure to cadmium. The daily oral intake is 10–35 µg. Smoking is a significant additional source of cadmium exposure.

Guideline value	0.003 mg/litre
Occurrence	Levels in drinking-water usually less than 1 µg/litre
PTWI	7 µg/kg of body weight, on the basis that if levels of cadmium in the renal cortex are not to exceed 50 mg/kg, total intake of cadmium (assuming an absorption rate for dietary cadmium of 5% and a daily excretion rate of 0.005% of body burden) should not exceed 1 µg/kg of body weight per day
Limit of detection	0.01 µg/litre by ICP/MS; 2 µg/litre by FAAS
Treatment achievability	0.002 mg/litre should be achievable using coagulation or precipitation softening

Guideline derivation • allocation to water • weight • consumption	10% of PTWI 60-kg adult 2 litres/day
Additional comments	• Although new information indicates that a proportion of the general population may be at increased risk for tubular dysfunction when exposed at the current PTWI, the risk estimates that can be made at present are imprecise. • It is recognized that the margin between the PTWI and the actual weekly intake of cadmium by the general population is small, less than 10-fold, and that this margin may be even smaller in smokers.

Toxicological review

Absorption of cadmium compounds is dependent on the solubility of the compounds. Cadmium accumulates primarily in the kidneys and has a long biological half-life in humans of 10–35 years. There is evidence that cadmium is carcinogenic by the inhalation route, and IARC has classified cadmium and cadmium compounds in Group 2A. However, there is no evidence of carcinogenicity by the oral route and no clear evidence for the genotoxicity of cadmium. The kidney is the main target organ for cadmium toxicity. The critical cadmium concentration in the renal cortex that would produce a 10% prevalence of low-molecular-weight proteinuria in the general population is about 200 mg/kg and would be reached after a daily dietary intake of about 175 μg per person for 50 years.

History of guideline development

The 1958 WHO *International Standards for Drinking-water* did not refer to cadmium. The 1963 International Standards recommended a maximum allowable concentration of 0.01 mg/litre, based on health concerns. This value was retained in the 1971 International Standards as a tentative upper concentration limit, based on the lowest concentration that could be conveniently measured. In the first edition of the *Guidelines for Drinking-water Quality*, published in 1984, a guideline value of 0.005 mg/litre was recommended for cadmium in drinking-water. This value was lowered to 0.003 mg/litre in the 1993 Guidelines, based on the PTWI set by JECFA.

Assessment date

The risk assessment was conducted in 2003.

Principal references

JECFA (2000) *Summary and conclusions of the fifty-fifth meeting, Geneva, 6–15 June 2000*. Geneva, World Health Organization, Joint FAO/WHO Expert Committee on Food Additives.

WHO (2003) *Cadmium in drinking-water. Background document for preparation of WHO Guidelines for drinking-water quality.* Geneva, World Health Organization (WHO/SDE/WSH/03.04/80).

12.18 Carbofuran

Carbofuran (CAS No. 1563-66-2) is used worldwide as a pesticide for many crops. Residues in treated crops are generally very low or not detectable. The physical and chemical properties of carbofuran and the few data on occurrence indicate that drinking-water from both groundwater and surface water sources is potentially the major route of exposure.

Guideline value	0.007 mg/litre
Occurrence	Has been detected in surface water, groundwater and drinking-water, generally at levels of a few micrograms per litre or lower; highest concentration (30 µg/litre) measured in groundwater
ADI	0.002 mg/kg of body weight based on a NOAEL of 0.22 mg/kg of body weight per day for acute (reversible) effects in dogs in a short-term (4-week) study conducted as an adjunct to a 13-week study in which inhibition of erythrocyte acetylcholinesterase activity was observed, and using an uncertainty factor of 100
Limit of detection	0.1 µg/litre by GC with a nitrogen–phosphorus detector; 0.9 µg/litre by reverse-phase HPLC with a fluorescence detector
Treatment achievability	1 µg/litre should be achievable using GAC
Guideline derivation • allocation to water • weight • consumption	 10% of ADI 60-kg adult 2 litres/day
Additional comments	Use of a 4-week study was considered appropriate because the NOAEL is based on a reversible acute effect; the NOAEL will also be protective for chronic effects.

Toxicological review

Carbofuran is highly toxic after acute oral administration. The main systemic effect of carbofuran poisoning in short- and long-term toxicity studies appears to be cholinesterase inhibition. No evidence of teratogenicity has been found in reproductive toxicity studies. On the basis of available studies, carbofuran does not appear to be carcinogenic or genotoxic.

History of guideline development

The 1958 and 1963 WHO *International Standards for Drinking-water* did not refer to carbofuran, but the 1971 International Standards suggested that pesticide residues that may occur in community water supplies make only a minimal contribution to the total daily intake of pesticides for the population served. Carbofuran was not evaluated in the first edition of the *Guidelines for Drinking-water Quality*, published in 1984, but a health-based guideline value of 0.005 mg/litre was established for carbofuran in the 1993 Guidelines, based on human data and supported by observations in laboratory animals. This value was amended to 0.007 mg/litre in the addendum to the Guidelines published in 1998, on the basis of the ADI established by JMPR in 1996.

Assessment date
The risk assessment was originally conducted in 1998. The Final Task Force Meeting in 2003 agreed that this risk assessment be brought forward to this edition of the *Guidelines for Drinking-water Quality*.

Principal references
FAO/WHO (1997) *Pesticide residues in food – 1996. Evaluations – 1996. Part II – Toxicological*. Geneva, World Health Organization, Joint FAO/WHO Meeting on Pesticide Residues (WHO/PCS/97.1).
WHO (2003) *Carbofuran in drinking-water. Background document for preparation of WHO Guidelines for drinking-water quality*. Geneva, World Health Organization (WHO/SDE/WSH/03.04/81).

12.19 Carbon tetrachloride

Carbon tetrachloride is used mainly in the production of chlorofluorocarbon refrigerants, foam-blowing agents and solvents. However, since the Montreal Protocol on Substances that Deplete the Ozone Layer (1987) and its amendments (1990 and 1992) established a timetable for the phase-out of the production and consumption of carbon tetrachloride, manufacture and use have dropped and will continue to drop. Carbon tetrachloride is released mostly into the atmosphere but also into industrial wastewater. Although it readily migrates from surface water to the atmosphere, levels in anaerobic groundwater may remain elevated for months or even years. Although available data on concentrations in food are limited, the intake from air is expected to be much greater than that from food or drinking-water.

Guideline value	0.004 mg/litre
Occurrence	Concentrations in drinking-water generally less than 5 µg/litre
TDI	1.4 µg/kg of body weight, based on a NOAEL of 1 mg/kg of body weight per day for hepatotoxic effects in a 12-week oral gavage study in rats, incorporating a conversion factor of 5/7 for daily dosing and applying an uncertainty factor of 500 (100 for inter- and intraspecies variation, 10 for the duration of the study and a modifying factor of 0.5 because it was a bolus study)
Limit of detection	0.1–0.3 µg/litre by GC with ECD or MS
Treatment achievability	0.001 mg/litre should be achievable using air stripping

Guideline derivation • allocation to water • weight • consumption	10% of TDI 60-kg adult 2 litres/day
Additional comments	The guideline value is lower than the range of values associated with upper-bound lifetime excess cancer risks of 10^{-4}, 10^{-5} and 10^{-6} calculated by linear extrapolation.

Toxicological review

The primary targets for carbon tetrachloride toxicity are liver and kidney. In experiments with mice and rats, carbon tetrachloride proved to be capable of inducing hepatomas and hepatocellular carcinomas. The doses inducing hepatic tumours were higher than those inducing cell toxicity. It is likely that the carcinogenicity of carbon tetrachloride is secondary to its hepatotoxic effects. On the basis of available data, carbon tetrachloride can be considered to be a non-genotoxic compound. Carbon tetrachloride is classified by IARC as being possibly carcinogenic to humans (Group 2B): there is sufficient evidence that carbon tetrachloride is carcinogenic in laboratory animals, but inadequate evidence in humans.

History of guideline development

The 1958, 1963 and 1971 WHO *International Standards for Drinking-water* did not refer to carbon tetrachloride. In the first edition of the *Guidelines for Drinking-water Quality*, published in 1984, a tentative guideline value of 0.003 mg/litre was recommended; the guideline was designated as tentative because reliable evidence on which to calculate a guideline value based on carcinogenicity was available in only one animal species, because of the good qualitative supporting data and because of its frequency of occurrence in water. The 1993 Guidelines established a health-based guideline value of 0.002 mg/litre for carbon tetrachloride.

Assessment date

The risk assessment was conducted in 2003.

Principal references

IPCS (1999) *Carbon tetrachloride*. Geneva, World Health Organization, International Programme on Chemical Safety (Environmental Health Criteria 208).

WHO (2003) *Carbon tetrachloride in drinking-water. Background document for preparation of WHO Guidelines for drinking-water quality*. Geneva, World Health Organization (WHO/SDE/WSH/03.04/82).

12.20 Chloral hydrate (trichloroacetaldehyde)

Chloral hydrate is formed as a by-product of chlorination when chlorine reacts with humic acids. It has been widely used as a sedative or hypnotic drug in humans at oral doses of up to 14 mg/kg of body weight.

Provisional guideline value	0.01 mg/litre The guideline value is designated as provisional because of limitations of the available database.
Occurrence	Found in drinking-water at concentrations of up to 100 µg/litre
TDI	1.6 µg/kg of body weight, based on a LOAEL of 16 mg/kg of body weight per day for liver enlargement from a 90-day drinking-water study in mice, using an uncertainty factor of 10 000 to take into consideration intra- and intraspecies variation, the short duration of the study and the use of a LOAEL instead of a NOAEL
Quantification limit	1 µg/litre by GC with ECD; 3 µg/litre by GC/MS
Treatment achievability	Chloral hydrate concentrations in drinking-water are generally below 0.05 mg/litre. Chloral hydrate concentrations may be reduced by removal of precursor compounds, changes to disinfection practice or GAC treatment.
Guideline derivation • allocation to water • weight • consumption	 20% of TDI 60-kg adult 2 litres/day

Toxicological review

The information available on the toxicity of chloral hydrate is limited, but effects on the liver have been observed in 90-day studies in mice. Chloral hydrate has been shown to be genotoxic in some short-term tests *in vitro*, but it does not bind to DNA. It has been found to disrupt chromosome segregation in cell division.

History of guideline development

The 1958, 1963 and 1971 WHO *International Standards for Drinking-water* and the first edition of the *Guidelines for Drinking-water Quality*, published in 1984, did not refer to chloral hydrate. The 1993 Guidelines established a provisional health-based guideline value of 0.01 mg/litre for chloral hydrate in drinking-water. The guideline value was designated as provisional because of the limitations of the available database, necessitating the use of an uncertainty factor of 10 000.

Assessment date

The risk assessment was originally conducted in 1993. The Final Task Force Meeting in 2003 agreed that this risk assessment be brought forward to this edition of the *Guidelines for Drinking-water Quality*.

Principal reference

WHO (2003) *Chloral hydrate (trichloroacetaldehyde) in drinking-water. Background document for preparation of WHO Guidelines for drinking-water quality*. Geneva, World Health Organization (WHO/SDE/WSH/03.04/49).

12.21 Chlordane

Chlordane (CAS No. 57-47-9) is a broad-spectrum insecticide that has been used since 1947. Its use has recently been increasingly restricted in many countries, and it is now used mainly to destroy termites by subsurface injection into soil. Chlordane may be a low-level source of contamination of groundwater when applied by subsurface injection. Technical chlordane is a mixture of compounds, with the *cis* and *trans* forms of chlordane predominating. It is very resistant to degradation, is highly immobile in soil and it unlikely to migrate to groundwater, where it has only rarely been found. It is readily lost to the atmosphere. Although levels of chlordane in food have been decreasing, it is highly persistent and has a high bioaccumulation potential.

Guideline value	0.0002 mg/litre (0.2 µg/litre)
Occurrence	Has been detected in both drinking-water and groundwater, usually at levels below 0.1 µg/litre
PTDI	0.5 µg/kg of body weight based on a NOAEL of 50 µg/kg of body weight per day for increased liver weights, serum bilirubin levels and incidence of hepatocellular swelling, derived from a long-term dietary study in rats, and using an uncertainty factor of 100
Limit of detection	0.014 µg/litre by GC with an ECD
Treatment achievability	0.1 µg/litre should be achievable using GAC
Guideline derivation • allocation to water • weight • consumption	1% of PTDI 60-kg adult 2 litres/day
Additional comments	Chlordane is listed under the Stockholm Convention on Persistent Organic Pollutants. Hence, monitoring may occur in addition to that required by drinking-water guidelines.

Toxicological review

In experimental animals, prolonged exposure in the diet causes liver damage. Chlordane produces liver tumours in mice, but the weight of evidence indicates that it is not genotoxic. Chlordane can interfere with cell communication *in vitro*, a characteristic of many tumour promoters. IARC re-evaluated chlordane in 1991 and concluded that there is inadequate evidence for its carcinogenicity in humans and sufficient evidence for its carcinogenicity in animals, classifying it in Group 2B.

History of guideline development

The 1958 and 1963 WHO *International Standards for Drinking-water* did not refer to chlordane, but the 1971 International Standards suggested that pesticide residues that may occur in community water supplies make only a minimal contribution to the total daily intake of pesticides for the population served. In the first edition of the *Guidelines for Drinking-water Quality*, published in 1984, a health-based guideline value of 0.3 µg/litre was recommended for chlordane (total isomers), based on the

ADI recommended by JMPR in 1977. The 1993 Guidelines established a health-based guideline value of 0.2 µg/litre for chlordane in drinking-water, based on an ADI established by JMPR in 1986.

Assessment date
The risk assessment was conducted in 2003.

Principal references
FAO/WHO (1995) *Pesticide residues in food – 1994. Report of the Joint Meeting of the FAO Panel of Experts on Pesticide Residues in Food and the Environment and WHO Toxicological and Environmental Core Assessment Groups*. Rome, Food and Agriculture Organization of the United Nations (FAO Plant Production and Protection Paper 127).

WHO (2003) *Chlordane in drinking-water. Background document for preparation of WHO Guidelines for drinking-water quality*. Geneva, World Health Organization (WHO/SDE/WSH/03.04/84).

12.22 Chloride

Chloride in drinking-water originates from natural sources, sewage and industrial effluents, urban runoff containing de-icing salt and saline intrusion.

The main source of human exposure to chloride is the addition of salt to food, and the intake from this source is usually greatly in excess of that from drinking-water.

Excessive chloride concentrations increase rates of corrosion of metals in the distribution system, depending on the alkalinity of the water. This can lead to increased concentrations of metals in the supply.

No health-based guideline value is proposed for chloride in drinking-water. However, chloride concentrations in excess of about 250 mg/litre can give rise to detectable taste in water (see chapter 10).

History of guideline development

The 1958 WHO *International Standards for Drinking-water* suggested that concentrations of chloride greater than 600 mg/litre would markedly impair the potability of the water. The 1963 and 1971 International Standards retained this value as a maximum allowable or permissible concentration. In the first edition of the *Guidelines for Drinking-water Quality*, published in 1984, a guideline value of 250 mg/litre was established for chloride, based on taste considerations. No health-based guideline value for chloride in drinking-water was proposed in the 1993 Guidelines, although it was confirmed that chloride concentrations in excess of about 250 mg/litre can give rise to detectable taste in water.

Assessment date
The risk assessment was originally conducted in 1993. The Final Task Force Meeting in 2003 agreed that this risk assessment be brought forward to this edition of the *Guidelines for Drinking-water Quality*.

Principal reference
WHO (2003) *Chloride in drinking-water. Background document for preparation of WHO Guidelines for drinking-water quality.* Geneva, World Health Organization (WHO/SDE/WSH/03.04/3).

12.23 Chlorine
Chlorine is produced in large amounts and widely used both industrially and domestically as an important disinfectant and bleach. In particular, it is widely used in the disinfection of swimming pools and is the most commonly used disinfectant and oxidant in drinking-water treatment. In water, chlorine reacts to form hypochlorous acid and hypochlorites.

Guideline value	5 mg/litre
Occurrence	Present in most disinfected drinking-water at concentrations of 0.2–1 mg/litre
TDI	150 µg/kg of body weight, derived from a NOAEL for the absence of toxicity in rodents ingesting chlorine in drinking-water for 2 years
Limit of detection	0.01 µg/litre following pre-column derivatization to 4-bromoacetanilide by HPLC; 10 µg/litre as free chlorine by colorimetry; 0.2 mg/litre by ion chromatography
Treatment achievability	It is possible to reduce the concentration of chlorine effectively to zero (< 0.1 mg/litre) by reduction. However, it is normal practice to supply water with a chlorine residual of a few tenths of a milligram per litre to act as a preservative during distribution.
Guideline derivation • allocation to water • weight • consumption	 100% of TDI 60-kg adult 2 litres/day
Additional comments	• The guideline value is conservative, as no adverse effect level was identified in the critical study. • Most individuals are able to taste chlorine at the guideline value.

Toxicological review
In humans and animals exposed to chlorine in drinking-water, no specific adverse treatment-related effects have been observed. IARC has classified hypochlorite in Group 3.

History of guideline development
The 1958, 1963 and 1971 WHO *International Standards for Drinking-water* and the first edition of the *Guidelines for Drinking-water Quality*, published in 1984, did not refer to chlorine. The 1993 Guidelines established a guideline value of 5 mg/litre for free chlorine in drinking-water, but noted that this value is conservative, as no adverse effect level was identified in the study used. It was also noted that most individuals are able to taste chlorine at the guideline value.

Assessment date
The risk assessment was originally conducted in 1993. The Final Task Force Meeting in 2003 agreed that this risk assessment be brought forward to this edition of the *Guidelines for Drinking-water Quality*.

Principal reference
WHO (2003) *Chlorine in drinking-water. Background document for preparation of WHO Guidelines for drinking-water quality*. Geneva, World Health Organization (WHO/SDE/WSH/03.04/45).

12.24 Chlorite and chlorate
Chlorite and chlorate are DBPs resulting from the use of chlorine dioxide as a disinfectant and for odour/taste control in water. Chlorine dioxide is also used as a bleaching agent for cellulose, paper pulp, flour and oils. Sodium chlorite and sodium chlorate are both used in the production of chlorine dioxide as well as for other commercial purposes. Chlorine dioxide rapidly decomposes into chlorite, chlorate and chloride ions in treated water, chlorite being the predominant species; this reaction is favoured by alkaline conditions. The major route of environmental exposure to chlorine dioxide, sodium chlorite and sodium chlorate is through drinking-water.

Provisional guideline values	
Chlorite	0.7 mg/litre
Chlorate	0.7 mg/litre The guideline values for chlorite and chlorate are designated as provisional because use of chlorine dioxide as a disinfectant may result in the chlorite and chlorate guideline values being exceeded, and difficulties in meeting the guideline value must never be a reason for compromising adequate disinfection.
Occurrence	Levels of chlorite in water reported in one study ranged from 3.2 to 7.0 mg/litre; however, the combined levels will not exceed the dose of chlorine dioxide applied.

TDIs	
Chlorite	30 μg/kg of body weight based on a NOAEL of 2.9 mg/kg of body weight per day identified in a two-generation study in rats, based on lower startle amplitude, decreased absolute brain weight in the F_1 and F_2 generations and altered liver weights in two generations, using an uncertainty factor of 100 (10 each for inter- and intraspecies variation)
Chlorate	30 μg/kg of body weight based on a NOAEL of 30 mg/kg of body weight per day in a recent well conducted 90-day study in rats, based on thyroid gland colloid depletion at the next higher dose, and using an uncertainty factor of 1000 (10 each for inter- and intraspecies variation and 10 for the short duration of the study)
Limit of detection	5 μg/litre by ion chromatography with suppressed conductivity detection for chlorate
Treatment achievability	It is possible to reduce the concentration of chlorine dioxide effectively to zero (< 0.1 mg/litre) by reduction; however, it is normal practice to supply water with a chlorine dioxide residual of a few tenths of a milligram per litre to act as a preservative during distribution. Chlorate concentrations arising from the use of sodium hypochlorite are generally around 0.1 mg/litre, although concentrations above 1 mg/litre have been reported. With chlorine dioxide disinfection, the concentration of chlorate depends heavily on process conditions (in both the chlorine dioxide generator and the water treatment plant) and applied dose of chlorine dioxide. As there is no viable option for reducing chlorate concentrations, control of chlorate concentration must rely on preventing its addition (from sodium hypochlorite) or formation (from chlorine dioxide). Chlorite ion is an inevitable by-product arising from the use of chlorine dioxide. When chlorine dioxide is used as the final disinfectant at typical doses, the resulting chlorite concentration should be <0.2 mg/litre. If chlorine dioxide is used as a pre-oxidant, the resulting chlorite concentration may need to be reduced using ferrous iron or activated carbon.
Guideline derivation • allocation to water • weight • consumption	 80% of TDI 60-kg adult 2 litres/day

Toxicological review

Chlorine dioxide

Chlorine dioxide has been shown to impair neurobehavioural and neurological development in rats exposed perinatally. Significant depression of thyroid hormones has also been observed in rats and monkeys exposed to it in drinking-water studies. A guideline value has not been established for chlorine dioxide because of its rapid hydrolysis to chlorite and because the chlorite provisional guideline value is adequately protective for potential toxicity from chlorine dioxide. The taste and odour threshold for this compound is 0.4 mg/litre.

Chlorite

IARC has concluded that chlorite is not classifiable as to its carcinogenicity to humans. The primary and most consistent finding arising from exposure to chlorite is oxidative stress resulting in changes in the red blood cells. This end-point is seen in laboratory animals and, by analogy with chlorate, in humans exposed to high doses in poisoning incidents. Studies with human volunteers for up to 12 weeks did not identify any effect on blood parameters at the highest dose tested, 36 μg/kg of body weight per day.

Chlorate

Like chlorite, the primary concern with chlorate is oxidative damage to red blood cells. Also like chlorite, a chlorate dose of 36 μg/kg of body weight per day for 12 weeks did not result in any adverse effects in human volunteers. Although the database for chlorate is less extensive than that for chlorite, a recent well conducted 90-day study in rats is available. A long-term study is in progress, which should provide more information on chronic exposure to chlorate.

History of guideline development

The 1958, 1963 and 1971 WHO *International Standards for Drinking-water* and the first edition of the *Guidelines for Drinking-water Quality*, published in 1984, did not refer to chlorine dioxide, chlorate or chlorite. The 1993 Guidelines established a provisional health-based guideline value of 0.2 mg/litre for chlorite in drinking-water. The guideline value was designated as provisional because use of chlorine dioxide as a disinfectant may result in the chlorite guideline value being exceeded, and difficulties in meeting the guideline value must never be a reason for compromising disinfection. The 1993 Guidelines did not establish a health-based guideline value for chlorine dioxide in drinking-water because of its rapid breakdown and because the provisional guideline value for chlorite is adequately protective for potential toxicity from chlorine dioxide. The 1993 Guidelines concluded that available data on the effects of chlorate in humans and experimental animals are insufficient to permit development of a guideline value and recommended that further research was needed to characterize the non-lethal effects of chlorate. It was noted that the taste and odour threshold for chlorine dioxide is 0.4 mg/litre.

Assessment date

The risk assessment was conducted in 2003.

Principal references

IPCS (2000) *Disinfectants and disinfectant by-products.* Geneva, World Health Organization, International Programme on Chemical Safety (Environmental Health Criteria 216).

WHO (2003) *Chlorite and chlorate in drinking-water. Background document for preparation of WHO Guidelines for drinking-water quality.* Geneva, World Health Organization (WHO/SDE/WSH/03.04/86).

12.25 Chloroacetones

1,1-Dichloroacetone is formed from the reaction between chlorine and organic precursors and has been detected in chlorinated drinking-water. Concentrations are estimated to be less than 10 µg/litre and usually less than 1 µg/litre.

The toxicological data on 1,1-dichloroacetone are very limited, although studies with single doses indicate that it affects the liver.

There are insufficient data at present to permit the proposal of guideline values for 1,1-dichloroacetone or any of the other chloroacetones.

History of guideline development

The 1958, 1963 and 1971 WHO *International Standards for Drinking-water* and the first edition of the *Guidelines for Drinking-water Quality*, published in 1984, did not refer to chloroacetones. The 1993 Guidelines concluded that there were insufficient data available to permit the proposal of guideline values for any of the chloroacetones.

Assessment date

The risk assessment was originally conducted in 1993. The Final Task Force Meeting in 2003 agreed that this risk assessment be brought forward to this edition of the *Guidelines for Drinking-water Quality*.

Principal reference

WHO (2003) *Chloroacetones in drinking-water. Background document for preparation of WHO Guidelines for drinking-water quality.* Geneva, World Health Organization (WHO/SDE/WSH/03.04/50).

12.26 Chlorophenols (2-chlorophenol, 2,4-dichlorophenol, 2,4,6-trichlorophenol)

Chlorophenols are present in drinking-water as a result of the chlorination of phenols, as by-products of the reaction of hypochlorite with phenolic acids, as biocides or as degradation products of phenoxy herbicides. Those most likely to occur in drinking-water as by-products of chlorination are 2-chlorophenol, 2,4-dichlorophenol and 2,4,6-trichlorophenol. The taste thresholds for chlorophenols in drinking-water are low.

Guideline value for 2,4,6-trichlorophenol	0.2 mg/litre
Occurrence	Concentrations of chlorophenols in drinking-water are usually less than 1 µg/litre.
Basis of guideline derivation	Applying the linearized multistage model to leukaemias in male rats observed in a 2-year feeding study (hepatic tumours found in this study were not used for risk estimation because of the possible role of contaminants in their induction)
Limit of detection	0.5–5 µg/litre by formation of pentafluorobenzyl ether derivatives; 1–10 µg/litre (monochlorophenols), 0.5 µg/litre (dichlorophenols) and 0.01 µg/litre (trichlorophenols) using GC with ECD
Treatment achievability	2,4,6-Trichlorophenol concentrations are generally less than 1 µg/litre. If necessary, 2,4,6-trichlorophenol concentrations can be reduced using GAC.
Additional comments	The guideline value for 2,4,6-trichlorophenol exceeds its lowest reported taste threshold.

Toxicological review
2-Chlorophenol

Data on the toxicity of 2-chlorophenol are limited. Therefore, no health-based guideline value has been derived.

2,4-Dichlorophenol

Data on the toxicity of 2,4-dichlorophenol are limited. Therefore, no health-based guideline value has been derived.

2,4,6-Trichlorophenol

2,4,6-Trichlorophenol has been reported to induce lymphomas and leukaemias in male rats and hepatic tumours in male and female mice. The compound has not been shown to be mutagenic in the Ames test but has shown weak mutagenic activity in other *in vitro* and *in vivo* studies. IARC has classified 2,4,6-trichlorophenol in Group 2B.

History of guideline development
The 1958, 1963 and 1971 WHO *International Standards for Drinking-water* did not refer to chlorophenols. In the first edition of the *Guidelines for Drinking-water Quality*, published in 1984, no guideline values for 2-chlorophenol, 4-chlorophenol, 2,4-dichlorophenol, 2,6-dichlorophenol or 2,4,5-trichlorophenol were recommended after a detailed evaluation of the compounds, although it was suggested that individual chlorophenols should not be present in drinking-water at a level above 0.0001 mg/litre for organoleptic reasons (and the total phenol content of water to be chlorinated should be kept below 0.001 mg/litre). In the same edition, a health-based guideline value of 0.01 mg/litre was recommended for 2,4,6-trichlorophenol, while noting

that the linear multistage extrapolation model appropriate for chemical carcinogens that was used in its derivation involved considerable uncertainty. It was also noted that 2,4,6-trichlorophenol may be detected by its taste and odour at a concentration of 0.0001 mg/litre. No health-based guidelines for 2-chlorophenol or 2,4-dichlorophenol were derived in the 1993 Guidelines, as data on their toxicity were limited. A guideline value of 0.2 mg/litre, associated with a 10^{-5} upper-bound excess lifetime cancer risk, was calculated for 2,4,6-trichlorophenol. This concentration exceeds the lowest reported taste threshold for the chemical (0.002 mg/litre).

Assessment date
The risk assessment was originally conducted in 1993. The Final Task Force Meeting in 2003 agreed that this risk assessment be brought forward to this edition of the *Guidelines for Drinking-water Quality*.

Principal reference
WHO (2003) *Chlorophenols in drinking-water. Background document for preparation of WHO Guidelines for drinking-water quality*. Geneva, World Health Organization (WHO/SDE/WSH/03.04/47).

12.27 Chloropicrin

Chloropicrin, or trichloronitromethane, is formed by the reaction of chlorine with humic and amino acids and with nitrophenols. Its formation is increased in the presence of nitrates. Limited data from the USA indicate that concentrations in drinking-water are usually less than 5 µg/litre.

Decreased survival and body weights have been reported following long-term oral exposure in laboratory animals. Chloropicrin has been shown to be mutagenic in bacterial tests and in *in vitro* assays in lymphocytes. Because of the high mortality in a carcinogenesis bioassay and the limited number of end-points examined in the 78-week toxicity study, the available data were considered inadequate to permit the establishment of a guideline value for chloropicrin.

History of guideline development
The 1958, 1963 and 1971 WHO *International Standards for Drinking-water* and the first edition of the *Guidelines for Drinking-water Quality*, published in 1984, did not refer to chloropicrin. The 1993 Guidelines considered the available data to be inadequate to permit the establishment of a guideline value for chloropicrin in drinking-water.

Assessment date
The risk assessment was originally conducted in 1993. The Final Task Force Meeting in 2003 agreed that this risk assessment be brought forward to this edition of the *Guidelines for Drinking-water Quality*.

Principal reference
WHO (2003) *Chloropicrin in drinking-water. Background document for preparation of WHO Guidelines for drinking-water quality.* Geneva, World Health Organization (WHO/SDE/WSH/03.04/52).

12.28 Chlorotoluron

Chlorotoluron (CAS No. 15545-48-9) is a pre- or early post-emergence herbicide that is slowly biodegradable and mobile in soil. There is only very limited exposure to this compound from food.

Guideline value	0.03 mg/litre
Occurrence	Detected in drinking-water at concentrations of less than 1 µg/litre
TDI	11.3 µg/kg of body weight, derived from a NOAEL of 11.3 mg/kg of body weight per day for systemic effects in a 2-year feeding study in mice using an uncertainty factor of 1000 (100 for inter- and intraspecies variation and 10 for evidence of carcinogenicity)
Limit of detection	0.1 µg/litre by separation by reverse-phase HPLC followed by UV and electrochemical detection
Treatment achievability	0.1 µg/litre should be achievable using GAC
Guideline derivation • allocation to water • weight • consumption	 10% of TDI 60-kg adult 2 litres/day

Toxicological review
Chlorotoluron is of low toxicity in single, short-term and long-term exposures in animals, but it has been shown to cause an increase in adenomas and carcinomas of the kidneys of male mice given high doses for 2 years. As no carcinogenic effects were reported in a 2-year study in rats, it has been suggested that chlorotoluron has a carcinogenic potential that is both species- and sex-specific. Chlorotoluron and its metabolites have shown no evidence of genotoxicity.

History of guideline development
The 1958 and 1963 WHO *International Standards for Drinking-water* did not refer to chlorotoluron, but the 1971 International Standards suggested that pesticide residues that may occur in community water supplies make only a minimal contribution to the total daily intake of pesticides for the population served. Chlorotoluron was not evaluated in the first edition of the *Guidelines for Drinking-water Quality*, published in 1984, but the 1993 Guidelines established a health-based guideline value of 0.03 mg/litre for chlorotoluron in drinking-water.

Assessment date
The risk assessment was originally conducted in 1993. The Final Task Force Meeting in 2003 agreed that this risk assessment be brought forward to this edition of the *Guidelines for Drinking-water Quality*.

Principal reference
WHO (2003) *Chlorotoluron in drinking-water. Background document for preparation of WHO Guidelines for drinking-water quality*. Geneva, World Health Organization (WHO/SDE/WSH/03.04/33).

12.29 Chlorpyrifos
Chlorpyrifos (CAS No. 2921-88-2) is a broad-spectrum organophosphorus insecticide used for the control of mosquitos, flies, various crop pests in soil and on foliage, household pests and aquatic larvae. Athough it is not recommended for addition to water for public health purposes by WHOPES, it may be used in some countries as an aquatic larvicide for the control of mosquito larvae. Chlorpyrifos is strongly absorbed by soil and does not readily leach from it, degrading slowly by microbial action. It has a low solubility in water and great tendency to partition from aqueous into organic phases in the environment.

Guideline value	0.03 mg/litre
Occurrence	Detected in surface waters in USA, usually at concentrations below 0.1 µg/litre; also detected in groundwater in less than 1% of the wells tested, usually at concentrations below 0.01 µg/litre
ADI	0.01 mg/kg of body weight on the basis of a NOAEL of 1 mg/kg of body weight per day for inhibition of brain acetylcholinesterase activity in studies in mice, rats and dogs, using a 100-fold uncertainty factor, and on the basis of a NOAEL of 0.1 mg/kg of body weight per day for inhibition of erythrocyte acetylcholinesterase activity in a study of human subjects exposed for 9 days, using a 10-fold uncertainty factor
Limit of detection	1 µg/litre by GC using an ECD or flame photometric detection
Treatment achievability	No data available; should be amenable to treatment by coagulation (10–20% removal), activated carbon adsorption and ozonation
Guideline derivation • allocation to water • weight • consumption	 10% of ADI 60-kg adult 2 litres/day

Toxicological review
JMPR concluded that chlorpyrifos is unlikely to pose a carcinogenic risk to humans. Chlorpyrifos was not genotoxic in an adequate range of studies *in vitro* and *in vivo*. In long-term studies, inhibition of cholinesterase activity was the main toxicological finding in all species.

History of guideline development
The 1958 and 1963 WHO *International Standards for Drinking-water* did not refer to chlorpyrifos, but the 1971 International Standards suggested that pesticide residues that may occur in community water supplies make only a minimal contribution to the total daily intake of pesticides for the population served. Chlorpyrifos was not evaluated in the first edition of the *Guidelines for Drinking-water Quality*, published in 1984, in the second edition, published in 1993, or in the addendum to the second edition, published in 1998.

Assessment date
The risk assessment was conducted in 2003.

Principal references
FAO/WHO (2000) *Pesticide residues in food – 1999 evaluations. Part II – Toxicological.* Geneva, World Health Organization, Joint FAO/WHO Meeting on Pesticide Residues (WHO/PCS/00.4).

WHO (2003) *Chlorpyrifos in drinking-water. Background document for preparation of WHO Guidelines for drinking-water quality.* Geneva, World Health Organization (WHO/SDE/WSH/03.04/87).

12.30 Chromium
Chromium is widely distributed in the Earth's crust. It can exist in valences of +2 to +6. In general, food appears to be the major source of intake.

Provisional guideline value	0.05 mg/litre for total chromium The guideline value is designated as provisional because of uncertainties in the toxicological database.
Occurrence	Total chromium concentrations in drinking-water are usually less than 2 µg/litre, although concentrations as high as 120 µg/litre have been reported.
Basis of guideline value derivation	There are no adequate toxicity studies available to provide a basis for a NOAEL. The guideline value was first proposed in 1958 for hexavalent chromium, based on health concerns, but was later changed to a guideline for total chromium because of difficulties in analysing for the hexavalent form only.
Limit of detection	0.05–0.2 µg/litre for total chromium by AAS
Treatment achievability	0.015 mg/litre should be achievable using coagulation

Toxicological review
In a long-term carcinogenicity study in rats given chromium(III) by the oral route, no increase in tumour incidence was observed. In rats, chromium(VI) is a carcinogen via the inhalation route, although the limited data available do not show evidence

for carcinogenicity via the oral route. In epidemiological studies, an association has been found between exposure to chromium(VI) by the inhalation route and lung cancer. IARC has classified chromium(VI) in Group 1 (human carcinogen) and chromium(III) in Group 3. Chromium(VI) compounds are active in a wide range of *in vitro* and *in vivo* genotoxicity tests, whereas chromium(III) compounds are not.

History of guideline development
The 1958 WHO *International Standards for Drinking-water* recommended a maximum allowable concentration of 0.05 mg/litre for chromium (hexavalent), based on health concerns. This value was retained in the 1963 International Standards. Chromium was not evaluated in the 1971 International Standards. In the first edition of the *Guidelines for Drinking-water Quality*, published in 1984, the guideline value of 0.05 mg/litre for total chromium was retained; total chromium was specified because of difficulties in analysing for the hexavalent form only. The 1993 Guidelines questioned the guideline value of 0.05 mg/litre because of the carcinogenicity of hexavalent chromium by the inhalation route and its genotoxicity, although the available toxicological data did not support the derivation of a new value. As a practical measure, 0.05 mg/litre, which is considered to be unlikely to give rise to significant health risks, was retained as the provisional guideline value until additional information becomes available and chromium can be re-evaluated.

Assessment date
The risk assessment was originally conducted in 1993. The Final Task Force Meeting in 2003 agreed that this risk assessment be brought forward to this edition of the *Guidelines for Drinking-water Quality*.

Principal reference
WHO (2003) *Chromium in drinking-water. Background document for preparation of WHO Guidelines for drinking-water quality*. Geneva, World Health Organization (WHO/SDE/WSH/03.04/4).

12.31 Copper
Copper is both an essential nutrient and a drinking-water contaminant. It has many commercial uses. It is used to make pipes, valves and fittings and is present in alloys and coatings. Copper sulfate pentahydrate is sometimes added to surface water for the control of algae. Copper concentrations in drinking-water vary widely, with the primary source most often being the corrosion of interior copper plumbing. Levels in running or fully flushed water tend to be low, whereas those in standing or partially flushed water samples are more variable and can be substantially higher (frequently > 1 mg/litre). Copper concentrations in treated water often increase during distribution, especially in systems with an acid pH or high-carbonate waters with an alkaline pH. Food and water are the primary sources of copper exposure in developed

countries. Consumption of standing or partially flushed water from a distribution system that includes copper pipes or fittings can considerably increase total daily copper exposure, especially for infants fed formula reconstituted with tap water.

Guideline value	2 mg/litre
Occurrence	Concentrations in drinking-water range from ≤0.005 to >30 mg/litre, primarily as a result of the corrosion of interior copper plumbing.
Basis of guideline derivation	To be protective against acute gastrointestinal effects of copper and provide an adequate margin of safety in populations with normal copper homeostasis
Limit of detection	0.02–0.1 µg/litre by ICP/MS; 0.3 µg/litre by ICP/optical emission spectroscopy; 0.5 µg/litre by FAAS
Treatment achievability	Copper is not removed by conventional treatment processes. However, copper is not normally a raw water contaminant.
Additional comments	• For adults with normal copper homeostasis, the guideline value should permit consumption of 2 or 3 litres of water per day, use of a nutritional supplement and copper from foods without exceeding the tolerable upper intake level of 10 mg/day or eliciting an adverse gastrointestinal response. • Staining of laundry and sanitary ware occurs at copper concentrations above 1 mg/litre. At levels above 2.5 mg/litre, copper imparts an undesirable bitter taste to water; at higher levels, the colour of water is also impacted. • In most instances where copper tubing is used as a plumbing material, concentrations of copper will be below the guideline value. However, there are some conditions, such as highly acidic or aggressive waters, that will give rise to much higher copper concentrations, and the use of copper tubing may not be appropriate in such circumstances.

Toxicological review

IPCS concluded that the upper limit of the acceptable range of oral intake in adults is uncertain but is most likely in the range of several (more than 2 or 3) but not many milligrams per day in adults. This evaluation was based solely on studies of gastrointestinal effects of copper-contaminated drinking-water. The available data on toxicity in animals were not considered helpful in establishing the upper limit of the acceptable range of oral intake due to uncertainty about an appropriate model for humans, but they help to establish a mode of action for the response. The data on the gastrointestinal effects of copper must be used with caution, since the effects observed are influenced by the concentration of ingested copper to a greater extent than the total mass or dose ingested in a 24-h period. Recent studies have delineated the threshold for the effects of copper in drinking-water on the gastrointestinal tract, but there is still some uncertainty regarding the long-term effects of copper on sensitive populations, such as carriers of the gene for Wilson disease and other metabolic disorders of copper homeostasis.

History of guideline development
The 1958 WHO *International Standards for Drinking-water* suggested that concentrations of copper greater than 1.5 mg/litre would markedly impair the potability of the water. The 1963 and 1971 International Standards retained this value as a maximum allowable or permissible concentration. In the first edition of the *Guidelines for Drinking-water Quality*, published in 1984, a guideline value of 1.0 mg/litre was established for copper, based on its laundry and other staining properties. The 1993 Guidelines derived a provisional health-based guideline value of 2 mg/litre for copper from the PMTDI proposed by JECFA, based on a rather old study in dogs that did not take into account differences in copper metabolism between infants and adults. The guideline value was considered provisional because of the uncertainties regarding copper toxicity in humans. This guideline value was retained in the addendum to the Guidelines published in 1998 and remained provisional as a result of uncertainties in the dose–response relationship between copper in drinking-water and acute gastrointestinal effects in humans. It was stressed that the outcome of epidemiological studies in progress in Chile, Sweden and the USA may permit more accurate quantification of effect levels for copper-induced toxicity in humans, including sensitive subpopulations. Copper can also give rise to taste problems at concentrations above 5 mg/litre and can stain laundry and sanitary ware at concentrations above 1 mg/litre.

Assessment date
The risk assessment was conducted in 2003.

Principal references
IPCS (1998) *Copper*. Geneva, World Health Organization, International Programme on Chemical Safety (Environmental Health Criteria 200).

WHO (2003) *Copper in drinking-water. Background document for preparation of WHO Guidelines for drinking-water quality*. Geneva, World Health Organization (WHO/SDE/WSH/03.04/88).

12.32 Cyanazine
Cyanazine (CAS No. 21725-46-2) is a member of the triazine family of herbicides. It is used as a pre- and post-emergence herbicide for the control of annual grasses and broadleaf weeds. It can be degraded in soil and water by microorganisms and by hydrolysis.

Guideline value	0.0006 mg/litre (0.6 µg/litre)
Occurrence	Has been detected in surface water and groundwater, usually at concentrations of a few micrograms per litre, although levels as high as 1.3 and 3.5 mg/litre have been measured in surface water and groundwater, respectively

TDI	0.198 µg/kg of body weight based on a NOAEL of 0.198 mg/kg of body weight for hyperactivity in male rats in a 2-year toxicity/carcinogenicity study, with an uncertainty factor of 1000 (100 for inter- and intraspecies variation and 10 for limited evidence of carcinogenicity)
Limit of detection	0.01 µg/litre by GC with MS
Treatment achievability	0.1 µg/litre should be achievable using GAC
Guideline derivation • allocation to water • weight • consumption	10% of TDI 60-kg adult 2 litres/day

Toxicological review

On the basis of the available mutagenicity data on cyanazine, evidence for genotoxicity is equivocal. Cyanazine causes mammary gland tumours in Sprague-Dawley rats but not in mice. The mechanism of mammary gland tumour development in Sprague-Dawley rats is currently under investigation and may prove to be hormonal (cf. atrazine). Cyanazine is also teratogenic in Fischer 344 rats at dose levels of 25 mg/kg of body weight per day and higher.

History of guideline development

The 1958 and 1963 WHO *International Standards for Drinking-water* did not refer to cyanazine, but the 1971 International Standards suggested that pesticide residues that may occur in community water supplies make only a minimal contribution to the total daily intake of pesticides for the population served. In the first edition of the *Guidelines for Drinking-water Quality*, published in 1984, no guideline value for triazine herbicides, which include cyanazine, was recommended after a detailed evaluation of the compounds. Cyanazine was not evaluated in the second edition of the *Guidelines for Drinking-water Quality*, published in 1993. In the addendum to the second edition of these Guidelines, published in 1998, a health-based guideline value of 0.6 µg/litre was established for cyanazine in drinking-water.

Assessment date

The risk assessment was originally conducted in 1998. The Final Task Force Meeting in 2003 agreed that this risk assessment be brought forward to this edition of the *Guidelines for Drinking-water Quality*.

Principal reference

WHO (2003) *Cyanazine in drinking-water. Background document for preparation of WHO Guidelines for drinking-water quality.* Geneva, World Health Organization (WHO/SDE/WSH/03.04/60).

12.33 Cyanide

Cyanides can be found in some foods, particularly in some developing countries, and they are occasionally found in drinking-water, primarily as a consequence of industrial contamination.

Guideline value	0.07 mg/litre
Occurrence	Occasionally found in drinking-water
TDI	12 µg/kg of body weight, based on a LOAEL of 1.2 mg/kg of body weight per day for effects on behavioural patterns and serum biochemistry in a 6-month study in pigs, using an uncertainty factor of 100 for inter- and intraspecies variation (no additional factor for use of a LOAEL instead of a NOAEL was considered necessary because of doubts over the biological significance of the observed changes)
Limit of detection	2 µg/litre by titrimetric and photometric techniques
Treatment achievability	Cyanide is removed from water by high doses of chlorine.
Guideline derivation • allocation to water • weight • consumption	20% of TDI (because exposure to cyanide from other sources is normally small and because exposure from water is only intermittent) 60-kg adult 2 litres/day
Additional considerations	The guideline value is considered to be protective for acute and long-term exposure.

Toxicological review

The acute toxicity of cyanides is high. Effects on the thyroid and particularly the nervous system were observed in some populations as a consequence of the long-term consumption of inadequately processed cassava containing high levels of cyanide.

History of guideline development

The 1958 WHO *International Standards for Drinking-water* recommended a maximum allowable concentration of 0.01 mg/litre for cyanide, based on health concerns. This value was raised to 0.2 mg/litre in the 1963 International Standards. The tentative upper concentration limit was lowered to 0.05 mg/litre in the 1971 International Standards upon consideration of the ADI of hydrogen cyanide residues in some fumigated foods of 0.05 mg/kg of body weight and to ensure that the water source is not too highly contaminated by industrial effluents and that water treatment has been adequate. In the first edition of the *Guidelines for Drinking-water Quality*, published in 1984, it was determined that a guideline value of 0.1 mg/litre would be a reasonable level for the protection of public health. A health-based guideline value of 0.07 mg/litre, which was considered to be protective for both acute and long-term exposure, was derived in the 1993 Guidelines.

Assessment date

The risk assessment was originally conducted in 1993. The Final Task Force Meeting in 2003 agreed that this risk assessment be brought forward to this edition of the *Guidelines for Drinking-water Quality*.

Principal reference

WHO (2003) *Cyanide in drinking-water. Background document for preparation of WHO Guidelines for drinking-water quality*. Geneva, World Health Organization (WHO/SDE/WSH/03.04/5).

12.34 Cyanogen chloride

Cyanogen chloride is a by-product of chloramination. It is a reaction product of organic precursors with hypochlorous acid in the presence of ammonium ion. Concentrations detected in drinking-water treated with chlorine and chloramine were 0.4 and 1.6 µg/litre, respectively.

Cyanogen chloride is rapidly metabolized to cyanide in the body. There are few data on the oral toxicity of cyanogen chloride, and the guideline value is based, therefore, on cyanide. The guideline value is 70 µg/litre for cyanide as total cyanogenic compounds (see Cyanide in section 12.33).

History of guideline development

The 1958, 1963 and 1971 WHO *International Standards for Drinking-water* and the first edition of the *Guidelines for Drinking-water Quality*, published in 1984, did not refer to cyanogen chloride. The 1993 Guidelines derived a health-based guideline value for cyanogen chloride based on cyanide, as cyanogen chloride is rapidly metabolized to cyanide in the body and as there are few data on the oral toxicity of cyanogen chloride. The guideline value is 0.07 mg/litre for cyanide as total cyanogenic compounds (see Cyanide in section 12.33).

Assessment date

The risk assessment was originally conducted in 1993. The Final Task Force Meeting in 2003 agreed that this risk assessment be brought forward to this edition of the *Guidelines for Drinking-water Quality*.

Principal reference

WHO (2003) *Cyanogen chloride in drinking-water. Background document for preparation of WHO Guidelines for drinking-water quality*. Geneva, World Health Organization (WHO/SDE/WSH/03.04/51).

12.35 2,4-D (2,4-dichlorophenoxyacetic acid)

The term 2,4-D is used here to refer to the free acid, 2,4-dichlorophenoxyacetic acid (CAS No. 94-75-7). Commercial 2,4-D products are marketed as the free acid, alkali and amine salts, and ester formulations. 2,4-D itself is chemically stable, but its esters are rapidly hydrolysed to the free acid. 2,4-D is a systemic herbicide used for control of broad-leaved weeds, including aquatic weeds. 2,4-D is rapidly biodegraded in the environment. Residues of 2,4-D in food rarely exceed a few tens of micrograms per kilogram.

Guideline value	0.03 mg/litre
Occurrence	Levels in water usually below 0.5 µg/litre, although concentrations as high as 30 µg/litre have been measured
ADI	0.01 mg/kg of body weight for the sum of 2,4-D and its salts and esters, expressed as 2,4-D, on the basis of a NOAEL of 1 mg/kg of body weight per day in a 1-year study of toxicity in dogs (for a variety of effects, including histopathological lesions in kidneys and liver) and a 2-year study of toxicity and carcinogenicity in rats (for renal lesions)
Limit of detection	0.1 µg/litre by gas–liquid chromatography with electrolytic conductivity detection
Treatment achievability	1 µg/litre should be achievable using GAC
Guideline derivation • allocation to water • weight • consumption	 10% of ADI 60-kg adult 2 litres/day
Additional comments	The guideline value applies to 2,4-D, as salts and esters of 2,4-D are rapidly hydrolysed to the free acid in water

Toxicological review

Epidemiological studies have suggested an association between exposure to chlorophenoxy herbicides, including 2,4-D, and two forms of cancer in humans: soft-tissue sarcomas and non-Hodgkin lymphoma. The results of these studies, however, are inconsistent; the associations found are weak, and conflicting conclusions have been reached by the investigators. Most of the studies did not provide information on exposure specifically to 2,4-D, and the risk was related to the general category of chlorophenoxy herbicides, a group that includes 2,4,5-trichlorophenoxyacetic acid (2,4,5-T), which was potentially contaminated with dioxins. JMPR concluded that it was not possible to evaluate the carcinogenic potential of 2,4-D on the basis of the available epidemiological studies. JMPR has also concluded that 2,4-D and its salts and esters are not genotoxic. The toxicity of the salts and esters of 2,4-D is comparable to that of the acid.

History of guideline development

The 1958 and 1963 WHO *International Standards for Drinking-water* did not refer to 2,4-D, but the 1971 International Standards suggested that pesticide residues that may occur in community water supplies make only a minimal contribution to the total daily intake of pesticides for the population served. In the first edition of the *Guide-*

lines for Drinking-water Quality, published in 1984, a health-based guideline value of 0.1 mg/litre was recommended for 2,4-D, based on the ADI recommended by WHO in 1976, but it was noted that some individuals may be able to detect 2,4-D by taste and odour at levels exceeding 0.05 mg/litre. The 1993 Guidelines established a health-based guideline value of 0.03 mg/litre for 2,4-D in drinking-water. This guideline value was retained in the addendum to these Guidelines, published in 1998, but was based on the more recent (1996) toxicological evaluation conducted by JMPR. This guideline value applies to 2,4-D, as salts and esters of 2,4-D are rapidly hydrolysed to the free acid in water.

Assessment date
The risk assessment was originally conducted in 1998. The Final Task Force Meeting in 2003 agreed that this risk assessment be brought forward to this edition of the *Guidelines for Drinking-water Quality*.

Principal references
FAO/WHO (1997) *Pesticide residues in food – 1996. Evaluations 1996. Part II – Toxicological.* Geneva, World Health Organization, Joint FAO/WHO Meeting on Pesticide Residues (WHO/PCS/97.1).

WHO (2003) *2,4-D in drinking-water. Background document for preparation of WHO Guidelines for drinking-water quality.* Geneva, World Health Organization (WHO/SDE/WSH/03.04/70).

12.36 2,4-DB
The half-lives for degradation of chlorophenoxy herbicides, including 2,4-DB (CAS No. 94-82-6), in the environment are in the order of several days. Chlorophenoxy herbicides are not often found in food.

Guideline value	0.09 mg/litre
Occurrence	Chlorophenoxy herbicides not frequently found in drinking- water; when detected, concentrations are usually no greater than a few micrograms per litre
TDI	30 µg/kg of body weight, based on a NOAEL of 3 mg/kg of body weight per day for effects on body and organ weights, blood chemistry and haematological parameters in a 2-year study in rats, with an uncertainty factor of 100 (for inter- and intraspecies variation)
Limit of detection	1 µg/litre to 1 mg/litre for various methods commonly used for the determination of chlorophenoxy herbicides in water, including solvent extraction, separation by GC, gas–liquid chromatography, thin-layer chromatography or HPLC, with ECD or UV detection
Treatment achievability	0.1 µg/litre should be achievable using GAC

Guideline derivation	
• allocation to water	10% of TDI
• weight	60-kg adult
• consumption	2 litres/day
Additional considerations	The NOAEL used in the guideline value derivation is similar to the NOAEL of 2.5 mg/kg of body weight per day obtained in a short-term study in beagle dogs and the NOAEL for hepatocyte hypertrophy of 5 mg/kg of body weight per day obtained in a 3-month study in rats.

Toxicological review

Chlorophenoxy herbicides, as a group, have been classified in Group 2B by IARC. However, the available data from studies in exposed populations and animals do not permit assessment of the carcinogenic potential to humans of any specific chlorophenoxy herbicide. Therefore, drinking-water guidelines for these compounds are based on a threshold approach for other toxic effects.

History of guideline development

The 1958 and 1963 WHO *International Standards for Drinking-water* did not refer to chlorophenoxy herbicides, including 2,4-DB, but the 1971 International Standards suggested that pesticide residues that may occur in community water supplies make only a minimal contribution to the total daily intake of pesticides for the population served. 2,4-DB was not evaluated in the first edition of the *Guidelines for Drinking-water Quality*, published in 1984, but the 1993 Guidelines established a health-based guideline value of 0.09 mg/litre for 2,4-DB.

Assessment date

The risk assessment was originally conducted in 1993. The Final Task Force Meeting in 2003 agreed that this risk assessment be brought forward to this edition of the *Guidelines for Drinking-water Quality*.

Principal reference

WHO (2003) *Chlorophenoxy herbicides (excluding 2,4-D and MCPA) in drinking-water. Background document for preparation of WHO Guidelines for drinking-water quality*. Geneva, World Health Organization (WHO/SDE/WSH/03.04/44).

12.37 DDT and metabolites

The structure of DDT (CAS No. 107917-42-0) permits several different isomeric forms, and commercial products consist predominantly of p,p'-DDT. Its use has been restricted or banned in several countries, although DDT is still used in some countries for the control of vectors that transmit yellow fever, sleeping sickness, typhus, malaria and other insect-transmitted diseases. DDT and its metabolites are persistent in the environment and resistant to complete degradation by microorganisms. Food is the major source of intake of DDT and related compounds for the general population.

Guideline value	0.001 mg/litre
Occurrence	Detected in surface water at concentrations below 1 µg/litre; also detected in drinking-water at 100-fold lower concentrations
PTDI	0.01 mg/kg of body weight based on a NOAEL of 1 mg/kg of body weight per day for developmental toxicity in rats, applying an uncertainty factor of 100
Limit of detection	0.011 µg/litre by GC using an ECD
Treatment achievability	0.1 µg/litre should be achievable using coagulation or GAC
Guideline derivation • allocation to water • weight • consumption	 1% of PTDI 10-kg child 1 litre/day
Additional comments	• DDT is listed under the Stockholm Convention on Persistent Organic Pollutants. Hence, monitoring may occur in addition to that required by drinking-water guidelines. • The guideline value is derived on the basis of a 10-kg child consuming 1 litre of drinking-water per day, because infants and children may be exposed to greater amounts of chemicals in relation to their body weight and because of concern over the bioaccumulation of DDT. • It should be emphasized that the benefits of DDT use in malaria and other vector control programmes outweigh any health risk from the presence of DDT in drinking-water.

Toxicological review

A working group convened by IARC classified the DDT complex as a non-genotoxic carcinogen in rodents and a potent promoter of liver tumours. IARC has concluded that there is insufficient evidence in humans and sufficient evidence in experimental animals for the carcinogenicity of DDT (Group 2B) based upon liver tumours observed in rats and mice. The results of epidemiological studies of pancreatic cancer, multiple myeloma, non-Hodgkin lymphoma and uterine cancer did not support the hypothesis of an association with environmental exposure to the DDT complex. Conflicting data were obtained with regard to some genotoxic end-points. In most studies, DDT did not induce genotoxic effects in rodent or human cell systems, nor was it mutagenic to fungi or bacteria. The US Agency for Toxic Substances and Disease Registry concluded that the DDT complex could impair reproduction and/or development in several species. Hepatic effects of DDT in rats include increased liver weights, hypertrophy, hyperplasia, induction of microsomal enzymes, including cytochrome P450, cell necrosis, increased activity of serum liver enzymes and mitogenic effects, which might be related to a regenerative liver response to DDT.

12. CHEMICAL FACT SHEETS

History of guideline development
The 1958 and 1963 WHO *International Standards for Drinking-water* did not refer to DDT, but the 1971 International Standards suggested that pesticide residues that may occur in community water supplies make only a minimal contribution to the total daily intake of pesticides for the population served. In the first edition of the *Guidelines for Drinking-water Quality*, published in 1984, a health-based guideline value of 0.001 mg/litre was recommended for DDT (total isomers), based on the ADI recommended by JMPR in 1969. The 1993 Guidelines established a health-based guideline value of 0.002 mg/litre for DDT and its metabolites in drinking-water, derived from the ADI recommended by JMPR in 1984 and taking into consideration the fact that infants and children may be exposed to greater amounts of chemicals in relation to their body weight, concern over the bioaccumulation of DDT and the significant exposure to DDT by routes other than water. It was noted that the guideline value exceeds the water solubility of DDT of 0.001 mg/litre, but that some DDT may be adsorbed onto the small amount of particulate matter present in drinking-water, so the guideline value could be reached under certain circumstances. It was also emphasized that the benefits of DDT use in malaria and other vector control programmes far outweigh any health risk from the presence of DDT in drinking-water.

Assessment date
The risk assessment was conducted in 2003.

Principal references
FAO/WHO (2001) *Pesticide residues in food – 2000. Evaluations – 2000. Part II – Toxicology*. Geneva, World Health Organization, Joint FAO/WHO Meeting on Pesticide Residues (WHO/PCS/01.3).

WHO (2003) *DDT and its derivatives in drinking-water. Background document for preparation of WHO Guidelines for drinking-water quality*. Geneva, World Health Organization (WHO/SDE/WSH/03.04/89).

12.38 Dialkyltins

The group of chemicals known as the organotins is composed of a large number of compounds with differing properties and applications. The most widely used of the organotins are the disubstituted compounds, which are employed as stabilizers in plastics, including polyvinyl chloride (PVC) water pipes, and the trisubstituted compounds, which are widely used as biocides.

The disubstituted compounds that may leach from PVC water pipes at low concentrations for a short time after installation are primarily immunotoxins, although they appear to be of low general toxicity. The data available are insufficient to permit the proposal of guideline values for individual dialkyltins.

History of guideline development
The 1958, 1963 and 1971 WHO *International Standards for Drinking-water* and the first edition of the *Guidelines for Drinking-water Quality*, published in 1984, did not refer to dialkyltins. The 1993 Guidelines concluded that the data available were insufficient to permit the proposal of guideline values for individual dialkyltins.

Assessment date
The risk assessment was conducted in 2003.

Principal reference
WHO (2003) *Dialkyltins in drinking-water. Background document for preparation of WHO Guidelines for drinking-water quality*. Geneva, World Health Organization (WHO/SDE/WSH/03.04/109).

12.39 1,2-Dibromo-3-chloropropane (DBCP)
1,2-Dibromo-3-chloropropane (CAS No. 96-12-8) is a soil fumigant that is highly soluble in water. It has a taste and odour threshold in water of 10 µg/litre. DBCP was detected in vegetables grown in treated soils, and low levels have been detected in air.

Guideline value	0.001 mg/litre
Occurrence	Limited survey found levels of up to a few micrograms per litre in drinking-water
Basis of guideline derivation	Linearized multistage model was applied to the data on the incidence of stomach, kidney and liver tumours in the male rat in a 104-week dietary study
Limit of detection	0.02 µg/litre by GC with ECD
Treatment achievability	1 µg/litre should be achievable using air stripping followed by GAC
Additional comments	The guideline value of 1 µg/litre should be protective for the reproductive toxicity of DBCP.

Toxicological review
On the basis of animal data from different strains of rats and mice, DBCP was determined to be carcinogenic in both sexes by the oral, inhalation and dermal routes. DBCP was also determined to be a reproductive toxicant in humans and several species of laboratory animals. DBCP was found to be genotoxic in a majority of *in vitro* and *in vivo* assays. IARC has classified DBCP in Group 2B based upon sufficient evidence of carcinogenicity in animals. Recent epidemiological evidence suggests an increase in cancer mortality in individuals exposed to high levels of DBCP.

History of guideline development
The 1958 and 1963 WHO *International Standards for Drinking-water* did not refer to DBCP, but the 1971 International Standards suggested that pesticide residues that may

occur in community water supplies make only a minimal contribution to the total daily intake of pesticides for the population served. DBCP was not evaluated in the first edition of the *Guidelines for Drinking-water Quality*, published in 1984, but the 1993 Guidelines calculated a guideline value of 0.001 mg/litre for DBCP in drinking-water, corresponding to an upper-bound excess lifetime cancer risk of 10^{-5} and sufficiently protective for the reproductive toxicity of the pesticide. It was noted that for a contaminated water supply, extensive treatment would be required to reduce the level of DBCP to the guideline value.

Assessment date
The risk assessment was originally conducted in 1993. The Final Task Force Meeting in 2003 agreed that this risk assessment be brought forward to this edition of the *Guidelines for Drinking-water Quality*.

Principal reference
WHO (2003) *1,2-Dibromo-3-chloropropane in drinking-water. Background document for preparation of WHO Guidelines for drinking-water quality*. Geneva, World Health Organization (WHO/SDE/WSH/03.04/34).

12.40 1,2-Dibromoethane (ethylene dibromide)

1,2-Dibromoethane (CAS No. 106-93-4) is used as a lead scavenger in tetra-alkyl lead petrol and antiknock preparations and as a fumigant for soils, grains and fruits. However, with the phasing out of leaded petrol and of the use of 1,2-dibromoethane in agricultural applications in many countries, use of this substance has declined significantly. In addition to its continued use as a petrol additive in some countries, 1,2-dibromoethane is currently used principally as a solvent and as an intermediate in the chemical industry.

Provisional guideline value	0.0004 mg/litre (0.4 µg/litre) The guideline value is provisional due to serious limitations of the critical studies.
Occurrence	Detected in groundwater following its use as a soil fumigant at concentrations as high as 100 µg/litre
Basis of guideline derivation	Lower end of the range (and thus more conservative estimate) of lifetime low-dose cancer risks calculated by linearized multistage modelling of the incidences of haemangiosarcomas and tumours in the stomach, liver, lung and adrenal cortex (adjusted for the observed high early mortality, where appropriate, and corrected for the expected rate of increase in tumour formation in rodents in a standard bioassay of 104 weeks) of rats and/or mice exposed to 1,2-dibromoethane by gavage

Limit of detection	0.01 μg/litre by microextraction GC/MS; 0.03 μg/litre by purge and trap GC with halogen-specific detector; 0.8 μg/litre by purge-and-trap capillary column GC with photoionization and electrolytic conductivity detectors in series
Treatment achievability	0.1 μg/litre should be achievable using GAC

Toxicological review

1,2-Dibromoethane has induced an increased incidence of tumours at several sites in all carcinogenicity bioassays identified in which rats or mice were exposed to the compound by gavage, ingestion in drinking-water, dermal application and inhalation. However, many of these studies were characterized by high early mortality, limited histopathological examination, small group sizes or use of only one exposure level. The substance acted as an initiator of liver foci in an initiation/promotion assay but did not initiate skin tumour development. 1,2-Dibromoethane was consistently genotoxic in *in vitro* assays, although results of *in vivo* assays were mixed. Biotransformation to active metabolites, which have been demonstrated to bind to DNA, is probably involved in the induction of tumours. Available data do not support the existence of a non-genotoxic mechanism of tumour induction. The available data thus indicate that 1,2-dibromoethane is a genotoxic carcinogen in rodents. Data on the potential carcinogenicity in humans are inadequate; however, it is likely that 1,2-dibromoethane is metabolized similarly in rodent species and in humans (although there may be varying potential for the production of active metabolites in humans, owing to genetic polymorphism). IARC classified 1,2-dibromoethane in Group 2A (the agent is probably carcinogenic to humans).

History of guideline development

The 1958 and 1963 WHO *International Standards for Drinking-water* did not refer to 1,2-dibromoethane, but the 1971 International Standards suggested that pesticide residues that may occur in community water supplies make only a minimal contribution to the total daily intake of pesticides for the population served. 1,2-Dibromoethane was not evaluated in the first edition of the *Guidelines for Drinking-water Quality*, published in 1984, but the 1993 Guidelines noted that 1,2-dibromoethane appears to be a genotoxic carcinogen. However, as the studies to date were inadequate for mathematical risk extrapolation, a guideline value for 1,2-dibromoethane was not derived. The Guidelines recommended that 1,2-dibromoethane be re-evaluated as soon as new data became available. In the addendum to these Guidelines, published in 1998, the guideline value that corresponds to an upper-bound excess lifetime cancer risk for various tumour types of 10^{-5} was calculated to be in the range 0.0004–0.015 mg/litre. This guideline value was considered to be provisional because of the serious limitations of the critical studies.

Assessment date
The risk assessment was conducted in 2003.

Principal references
IPCS (1995) *Report of the 1994 meeting of the Core Assessment Group*. Geneva, World Health Organization, International Programme on Chemical Safety, Joint Meeting on Pesticides (WHO/PCS/95.7).

IPCS (1996) *1,2-Dibromoethane*. Geneva, World Health Organization, International Programme on Chemical Safety (Environmental Health Criteria 177).

WHO (2003) *1,2-Dibromoethane in drinking-water. Background document for preparation of WHO Guidelines for drinking-water quality*. Geneva, World Health Organization (WHO/SDE/WSH/03.04/66).

12.41 Dichloroacetic acid
Chlorinated acetic acids are formed from organic material during water chlorination.

Provisional guideline value	0.05 mg/litre
	The guideline value is designated as provisional because the data are insufficient to ensure that the value is technically achievable. Difficulties in meeting a guideline value must never be a reason to compromise adequate disinfection.
Occurrence	Found in finished chlorinated water at concentrations up to about 100 µg/litre, but in most cases at concentrations less than 50 µg/litre
TDI	7.6 µg/kg of body weight, based on a study in which no effects were seen in the livers of mice exposed to dichloroacetate at 7.6 mg/kg of body weight per day for 75 weeks and incorporating an uncertainty factor of 1000 (100 for intra- and interspecies variation and 10 for possible carcinogenicity)
Limit of detection	1 µg/litre by GC with ECD; 2 µg/litre by GC/MS
Treatment achievability	Concentrations may be reduced by installing or optimizing coagulation to remove precursors and/or by controlling the pH during chlorination.
Guideline derivation • allocation to water • weight • consumption	 20% of TDI 60-kg adult 2 litres/day

Toxicological review
In several bioassays, dichloroacetate has been shown to induce hepatic tumours in mice. However, the evidence for the carcinogenicity of dichloroacetate is insufficient to derive a guideline value based on carcinogenicity. No adequate data on genotoxicity are available.

History of guideline development

The 1958, 1963 and 1971 WHO *International Standards for Drinking-water* and the first edition of the *Guidelines for Drinking-water Quality*, published in 1984, did not refer to dichloroacetic acid. In the 1993 Guidelines, a provisional guideline value of 0.05 mg/litre was derived for dichloroacetic acid; the guideline value was designated as provisional because the data were insufficient to ensure that the value was technically achievable.

Assessment date

The risk assessment was originally conducted in 1993. The Final Task Force Meeting in 2003 agreed that this risk assessment be brought forward to this edition of the *Guidelines for Drinking-water Quality*.

Principal reference

WHO (2003) *Dichloroacetic acid in drinking-water. Background document for preparation of WHO Guidelines for drinking-water quality*. Geneva, World Health Organization (WHO/SDE/WSH/03.04/121).

12.42 Dichlorobenzenes (1,2-dichlorobenzene, 1,3-dichlorobenzene, 1,4-dichlorobenzene)

The dichlorobenzenes (DCBs) are widely used in industry and in domestic products such as odour-masking agents, chemical dyestuffs and pesticides. Sources of human exposure are predominantly air and food.

Guideline values	
1,2-Dichlorobenzene	1 mg/litre
1,4-Dichlorobenzene	0.3 mg/litre
Occurrence	Have been found in raw water sources at levels as high as 10 µg/litre and in drinking-water at concentrations up to 3 µg/litre; much higher concentrations (up to 7 mg/litre) present in contaminated groundwater
TDIs	
1,2-Dichlorobenzene	429 µg/kg of body weight, based on a NOAEL of 60 mg/kg of body weight per day for tubular degeneration of the kidney identified in a 2-year mouse gavage study, correcting for 5 days per week dosing and using an uncertainty factor of 100 (for inter- and intraspecies variation)
1,4-Dichlorobenzene	107 µg/kg of body weight, based on a LOAEL of 150 mg/kg of body weight per day for kidney effects identified in a 2-year rat study, correcting for 5 days per week dosing and using an uncertainty factor of 1000 (100 for inter- and intraspecies variation and 10 for the use of a LOAEL instead of a NOAEL and the carcinogenicity end-point)

Limit of detection	0.01–0.25 µg/litre by gas–liquid chromatography with ECD; 3.5 µg/litre by GC using a photoionization detector
Treatment achievability	0.01 mg/litre should be achievable using air stripping
Guideline derivation • allocation to water • weight • consumption	 10% of TDI 60-kg adult 2 litres/day
Additional comments	Guideline values for both 1,2- and 1,4-DCB far exceed their lowest reported taste thresholds in water of 1 and 6 µg/litre, respectively.

Toxicological review

1,2-Dichlorobenzene

1,2-DCB is of low acute toxicity by the oral route of exposure. Oral exposure to high doses of 1,2-DCB affects mainly the liver and kidneys. The balance of evidence suggests that 1,2-DCB is not genotoxic, and there is no evidence for its carcinogenicity in rodents.

1,3-Dichlorobenzene

There are insufficient toxicological data on this compound to permit a guideline value to be proposed, but it should be noted that it is rarely found in drinking-water.

1,4-Dichlorobenzene

1,4-DCB is of low acute toxicity, but there is evidence that it increases the incidence of renal tumours in rats and of hepatocellular adenomas and carcinomas in mice after long-term exposure. IARC has placed 1,4-DCB in Group 2B. 1,4-DCB is not considered to be genotoxic, and the relevance for humans of the tumours observed in animals is doubtful.

History of guideline development

The 1958, 1963 and 1971 WHO *International Standards for Drinking-water* did not refer to DCBs. In the first edition of the *Guidelines for Drinking-water Quality*, published in 1984, no guideline value was recommended for 1,2- or 1,4-DCB after a detailed evaluation of the compounds. Toxicological limits for drinking-water of 0.005–0.05 mg/litre were derived based on an ADI; given that the threshold odour concentrations are 0.003 mg/litre for 1,2-DCB and 0.001 mg/litre for 1,4-DCB, 10% of each of these values was recommended as a level unlikely to give rise to taste and odour problems in drinking-water supplies. The 1993 Guidelines calculated a health-based guideline value of 1 mg/litre for 1,2-DCB, which far exceeds the lowest reported taste threshold of 1,2-DCB in water (0.001 mg/litre). There were insufficient toxicological data on 1,3-DCB to permit a guideline value to be proposed, but the 1993 Guidelines noted that it is rarely found in drinking-water. A health-based guideline value of 0.3 mg/litre was proposed for 1,4-DCB, which far exceeds the lowest reported odour threshold of 1,4-DCB in water (0.0003 mg/litre).

Assessment date

The risk assessment was originally conducted in 1993. The Final Task Force Meeting in 2003 agreed that this risk assessment be brought forward to this edition of the *Guidelines for Drinking-water Quality*.

Principal reference

WHO (2003) *Dichlorobenzenes in drinking-water. Background document for preparation of WHO Guidelines for drinking-water quality*. Geneva, World Health Organization (WHO/SDE/WSH/03.04/28).

12.43 1,1-Dichloroethane

1,1-Dichloroethane is used as a chemical intermediate and solvent. There are limited data showing that it can be present at concentrations of up to 10 µg/litre in drinking-water. However, because of the widespread use and disposal of this chemical, its occurrence in groundwater may increase.

1,1-Dichloroethane is rapidly metabolized by mammals to acetic acid and a variety of chlorinated compounds. It is of relatively low acute toxicity, and limited data are available on its toxicity from short- and long-term studies. There is limited *in vitro* evidence of genotoxicity. One carcinogenicity study by gavage in mice and rats provided no conclusive evidence of carcinogenicity, although there was some evidence of an increased incidence of haemangiosarcomas in treated animals.

In view of the very limited database on toxicity and carcinogenicity, it was concluded that no guideline value should be proposed.

History of guideline development

The 1958, 1963 and 1971 WHO *International Standards for Drinking-water* and the first edition of the *Guidelines for Drinking-water Quality*, published in 1984, did not refer to 1,1-dichloroethane. In view of the very limited database on toxicity and carcinogenicity, the 1993 Guidelines concluded that no guideline value for 1,1-dichloroethane should be proposed.

Assessment date

The risk assessment was originally conducted in 1993. The Final Task Force Meeting in 2003 agreed that this risk assessment be brought forward to this edition of the *Guidelines for Drinking-water Quality*.

Principal reference

WHO (2003) *1,1-Dichloroethane in drinking-water. Background document for preparation of WHO Guidelines for drinking-water quality*. Geneva, World Health Organization (WHO/SDE/WSH/03.04/19).

12.44 1,2-Dichloroethane

1,2-Dichloroethane is used mainly as an intermediate in the production of vinyl chloride and other chemicals and to a lesser extent as a solvent. It may enter surface waters via effluents from industries that manufacture or use the substance. It may also enter groundwater, where it may persist for long periods, following disposal in waste sites. It is found in urban air.

Guideline value	0.030 mg/litre
Occurrence	Has been found in drinking-water at levels of up to a few micrograms per litre
Basis of guideline derivation	Applying the linearized multistage model to haemangiosarcomas observed in male rats in a 78-week gavage study
Limit of detection	0.06–2.8 µg/litre by GC/MS; 0.03–0.2 µg/litre by GC with electrolytic conductivity detector; 5 µg/litre by GC with FID; 0.03 µg/litre by GC with photoionization detection
Treatment achievability	0.0001 mg/litre should be achievable using GAC
Additional considerations	The guideline value of 0.030 mg/litre is consistent with the value derived from IPCS (1998), based on a 10^{-5} risk level.

Toxicological review

IARC has classified 1,2-dichloroethane in Group 2B (possible human carcinogen). It has been shown to produce statistically significant increases in a number of tumour types in laboratory animals, including the relatively rare haemangiosarcoma, and the balance of evidence indicates that it is potentially genotoxic. Targets of 1,2-dichloroethane toxicity in orally exposed animals included the immune system, central nervous system, liver and kidney. Data indicate that 1,2-dichloroethane is less potent when inhaled.

History of guideline development

The 1958, 1963 and 1971 WHO *International Standards for Drinking-water* did not refer to 1,2-dichloroethane. In the first edition of the *Guidelines for Drinking-water Quality*, published in 1984, a health-based guideline value of 0.01 mg/litre was recommended for 1,2-dichloroethane, while noting that the mathematical model appropriate to chemical carcinogens that was used in its derivation involved considerable uncertainty. The 1993 Guidelines calculated a guideline value of 0.03 mg/litre for 1,2-dichloroethane on the basis of haemangiosarcomas observed in male rats, corresponding to an upper-bound excess lifetime cancer risk of 10^{-5}.

Assessment date

The risk assessment was conducted in 2003.

Principal references

IPCS (1995) *1,2-Dichloroethane*, 2nd ed. Geneva, World Health Organization, International Programme on Chemical Safety (Environmental Health Criteria 176).

IPCS (1998) *1,2-Dichloroethane*. Geneva, World Health Organization, International Programme on Chemical Safety (Concise International Chemical Assessment Document 1).

WHO (2003) *1,2-Dichloroethane in drinking-water. Background document for preparation of WHO Guidelines for drinking-water quality*. Geneva, World Health Organization (WHO/SDE/WSH/03.04/67).

12.45 1,1-Dichloroethene

1,1-Dichloroethene, or vinylidene chloride, is used mainly as a monomer in the production of polyvinylidene chloride co-polymers and as an intermediate in the synthesis of other organic chemicals. It is an occasional contaminant of drinking-water, usually being found together with other chlorinated hydrocarbons. There are no data on levels in food, but levels in air are generally less than 40 ng/m^3 except at some manufacturing sites.

Guideline value	0.03 mg/litre
Occurrence	Detected in finished drinking-water taken from groundwater sources at median concentrations of 0.28–1.2 µg/litre and in public drinking-water supplies at concentrations ranging from ≤0.2 to 0.5 µg/litre
TDI	9 µg/kg of body weight, based on a LOAEL (for increased incidence of hepatic lesions in females) of 9 mg/kg of body weight per day in a 2-year drinking-water study in rats, using an uncertainty factor of 1000 (100 for intra- and interspecies variation and 10 for the use of a LOAEL instead of a NOAEL and the potential for carcinogenicity)
Limit of detection	0.025 µg/litre by capillary GC with ECD; 0.07 µg/litre by purge-and-trap packed column GC with ECD or microcoulometric detector; 4.7 µg/litre by purge-and-trap packed column GC/MS
Treatment achievability	0.01 mg/litre should be achievable using GAC or air stripping
Guideline derivation • allocation to water • weight • consumption	 10% of TDI 60-kg adult 2 litres/day

Toxicological review

1,1-Dichloroethene is a central nervous system depressant and may cause liver and kidney toxicity in occupationally exposed humans. It causes liver and kidney damage in laboratory animals. IARC has placed 1,1-dichloroethene in Group 3. It was found to be genotoxic in a number of test systems *in vitro* but was not active in the dominant lethal and micronucleus assays *in vivo*. It induced kidney tumours in mice in one inhalation study but was reported not to be carcinogenic in a number of other studies, including several in which it was given in drinking-water.

History of guideline development
The 1958, 1963 and 1971 WHO *International Standards for Drinking-water* did not refer to 1,1-dichloroethene. In the first edition of the *Guidelines for Drinking-water Quality*, published in 1984, a health-based guideline value of 0.0003 mg/litre was recommended for 1,1-dichloroethene, while noting that the mathematical model appropriate to chemical carcinogens that was used in its derivation involved considerable uncertainty. A health-based guideline value of 0.03 mg/litre for 1,1-dichloroethene was recommended in the 1993 Guidelines.

Assessment date
The risk assessment was originally conducted in 1993. The Final Task Force Meeting in 2003 agreed that this risk assessment be brought forward to this edition of the *Guidelines for Drinking-water Quality*.

Principal reference
WHO (2003) *1,1-Dichloroethene in drinking-water. Background document for preparation of WHO Guidelines for drinking-water quality*. Geneva, World Health Organization (WHO/SDE/WSH/03.04/20).

12.46 1,2-Dichloroethene

1,2-Dichloroethene exists in a *cis* and a *trans* form. The *cis* form is more frequently found as a water contaminant. The presence of these two isomers, which are metabolites of other unsaturated halogenated hydrocarbons in wastewater and anaerobic groundwater, may indicate the simultaneous presence of more toxic organochlorine chemicals, such as vinyl chloride. Accordingly, their presence indicates that more intensive monitoring should be conducted. There are no data on exposure from food. Concentrations in air are low, with higher concentrations, in the microgram per cubic metre range, near production sites. The *cis* isomer was previously used as an anaesthetic.

Guideline value	0.05 mg/litre
Occurrence	Has been found in drinking-water supplies derived from groundwater at levels up to 120 µg/litre
TDI	17 µg/kg of body weight, based on a NOAEL (for increases in serum alkaline phosphatase levels and increased thymus weight) of 17 mg/kg of body weight from a 90-day study in mice administered *trans*-1,2-dichloroethene in drinking-water, using an uncertainty factor of 1000 (100 for inter- and intraspecies variation and 10 for the short duration of the study)
Limit of detection	0.17 µg/litre by GC with MS
Treatment achievability	0.01 mg/litre should be achievable using GAC or air stripping

Guideline derivation • allocation to water • weight • consumption	10% of TDI 60-kg adult 2 litres/day
Additional comments	Data on the *trans* isomer were used to calculate a joint guideline value for both isomers because toxicity for the *trans* isomer occurred at a lower dose than for the *cis* isomer and because data suggest that the mouse is a more sensitive species than the rat.

Toxicological review

There is little information on the absorption, distribution and excretion of 1,2-dichloroethene. However, by analogy with 1,1-dichloroethene, it would be expected to be readily absorbed, distributed mainly to the liver, kidneys and lungs and rapidly excreted. The *cis* isomer is more rapidly metabolized than the *trans* isomer in *in vitro* systems. Both isomers have been reported to cause increased serum alkaline phosphatase levels in rodents. In a 3-month study in mice given the *trans* isomer in drinking-water, there was a reported increase in serum alkaline phosphatase and reduced thymus and lung weights. Transient immunological effects were also reported, the toxicological significance of which is unclear. *Trans*-1,2-dichloroethene also caused reduced kidney weights in rats, but at higher doses. Only one rat toxicity study is available for the *cis* isomer, which produced toxic effects in rats similar in magnitude to those induced by the *trans* isomer in mice, but at higher doses. There are limited data to suggest that both isomers may possess some genotoxic activity. There is no information on carcinogenicity.

History of guideline development

The 1958, 1963 and 1971 WHO *International Standards for Drinking-water* did not refer to 1,2-dichloroethene. In the first edition of the *Guidelines for Drinking-water Quality*, published in 1984, no guideline value was recommended after a detailed evaluation of the compound. In the 1993 Guidelines, a joint guideline value of 0.05 mg/litre was calculated for both 1,2-dichloroethene isomers using toxicity data on the *trans* isomer.

Assessment date

The risk assessment was originally conducted in 1993. The Final Task Force Meeting in 2003 agreed that this risk assessment be brought forward to this edition of the *Guidelines for Drinking-water Quality*.

Principal reference

WHO (2003) *1,2-Dichloroethene in drinking-water. Background document for preparation of WHO Guidelines for drinking-water quality*. Geneva, World Health Organization (WHO/SDE/WSH/03.04/72).

12.47 Dichloromethane

Dichloromethane, or methylene chloride, is widely used as a solvent for many purposes, including coffee decaffeination and paint stripping. Exposure from drinking-water is likely to be insignificant compared with that from other sources.

Guideline value	0.02 mg/litre
Occurrence	Dichloromethane has been found in surface water samples at concentrations ranging from 0.1 to 743 µg/litre. Levels are usually higher in groundwater because volatilization is restricted; concentrations as high as 3600 µg/litre have been reported. Mean concentrations in drinking-water were less than 1 µg/litre.
TDI	6 µg/kg of body weight, derived from a NOAEL of 6 mg/kg of body weight per day for hepatotoxic effects in a 2-year drinking-water study in rats, using an uncertainty factor of 1000 (100 for inter- and intraspecies variation and 10 for concern about carcinogenic potential)
Limit of detection	0.3 µg/litre by purge-and-trap GC with MS detection (note that dichloromethane vapour readily penetrates tubing during the procedure)
Treatment achievability	20 µg/litre should be achievable using air stripping
Guideline derivation • allocation to water • weight • consumption	 10% of TDI 60-kg adult 2 litres/day

Toxicological review

Dichloromethane is of low acute toxicity. An inhalation study in mice provided conclusive evidence of carcinogenicity, whereas drinking-water studies in rats and mice provided only suggestive evidence. IARC has placed dichloromethane in Group 2B; however, the balance of evidence suggests that it is not a genotoxic carcinogen and that genotoxic metabolites are not formed in relevant amounts *in vivo*.

History of guideline development

The 1958, 1963 and 1971 WHO *International Standards for Drinking-water* did not refer to dichloromethane. In the first edition of the *Guidelines for Drinking-water Quality*, published in 1984, no guideline value was recommended after a detailed evaluation of the compound. The 1993 Guidelines established a health-based guideline value of 0.02 mg/litre for dichloromethane, noting that widespread exposure from other sources is possible.

Assessment date

The risk assessment was originally conducted in 1993. The Final Task Force Meeting in 2003 agreed that this risk assessment be brought forward to this edition of the *Guidelines for Drinking-water Quality*.

Principal reference
WHO (2003) *Dichloromethane in drinking-water. Background document for preparation of WHO Guidelines for drinking-water quality*. Geneva, World Health Organization (WHO/SDE/WSH/03.04/18).

12.48 1,2-Dichloropropane (1,2-DCP)

1,2-Dichloropropane (CAS No. 78-87-5) is used as an insecticide fumigant on grain and soil and to control peach tree borers. It is also used as an intermediate in the production of perchloroethylene and other chlorinated products and as a solvent. 1,2-DCP is relatively resistant to hydrolysis, is poorly adsorbed onto soil and can migrate into groundwater.

Provisional guideline value	0.04 mg/litre The guideline value is provisional owing to limitations of the toxicological database.
Occurrence	Detected in groundwater and drinking-water, usually at concentrations below 20 µg/litre, although levels as high as 440 µg/litre have been measured in well water
TDI	14 µg/kg of body weight based on a LOAEL of 71.4 mg/kg of body weight per day (100 mg/kg of body weight per day corrected for 5 days per week dosing) for changes in haematological parameters in a 13-week study in male rats, with an uncertainty factor of 5000 (100 for inter- and intraspecies variation, 10 for use of a LOAEL and 5 to reflect limitations of the database, including the limited data on *in vivo* genotoxicity and use of a subchronic study)
Limit of detection	0.02 µg/litre by a purge-and-trap GC method with an electrolytic conductivity detector or GC/MS
Treatment achievability	1 µg/litre should be achievable using GAC
Guideline derivation • allocation to water • weight • consumption	 10% of TDI 60-kg adult 2 litres/day

Toxicological review

1,2-DCP was evaluated by IARC in 1986 and 1987. The substance was classified in Group 3 (not classifiable as to its carcinogenicity to humans) on the basis of limited evidence for its carcinogenicity in experimental animals and insufficient data with which to evaluate its carcinogenicity in humans. Results from *in vitro* assays for mutagenicity were mixed. The *in vivo* studies, which were limited in number and design, were negative. In accordance with the IARC evaluation, the evidence from the long-term carcinogenicity studies in mice and rats was considered limited, and it was concluded that the use of a threshold approach for the toxicological evaluation of 1,2-DCP was appropriate.

12. CHEMICAL FACT SHEETS

History of guideline development
The 1958 and 1963 WHO *International Standards for Drinking-water* did not refer to 1,2-DCP, but the 1971 International Standards suggested that pesticide residues that may occur in community water supplies make only a minimal contribution to the total daily intake of pesticides for the population served. 1,2-DCP was not evaluated in the first edition of the *Guidelines for Drinking-water Quality*, published in 1984, but the 1993 Guidelines proposed a provisional health-based guideline value of 0.02 mg/litre for 1,2-DCP in drinking-water. The value was provisional because an uncertainty factor of 10 000 was used in its derivation. This guideline value was amended to 0.04 mg/litre in the addendum to these Guidelines, published in 1998, using a lower uncertainty factor. This guideline value was considered to be provisional owing to the magnitude of the uncertainty factor and the fact that the database had not changed since the previous guideline value had been derived.

Assessment date
The risk assessment was originally conducted in 1998. The Final Task Force Meeting in 2003 agreed that this risk assessment be brought forward to this edition of the *Guidelines for Drinking-water Quality*.

Principal reference
WHO (2003) *1,2-Dichloropropane (1,2-DCP) in drinking-water. Background document for preparation of WHO Guidelines for drinking-water quality*. Geneva, World Health Organization (WHO/SDE/WSH/03.04/61).

12.49 1,3-Dichloropropane

1,3-Dichloropropane (CAS No. 142-28-9) has several industrial uses and may be found as a contaminant of soil fumigants containing 1,3-dichloropropene. It is rarely found in water.

1,3-Dichloropropane is of low acute toxicity. There is some indication that it may be genotoxic in bacterial systems. No short-term, long-term, reproductive or developmental toxicity data pertinent to exposure via drinking-water could be located in the literature. The available data are considered insufficient to permit recommendation of a guideline value.

History of guideline development
The 1958 and 1963 WHO *International Standards for Drinking-water* did not refer to 1,3-dichloropropane, but the 1971 International Standards suggested that pesticide residues that may occur in community water supplies make only a minimal contribution to the total daily intake of pesticides for the population served. 1,3-Dichloropropane was not evaluated in the first edition of the *Guidelines for Drinking-water Quality*, published in 1984, but the 1993 Guidelines concluded that the available data

were insufficient to permit recommendation of a guideline value for 1,3-dichloropropane in drinking-water.

Assessment date
The risk assessment was originally conducted in 1993. The Final Task Force Meeting in 2003 agreed that this risk assessment be brought forward to this edition of the *Guidelines for Drinking-water Quality*.

Principal reference
WHO (2003) *1,3-Dichloropropane in drinking-water. Background document for preparation of WHO Guidelines for drinking-water quality*. Geneva, World Health Organization (WHO/SDE/WSH/03.04/35).

12.50 1,3-Dichloropropene

1,3-Dichloropropene (CAS Nos. 542-75-6 isomer mixture; 10061-01-5 *cis* isomer; 10061-02-6 *trans* isomer) is a soil fumigant, the commercial product being a mixture of *cis* and *trans* isomers. It is used to control a wide variety of soil pests, particularly nematodes in sandy soils. Notwithstanding its high vapour pressure, it is soluble in water at the gram per litre level and can be considered a potential water contaminant.

Guideline value	0.02 mg/litre
Occurrence	Has been found in surface water and groundwater at concentrations of a few micrograms per litre
Basis of guideline derivation	Calculated by applying the linearized multistage model to the observation of lung and bladder tumours in female mice in a 2-year gavage study
Limit of detection	0.34 and 0.20 µg/litre by purge-and-trap packed column GC using an electrolytic conductivity detector or microcoulometric detector for *cis*-1,3-dichloropropene and *trans*-1,3- dichloropropene, respectively
Treatment achievability	No information found on removal from water

Toxicological review
1,3-Dichloropropene is a direct-acting mutagen that has been shown to produce forestomach tumours following long-term oral gavage exposure in rats and mice. Tumours have also been found in the bladder and lungs of female mice and the liver of male rats. Long-term inhalation studies in the rat have proved negative, whereas some benign lung tumours have been reported in inhalation studies in mice. IARC has classified 1,3-dichloropropene in Group 2B (possible human carcinogen).

History of guideline development
The 1958 and 1963 WHO *International Standards for Drinking-water* did not refer to 1,3-dichloropropene, but the 1971 International Standards suggested that pesticide

residues that may occur in community water supplies make only a minimal contribution to the total daily intake of pesticides for the population served. 1,3-Dichloropropene was not evaluated in the first edition of the *Guidelines for Drinking-water Quality*, published in 1984, but the 1993 Guidelines calculated a guideline value of 0.02 mg/litre for 1,3-dichloropropene in drinking-water, corresponding to an upper-bound excess lifetime cancer risk of 10^{-5}.

Assessment date
The risk assessment was originally conducted in 1993. The Final Task Force Meeting in 2003 agreed that this risk assessment be brought forward to this edition of the *Guidelines for Drinking-water Quality*.

Principal reference
WHO (2003) *1,3-Dichloropropene in drinking-water. Background document for preparation of WHO Guidelines for drinking-water quality*. Geneva, World Health Organization (WHO/SDE/WSH/03.04/36).

12.51 Dichlorprop (2,4-DP)
The half-lives for degradation of chlorophenoxy herbicides, including dichlorprop (CAS No. 120-36-5), in the environment are in the order of several days. Chlorophenoxy herbicides are not often found in food.

Guideline value	0.1 mg/litre
Occurrence	Chlorophenoxy herbicides not frequently found in drinking- water; when detected, concentrations are usually no greater than a few micrograms per litre
TDI	36.4 µg/kg of body weight, based on a NOAEL for renal toxicity in a 2-year study in rats of 100 mg/kg of diet, equal to 3.64 mg/kg of body weight per day, applying an uncertainty factor of 100 (for intra- and interspecies variation)
Limit of detection	1 µg/litre to 1 mg/litre for various methods commonly used for the determination of chlorophenoxy herbicides in water, including solvent extraction, separation by GC, gas–liquid chromatography, thin-layer chromatography or HPLC, with ECD or UV detection
Treatment achievability	No data available
Guideline derivation • allocation to water • weight • consumption	 10% of TDI 60-kg adult 2 litres/day

Toxicological review
Chlorophenoxy herbicides, as a group, have been classified in Group 2B by IARC. However, the available data from studies in exposed populations and animals do not

permit assessment of the carcinogenic potential to humans of any specific chlorophenoxy herbicide. Therefore, drinking-water guidelines for these compounds are based on a threshold approach for other toxic effects. In dietary studies in rats, slight liver hypertrophy was observed in a 3-month study, and effects in a 2-year study included hepatocellular swelling, mild anaemia, increased incidence of brown pigment in the kidneys (possibly indicative of slight degeneration of the tubular epithelium) and decreased urinary specific gravity and protein.

History of guideline development
The 1958 and 1963 WHO *International Standards for Drinking-water* did not refer to chlorophenoxy herbicides, including dichlorprop, but the 1971 International Standards suggested that pesticide residues that may occur in community water supplies make only a minimal contribution to the total daily intake of pesticides for the population served. Dichlorprop was not evaluated in the first edition of the *Guidelines for Drinking-water Quality*, published in 1984, but the 1993 Guidelines established a health-based guideline value of 0.1 mg/litre for dichlorprop.

Assessment date
The risk assessment was originally conducted in 1993. The Final Task Force Meeting in 2003 agreed that this risk assessment be brought forward to this edition of the *Guidelines for Drinking-water Quality*.

Principal reference
WHO (2003) *Chlorophenoxy herbicides (excluding 2,4-D and MCPA) in drinking-water. Background document for preparation of WHO Guidelines for drinking-water quality*. Geneva, World Health Organization (WHO/SDE/WSH/03.04/44).

12.52 Di(2-ethylhexyl)adipate

Di(2-ethylhexyl)adipate (DEHA) is used mainly as a plasticizer for synthetic resins such as PVC. Reports of the presence of DEHA in surface water and drinking-water are scarce, but DEHA has occasionally been identified in drinking-water at levels of a few micrograms per litre. As a consequence of its use in PVC films, food is the most important source of human exposure (up to 20 mg/day).

DEHA is of low short-term toxicity; however, dietary levels above 6000 mg/kg of feed induce peroxisomal proliferation in the liver of rodents. This effect is often associated with the development of liver tumours. DEHA induced liver carcinomas in female mice at very high doses but not in male mice or rats. It is not genotoxic. IARC has placed DEHA in Group 3.

A health-based value of 80 µg/litre can be calculated for DEHA on the basis of a TDI of 280 µg/kg of body weight, based on fetotoxicity in rats, and allocating 1% of the TDI to drinking-water. However, because DEHA occurs at concentrations well below those at which toxic effects are observed, it is not considered necessary to derive a health-based guideline value.

History of guideline development
The 1958, 1963 and 1971 WHO *International Standards for Drinking-water* and the first edition of the *Guidelines for Drinking-water Quality*, published in 1984, did not refer to DEHA. The 1993 Guidelines proposed a health-based guideline value of 0.08 mg/litre for DEHA in drinking-water.

Assessment date
The risk assessment was conducted in 2003.

Principal reference
WHO (2003) *Di(2-ethylhexyl)adipate in drinking-water. Background document for preparation of WHO Guidelines for drinking-water quality.* Geneva, World Health Organization (WHO/SDE/WSH/03.04/68).

12.53 Di(2-ethylhexyl)phthalate
Di(2-ethylhexyl)phthalate (DEHP) is used primarily as a plasticizer. Exposure among individuals may vary considerably because of the broad nature of products into which DEHP is incorporated. In general, food will be the main exposure route.

Guideline value	0.008 mg/litre
Occurrence	Found in surface water, groundwater and drinking-water in concentrations of a few micrograms per litre; in polluted surface water and groundwater, concentrations of hundreds of micrograms per litre have been reported
TDI	25 µg/kg of body weight, based on a NOAEL of 2.5 mg/kg of body weight per day for peroxisomal proliferation in the liver in rats, using an uncertainty factor of 100 for inter- and Intraspecies variation
Limit of detection	0.1 µg/litre by GC/MS
Treatment achievability	No data available
Guideline derivation • allocation to water • weight • consumption	 1% of TDI 60-kg adult 2 litres/day
Additional comments	The reliability of some data on environmental water samples is questionable because of secondary contamination during sampling and working-up procedures. Concentrations that exceed the solubility more than 10-fold have been reported.

Toxicological review
In rats, DEHP is readily absorbed from the gastrointestinal tract. In primates (including humans), absorption after ingestion is lower. Species differences are also observed

in the metabolic profile. Most species excrete primarily the conjugated mono-ester in urine. Rats, however, predominantly excrete terminal oxidation products. DEHP is widely distributed in the body, with highest levels in liver and adipose tissue, without showing significant accumulation. The acute oral toxicity is low. The most striking effect in short-term toxicity studies is the proliferation of hepatic peroxisomes, indicated by increased peroxisomal enzyme activity and histopathological changes. The available information suggests that primates, including humans, are far less sensitive to this effect than rodents. In long-term oral carcinogenicity studies, hepatocellular carcinomas were found in rats and mice. IARC has concluded that DEHP is possibly carcinogenic to humans (Group 2B). In 1988, JECFA evaluated DEHP and recommended that human exposure to this compound in food be reduced to the lowest level attainable. The Committee considered that this might be achieved by using alternative plasticizers or alternatives to plastic material containing DEHP. In a variety of *in vitro* and *in vivo* studies, DEHP and its metabolites have shown no evidence of genotoxicity, with the exception of induction of aneuploidy and cell transformation.

History of guideline development
The 1958, 1963 and 1971 WHO *International Standards for Drinking-water* and the first edition of the *Guidelines for Drinking-water Quality*, published in 1984, did not refer to DEHP. The 1993 Guidelines established a health-based guideline value of 0.008 mg/litre for DEHP in drinking-water.

Assessment date
The risk assessment was originally conducted in 1993. The Final Task Force Meeting in 2003 agreed that this risk assessment be brought forward to this edition of the *Guidelines for Drinking-water Quality*.

Principal reference
WHO (2003) *Di(2-ethylhexyl)phthalate in drinking-water. Background document for preparation of WHO Guidelines for drinking-water quality*. Geneva, World Health Organization (WHO/SDE/WSH/03.04/29).

12.54 Dimethoate

Dimethoate (CAS No. 60-51-5) is an organophosphorus insecticide used to control a broad range of insects in agriculture, as well as the housefly. It has a half-life of 18 h to 8 weeks and is not expected to persist in water, although it is relatively stable at pH 2–7. A total daily intake from food of 0.001 µg/kg of body weight has been estimated.

Guideline value	0.006 mg/litre
Occurrence	Detected at trace levels in a private well in Canada, but not detected in a Canadian survey of surface water or drinking-water supplies
ADI	0.002 mg/kg of body weight based on an apparent NOAEL of 1.2 mg/kg of body weight per day for reproductive performance in a study of reproductive toxicity in rats, applying an uncertainty factor of 500 to take into consideration concern regarding whether this could be a LOAEL
Limit of detection	0.05 µg/litre by GC/MS
Treatment achievability	1 µg/litre should be achievable using GAC and chlorination
Guideline derivation • allocation to water • weight • consumption	10% of ADI 60-kg adult 2 litres/day

Toxicological review

In studies with human volunteers, dimethoate has been shown to be a cholinesterase inhibitor and a skin irritant. Dimethoate is not carcinogenic to rodents. JMPR concluded that although *in vitro* studies indicate that dimethoate has mutagenic potential, this potential does not appear to be expressed in vivo. In a multigeneration study of reproductive toxicity in rats, the NOAEL appeared to be 1.2 mg/kg of body weight per day, but there was some indication that reproductive performance may have been affected at lower doses. No data were available to assess whether the effects on reproductive performance were secondary to inhibition of cholinesterase. JMPR concluded that it was not appropriate to base the ADI on the results of the studies of volunteers, since the crucial end-point (reproductive performance) has not been assessed in humans. It was suggested that there may be a need to re-evaluate the toxicity of dimethoate after the periodic review of the residue and analytical aspects of dimethoate has been completed if it is determined that omethoate is a major residue.

History of guideline development

The 1958 and 1963 WHO *International Standards for Drinking-water* did not refer to dimethoate, but the 1971 International Standards suggested that pesticide residues that may occur in community water supplies make only a minimal contribution to the total daily intake of pesticides for the population served. Dimethoate was not evaluated in the first edition of the *Guidelines for Drinking-water Quality*, published in 1984, in the second edition, published in 1993, or in the addendum to the second edition, published in 1998.

Assessment date

The risk assessment was conducted in 2003.

Principal references

FAO/WHO (1997) *Pesticide residues in food – 1996 evaluations. Part II – Toxicological*. Geneva, World Health Organization, Joint FAO/WHO Meeting on Pesticide Residues (WHO/PCS/97.1).

WHO (2003) *Dimethoate in drinking-water. Background document for preparation of WHO Guidelines for drinking-water quality*. Geneva, 1World Health Organization (WHO/SDE/WSH/03.04/90).

12.55 Diquat

Diquat (CAS No. 2764-72-9) is a non-selective contact herbicide and crop desiccant. Diquat may also be used (at or below 1 mg/litre) as an aquatic herbicide for the control of free-floating and submerged aquatic weeds in ponds, lakes and irrigation ditches. Because of its rapid degradation in water and strong adsorption onto sediments, diquat has rarely been found in drinking-water.

Diquat does not appear to be carcinogenic or genotoxic. The main toxicological finding in experimental animals is cataract formation. A health-based value of 6 µg/litre for diquat ion can be calculated on the basis of an ADI of 0.002 mg of diquat ion per kg of body weight, based on cataract formation at the next higher dose in a 2-year study in rats. However, because diquat has rarely been found in drinking-water, it is not considered necessary to derive a guideline value. It should also be noted that the limit of detection of diquat in water is 0.001 mg/litre, and its practical quantification limit is about 0.01 mg/litre.

History of guideline development

The 1958 and 1963 WHO *International Standards for Drinking-water* did not refer to diquat, but the 1971 International Standards suggested that pesticide residues that may occur in community water supplies make only a minimal contribution to the total daily intake of pesticides for the population served. Diquat was not evaluated in the first two editions of the *Guidelines for Drinking-water Quality*, published in 1984 and 1993. In the addendum to the second edition of these Guidelines, published in 1998, a health-based value of 0.006 mg/litre was calculated for the diquat ion using the ADI established by JMPR in 1993. However, the limit of detection of diquat in water is 0.001 mg/litre, and its practical quantification limit is about 0.01 mg/litre. A provisional guideline value of 0.01 mg/litre was therefore established for diquation.

Assessment date

The risk assessment was conducted in 2003.

Principal references

FAO/WHO (1994) *Pesticide residues in food – 1993. Evaluations – 1993. Part II – Toxicology*. Geneva, World Health Organization, Joint FAO/WHO Meeting on Pesticide Residues (WHO/PCS/94.4).

WHO (2003) *Diquat in drinking-water. Background document for preparation of WHO Guidelines for drinking-water quality.* Geneva, World Health Organization (WHO/SDE/WSH/03.04/91).

12.56 Edetic acid (EDTA)

Human exposure to EDTA arises directly from its use in food additives, medicines, and personal care and hygiene products. Exposure to EDTA from drinking-water is probably very small in comparison with that from other sources. Once EDTA is present in the aquatic environment, its speciation will depend on the water quality and the presence of trace metals with which it will combine. The removal of EDTA from communal wastewater by biodegradation in sewage purification plants is very limited.

Guideline value	0.6 mg/litre (for EDTA as the free acid)
Occurrence	Present in surface waters generally at concentrations below 70 µg/litre, although higher concentrations (900 µg/litre) have been measured; detected in drinking-water prepared from surface waters at concentrations of 10–30 µg/litre
ADI	1.9 mg/kg of body weight as the free acid (ADI of 2.5 mg/kg of body weight proposed by JECFA for calcium disodium edetate as a food additive)
Limit of detection	1 µg/litre by potentiometric stripping analyis
Treatment achievability	0.01 mg/litre using GAC plus ozonation
Guideline derivation • allocation to water • weight • consumption	 1% of ADI 60-kg adult 2 litres/day
Additional comments	Concern has been expressed over the ability of EDTA to complex, and therefore reduce the availability of, zinc. However, this is of significance only at elevated doses substantially in excess of those encountered in the environment.

Toxicological review

Calcium disodium edetate is poorly absorbed from the gut. The long-term toxicity of EDTA is complicated by its ability to chelate essential and toxic metals. Those toxicological studies that are available indicate that the apparent toxicological effects of EDTA have in fact been due to zinc deficiency as a consequence of complexation. EDTA does not appear to be teratogenic or carcinogenic in animals. The vast clinical experience of the use of EDTA in the treatment of metal poisoning has demonstrated its safety in humans.

History of guideline development

The 1958, 1963 and 1971 WHO *International Standards for Drinking-water* and the first edition of the *Guidelines for Drinking-water Quality*, published in 1984, did not refer to edetic acid. The 1993 Guidelines proposed a provisional health-based guideline value of 0.2 mg/litre for edetic acid, based on an ADI for calcium disodium edetate as a food additive proposed by JECFA in 1973 and assuming that a 10-kg child consumes 1 litre of water per day, in view of the possibility of zinc complexation. The value was considered provisional to reflect the fact that the JECFA ADI had not been considered since 1973. JECFA further evaluated the toxicological studies available on EDTA in 1993 and was unable to add any further important information regarding the toxicity of EDTA and its calcium and sodium salts to the 1973 evaluation. In the addendum to the second edition of the Guidelines, published in 1998, a guideline value of 0.6 mg/litre was derived for EDTA (free acid), using different assumptions from those used in the derivation of the provisional guideline value in the 1993 Guidelines. In particular, it was noted that the ability of EDTA to complex, and therefore reduce the availability of, zinc was of significance only at elevated doses substantially in excess of those encountered in the environment.

Assessment date

The risk assessment was originally conducted in 1998. The Final Task Force Meeting in 2003 agreed that this risk assessment be brought forward to this edition of the *Guidelines for Drinking-water Quality*.

Principal reference

WHO (2003) *Edetic acid (EDTA) in drinking-water. Background document for preparation of WHO Guidelines for drinking-water quality*. Geneva, World Health Organization (WHO/SDE/WSH/03.04/58).

12.57 Endosulfan

Endosulfan (CAS No. 115-29-7) is an insecticide used in countries throughout the world to control pests on fruit, vegetables and tea and on non-food crops such as tobacco and cotton. In addition to its agricultural use, it is used in the control of the tsetse fly, as a wood preservative and for the control of home garden pests. Endosulfan contamination does not appear to be widespread in the aquatic environment, but the chemical has been found in agricultural runoff and rivers in industrialized areas where it is manufactured or formulated, as well as in surface water and groundwater samples collected from hazardous waste sites in the USA. Surface water samples in the USA generally contain less than 1 µg/litre. The main source of exposure of the general population is food, but residues have generally been found to be well below the FAO/WHO maximum residue limits. Another important route of exposure to endosulfan for the general population is the use of tobacco products.

JMPR concluded that endosulfan is not genotoxic, and no carcinogenic effects were noted in long-term studies using mice and rats. The kidney is the target organ for toxicity. Several recent studies have shown that endosulfan, alone or in combination with other pesticides, may bind to estrogen receptors and perturb the endocrine system. A health-based value of 20 µg/litre can be calculated for endosulfan on the basis of an ADI of 0.006 mg/kg of body weight, based on results from a 2-year dietary study of toxicity in rats, and supported by a 78-week study in mice, a 1-year study in dogs and a developmental toxicity study in rats. However, because endosulfan occurs at concentrations well below those at which toxic effects are observed, it is not considered necessary to derive a guideline value.

History of guideline development
The 1958 and 1963 WHO *International Standards for Drinking-water* did not refer to endosulfan, but the 1971 International Standards suggested that pesticide residues that may occur in community water supplies make only a minimal contribution to the total daily intake of pesticides for the population served. Endosulfan was not evaluated in the first edition of the *Guidelines for Drinking-water Quality*, published in 1984, in the second edition, published in 1993, or in the addendum to the second edition, published in 1998.

Assessment date
The risk assessment was conducted in 2003.

Principal references
FAO/WHO (1999) *Pesticide residues in food – 1998 evaluations. Part II – Toxicological.* Geneva, World Health Organization, Joint FAO/WHO Meeting on Pesticide Residues (WHO/PCS/99.18).

WHO (2003) *Endosulfan in drinking-water. Background document for preparation of WHO Guidelines for drinking-water quality.* Geneva, World Health Organization (WHO/SDE/WSH/03.04/92).

12.58 Endrin
Endrin (CAS No. 72-20-8) is a broad-spectrum foliar insecticide that acts against a wide range of agricultural pests. It is also used as a rodenticide. Small amounts of endrin are present in food, but the total intake from food appears to be decreasing.

Guideline value	0.0006 mg/litre (0.6 µg/litre)
Occurrence	Traces of endrin found in the drinking-water supplies of several countries
PTDI	0.0002 mg/kg of body weight, based on a NOAEL of 0.025 mg/kg of body weight per day in a 2-year study in dogs and applying an uncertainty factor of 100
Limit of detection	0.002 µg/litre by GC with ECD
Treatment achievability	0.2 µg/litre should be achievable using GAC

Guideline derivation	
• allocation to water	10% of PTDI
• weight	60-kg adult
• consumption	2 litres/day
Additional comments	Endrin is listed under the Stockholm Convention on Persistent Organic Pollutants. Hence, monitoring may occur in addition to that required by drinking-water guidelines.

Toxicological review
Toxicological data are insufficient to indicate whether endrin is a carcinogenic hazard to humans. The primary site of action of endrin is the central nervous system.

History of guideline development
The 1958 and 1963 WHO *International Standards for Drinking-water* did not refer to endrin, but the 1971 International Standards suggested that pesticide residues that may occur in community water supplies make only a minimal contribution to the total daily intake of pesticides for the population served. Endrin was not evaluated in the first edition of the *Guidelines for Drinking-water Quality*, published in 1984, in the second edition, published in 1993, or in the addendum to the second edition, published in 1998.

Assessment date
The risk assessment was conducted in 2003.

Principal references
FAO/WHO (1995) *Pesticide residues in food – 1994. Report of the Joint Meeting of the FAO Panel of Experts on Pesticide Residues in Food and the Environment and WHO Toxicological and Environmental Core Assessment Groups*. Rome, Food and Agriculture Organization of the United Nations (FAO Plant Production and Protection Paper 127).

IPCS (1992) *Endrin*. Geneva, World Health Organization, International Programme on Chemical Safety (Environmental Health Criteria 130).

WHO (2003) *Endrin in drinking-water. Background document for preparation of WHO Guidelines for drinking-water quality*. Geneva, World Health Organization (WHO/SDE/WSH/03.04/93).

12.59 Epichlorohydrin
Epichlorohydrin is used for the manufacture of glycerol, unmodified epoxy resins and water treatment resins. No quantitative data are available on its occurrence in food or drinking-water. Epichlorohydrin is hydrolysed in aqueous media.

Provisional guideline value	0.0004 mg/litre (0.4 µg/litre) The guideline value is considered to be provisional because of the uncertainties surrounding the toxicity of epichlorohydrin and the use of a large uncertainty factor in deriving the guideline value.
Occurrence	No quantitative data available
TDI	0.14 µg/kg of body weight, on the basis of a LOAEL of 2 mg/kg of body weight per day for forestomach hyperplasia observed in a 2-year gavage study in rats, correcting for 5 days per week dosing and using an uncertainty factor of 10 000 to take into consideration inter- and intraspecies variation (100), the use of a LOAEL instead of a NOAEL (10) and carcinogenicity (10)
Limit of detection	0.01 µg/litre by GC with ECD; 0.1 and 0.5 µg/litre by GC/MS; 0.01 mg/litre by GC with FID
Treatment achievability	Conventional treatment processes do not remove epichlorohydrin. Epichlorohydrin concentrations in drinking-water are controlled by limiting either the epichlorohydrin content of polyamine flocculants or the dose used, or both.
Guideline derivation • allocation to water • weight • consumption	 10% of TDI 60-kg adult 2 litres/day
Additional comments	Although epichlorohydrin is a genotoxic carcinogen, the use of the linearized multistage model for estimating cancer risk was considered inappropriate because tumours are seen only at the site of administration, where epichlorohydrin is highly irritating.

Toxicological review

Epichlorohydrin is rapidly and extensively absorbed following oral, inhalation or dermal exposure. It binds easily to cellular components. Major toxic effects are local irritation and damage to the central nervous system. It induces squamous cell carcinomas in the nasal cavity by inhalation and forestomach tumours by the oral route. It has been shown to be genotoxic *in vitro* and *in vivo*. IARC has placed epichlorohydrin in Group 2A (probably carcinogenic to humans).

History of guideline development

The 1958, 1963 and 1971 WHO *International Standards for Drinking-water* and the first edition of the *Guidelines for Drinking-water Quality*, published in 1984, did not refer to epichlorohydrin. The 1993 Guidelines proposed a provisional health-based guideline value of 0.0004 mg/litre for epichlorohydrin. The value was provisional because it was derived using an uncertainty factor of 10 000. It was noted that a practical quantification level for epichlorohydrin is of the order of 0.03 mg/litre, but concentrations in drinking-water can be controlled by specifying the epichlorohydrin content of products coming into contact with it.

Assessment date
The risk assessment was conducted in 2003.

Principal reference
WHO (2003) *Epichlorohydrin in drinking-water. Background document for preparation of WHO Guidelines for drinking-water quality*. Geneva, World Health Organization (WHO/SDE/WSH/03.04/94).

12.60 Ethylbenzene

The primary sources of ethylbenzene in the environment are the petroleum industry and the use of petroleum products. Because of its physical and chemical properties, more than 96% of ethylbenzene in the environment can be expected to be present in air. Values of up to 26 µg/m^3 in air have been reported. Ethylbenzene is found in trace amounts in surface water, groundwater, drinking-water and food.

Guideline value	0.3 mg/litre
Occurrence	Concentrations in drinking-water are generally below 1 µg/litre; levels up to 300 µg/litre have been reported in groundwater contaminated by point emissions.
TDI	97.1 µg/kg of body weight, based on a NOAEL of 136 mg/kg of body weight per day for hepatotoxicity and nephrotoxicity observed in a limited 6-month study in rats, correcting for 5 days per week dosing and using an uncertainty factor of 1000 (100 for inter- and intraspecies variation and 10 for the limited database and short duration of the study)
Limit of detection	0.002–0.005 µg/litre by GC with photoionization detector; 0.03–0.06 µg/litre by GC/MS
Treatment achievability	0.001 mg/litre should be achievable using air stripping
Guideline derivation • allocation to water • weight • consumption	 10% of TDI 60-kg adult 2 litres/day
Additional comments	The guideline value exceeds the lowest reported odour threshold for ethylbenzene in drinking-water (0.002 mg/litre).

Toxicological review
Ethylbenzene is readily absorbed by oral, inhalation or dermal routes. In humans, storage in fat has been reported. Ethylbenzene is almost completely converted to soluble metabolites, which are excreted rapidly in urine. The acute oral toxicity is low. No definite conclusions can be drawn from limited teratogenicity data. No data on reproduction, long-term toxicity or carcinogenicity are available. Ethylbenzene has shown no evidence of genotoxicity in *in vitro* or in *in vivo* systems.

History of guideline development
The 1958, 1963 and 1971 WHO *International Standards for Drinking-water* and the first edition of the *Guidelines for Drinking-water Quality*, published in 1984, did not refer to ethylbenzene. The 1993 Guidelines proposed a health-based guideline value of 0.3 mg/litre for ethylbenzene, noting that this value exceeds the lowest reported odour threshold for ethylbenzene in drinking-water (0.002 mg/litre).

Assessment date
The risk assessment was originally conducted in 1993. The Final Task Force Meeting in 2003 agreed that this risk assessment be brought forward to this edition of the *Guidelines for Drinking-water Quality*.

Principal reference
WHO (2003) *Ethylbenzene in drinking-water. Background document for preparation of WHO Guidelines for drinking-water quality*. Geneva, World Health Organization (WHO/SDE/WSH/03.04/26).

12.61 Fenitrothion
Fenitrothion (CAS No. 122-14-5) is mainly used in agriculture for controlling insects on rice, cereals, fruits, vegetables, stored grains and cotton and in forest areas. It is also used for the control of flies, mosquitos and cockroaches in public health programmes and/or indoor use. Fenitrothion is stable in water only in the absence of sunlight or microbial contamination. In soil, biodegradation is the primary route of degradation, although photolysis may also play a role. Fenitrothion residues detected in water were low (maximum 1.30 µg/litre) during the spruce budworm spray programme. Following the spraying of forests to control spruce budworm, water samples did not contain detectable amounts of fenitrothion; post-spray samples contained <0.01 µg/litre. Levels of fenitrothion residues in fruits, vegetables and cereal grains decline rapidly after treatment, with a half-life of 1–2 days. Intake of fenitrothion appears to be primarily (95%) from food.

On the basis of testing in an adequate range of studies *in vitro* and *in vivo*, JMPR concluded that fenitrothion is unlikely to be genotoxic. It also concluded that fenitrothion is unlikely to pose a carcinogenic risk to humans. In long-term studies of toxicity, inhibition of cholinesterase activity was the main toxicological finding in all species. A health-based value of 8 µg/litre can be calculated for fenitrothion on the basis of an ADI of 0.005 mg/kg of body weight, based on a NOAEL of 0.5 mg/kg of body weight per day for inhibition of brain and erythrocyte cholinesterase activity in a 2-year study of toxicity in rats and supported by a NOAEL of 0.57 mg/kg of body weight per day for inhibition of brain and erythrocyte cholinesterase activity in a 3-month study of ocular toxicity in rats and a NOAEL of 0.65 mg/kg of body weight per day for reduced food consumption and body weight gain in a study of reproductive toxicity in rats, and allocating 5% of the ADI to drinking-water. However, because fenitrothion occurs at concentrations well below those at which toxic effects are observed, it is not considered necessary to derive a guideline value.

History of guideline development
The 1958 and 1963 WHO *International Standards for Drinking-water* did not refer to fenitrothion, but the 1971 International Standards suggested that pesticide residues that may occur in community water supplies make only a minimal contribution to the total daily intake of pesticides for the population served. Fenitrothion was not evaluated in the first edition of the *Guidelines for Drinking-water Quality*, published in 1984, in the second edition, published in 1993, or in the addendum to the second edition, published in 1998.

Assessment date
The risk assessment was conducted in 2003.

Principal references
FAO/WHO (2001) *Pesticide residues in food – 2000 evaluations. Part II – Toxicological*. Geneva, World Health Organization, Joint FAO/WHO Meeting on Pesticide Residues (WHO/PCS/01.3).

WHO (2003) *Fenitrothion in drinking-water. Background document for preparation of WHO Guidelines for drinking-water quality*. Geneva, World Health Organization (WHO/SDE/WSH/03.04/95).

12.62 Fenoprop (2,4,5-TP; 2,4,5-trichlorophenoxy propionic acid)
The half-lives for degradation of chlorophenoxy herbicides, including fenoprop (CAS No. 93-72-1), in the environment are in the order of several days. Chlorophenoxy herbicides are not often found in food.

Guideline value	0.009 mg/litre
Occurrence	Chlorophenoxy herbicides not frequently found in drinking-water; when detected, concentrations are usually no greater than a few micrograms per litre
TDI	3 µg/kg of body weight, based on a NOAEL of 0.9 mg/kg of body weight for adverse effects on the liver in a study in which beagle dogs were administered fenoprop in the diet for 2 years, with an uncertainty factor of 300 (100 for inter- and intraspecies variation and 3 for limitations in the database)
Limit of detection	0.2 µg/litre by either packed or capillary column GC with ECD
Treatment achievability	No data found; 0.001 mg/litre should be achievable using GAC
Guideline derivation • allocation to water • weight • consumption	 10% of TDI 60-kg adult 2 litres/day

12. CHEMICAL FACT SHEETS

Toxicological review

Chlorophenoxy herbicides, as a group, have been classified in Group 2B by IARC. However, the available data from studies in exposed populations and animals do not permit assessment of the carcinogenic potential to humans of any specific chlorophenoxy herbicide. Therefore, drinking-water guidelines for these compounds are based on a threshold approach for other toxic effects. Effects observed in long-term studies with beagle dogs given fenoprop in the diet include mild degeneration and necrosis of hepatocytes and fibroblastic proliferation in one study and severe liver pathology in another study. In rats, increased kidney weight was observed in two long-term dietary studies.

History of guideline development

The 1958 and 1963 WHO *International Standards for Drinking-water* did not refer to chlorophenoxy herbicides, including fenoprop, but the 1971 International Standards suggested that pesticide residues that may occur in community water supplies make only a minimal contribution to the total daily intake of pesticides for the population served. Fenoprop was not evaluated in the first edition of the *Guidelines for Drinking-water Quality*, published in 1984, but the 1993 Guidelines established a health-based guideline value of 0.009 mg/litre for fenoprop.

Assessment date

The risk assessment was originally conducted in 1993. The Final Task Force Meeting in 2003 agreed that this risk assessment be brought forward to this edition of the *Guidelines for Drinking-water Quality*.

Principal reference

WHO (2003) *Chlorophenoxy herbicides (excluding 2,4-D and MCPA) in drinking-water. Background document for preparation of WHO Guidelines for drinking-water quality*. Geneva, World Health Organization (WHO/SDE/WSH/03.04/44).

12.63 Fluoride

Fluoride accounts for about 0.3 g/kg of the Earth's crust and exists in the form of fluorides in a number of minerals. The most important source of fluoride in drinking-water is naturally occurring. Inorganic fluoride-containing minerals are used widely in industry for a wide range of purposes, including aluminium production. Fluorides can be released to the environment from the phosphate-containing rock used to produce phosphate fertilizers; these phosphate deposits contain about 4% fluorine. Fluorosilicic acid, sodium hexafluorosilicate and sodium fluoride are used in municipal water fluoridation schemes. Daily exposure to fluoride depends mainly on the geographical area. In most circumstances, food seems to be the primary source of fluoride intake, with lesser contributions from drinking-water and from toothpaste. In areas with relatively high concentrations, particularly in groundwater, drinking-water becomes increasingly important as a source of fluoride. Intakes in areas where high-fluoride coal is used indoors may also be significant.

Guideline value	1.5 mg/litre
Occurrence	In groundwater, concentrations vary with the type of rock the water flows through but do not usually exceed 10 mg/litre; the highest natural level reported is 2800 mg/litre.
Basis of guideline derivation	Epidemiological evidence that concentrations above this value carry an increasing risk of dental fluorosis, and progressively higher concentrations lead to increasing risks of skeletal fluorosis. The value is higher than that recommended for artificial fluoridation of water supplies, which is usually 0.5–1.0 mg/litre.
Limit of detection	0.01 mg/litre by ion chromatography; 0.1 mg/litre by ion-selective electrodes or the SPADNS (sulfo phenyl azo dihydroxy naphthalene disulfonic acid) colorimetric method
Treatment achievability	1 mg/litre should be achievable using activated alumina (not a "conventional" treatment process, but relatively simple to install filters)
Additional comments	• A management guidance document on fluoride is available. • In setting national standards for fluoride or in evaluating the possible health consequences of exposure to fluoride, it is essential to consider the intake of water by the population of interest and the intake of fluoride from other sources (e.g., from food, air and dental preparations). Where the intakes from other sources are likely to approach, or be greater than, 6 mg/day, it would be appropriate to consider setting standards at a lower concentration than the guideline value. • In areas with high natural fluoride levels in drinking-water, the guideline value may be difficult to achieve, in some circumstances, with the treatment technology available.

Toxicological review

Many epidemiological studies of possible adverse effects of the long-term ingestion of fluoride via drinking-water have been carried out. These studies clearly establish that fluoride primarily produces effects on skeletal tissues (bones and teeth). In many regions with high fluoride exposure, fluoride is a significant cause of morbidity. Low concentrations provide protection against dental caries, especially in children. The pre- and post-eruptive protective effects of fluoride (involving the incorporation of fluoride into the matrix of the tooth during its formation, the development of shallower tooth grooves, which are consequently less prone to decay, and surface contact with enamel) increase with fluoride concentration up to about 2 mg/litre of drinking-water; the minimum concentration of fluoride in drinking-water required to produce it is approximately 0.5 mg/litre. However, fluoride can also have an adverse effect on tooth enamel and may give rise to mild dental fluorosis at drinking-water concentrations between 0.9 and 1.2 mg/litre, depending on intake. Elevated fluoride intakes can also have more serious effects on skeletal tissues. It has been concluded that there is a clear excess risk of adverse skeletal effects for a total intake of 14 mg/day and suggestive evidence of an increased risk of effects on the skeleton at total fluoride intakes above about 6 mg/day.

History of guideline development
The 1958 and 1963 WHO *International Standards for Drinking-water* referred to fluoride, stating that concentrations in drinking-water in excess of 1.0–1.5 mg of fluorine per litre may give rise to dental fluorosis in some children, and much higher concentrations may eventually result in skeletal damage in both children and adults. To prevent the development of dental caries in children, a number of communal water supplies are fluoridated to bring the fluorine concentration to 1.0 mg/litre. The 1971 International Standards recommended control limits for fluorides in drinking-water for various ranges of the annual average of maximum daily air temperatures; control limits ranged from 0.6–0.8 mg/litre for temperatures of 26.3–32.6 °C to 0.9–1.7 mg/litre for temperatures of 10–12 °C. In the first edition of the *Guidelines for Drinking-water Quality*, published in 1984, a guideline value of 1.5 mg/litre was established for fluoride, as mottling of teeth has been reported very occasionally at higher levels. It was also noted that local application of the guideline value must take into account climatic conditions and higher levels of water intake. The 1993 Guidelines concluded that there was no evidence to suggest that the guideline value of 1.5 mg/litre set in 1984 needed to be revised. It was also recognized that in areas with high natural fluoride levels, the guideline value may be difficult to achieve in some circumstances with the treatment technology available. It was also emphasized that in setting national standards for fluoride, it is particularly important to consider climatic conditions, volume of water intake and intake of fluoride from other sources.

Assessment date
The risk assessment was conducted in 2003.

Principal references
IPCS (2002) *Fluorides*. Geneva, World Health Organization, International Programme on Chemical Safety (Environmental Health Criteria 227).

WHO (2003) *Fluoride in drinking-water. Background document for preparation of WHO Guidelines for drinking-water quality*. Geneva, World Health Organization (WHO/SDE/WSH/03.04/96).

12.64 Formaldehyde
Formaldehyde occurs in industrial effluents and is emitted into air from plastic materials and resin glues. Formaldehyde in drinking-water results primarily from the oxidation of natural organic matter during ozonation and chlorination. It is also found in drinking-water as a result of release from polyacetal plastic fittings.

Guideline value	0.9 mg/litre
Occurrence	Concentrations of up to 30 µg/litre have been found in ozonated drinking-water.
TDI	150 µg/kg of body weight, derived from a NOAEL (for a variety of effects, including increased relative kidney weights in females and an increased incidence of renal papillary necrosis in both sexes) of 15 mg/kg of body weight per day in a 2-year study in rats, incorporating an uncertainty factor of 100 (for intra- and interspecies variation); no account was taken of potential carcinogenicity from the inhalation of formaldehyde from various indoor water uses, such as showering
Limit of detection	6.2 µg/litre by HPLC following derivatization with 2,4-dinitrophenylhydrazine and liquid–solid extraction
Treatment achievability	<0.03 mg/litre by process control/modification
Guideline derivation • allocation to water • weight • consumption	 20% of TDI 60-kg adult 2 litres/day

Toxicological review

Rats and mice exposed to formaldehyde by inhalation exhibited an increased incidence of carcinomas of the nasal cavity at doses that caused irritation of the nasal epithelium. Ingestion of formaldehyde in drinking-water for 2 years caused stomach irritation in rats. Papillomas of the stomach associated with severe tissue irritation were observed in one study. IARC has classified formaldehyde in Group 2A. The weight of evidence indicates that formaldehyde is not carcinogenic by the oral route.

History of guideline development

The 1958, 1963 and 1971 WHO *International Standards for Drinking-water* and the first edition of the *Guidelines for Drinking-water Quality*, published in 1984, did not refer to formaldehyde. The 1993 Guidelines established a health-based guideline value of 0.9 mg/litre for formaldehyde in drinking-water.

Assessment date

The risk assessment was originally conducted in 1993. The Final Task Force Meeting in 2003 agreed that this risk assessment be brought forward to this edition of the *Guidelines for Drinking-water Quality*.

Principal reference

WHO (2003) *Formaldehyde in drinking-water. Background document for preparation of WHO Guidelines for drinking-water quality*. Geneva, World Health Organization (WHO/SDE/WSH/03.04/48).

12.65 Glyphosate and AMPA

Glyphosate (CAS No. 1071-83-6) is a broad-spectrum herbicide used in both agriculture and forestry and for aquatic weed control. Microbial biodegradation of glyphosate occurs in soil, aquatic sediment and water, the major metabolite being aminomethylphosphonic acid (AMPA) (CAS No. 1066-51-9). Glyphosate is chemically stable in water and is not subject to photochemical degradation. The low mobility of glyphosate in soil indicates minimal potential for the contamination of groundwater. Glyphosate can, however, enter surface and subsurface waters after direct use near aquatic environments or by runoff or leaching from terrestrial applications.

Glyphosate and AMPA have similar toxicological profiles, and both are considered to exhibit low toxicity. A health-based value of 0.9 mg/litre can be derived based on the group ADI for AMPA alone or in combination with glyphosate of 0.3 mg/kg of body weight, based upon a NOAEL of 32 mg/kg of body weight per day, the highest dose tested, identified in a 26-month study of toxicity in rats fed technical-grade glyphosate and using an uncertainty factor of 100.

Because of their low toxicity, the health-based value derived for AMPA alone or in combination with glyphosate is orders of magnitude higher than concentrations of glyphosate or AMPA normally found in drinking-water. Under usual conditions, therefore, the presence of glyphosate and AMPA in drinking-water does not represent a hazard to human health. For this reason, the establishment of a guideline value for glyphosate and AMPA is not deemed necessary.

History of guideline development
The 1958 and 1963 WHO *International Standards for Drinking-water* did not refer to glyphosate, but the 1971 International Standards suggested that pesticide residues that may occur in community water supplies make only a minimal contribution to the total daily intake of pesticides for the population served. Glyphosate was not evaluated in the first two editions of the *Guidelines for Drinking-water Quality*, published in 1984 and 1993. In the addendum to these Guidelines, published in 1998, a health-based value of 5 mg/litre was derived for glyphosate using the ADI derived in the EHC monograph for glyphosate published in 1994. However, the health-based value is orders of magnitude higher than the concentrations normally found in drinking-water. Under usual conditions, therefore, the presence of glyphosate in drinking-water does not represent a hazard to human health, and it was not deemed necessary to establish a guideline value for glyphosate. It was noted that most AMPA, the major metabolite of glyphosate, found in water comes from sources other than glyphosate degradation.

Assessment date
The risk assessment was conducted in 2003.

Principal references

FAO/WHO (1998) *Pesticide residues in food – 1997 evaluations. Part II – Toxicological and environmental.* Geneva, World Health Organization, Joint FAO/WHO Meeting on Pesticide Residues (WHO/PCS/98.6).

IPCS (1994) *Glyphosate.* Geneva, World Health Organization, International Programme on Chemical Safety (Environmental Health Criteria 159).

WHO (2003) *Glyphosate and AMPA in drinking-water. Background document for preparation of WHO Guidelines for drinking-water quality.* Geneva, World Health Organization (WHO/SDE/WSH/03.04/97).

12.66 Halogenated acetonitriles (dichloroacetonitrile, dibromoacetonitrile, bromochloroacetonitrile, trichloroacetonitrile)

Halogenated acetonitriles are produced during water chlorination or chloramination from naturally occurring substances, including algae, fulvic acid and proteinaceous material. In general, increasing temperature and/or decreasing pH have been associated with increasing concentrations of halogenated acetonitriles. Ambient bromide levels appear to influence, to some degree, the speciation of halogenated acetonitrile compounds. Dichloroacetonitrile is by far the most predominant halogenated acetonitrile species detected in drinking-water.

Provisional guideline value for dichloroacetonitrile	0.02 mg/litre The guideline value for dichloroacetonitrile is provisional due to limitations of the toxicological database.
Guideline value for dibromoacetonitrile	0.07 mg/litre
Occurrence	Halogenated acetonitriles have been found in surface water and groundwater distribution systems at concentrations generally below 10 µg/litre and usually below 1 µg/litre.
TDIs	
Dichloroacetonitrile	2.7 µg/kg of body weight based on a LOAEL of 8 mg/kg of body weight per day for increased relative liver weight in male and female rats in a 90-day study, using an uncertainty factor of 3000 (taking into consideration intra- and interspecies variation, the short duration of the study, the use of a minimal LOAEL and database deficiencies)
Dibromoacetonitrile	11 µg/kg of body weight, based on a NOAEL of 11.3 mg/kg of body weight per day for decreased body weight in male F344 rats in a 90-day drinking-water study and an uncertainty factor of 1000 (accounting for inter- and intraspecies variation, subchronic to chronic extrapolation and database insufficiencies)
Limit of detection	0.03 µg/litre by GC with an ECD

Treatment achievability	Concentrations of individual halogenated acetonitriles can exceed 0.01 mg/litre, although levels of 0.002 mg/litre or less are more usual. Trichloroacetonitrile concentrations are likely to be much less than 0.001 mg/litre. Reduction of organic precursors will reduce their formation.
Guideline derivation • allocation to water • weight • consumption	 20% of TDI 60-kg adult 2 litres/day

Toxicological review

IARC has concluded that dichloro-, dibromo-, bromochloro- and trichloroacetonitrile are not classifiable as to their carcinogenicity in humans. Dichloroacetonitrile and bromochloroacetonitrile have been shown to be mutagenic in bacterial assays, whereas results for dibromoacetonitrile and trichloroacetonitrile were negative. All four of these halogenated acetonitriles induced sister chromatid exchange and DNA strand breaks and adducts in mammalian cells *in vitro* but were negative in the mouse micronucleus test.

The majority of reproductive and developmental toxicity studies of the halogenated acetonitriles were conducted using tricaprylin as a vehicle for gavage administration of the compound under study. As tricaprylin was subsequently demonstrated to be a developmental toxicant that potentiated the effects of trichloroacetonitrile and, presumably, other halogenated acetonitriles, results reported for developmental studies using tricaprylin as the gavage vehicle are likely to overestimate the developmental toxicity of these halogenated acetonitriles.

Dichloroacetonitrile

Dichloroacetonitrile induced decreases in body weight and increases in relative liver weight in short-term studies. Although developmental toxicity has been demonstrated, the studies used tricaprylin as the vehicle for gavage administration.

Dibromoacetonitrile

Dibromoacetonitrile is currently under test for chronic toxicity in mice and rats. None of the available reproductive or developmental studies were adequate to use in the quantitative dose–response assessment. The data gap may be particularly relevant since cyanide, a metabolite of dibromoacetonitrile, induces male reproductive system toxicity, and due to uncertainty regarding the significance of the testes effects observed in the 14-day National Toxicology Program (NTP) rat study.

Bromochloroacetonitrile

Available data are insufficient to serve as a basis for derivation of a guideline value for bromochloroacetonitrile.

Trichloroacetonitrile
Available data are also insufficient to serve as a basis for derivation of a guideline value for trichloroacetonitrile. The previous provisional guideline value of 1 µg/litre was based on a developmental toxicity study in which trichloroacetonitrile was administered by gavage in tricaprylin vehicle, and a recent re-evaluation judged this study to be unreliable in light of the finding in a more recent study that tricaprylin potentiates the developmental and teratogenic effects of halogenated acetonitriles and alters the spectrum of malformations in the fetuses of treated dams.

History of guideline development
The 1958, 1963 and 1971 WHO *International Standards for Drinking-water* and the first edition of the *Guidelines for Drinking-water Quality*, published in 1984, did not refer to halogenated acetonitriles. The 1993 Guidelines established provisional health-based guideline values of 0.09 mg/litre for dichloroacetonitrile, 0.1 mg/litre for dibromoacetonitrile and 0.001 mg/litre for trichloroacetonitrile. The guideline values were designated as provisional because of the limitations of the databases (i.e., lack of long-term toxicity and carcinogenicity bioassays). Available data were insufficient to serve as a basis for derivation of a guideline value for bromochloroacetonitrile.

Assessment date
The risk assessment was conducted in 2003.

Principal references
IPCS (2000) *Disinfectants and disinfectant by-products.* Geneva, World Health Organization, International Programme on Chemical Safety (Environmental Health Criteria 216).
WHO (2003) *Halogenated acetonitriles in drinking-water. Background document for preparation of WHO Guidelines for drinking-water quality.* Geneva, World Health Organization (WHO/SDE/WSH/03.04/98).

12.67 Hardness
Hardness in water is caused by dissolved calcium and, to a lesser extent, magnesium. It is usually expressed as the equivalent quantity of calcium carbonate.

Depending on pH and alkalinity, hardness above about 200 mg/litre can result in scale deposition, particularly on heating. Soft waters with a hardness of less than about 100 mg/litre have a low buffering capacity and may be more corrosive to water pipes.

A number of ecological and analytical epidemiological studies have shown a statistically significant inverse relationship between hardness of drinking-water and cardiovascular disease. There is some indication that very soft waters may have an adverse effect on mineral balance, but detailed studies were not available for evaluation.

No health-based guideline value is proposed for hardness. However, the degree of hardness in water may affect its acceptability to the consumer in terms of taste and scale deposition (see chapter 10).

History of guideline development
The 1958 and 1963 WHO *International Standards for Drinking-water* did not refer to hardness. The 1971 International Standards stated that the maximum permissible level of hardness in drinking-water was 10 mEq/litre (500 mg calcium carbonate/litre), based on the acceptability of water for domestic use. In the first edition of the *Guidelines for Drinking-water Quality*, published in 1984, it was concluded that there was no firm evidence that drinking hard water causes any adverse effects on human health and that no recommendation on the restriction of municipal water softening or on the maintenance of a minimum residual calcium or magnesium level was warranted. A guideline value of 500 mg/litre (as calcium carbonate) was established for hardness, based on taste and household use considerations. No health-based guideline value for hardness was proposed in the 1993 Guidelines, although hardness above approximately 200 mg/litre may cause scale deposition in the distribution system. Public acceptability of the degree of hardness may vary considerably from one community to another, depending on local conditions, and the taste of water with hardness in excess of 500 mg/litre is tolerated by consumers in some instances.

Assessment date
The risk assessment was originally conducted in 1993. The Final Task Force Meeting in 2003 agreed that this risk assessment be brought forward to this edition of the *Guidelines for Drinking-water Quality*.

Principal reference
WHO (2003) *Hardness in drinking-water. Background document for preparation of WHO Guidelines for drinking-water quality*. Geneva, World Health Organization (WHO/SDE/WSH/03.04/6).

12.68 Heptachlor and heptachlor epoxide

Heptachlor (CAS No. 76-44-8) is a broad-spectrum insecticide, the use of which has been banned or restricted in many countries. At present, the major use of heptachlor is for termite control by subsurface injection into soil. Heptachlor is quite persistent in soil, where it is mainly transformed to its epoxide. Heptachlor epoxide (CAS No. 1024-57-3) is very resistant to further degradation. Heptachlor and heptachlor epoxide bind to soil particles and migrate very slowly. Heptachlor and heptachlor epoxide have been found in drinking-water at levels of nanograms per litre. Diet is considered to represent the major source of exposure to heptachlor, although intake is decreasing.

Prolonged exposure to heptachlor has been associated with damage to the liver and central nervous system toxicity. In 1991, IARC reviewed the data on heptachlor and concluded that the evidence for carcinogenicity was sufficient in animals and inadequate in humans, classifying it in Group 2B. A health-based value of 0.03 µg/litre can be calculated for heptachlor and heptachlor epoxide on the basis of a PTDI of 0.1 µg/kg of body weight, based on a NOAEL for heptachlor of 0.025 mg/kg of body weight per day from two studies in the dog, taking into consideration inadequacies of the database and allocating 1% of the PTDI to drinking-water. However, because heptachlor and heptachlor epoxide occur at concentrations well below those at which toxic effects are observed, it is not considered necessary to derive a guideline value. It should also be noted that concentrations below 0.1 µg/litre are generally not achievable using conventional treatment technology.

History of guideline development
The 1958 and 1963 WHO *International Standards for Drinking-water* did not refer to heptachlor and heptachlor epoxide, but the 1971 International Standards suggested that pesticide residues that may occur in community water supplies make only a minimal contribution to the total daily intake of pesticides for the population served. In the first edition of the *Guidelines for Drinking-water Quality*, published in 1984, a health-based guideline value of 0.1 µg/litre was recommended for heptachlor and heptachlor epoxide, based on the ADI recommended by JMPR. It was noted that this guideline value was less than the value that would have been calculated by applying the multistage model at a projected incremental cancer risk of 1 per 100 000 per lifetime. The 1993 Guidelines established a health-based guideline value of 0.03 µg/litre for heptachlor, based on an ADI established by JMPR in 1991 and taking into consideration the fact that the main source of exposure seems to be food.

Assessment date
The risk assessment was conducted in 2003.

Principal references
FAO/WHO (1992) *Pesticide residues in food – 1991. Evaluations – 1991. Part II. Toxicology*. Geneva, World Health Organization, Joint FAO/WHO Meeting on Pesticide Residues (WHO/PCS/92.52).

FAO/WHO (1995) *Pesticide residues in food – 1994. Report of the Joint Meeting of the FAO Panel of Experts on Pesticide Residues in Food and the Environment and WHO Toxicological and Environmental Core Assessment Groups*. Rome, Food and Agriculture Organization of the United Nations (FAO Plant Production and Protection Paper 127).

WHO (2003) *Heptachlor and heptachlor epoxide in drinking-water. Background document for preparation of WHO Guidelines for drinking-water quality*. Geneva, World Health Organization (WHO/SDE/WSH/03.04/99).

12.69 Hexachlorobenzene (HCB)

The major agricultural application for HCB (CAS No. 118-74-1) was as a seed dressing for crops to prevent the growth of fungi, but its use is now uncommon. At present, it appears mainly as a by-product of several chemical processes or an impurity in some pesticides. HCB is distributed throughout the environment because it is mobile and resistant to degradation. It bioaccumulates in organisms because of its physicochemical properties and its slow elimination. HCB is commonly detected at low levels in food, and it is generally present at low concentrations in ambient air. It has been detected only infrequently, and at very low concentrations (below 0.1 μg/litre), in drinking-water supplies.

IARC has evaluated the evidence for the carcinogenicity of HCB in animals and humans and assigned it to Group 2B. HCB has been shown to induce tumours in three animal species and at a variety of sites. A health-based value of 1 μg/litre can be derived for HCB by applying the linearized multistage low-dose extrapolation model to liver tumours observed in female rats in a 2-year dietary study. Using an alternative (TD_{05}) approach, a health-based guidance value of 0.16 μg/kg body weight per day can be calculated, which corresponds to a drinking-water concentration of approximately 0.05 μg/litre, if one assumes a 1% allocation of the guidance value to drinking-water.

Because the health-based values derived from both of these approaches are considerably higher than the concentrations at which HCB is detected in drinking-water (i.e., sub-nanograms per litre), when it is detected, it is not considered necessary to establish a guideline value for HCB in drinking-water. Hexachlorobenzene is listed under the Stockholm Convention on Persistent Organic Pollutants.

History of guideline development

The 1958 and 1963 WHO *International Standards for Drinking-water* did not refer to HCB, but the 1971 International Standards suggested that pesticide residues that may occur in community water supplies make only a minimal contribution to the total daily intake of pesticides for the population served. In the first edition of the *Guidelines for Drinking-water Quality*, published in 1984, a health-based guideline value of 0.01 μg/litre was recommended for HCB, derived from the linear multistage extrapolation model for a cancer risk of less than 1 in 100 000 for a lifetime of exposure; it was noted that the mathematical model used involved considerable uncertainty. The 1993 Guidelines calculated a guideline value of 1 μg/litre for HCB in drinking-water, corresponding to an upper-bound excess lifetime cancer risk of 10^{-5}.

Assessment date

The risk assessment was conducted in 2003.

Principal references
IPCS (1997) *Hexachlorobenzene*. Geneva, World Health Organization, International Programme on Chemical Safety (Environmental Health Criteria 195).
WHO (2003) *Hexachlorobenzene in drinking-water. Background document for preparation of WHO Guidelines for drinking-water quality*. Geneva, World Health Organization (WHO/SDE/WSH/03.04/100).

12.70 Hexachlorobutadiene (HCBD)

HCBD is used as a solvent in chlorine gas production, a pesticide, an intermediate in the manufacture of rubber compounds and a lubricant. Concentrations of up to 6 µg/litre have been reported in the effluents from chemical manufacturing plants. It is also found in air and food.

Guideline value	0.0006 mg/litre (0.6 µg/litre)
Occurrence	Has been detected in surface water at concentrations of a few micrograms per litre and in drinking-water at concentrations below 0.5 µg/litre
TDI	0.2 µg/kg of body weight, based on a NOAEL of 0.2 mg/kg of body weight per day for renal toxicity in a 2-year feeding study in rats, using an uncertainty factor of 1000 (100 for inter- and intraspecies variation and 10 for limited evidence of carcinogenicity and genotoxicity of some metabolites)
Limit of detection	0.01 µg/litre by GC/MS; 0.18 µg/litre by GC with ECD
Treatment achievability	0.001 mg/litre should be achievable using GAC
Guideline derivation • allocation to water • weight • consumption	10% of TDI 60-kg adult 2 litres/day
Additional comments	The practical quantification level for HCBD is of the order of 2 µg/litre, but concentrations in drinking-water can be controlled by specifying the HCBD content of products coming into contact with it.

Toxicological review
HCBD is easily absorbed and metabolized via conjugation with glutathione. This conjugate can be further metabolized to a nephrotoxic derivative. Kidney tumours were observed in a long-term oral study in rats. HCBD has not been shown to be carcinogenic by other routes of exposure. IARC has placed HCBD in Group 3. Positive and negative results for HCBD have been obtained in bacterial assays for point mutation; however, several metabolites have given positive results.

History of guideline development
The 1958, 1963 and 1971 WHO *International Standards for Drinking-water* and the first edition of the *Guidelines for Drinking-water Quality*, published in 1984, did not

refer to HCBD. The 1993 Guidelines derived a health-based guideline value of 0.0006 mg/litre for HCBD, noting that although a practical quantification level for HCBD is of the order of 0.002 mg/litre, concentrations in drinking-water can be controlled by specifying the HCBD content of products coming into contact with it.

Assessment date
The risk assessment was conducted in in 2003.

Principal references
IPCS (1994) *Hexachlorobutadiene*. Geneva, World Health Organization, International Programme on Chemical Safety (Environmental Health Criteria 156).
WHO (2003) *Hexachlorobutadiene in drinking-water. Background document for preparation of WHO Guidelines for drinking-water quality*. Geneva, World Health Organization (WHO/SDE/WSH/03.04/101).

12.71 Hydrogen sulfide

Hydrogen sulfide is a gas with an offensive "rotten eggs" odour that is detectable at very low concentrations, below $0.8\,\mu g/m^3$ in air. It is formed when sulfides are hydrolysed in water. However, the level of hydrogen sulfide found in drinking-water will usually be low, because sulfides are readily oxidized in well aerated water.

The acute toxicity to humans of hydrogen sulfide following inhalation of the gas is high; eye irritation can be observed at concentrations of 15–30 mg/m^3. Although oral toxicity data are lacking, it is unlikely that a person could consume a harmful dose of hydrogen sulfide from drinking-water. Consequently, no guideline value is proposed. However, hydrogen sulfide should not be detectable in drinking-water by taste or odour (see chapter 10).

History of guideline development
The 1958, 1963 and 1971 WHO *International Standards for Drinking-water* did not refer to hydrogen sulfide. In the first edition of the *Guidelines for Drinking-water Quality*, published in 1984, it was recommended that hydrogen sulfide should not be detectable by the consumer, based on aesthetic considerations. A guideline value was not needed, since any contamination can be easily detected by the consumer. The 1993 Guidelines did not propose a health-based guideline value, as oral toxicity data are lacking; nevertheless, it is unlikely that a person could consume a harmful dose of hydrogen sulfide from drinking-water. The taste and odour thresholds of hydrogen sulfide in water are estimated to be between 0.05 and 0.1 mg/litre.

Assessment date
The risk assessment was originally conducted in 1993. The Final Task Force Meeting in 2003 agreed that this risk assessment be brought forward to this edition of the *Guidelines for Drinking-water Quality*.

Principal reference
WHO (2003) *Hydrogen sulfide in drinking-water. Background document for preparation of WHO Guidelines for drinking-water quality.* Geneva, World Health Organization (WHO/SDE/WSH/03.04/7).

12.72 Inorganic tin

Tin is used principally in the production of coatings used in the food industry. Food, particularly canned food, therefore represents the major route of human exposure to tin. For the general population, drinking-water is not a significant source of tin, and levels in drinking-water greater than 1–2 µg/litre are exceptional. However, there is increasing use of tin in solder, which may be used in domestic plumbing, and tin has been proposed for use as a corrosion inhibitor.

Tin and inorganic tin compounds are poorly absorbed from the gastrointestinal tract, do not accumulate in tissues and are rapidly excreted, primarily in the faeces.

No increased incidence of tumours was observed in long-term carcinogenicity studies conducted in mice and rats fed stannous chloride. Tin has not been shown to be teratogenic or fetotoxic in mice, rats or hamsters. In rats, the NOAEL in a long-term feeding study was 20 mg/kg of body weight per day.

The main adverse effect on humans of excessive levels of tin in canned beverages (above 150 mg/kg) or other canned foods (above 250 m/kg) has been acute gastric irritation. There is no evidence of adverse effects in humans associated with chronic exposure to tin.

In 1989, JECFA established a PTWI of 14 mg/kg of body weight from a TDI of 2 mg/kg of body weight on the basis that the problem with tin is associated with acute gastrointestinal irritancy, the threshold for which is about 200 mg/kg in food. This was reaffirmed by JECFA in 2000. In view of its low toxicity, the presence of tin in drinking-water does not, therefore, represent a hazard to human health. For this reason, the establishment of a guideline value for inorganic tin is not deemed necessary.

History of guideline development
The 1958 and 1963 WHO *International Standards for Drinking-water* did not refer to inorganic tin. The 1971 International Standards stated that tin should be controlled in drinking-water, but that insufficient information was available to enable a tentative limit to be established. In the first edition of the *Guidelines for Drinking-water Quality*, published in 1984, it was concluded that no action was required for tin. The establishment of a guideline value for inorganic tin was not deemed necessary in the 1993 Guidelines, as, because of the low toxicity of inorganic tin, a tentative guideline value could be derived 3 orders of magnitude higher than the normal tin concentration in drinking-water. Therefore, the presence of tin in drinking-water does not represent a hazard to human health.

Assessment date
The risk assessment was conducted in 2003.

Principal reference
WHO (2003) *Inorganic tin in drinking-water. Background document for preparation of WHO Guidelines for drinking-water quality.* Geneva, World Health Organization (WHO/SDE/WSH/03.04/115).

12.73 Iodine
Iodine occurs naturally in water in the form of iodide. Traces of iodine are produced by oxidation of iodide during water treatment. Iodine is occasionally used for water disinfection in the field or in emergency situations.

Iodine is an essential element for the synthesis of thyroid hormones. Estimates of the dietary requirement for adult humans range from 80 to 150 µg/day; in many parts of the world, there are dietary deficiencies in iodine. In 1988, JECFA set a PMTDI for iodine of 1 mg/day (17 µg/kg of body weight per day) from all sources, based primarily on data on the effects of iodide. However, recent data from studies in rats indicate that the effects of iodine in drinking-water on thyroid hormone concentrations in the blood differ from those of iodide.

Available data therefore suggest that derivation of a guideline value for iodine on the basis of information on the effects of iodide is inappropriate, and there are few relevant data on the effects of iodine. Because iodine is not recommended for long-term disinfection, lifetime exposure to iodine concentrations such as might occur from water disinfection is unlikely. For these reasons, a guideline value for iodine has not been established at this time. There is, however, a need for guidance concerning the use of iodine as a disinfectant in emergency situations and for travellers.

History of guideline development
The 1958, 1963 and 1971 WHO *International Standards for Drinking-water* and the first edition of the *Guidelines for Drinking-water Quality*, published in 1984, did not refer to iodine. The 1993 Guidelines did not establish a guideline value for iodine because available data suggest that derivation of a guideline value for iodine on the basis of information on the effects of iodide is inappropriate and there are few relevant data on the effects of iodine; also, because iodine is not recommended for long-term disinfection, lifetime exposure to iodine concentrations such as might occur from water disinfection is unlikely.

Assessment date
The risk assessment was originally conducted in 1993. The Final Task Force Meeting in 2003 agreed that this risk assessment be brought forward to this edition of the *Guidelines for Drinking-water Quality*.

Principal reference
WHO (2003) *Iodine in drinking-water. Background document for preparation of WHO Guidelines for drinking-water quality.* Geneva, World Health Organization (WHO/SDE/WSH/03.04/46).

12.74 Iron

Iron is one of the most abundant metals in the Earth's crust. It is found in natural fresh waters at levels ranging from 0.5 to 50 mg/litre. Iron may also be present in drinking-water as a result of the use of iron coagulants or the corrosion of steel and cast iron pipes during water distribution.

Iron is an essential element in human nutrition. Estimates of the minimum daily requirement for iron depend on age, sex, physiological status and iron bioavailability and range from about 10 to 50 mg/day.

As a precaution against storage in the body of excessive iron, in 1983 JECFA established a PMTDI of 0.8 mg/kg of body weight, which applies to iron from all sources except for iron oxides used as colouring agents and iron supplements taken during pregnancy and lactation or for specific clinical requirements. An allocation of 10% of this PMTDI to drinking-water gives a value of about 2 mg/litre, which does not present a hazard to health. The taste and appearance of drinking-water will usually be affected below this level (see chapter 10).

No guideline value for iron in drinking-water is proposed.

History of guideline development
The 1958 WHO *International Standards for Drinking-water* suggested that concentrations of iron greater than 1.0 mg/litre would markedly impair the potability of the water. The 1963 and 1971 International Standards retained this value as a maximum allowable or permissible concentration. In the first edition of the *Guidelines for Drinking-water Quality*, published in 1984, a guideline value of 0.3 mg/litre was established, as a compromise between iron's use in water treatment and aesthetic considerations. No health-based guideline value for iron in drinking-water was proposed in the 1993 Guidelines, but it was mentioned that a value of about 2 mg/litre can be derived from the PMTDI established in 1983 by JECFA as a precaution against storage in the body of excessive iron. Iron stains laundry and plumbing fixtures at levels above 0.3 mg/litre; there is usually no noticeable taste at iron concentrations below 0.3 mg/litre, and concentrations of 1–3 mg/litre can be acceptable for people drinking anaerobic well water.

Assessment date
The risk assessment was originally conducted in 1993. The Final Task Force Meeting in 2003 agreed that this risk assessment be brought forward to this edition of the *Guidelines for Drinking-water Quality*.

Principal reference
WHO (2003) *Iron in drinking-water. Background document for preparation of WHO Guidelines for drinking-water quality*. Geneva, World Health Organization (WHO/SDE/WSH/03.04/8).

12.75 Isoproturon

Isoproturon (CAS No. 34123-59-6) is a selective, systemic herbicide used in the control of annual grasses and broad-leaved weeds in cereals. It can be photodegraded, hydrolysed and biodegraded and persists for periods ranging from days to weeks. It is mobile in soil. There is evidence that exposure to this compound through food is low.

Guideline value	0.009 mg/litre
Occurrence	Has been detected in surface water and groundwater, usually at concentrations below 0.1 µg/litre; levels above 0.1 µg/litre have occasionally been detected in drinking-water
TDI	3 µg/kg of body weight based on a NOAEL of approximately 3 mg/kg of body weight in a 90-day study in dogs and a 2-year Feeding study in rats, with an uncertainty factor of 1000 (100 for inter- and intraspecies variation and 10 for evidence of non-genotoxic carcinogenicity in rats)
Limit of detection	10–100 ng/litre by reverse-phase HPLC followed by UV or electrochemical detection
Treatment achievability	0.1 µg/litre should be achievable using ozonation
Guideline derivation • allocation to water • weight • consumption	 10% of TDI 60-kg adult 2 litres/day

Toxicological review

Isoproturon is of low acute toxicity and low to moderate toxicity following short- and long-term exposures. It does not possess significant genotoxic activity, but it causes marked enzyme induction and liver enlargement. Isoproturon caused an increase in hepatocellular tumours in male and female rats, but this was apparent only at doses that also caused liver toxicity. Isoproturon appears to be a tumour promoter rather than a complete carcinogen.

History of guideline development

The 1958 and 1963 WHO *International Standards for Drinking-water* did not refer to isoproturon, but the 1971 International Standards suggested that pesticide residues that may occur in community water supplies make only a minimal contribution to the total daily intake of pesticides for the population served. Isoproturon was not evaluated in the first edition of the *Guidelines for Drinking-water Quality*, published in

1984, but the 1993 Guidelines calculated a health-based guideline value of 0.009 mg/litre for isoproturon in drinking-water.

Assessment date
The risk assessment was originally conducted in 1993. The Final Task Force Meeting in 2003 agreed that this risk assessment be brought forward to this edition of the *Guidelines for Drinking-water Quality*.

Principal reference
WHO (2003) *Isoproturon in drinking-water. Background document for preparation of WHO Guidelines for drinking-water quality*. Geneva, World Health Organization (WHO/SDE/WSH/03.04/37).

12.76 Lead

Lead is used principally in the production of lead-acid batteries, solder and alloys. The organolead compounds tetraethyl and tetramethyl lead have also been used extensively as antiknock and lubricating agents in petrol, although their use for these purposes in many countries is being phased out. Owing to the decreasing use of lead-containing additives in petrol and of lead-containing solder in the food processing industry, concentrations in air and food are declining, and intake from drinking-water constitutes a greater proportion of total intake. Lead is rarely present in tap water as a result of its dissolution from natural sources; rather, its presence is primarily from household plumbing systems containing lead in pipes, solder, fittings or the service connections to homes. The amount of lead dissolved from the plumbing system depends on several factors, including pH, temperature, water hardness and standing time of the water, with soft, acidic water being the most plumbosolvent.

Guideline value	0.01 mg/litre
Occurrence	Concentrations in drinking-water are generally below 5 µg/litre, although much higher concentrations (above 100 µg/litre) have been measured where lead fittings are present.
PTWI	25 µg/kg of body weight (equivalent to 3.5 µg/kg of body weight per day) for infants and children on the basis that lead is a cumulative poison and that there should be no accumulation of body burden of lead
Limit of detection	1 µg/litre by AAS
Treatment achievability	Not a raw water contaminant; treatment not applicable
Guideline derivation • allocation to water • weight • consumption	50% of PTWI 5-kg infant 0.75 litre/day

Additional comments	• As infants are considered to be the most sensitive subgroup of the population, this guideline value will also be protective for other age groups. • Lead is exceptional in that most lead in drinking-water arises from plumbing in buildings and the remedy consists principally of removing plumbing and fittings containing lead. This requires much time and money, and it is recognized that not all water will meet the guideline immediately. Meanwhile, all other practical measures to reduce total exposure to lead, including corrosion control, should be implemented.

Toxicological review

Placental transfer of lead occurs in humans as early as the 12th week of gestation and continues throughout development. Young children absorb 4–5 times as much lead as adults, and the biological half-life may be considerably longer in children than in adults. Lead is a general toxicant that accumulates in the skeleton. Infants, children up to 6 years of age and pregnant women are most susceptible to its adverse health effects. Inhibition of the activity of d-aminolaevulinic dehydratase (porphobilinogen synthase; one of the major enzymes involved in the biosynthesis of haem) in children has been observed at blood lead levels as low as $5\,\mu g/dl$, although adverse effects are not associated with its inhibition at this level. Lead also interferes with calcium metabolism, both directly and by interfering with vitamin D metabolism. These effects have been observed in children at blood lead levels ranging from 12 to $120\,\mu g/dl$, with no evidence of a threshold. Lead is toxic to both the central and peripheral nervous systems, inducing subencephalopathic neurological and behavioural effects. There is electrophysiological evidence of effects on the nervous system in children with blood lead levels well below $30\,\mu g/dl$. The balance of evidence from cross-sectional epidemiological studies indicates that there are statistically significant associations between blood lead levels of $30\,\mu g/dl$ and more and intelligence quotient deficits of about four points in children. Results from prospective (longitudinal) epidemiological studies suggest that prenatal exposure to lead may have early effects on mental development that do not persist to the age of 4 years. Research on primates has supported the results of the epidemiological studies, in that significant behavioural and cognitive effects have been observed following postnatal exposure resulting in blood lead levels ranging from 11 to $33\,\mu g/dl$. Renal tumours have been induced in experimental animals exposed to high concentrations of lead compounds in the diet, and IARC has classified lead and inorganic lead compounds in Group 2B (possible human carcinogen). However, there is evidence from studies in humans that adverse neurotoxic effects other than cancer may occur at very low concentrations of lead and that a guideline value derived on this basis would also be protective for carcinogenic effects.

History of guideline development

The 1958 WHO *International Standards for Drinking-water* recommended a maximum allowable concentration of 0.1 mg/litre for lead, based on health concerns.

This value was lowered to 0.05 mg/litre in the 1963 International Standards. The tentative upper concentration limit was increased to 0.1 mg/litre in the 1971 International Standards, because this level was accepted in many countries and the water had been consumed for many years without apparent ill effects, and it was difficult to reach a lower level in countries where lead pipes were used. In the first edition of the *Guidelines for Drinking-water Quality*, published in 1984, a health-based guideline value of 0.05 mg/litre was recommended. The 1993 Guidelines proposed a health-based guideline value of 0.01 mg/litre, using the PTWI established by JECFA for infants and children, on the basis that lead is a cumulative poison and that there should be no accumulation of body burden of lead. As infants are considered to be the most sensitive subgroup of the population, this guideline value would also be protective for other age groups. The Guidelines also recognized that lead is exceptional, in that most lead in drinking-water arises from plumbing, and the remedy consists principally of removing plumbing and fittings containing lead. As this requires much time and money, it is recognized that not all water will meet the guideline immediately. Meanwhile, all other practical measures to reduce total exposure to lead, including corrosion control, should be implemented. JECFA has reassessed lead and confirmed the previously derived PTWI.

Assessment date
The risk assessment was originally conducted in 1993. The Final Task Force Meeting in 2003 agreed that this risk assessment be brought forward to this edition of the *Guidelines for Drinking-water Quality*.

Principal reference
WHO (2003) *Lead in drinking-water. Background document for preparation of WHO Guidelines for drinking-water quality*. Geneva, World Health Organization (WHO/SDE/WSH/03.04/9).

12.77 Lindane
Lindane (γ-hexachlorocyclohexane, γ-HCH) (CAS No. 58-89-9) is used as an insecticide on fruit and vegetable crops, for seed treatment and in forestry. It is also used as a therapeutic pesticide in humans and animals. Several countries have restricted the use of lindane. Lindane can be degraded in soil and rarely leaches to groundwater. In surface waters, it can be removed by evaporation. Exposure of humans occurs mainly via food, but this is decreasing. There may also be exposure from its use in public health and as a wood preservative.

Guideline value	0.002 mg/litre
Occurrence	Has been detected in both surface water and groundwater, usually at concentrations below 0.1 µg/litre, although concentrations as high as 12 µg/litre have been measured in wastewater-contaminated rivers
ADI	0.005 mg/kg of body weight on the basis of a NOAEL of 0.47 mg/kg of body weight per day in a 2-year toxicity/carcinogenicity study in rats in which an increased incidence of periacinar hepatocellular hypertrophy, increased liver and spleen weights and increased mortality occurred at higher doses, using an uncertainty factor of 100
Limit of detection	0.01 µg/litre using GC
Treatment achievability	0.1 µg/litre should be achievable using GAC
Guideline derivation • allocation to water • weight • consumption	1% of ADI 60-kg adult 2 litres/day

Toxicological review

Lindane was toxic to the kidney and liver after administration orally, dermally or by inhalation in short-term and long-term studies of toxicity and reproductive toxicity in rats. The renal toxicity of lindane was specific to male rats and was considered not to be relevant to human risk assessment, since it is a consequence of accumulation of α_{2u}-globulin, a protein that is not found in humans. Hepatocellular hypertrophy was observed in a number of studies in mice, rats and rabbits and was reversed only partially after recovery periods of up to 6 weeks. Lindane did not induce a carcinogenic response in rats or dogs, but it caused an increased incidence of adenomas and carcinomas of the liver in agouti and pseudoagouti mice, but not in black or any other strains of mice, in a study of the role of genetic background in the latency and incidence of tumorigenesis. JMPR has concluded that there was no evidence of genotoxicity. In the absence of genotoxicity and on the basis of the weight of the evidence from the studies of carcinogenicity, JMPR has concluded that lindane is not likely to pose a carcinogenic risk to humans. Further, in an epidemiological study designed to assess the potential association between breast cancer and exposure to chlorinated pesticides, no correlation with lindane was found.

History of guideline development

The 1958 and 1963 WHO *International Standards for Drinking-water* did not refer to lindane, but the 1971 International Standards suggested that pesticide residues that may occur in community water supplies make only a minimal contribution to the total daily intake of pesticides for the population served. In the first edition of the *Guidelines for Drinking-water Quality*, published in 1984, a health-based guideline value of 3 µg/litre was recommended for lindane, based on the ADI recommended by JMPR. The 1993 Guidelines established a health-based guideline value of 2 µg/litre for lindane in drinking-water, on the basis of a study used to establish an ADI by JMPR in 1989 but using a compound intake estimate considered to be more appropriate in light of additional data and recognizing that there may be substantial exposure to lindane from its use in public health and as a wood preservative.

Assessment date
The risk assessment was conducted in 2003.

Principal references
FAO/WHO (2002) *Pesticide residues in food – 2002*. Rome, Food and Agriculture Organization of the United Nations, Joint FAO/WHO Meeting on Pesticide Residues (FAO Plant Production and Protection Paper 172).
WHO (2003) *Lindane in drinking-water. Background document for preparation of WHO Guidelines for drinking-water quality*. Geneva, World Health Organization (WHO/SDE/WSH/03.04/102).

12.78 Malathion

Malathion (CAS No. 121-75-5) is commonly used to control mosquitos and a variety of insects that attack fruits, vegetables, landscaping plants and shrubs. It can also be found in other pesticide products used indoors, on pets to control ticks and insects and to control human head and body lice. Under least favourable conditions (i.e., low pH and little organic content), malathion may persist in water with a half-life of months or even years. However, under most conditions, the half-life appears to be roughly 7–14 days. Malathion has been detected in surface water and drinking-water at concentrations below 2 µg/litre.

Malathion inhibits cholinesterase activity in mice, rats and human volunteers. It increased the incidence of liver adenomas in mice when administered in the diet. Most of the evidence indicates that malathion is not genotoxic, although some studies indicate that it can produce chromosomal aberrations and sister chromatid exchange *in vitro*. JMPR has concluded that malathion is not genotoxic.

A health-based value of 0.9 mg/litre can be calculated for malathion based on an allocation of 10% of the JMPR ADI – based on a NOAEL of 29 mg/kg of body weight per day in a 2-year study of toxicity and carcinogenicity in rats, using an uncertainty factor of 100 and supported by a NOAEL of 25 mg/kg of body weight per day in a developmental toxicity study in rabbits – to drinking-water. However, intake of malathion from all sources is generally low and well below the ADI. As the chemical occurs in drinking-water at concentrations much lower than the health-based value, the presence of malathion in drinking-water under usual conditions is unlikely to represent a hazard to human health. For this reason, it is considered unnecessary to derive a guideline value for malathion in drinking-water.

History of guideline development
The 1958 and 1963 WHO *International Standards for Drinking-water* did not refer to malathion, but the 1971 International Standards suggested that pesticide residues that may occur in community water supplies make only a minimal contribution to the total daily intake of pesticides for the population served. Malathion was not evaluated in the first edition of the *Guidelines for Drinking-water Quality*, published in 1984, in the second edition, published in 1993, or in the addendum to the second edition, published in 1998.

Assessment date
The risk assessment was conducted in 2003.

Principal references
FAO/WHO (1998) *Pesticide residues in food – 1997 evaluations. Part II – Toxicological and environmental*. Geneva, World Health Organization, Joint FAO/WHO Meeting on Pesticide Residues (WHO/PCS/98.6).

WHO (2003) *Malathion in drinking-water. Background document for preparation of WHO Guidelines for drinking-water quality*. Geneva, World Health Organization (WHO/SDE/WSH/03.04/103).

12.79 Manganese
Manganese is one of the most abundant metals in the Earth's crust, usually occurring with iron. It is used principally in the manufacture of iron and steel alloys, as an oxidant for cleaning, bleaching and disinfection as potassium permanganate and as an ingredient in various products. More recently, it has been used in an organic compound, MMT, as an octane enhancer in petrol in North America. Manganese greensands are used in some locations for potable water treatment. Manganese is an essential element for humans and other animals and occurs naturally in many food sources. The most important oxidative states for the environment and biology are Mn^{2+}, Mn^{4+} and Mn^{7+}. Manganese is naturally occurring in many surface water and groundwater sources, particularly in anaerobic or low oxidation conditions, and this is the most important source for drinking-water. The greatest exposure to manganese is usually from food.

Guideline value	0.4 mg/litre
Occurrence	Levels in fresh water typically range from 1 to 200 µg/litre, although levels as high as 10 mg/litre in acidic groundwater have been reported; higher levels in aerobic waters usually associated with industrial pollution
TDI	0.06 mg/kg of body weight, based on the upper range value of manganese intake of 11 mg/day, identified using dietary surveys, at which there are no observed adverse effects (i.e., considered a NOAEL), using an uncertainty factor of 3 to take into consideration the possible increased bioavailability of manganese from water
Limit of detection	0.01 µg/litre by AAS; 0.05 µg/litre by ICP/MS; 0.5 µg/litre by ICP/optical emission spectroscopy; 1 µg/litre by EAAS; 10 µg/litre by FAAS
Treatment achievability	0.05 mg/litre should be achievable using oxidation and filtration
Guideline derivation • allocation to water • weight • consumption	 20% of TDI (because manganese is essential trace element) 60-kg adult 2 litres/day
Additional comments	The presence of manganese in drinking-water will be objectionable to consumers if it is deposited in water mains and causes water discoloration. Concentrations below 0.05–0.1 mg/litre are usually acceptable to consumers but may sometimes still give rise to the deposition of black deposits in water mains over an extended period; this may vary with local circumstances.

Toxicological review

Manganese is an essential element for humans and other animals. Adverse effects can result from both deficiency and overexposure. Manganese is known to cause neurological effects following inhalation exposure, particularly in occupational settings, and there have been epidemiological studies that report adverse neurological effects following extended exposure to very high levels in drinking-water. However, there are a number of significant potential confounding factors in these studies, and a number of other studies have failed to observe adverse effects following exposure through drinking-water. Animal data, especially rodent data, are not desirable for human risk assessment because the physiological requirements for manganese vary among different species. Further, rodents are of limited value in assessing the neurobehavioural effects, because the neurological effects (e.g., tremor, gait disorders) seen in primates are often preceded or accompanied by psychological symptoms (e.g., irritability, emotional lability), which are not apparent in rodents. The only primate study is of limited use in a quantitative risk assessment because only one dose group was studied in a small number of animals and the manganese content in the basal diet was not provided.

History of guideline development

The 1958 WHO *International Standards for Drinking-water* suggested that concentrations of manganese greater than 0.5 mg/litre would markedly impair the potability of the water. The 1963 and 1971 International Standards retained this value as a maximum allowable or permissible concentration. In the first edition of the *Guidelines for Drinking-water Quality*, published in 1984, a guideline value of 0.1 mg/litre was established for manganese, based on its staining properties. The 1993 Guidelines concluded that although no single study is suitable for use in calculating a guideline value, the weight of evidence from actual daily intake and toxicity studies in laboratory animals given manganese in drinking-water supports the view that a provisional health-based guideline value of 0.5 mg/litre should be adequate to protect public health. It was also noted that concentrations below 0.1 mg/litre are usually acceptable to consumers, although this may vary with local circumstances.

Assessment date
The risk assessment was conducted in 2003.

Principal references
IPCS (1999) *Manganese and its compounds*. Geneva, World Health Organization, International Programme on Chemical Safety (Concise International Chemical Assessment Document 12).

WHO (2003) *Manganese in drinking-water. Background document for preparation of WHO Guidelines for drinking-water quality*. Geneva, World Health Organization (WHO/SDE/WSH/03.04/104).

12.80 MCPA [4-(2-methyl-4-chlorophenoxy)acetic acid]

MCPA (CAS No. 94-74-6) is a chlorophenoxy post-emergence herbicide that is very soluble, is highly mobile and can leach from the soil. It is metabolized by bacteria and can be photochemically degraded. MCPA has only limited persistence in water.

Guideline value	0.002 mg/litre
Occurrence	Not frequently detected in drinking-water; has been measured in surface water and groundwater at concentrations below 0.54 and 5.5 µg/litre, respectively
TDI	0.5 µg/kg of body weight, based on a NOAEL of 0.15 mg/kg of body weight for renal and liver toxicity observed at higher dose levels in a 1-year feeding study in dogs, with an uncertainty factor of 300 (100 for inter- and intraspecies variation and 3 for inadequacies in the database)
Limit of detection	0.01 µg/litre by GC/MS and by GC with ECD
Treatment achievability	0.1 µg/litre should be achievable using GAC or ozonation
Guideline derivation • allocation to water • weight • consumption	10% of TDI 60-kg adult 2 litres/day

Toxicological review

There are only limited and inconclusive data on the genotoxicity of MCPA. IARC evaluated MCPA in 1983 and concluded that the available data on humans and experimental animals were inadequate for an evaluation of carcinogenicity. Further evaluations by IARC on chlorophenoxy herbicides in 1986 and 1987 concluded that evidence for their carcinogenicity was limited in humans and inadequate in animals (Group 2B). Recent carcinogenicity studies on rats and mice did not indicate that MCPA was carcinogenic. No adequate epidemiological data on exposure to MCPA alone are available.

History of guideline development

The 1958 and 1963 WHO *International Standards for Drinking-water* did not refer to MCPA, but the 1971 International Standards suggested that pesticide residues that may occur in community water supplies make only a minimal contribution to the total daily intake of pesticides for the population served. MCPA was not evaluated in the first edition of the *Guidelines for Drinking-water Quality*, published in 1984, but the 1993 Guidelines established a health-based guideline value of 0.002 mg/litre for MCPA in drinking-water.

Assessment date

The risk assessment was originally conducted in 1993. The Final Task Force Meeting in 2003 agreed that this risk assessment be brought forward to this edition of the *Guidelines for Drinking-water Quality*.

Principal reference

WHO (2003) *MCPA in drinking-water. Background document for preparation of WHO Guidelines for drinking-water quality*. Geneva, World Health Organization (WHO/SDE/WSH/03.04/38).

12.81 Mecoprop (MCPP; [2(2-methyl-chlorophenoxy) propionic acid])

The half-lives for degradation of chlorophenoxy herbicides, including mecoprop (CAS No. 93-65-2; 7085-19-0 racemic mixture), in the environment are in the order of several days. Chlorophenoxy herbicides are not often found in food.

12. CHEMICAL FACT SHEETS

Guideline value	0.01 mg/litre
Occurrence	Chlorophenoxy herbicides not frequently found in drinking-water; when detected, concentrations are usually no greater than a few micrograms per litre
TDI	3.33 µg/kg of body weight, based on a NOAEL of 1 mg/kg of body weight for effects on kidney weight in 1- and 2-year studies in rats, with an uncertainty factor of 300 (100 for inter- and intraspecies variation and 3 for limitations in the database)
Limit of detection	0.01 µg/litre by GC/MS; 0.01–0.02 µg/litre by GC with ECD
Treatment achievability	0.1 µg/litre should be achievable using GAC or ozonation
Guideline derivation • allocation to water • weight • consumption	 10% of TDI 60-kg adult 2 litres/day

Toxicological review

Chlorophenoxy herbicides, as a group, have been classified in Group 2B by IARC. However, the available data from studies in exposed populations and animals do not permit assessment of the carcinogenic potential to humans of any specific chlorophenoxy herbicide. Therefore, drinking-water guidelines for these compounds are based on a threshold approach for other toxic effects. Effects of dietary administration of mecoprop in short- and long-term studies include decreased relative kidney weight (rats and beagle dogs), increased relative liver weight (rats), effects on blood parameters (rats and beagle dogs) and depressed body weight gain (beagle dogs).

History of guideline development

The 1958 and 1963 WHO *International Standards for Drinking-water* did not refer to chlorophenoxy herbicides, including mecoprop, but the 1971 International Standards suggested that pesticide residues that may occur in community water supplies make only a minimal contribution to the total daily intake of pesticides for the population served. Mecoprop was not evaluated in the first edition of the *Guidelines for Drinking-water Quality*, published in 1984, but the 1993 Guidelines established a health-based guideline value of 0.01 mg/litre for mecoprop.

Assessment date

The risk assessment was originally conducted in 1993. The Final Task Force Meeting in 2003 agreed that this risk assessment be brought forward to this edition of the *Guidelines for Drinking-water Quality*.

Principal reference

WHO (2003) *Chlorophenoxy herbicides (excluding 2,4-D and MCPA) in drinking-water. Background document for preparation of WHO Guidelines for drinking-water quality*. Geneva, World Health Organization (WHO/SDE/WSH/03.04/44).

12.82 Mercury

Mercury is used in the electrolytic production of chlorine, in electrical appliances, in dental amalgams and as a raw material for various mercury compounds. Methylation of inorganic mercury has been shown to occur in fresh water and in seawater, although almost all mercury in uncontaminated drinking-water is thought to be in the form of Hg^{2+}. Thus, it is unlikely that there is any direct risk of the intake of organic mercury compounds, especially of alkylmercurials, as a result of the ingestion of drinking-water. However, there is a real possibility that methylmercury will be converted into inorganic mercury. Food is the main source of mercury in non-occupationally exposed populations; the mean dietary intake of mercury in various countries ranges from 2 to 20 µg/day per person.

Guideline value	0.001 mg/litre for total mercury
Occurrence	Mercury is present in the inorganic form in surface water and groundwater at concentrations usually below 0.5 µg/litre, although local mineral deposits may produce higher levels in groundwater.
PTWI	5 µg/kg of body weight for total mercury for the general population, of which no more than 3.3 µg/kg of body weight should be present as methylmercury
Limit of detection	0.001 µg/litre by atomic fluorescence spectrometry; 0.05 µg/litre by cold vapour AAS; 0.6 µg/litre by ICP; 5 µg/litre by FAAS
Treatment achievability	0.1 µg/litre should be achievable using coagulation
Guideline derivation • allocation to water • weight • consumption	 10% of PTWI 60-kg adult 2 litres/day
Additional considerations	Pregnant women and nursing mothers are likely to be at greater risk than the general population from the adverse effects of methylmercury.

Toxicological review

The toxic effects of inorganic mercury compounds are seen mainly in the kidney. Methylmercury affects mainly the central nervous system.

History of guideline development

The 1958 and 1963 WHO *International Standards for Drinking-water* did not mention mercury. Mercury was first mentioned in the 1971 International Standards, which gave the tentative upper concentration limit for mercury as 0.001 mg/litre (total mercury), based on health concerns. It was noted that this figure was related to levels found in natural water. In the first edition of the *Guidelines for Drinking-water Quality*, published in 1984, the guideline value of 0.001 mg/litre, which applied to all chemical forms of mercury, was retained. The 1993 Guidelines also retained the guideline value for total mercury of 0.001 mg/litre, based on the PTWI for methylmercury established by JECFA in 1972 and reaffirmed by JECFA in 1988.

Assessment date
The risk assessment was originally conducted in 1993. The Final Task Force Meeting in 2003 agreed that this risk assessment be brought forward to this edition of the *Guidelines for Drinking-water Quality*.

Principal reference
WHO (2003) *Mercury in drinking-water. Background document for preparation of WHO Guidelines for drinking-water quality*. Geneva, World Health Organization (WHO/SDE/WSH/03.04/10).

12.83 Methoxychlor

Methoxychlor (CAS No. 72-43-5) is an insecticide used on vegetables, fruit, trees, fodder and farm animals. It is poorly soluble in water and highly immobile in most agricultural soils. Under normal conditions of use, methoxychlor does not seem to be of environmental concern. Daily intake from food and air is expected to be below 1 µg per person. Environmental metabolites are formed preferentially under anaerobic rather than aerobic conditions and include mainly the dechlorinated and demethylated products. There is some potential for the accumulation of the parent compound and its metabolites in surface water sediments.

Guideline value	0.02 mg/litre
Occurrence	Detected occasionally in drinking-water, at concentrations as high as 300 µg/litre in rural areas
TDI	5 µg/kg of body weight, based on a systemic NOAEL of 5 mg/kg of body weight in a teratology study in rabbits, with an uncertainty factor of 1000 (100 for inter- and intraspecies variation and 10 reflecting concern for threshold carcinogenicity and the limited database)
Limit of detection	0.001–0.01 µg/litre by GC
Treatment achievability	0.1 µg/litre should be achievable using GAC
Guideline derivation • allocation to water • weight • consumption	 10% of TDI 60-kg adult 2 litres/day

Toxicological review
The genotoxic potential of methoxychlor appears to be negligible. In 1979, IARC assigned methoxychlor to Group 3. Subsequent data suggest a carcinogenic potential of methoxychlor for liver and testes in mice. This may be due to the hormonal activity of proestrogenic mammalian metabolites of methoxychlor and may therefore have

a threshold. The study, however, was inadequate because only one dose was used and because this dose may have been above the maximum tolerated dose. The database for studies on long-term, short-term and reproductive toxicity is inadequate. A teratology study in rabbits reported a systemic NOAEL of 5 mg/kg of body weight per day, which is lower than the LOAELs and NOAELs from other studies. This NOAEL was therefore selected for use in the derivation of a TDI.

History of guideline development
The 1958 and 1963 WHO *International Standards for Drinking-water* did not refer to methoxychlor, but the 1971 International Standards suggested that pesticide residues that may occur in community water supplies make only a minimal contribution to the total daily intake of pesticides for the population served. In the first edition of the *Guidelines for Drinking-water Quality*, published in 1984, a health-based guideline value of 0.03 mg/litre was recommended for methoxychlor, based on the ADI recommended by JMPR in 1965 and reaffirmed in 1977. The 1993 Guidelines established a health-based guideline value of 0.02 mg/litre for methoxychlor in drinking-water.

Assessment date
The risk assessment was originally conducted in 1993. The Final Task Force Meeting in 2003 agreed that this risk assessment be brought forward to this edition of the *Guidelines for Drinking-water Quality*.

Principal reference
WHO (2003) *Methoxychlor in drinking-water. Background document for preparation of WHO Guidelines for drinking-water quality*. Geneva, World Health Organization (WHO/SDE/WSH/03.04/105).

12.84 Methyl parathion
Methyl parathion (CAS No. 298-00-0) is a non-systemic insecticide and acaricide that is produced throughout the world and has been registered for use on many crops, in particular cotton. It partitions mainly to air and soil in the environment. There is virtually no movement through soil, and neither the parent compound nor its breakdown products will reach groundwater. By far the most important route for the environmental degradation of methyl parathion is microbial degradation. Half-lives of methyl parathion in water are in the order of weeks to months. Concentrations of methyl parathion in natural waters of agricultural areas in the USA ranged up to 0.46 µg/litre, with highest levels in summer. The general population can come into contact with methyl parathion via air, water or food.

A NOAEL of 0.3 mg/kg of body weight per day was derived from the combined results of several studies conducted in humans, based on the depression of erythrocyte and plasma cholinesterase activities. Methyl parathion decreased cholinesterase activities in long-term studies in mice and rats, but did not induce carcinogenic effects. Methyl parathion was mutagenic in bacteria, but there was no evidence of genotoxicity in a limited range of studies in mammalian systems.

A health-based value of 9 µg/litre can be calculated for methyl parathion on the basis of an ADI of 0.003 mg/kg of body weight, based on a NOAEL of 0.25 mg/kg of body weight per day in a 2-year study in rats for retinal degeneration, sciatic nerve demyelination, reduced body weight, anaemia and decreased brain acetylcholinesterase activity, using an uncertainty factor of 100. Since the toxicological end-points seen in animals were other than acetylcholinesterase inhibition, it was considered more appropriate to use these data rather than the NOAEL derived for cholinesterase inhibition in humans.

Intake of methyl parathion from all sources is generally low and well below the ADI. As the health-based value is much higher than methyl parathion concentrations likely to be found in drinking-water, the presence of methyl parathion in drinking-water under usual conditions is unlikely to represent a hazard to human health. For this reason, the establishment of a guideline value for methyl parathion is not deemed necessary.

History of guideline development

The 1958 and 1963 WHO *International Standards for Drinking-water* did not refer to methyl parathion, but the 1971 International Standards suggested that pesticide residues that may occur in community water supplies make only a minimal contribution to the total daily intake of pesticides for the population served. Methyl parathion was not evaluated in the first edition of the *Guidelines for Drinking-water Quality*, published in 1984, in the second edition, published in 1993, or in the addendum to the second edition, published in 1998.

Assessment date

The risk assessment was conducted in 2003.

Principal references

FAO/WHO (1996) *Pesticide residues in food – 1995 evaluations. Part II – Toxicological and environmental.* Geneva, World Health Organization, Joint FAO/WHO Meeting on Pesticide Residues (WHO/PCS/96.48).

IPCS (1992) *Methyl parathion.* Geneva, World Health Organization, International Programme on Chemical Safety (Environmental Health Criteria 145).

WHO (2003) *Methyl parathion in drinking-water. Background document for preparation of WHO Guidelines for drinking-water quality.* Geneva, World Health Organization (WHO/SDE/WSH/03.04/106).

12.85 Metolachlor

Metolachlor (CAS No. 51218-45-2) is a selective pre-emergence herbicide used on a number of crops. It can be lost from the soil through biodegradation, photodegradation and volatilization. It is fairly mobile and under certain conditions can contaminate groundwater, but it is mostly found in surface water.

Guideline value	0.01 mg/litre
Occurrence	Detected in surface water and groundwater at concentrations that can exceed 10 µg/litre
TDI	3.5 µg/kg of body weight, based on a NOAEL of 3.5 mg/kg of body weight for an apparent decrease in kidney weight at the two highest dose levels in a 1-year dog study, with an uncertainty factor of 1000 (100 for inter- and intraspecies variation and 10 reflecting some concern regarding carcinogenicity)
Limit of detection	0.75–0.01 µg/litre by GC with nitrogen–phosphorus detection
Treatment achievability	0.1 µg/litre should be achievable using GAC
Guideline derivation • allocation to water • weight • consumption	 10% of TDI 60-kg adult 2 litres/day

Toxicological review

In a 1-year study in beagle dogs, administration of metolachlor resulted in decreased kidney weight at the two highest dose levels. In 2-year studies with rodents fed metolachlor in the diet, the only toxicological effects observed in albino mice were decreased body weight gain and decreased survival in females at the highest dose level, whereas rats showed decreased body weight gain and food consumption at the highest dose level. There is no evidence from available studies that metolachlor is carcinogenic in mice. In rats, an increase in liver tumours in females as well as a few nasal tumours in males have been observed. Metolachlor is not genotoxic.

History of guideline development

The 1958 and 1963 WHO *International Standards for Drinking-water* did not refer to metolachlor, but the 1971 International Standards suggested that pesticide residues that may occur in community water supplies make only a minimal contribution to the total daily intake of pesticides for the population served. Metolachlor was not evaluated in the first edition of the *Guidelines for Drinking-water Quality*, published in 1984, but the 1993 Guidelines established a health-based guideline value of 0.01 mg/litre for metolachlor in drinking-water.

Assessment date

The risk assessment was originally conducted in 1993. The Final Task Force Meeting in 2003 agreed that this risk assessment be brought forward to this edition of the *Guidelines for Drinking-water Quality*.

12. CHEMICAL FACT SHEETS

Principal reference
WHO (2003) *Metolachlor in drinking-water. Background document for preparation of WHO Guidelines for drinking-water quality*. Geneva, World Health Organization (WHO/SDE/WSH/03.04/39).

12.86 Microcystin-LR

Among the more than 80 microcystins identified to date, only a few occur frequently and in high concentrations. Microcystin-LR is among the most frequent and most toxic microcystin congeners. Frequently occurring cyanobacterial genera that contain these toxins are *Microcystis*, *Planktothrix* and *Anabaena*. Microcystins usually occur within the cells; substantial amounts are released to the surrounding water only in situations of cell rupture (i.e., lysis).

Provisional guideline value	0.001 mg/litre (for total microcystin-LR, free plus cell-bound) The guideline value is provisional, as it covers only microcystin-LR, the database is limited and new data for the toxicity of cyanobacterial toxins are being generated.
TDI	0.04 µg/kg of body weight, based on liver pathology observed in a 13-week study in mice and applying an uncertainty factor of 1000, taking into consideration limitations in the database, in particular lack of data on chronic toxicity and carcinogenicity
Limit of detection	0.1–1 µg/litre by HPLC following extraction of cells with 75% aqueous methanol or following concentration of microcystins from liquid samples on C-18; will allow differentiation between variants where standards are available. 0.1–0.5 µg/litre by commercially available immunoassay kits (ELISA) for microcystins dissolved in water or in aqueous extracts of cells; will detect most microcystins. These are less precise in quantification than HPLC, but useful for screening. 0.5–1.5 µg/litre by protein phosphatase assay for microcystins dissolved in water or in aqueous extracts of cells; will detect all microcystins. This assay is less precise in quantification and identification than HPLC, but useful for screening.
Monitoring	The preferred approach is visual monitoring (including microscopy for potentially microcystin-containing genera) of source water for evidence of increasing cyanobacterial cell density (blooms) or bloom-forming potential, and increased vigilance where such events occur. Chemical monitoring of microcystins is not the preferred focus.
Prevention and treatment	Actions to decrease the probability of bloom occurrence include catchment and source water management, such as reducing nutrient loading or changing reservoir stratification and mixing. Treatment effective for the removal of cyanobacteria includes filtration to remove intact cells. Treatment effective against free microcystins in water (as well as most other free cyanotoxins) includes oxidation through ozone or chlorine at sufficient concentrations and contact times, as well as GAC and some PAC applications.

Guideline derivation	
• allocation to water	80% of TDI
• weight	60-kg adult
• consumption	2 litres/day
Additional comments	While guideline values are derived where sufficient data exist, they are intended to inform the interpretation of monitoring data and not to indicate that there is a requirement for routine monitoring by chemical analysis.

Toxicological review

Microcystin-LR is a potent inhibitor of eukaryotic protein serine/threonine phosphatases 1 and 2A. The primary target for microcystin toxicity is the liver, as microcystins cross cell membranes chiefly through the bile acid transporter. Guideline derivation was based on an oral 13-week study with mice, supported by an oral 44-day study with pigs. A large number of poisonings of livestock and wildlife have been recorded. Evidence of tumour promotion has been published.

History of guideline development

Cyanobacterial toxins were not evaluated in the 1958, 1963 and 1971 WHO *International Standards for Drinking-water* or in the first two editions of the *Guidelines for Drinking-water Quality*, published in 1984 and 1993. In the addendum to the second edition of the Guidelines, published in 1998, it was concluded that there were insufficient data to allow a guideline value to be derived for any cyanobacterial toxins other than microcystin-LR. A health-based guideline value for total microcystin-LR (free plus cell-bound) of 0.001 mg/litre was derived, assuming significant exposure from drinking-water. The guideline value was designated as provisional, as it covers only microcystin-LR, the database is limited and new data for the toxicity of cyanobacterial toxins are being generated.

Assessment date

The risk assessment was conducted in 2003.

Principal references

Chorus I, Bartram J, eds. (1999) *Toxic cyanobacteria in water: A guide to their public health consequences, monitoring and management.* Published by E & FN Spon, London, on behalf of the World Health Organization, Geneva.

WHO (2003) *Cyanobacterial toxins: Microcystin-LR in drinking-water. Background document for preparation of WHO Guidelines for drinking-water quality.* Geneva, World Health Organization (WHO/SDE/WSH/03.04/57).

12.87 Molinate

Molinate (CAS No. 2212-67-1) is a herbicide used to control broad-leaved and grassy weeds in rice. The available data suggest that groundwater pollution by molinate is restricted to some rice-growing regions. Data on the occurrence of molinate in the environment are limited. Molinate is of low persistence in water and soil, with a half-life of about 5 days.

Guideline value	0.006 mg/litre
Occurrence	Concentrations in water rarely exceed 1 µg/litre.
TDI	2 µg/kg of body weight, based on a NOAEL for reproductive toxicity in the rat of 0.2 mg/kg of body weight, with an uncertainty factor of 100 (for inter- and intraspecies variation)
Limit of detection	0.01 µg/litre by GC/MS
Treatment achievability	0.001 mg/litre should be achievable using GAC
Guideline derivation • allocation to water • weight • consumption	 10% of TDI 60-kg adult 2 litres/day

Toxicological review

On the basis of the limited information available, molinate does not seem to be carcinogenic or mutagenic in animals. Evidence suggests that impairment of the reproductive performance of the male rat represents the most sensitive indicator of molinate exposure. However, epidemiological data based on the examination of workers involved in molinate production do not indicate any effect on human fertility.

History of guideline development

The 1958 and 1963 WHO *International Standards for Drinking-water* did not refer to molinate, but the 1971 International Standards suggested that pesticide residues that may occur in community water supplies make only a minimal contribution to the total daily intake of pesticides for the population served. Molinate was not evaluated in the first edition of the *Guidelines for Drinking-water Quality*, published in 1984, but the 1993 Guidelines established a health-based guideline value of 0.006 mg/litre for molinate in drinking-water.

Assessment date

The risk assessment was originally conducted in 1993. The Final Task Force Meeting in 2003 agreed that this risk assessment be brought forward to this edition of the *Guidelines for Drinking-water Quality*.

Principal reference

WHO (2003) *Molinate in drinking-water. Background document for preparation of WHO Guidelines for drinking-water quality*. Geneva, World Health Organization (WHO/SDE/WSH/03.04/40).

12.88 Molybdenum

Molybdenum is found naturally in soil and is used in the manufacture of special steels and in the production of tungsten and pigments, and molybdenum compounds are used as lubricant additives and in agriculture to prevent molybdenum deficiency in crops.

Guideline value	0.07 mg/litre
Occurrence	Concentrations in drinking-water are usually less than 0.01 mg/litre, although concentrations as high as 200 µg/litre have been reported in areas near mining sites.
NOAEL	0.2 mg/litre in a 2-year study of humans exposed through their drinking-water, using an uncertainty factor of 3 for intraspecies variation (because molybdenum is an essential element)
Limit of detection	0.25 µg/litre by graphite furnace AAS; 2 µg/litre by ICP/AES
Treatment achievability	Molybdenum is not removed from drinking-water.
Additional comments	The guideline value is within the range of that derived on the basis of results of toxicological studies in animal species and is consistent with the essential daily requirement.

Toxicological review

Molybdenum is considered to be an essential element, with an estimated daily requirement of 0.1–0.3 mg for adults. No data are available on the carcinogenicity of molybdenum by the oral route. Additional toxicological information is needed on the impact of molybdenum on bottle-fed infants.

History of guideline development

The 1958 and 1963 WHO *International Standards for Drinking-water* did not refer to molybdenum. The 1971 International Standards stated that molybdenum should be controlled in drinking-water, but that insufficient information was available to enable a tentative limit to be established. In the first edition of the *Guidelines for Drinking-water Quality*, published in 1984, it was concluded that no action was required for molybdenum. The 1993 Guidelines proposed a health-based guideline value of 0.07 mg/litre for molybdenum based on a 2-year study of humans exposed through their drinking-water. This value is within the range of that derived on the basis of results of toxicological studies in animal species and is consistent with the essential daily requirement.

Assessment date

The risk assessment was originally conducted in 1993. The Final Task Force Meeting in 2003 agreed that this risk assessment be brought forward to this edition of the *Guidelines for Drinking-water Quality*.

Principal reference
WHO (2003) *Molybdenum in drinking-water. Background document for preparation of WHO Guidelines for drinking-water quality*. Geneva, World Health Organization (WHO/SDE/WSH/03.04/11).

12.89 Monochloramine

Mono-, di- and trichloramines are considered by-products of drinking-water chlorination, being formed when ammonia is added to chlorinated water. Monochloramine may also be added to maintain residual disinfection activity in potable water distribution systems. The use of chloramines for disinfection instead of chlorine reduces the formation of THMs in drinking-water supplies. However, formation of other by-products, such as haloketones, chloropicrin, cyanogen chloride, haloacetic acids, haloacetonitriles, aldehydes and chlorophenols, has been reported. Monochloramine is recognized as a less effective disinfectant than chlorine. Only monochloramine, the most abundant chloramine, is considered here, as it has been the most extensively studied.

Guideline value	3 mg/litre
Occurrence	Typical chloramine concentrations of 0.5–2 mg/litre are found in drinking-water supplies where chloramine is used as a primary disinfectant or to provide a chlorine residual in the distribution system.
TDI	94 µg/kg of body weight, based on a NOAEL of 9.4 mg/kg of body weight per day, the highest dose administered to male rats in a 2-year NTP drinking-water study (although mean body weights of rats given the highest dose were lower than those of their respective control groups, it is probable that the lower body weights were caused by the unpalatability of the drinking-water)
Limit of detection	10 µg/litre by colorimetric methods
Treatment achievability	It is possible to reduce the concentration of chloramine effectively to zero (<0.1 mg/litre) by reduction; however, it is normal practice to supply water with a chloramine residual of a few tenths of a milligram per litre to act as a preservative during distribution.
Guideline derivation • allocation to water • weight • consumption	100% of TDI 60-kg adult 2 litres/day
Additional comments	• An additional uncertainty factor for possible carcinogenicity was not applied because equivocal cancer effects reported in the NTP study in only one species and in only one sex were within the range observed in historical controls. • Most individuals are able to taste chloramines at concentrations below 5 mg/litre, and some at levels as low as 0.3 mg/litre.

Toxicological review

Although monochloramine has been shown to be mutagenic in some *in vitro* studies, it has not been found to be genotoxic *in vivo*. IARC has classified chloramine in Group 3, and the US EPA has classified monochloramine in group D (not classifiable as to human carcinogenicity, as there is inadequate human and animal evidence). In the NTP bioassay in two species, the incidence of mononuclear cell leukaemias in female F344/N rats was increased, but no other increases in tumour incidence were observed. IPCS (2000) did not consider that the increase in mononuclear cell leukaemia was treatment-related.

History of guideline development

The 1958, 1963 and 1971 WHO *International Standards for Drinking-water* and the first edition of the *Guidelines for Drinking-water Quality*, published in 1984, did not refer to chloramines. The 1993 Guidelines established a health-based guideline value of 3 mg/litre for monochloramine in drinking-water. Available data were insufficient for the establishment of guideline values for dichloramine and trichloramine. It was noted that the odour thresholds for dichloramine and trichloramine are much lower than that for monochloramine.

Assessment date

The risk assessment was conducted in 2003.

Principal references

IPCS (2000) *Disinfectants and disinfectant by-products*. Geneva, World Health Organization, International Programme on Chemical Safety (Environmental Health Criteria 216).

WHO (2003) *Monochloramine in drinking-water. Background document for preparation of WHO Guidelines for drinking-water quality*. Geneva, World Health Organization (WHO/SDE/WSH/03.04/83).

12.90 Monochloroacetic acid

Chlorinated acetic acids are formed from organic material during water chlorination.

Guideline value	0.02 mg/litre
Occurrence	Present in surface water-derived drinking-water at <2–82 µg/litre (mean 2.1 µg/litre)
TDI	3.5 µg/kg of body weight, based on a LOAEL of 3.5 mg/kg of body weight per day from a study in which increased absolute and relative spleen weights were observed in male rats exposed to monochloroacetic acid in drinking-water for 2 years, and using an uncertainty factor of 1000 (100 for inter- and intraspecies variation and 10 for use of a minimal LOAEL instead of a NOAEL and database deficiencies, including the lack of a multigeneration reproductive toxicity study)
Limit of detection	2 µg/litre by GC with ECD; 5 µg/litre by GC/MS
Treatment achievability	No information available
Guideline derivation • allocation to water • weight • consumption	20% of TDI 60-kg adult 2 litres/day

Toxicological review
No evidence of carcinogenicity of monochloroacetate was found in 2-year gavage bioassays with rats and mice. Monochloroacetate has given mixed results in a limited number of mutagenicity assays and has been negative for clastogenicity in genotoxicity studies. IARC has not classified the carcinogenicity of monochloroacetic acid.

History of guideline development
The 1958, 1963 and 1971 WHO *International Standards for Drinking-water* and the first edition of the *Guidelines for Drinking-water Quality*, published in 1984, did not refer to monochloroacetic acid. The 1993 Guidelines did not establish a guideline value for monochloroacetic acid, as available toxicity data were considered insufficient.

Assessment date
The risk assessment was conducted in 2003.

Principal reference
WHO (2003) *Monochloroacetic acid in drinking-water. Background document for preparation of WHO Guidelines for drinking-water quality.* Geneva, World Health Organization (WHO/SDE/WSH/03.04/85).

12.91 Monochlorobenzene
Releases of monochlorobenzene (MCB) to the environment are thought to be mainly due to volatilization losses associated with its use as a solvent in pesticide formulations, as a degreasing agent and from other industrial applications. MCB has been

detected in surface water, groundwater and drinking-water; mean concentrations were less than 1 µg/litre in some potable water sources (maximum 5 µg/litre) in Canada. The major source of human exposure is probably air.

MCB is of low acute toxicity. Oral exposure to high doses of MCB affects mainly the liver, kidneys and haematopoietic system. There is limited evidence of carcinogenicity in male rats, with high doses increasing the occurrence of neoplastic nodules in the liver. The majority of evidence suggests that MCB is not mutagenic; although it binds to DNA *in vivo*, the level of binding is low.

A health-based value of 300 µg/litre can be calculated for MCB on the basis of a TDI of 85.7 µg/kg of body weight, based on neoplastic nodules identified in a 2-year rat study with dosing by gavage, and taking into consideration the limited evidence of carcinogenicity. However, because MCB occurs at concentrations well below those at which toxic effects are observed, it is not considered necessary to derive a guideline value. It should also be noted that the health-based value far exceeds the lowest reported taste and odour threshold for MCB in water.

History of guideline development

The 1958, 1963 and 1971 WHO *International Standards for Drinking-water* did not refer to MCB. In the first edition of the *Guidelines for Drinking-water Quality*, published in 1984, no guideline value for chlorobenzene was recommended after a detailed evaluation of the compound. Following consideration of the calculated toxicological limit for drinking-water of 0.005–0.05 mg/litre based on a tentative ADI and the fact that the threshold odour concentration of MCB in water is 0.03 mg/litre, no guideline value was recommended, and 0.003 mg/litre was recommended to avoid taste and odour problems in drinking-water. The 1993 Guidelines proposed a health-based guideline value of 0.3 mg/litre for MCB, noting that this value far exceeds the lowest reported taste and odour threshold for MCB in water (0.01 mg/litre).

Assessment date

The risk assessment was conducted in 2003.

Principal reference

WHO (2003) *Monochlorobenzene in drinking-water. Background document for preparation of WHO Guidelines for drinking-water quality*. Geneva, World Health Organization (WHO/SDE/WSH/03.04/107).

12.92 MX

MX, which is the common name for 3-chloro-4-dichloromethyl-5-hydroxy-2(5H)-furanone, is formed by the reaction of chlorine with complex organic matter in drinking-water. It has been identified in chlorinated humic acid solutions and drinking-water in Finland, the United Kingdom and the USA and was found to be present in 37 water sources at levels of 2–67 ng/litre. Five drinking-water samples from different Japanese cities contained MX at concentrations ranging from <3 to 9 ng/litre.

MX is a potent mutagen in bacteria and in cells *in vitro* and has undergone a lifetime study in rats in which some tumorigenic responses were observed. These data indicate that MX induces thyroid and bile duct tumours. IARC has classified MX in Group 2B on the basis of rat tumorigenicity and its strong mutagenicity.

A health-based value of 1.8 µg/litre can be calculated for MX on the basis of the increase in cholangiomas and cholangiocarcinomas in female rats using the linearized multistage model (without a body surface area correction). However, this is significantly above the concentrations that would be found in drinking-water, and, in view of the analytical difficulties in measuring this compound at such low concentrations, it is considered unnecessary to propose a formal guideline value for MX in drinking-water.

History of guideline development
The 1958, 1963 and 1971 WHO *International Standards for Drinking-water* and the first edition of the *Guidelines for Drinking-water Quality*, published in 1984, did not refer to MX. The 1993 Guidelines concluded that available data were inadequate to permit a guideline value for MX to be established.

Assessment date
The risk assessment was conducted in 2003.

Principal references
IPCS (2000) *Disinfectants and disinfectant by-products.* Geneva, World Health Organization, International Programme on Chemical Safety (Environmental Health Criteria 216).

WHO (2003) *MX in drinking-water. Background document for preparation of WHO Guidelines for drinking-water quality.* Geneva, World Health Organization (WHO/SDE/WSH/03.04/108).

12.93 Nickel
Nickel is used mainly in the production of stainless steel and nickel alloys. Food is the dominant source of nickel exposure in the non-smoking, non-occupationally exposed population; water is generally a minor contributor to the total daily oral intake. However, where there is heavy pollution or use of certain types of kettles, of non-resistant material in wells or of water that has come into contact with nickel- or chromium-plated taps, the nickel contribution from water may be significant.

Provisional guideline value	0.02 mg/litre
	The guideline value is considered provisional owing to uncertainties about the effect level for perinatal mortality.
Occurrence	The concentration of nickel in drinking-water is normally less than 0.02 mg/litre, although nickel released from taps and fittings may contribute up to 1 mg/litre. In special cases of release from natural or industrial nickel deposits in the ground, the nickel concentrations in drinking-water may be even higher.
TDI	5 µg/kg of body weight, derived from a NOAEL of 5 mg/kg of body weight per day from a dietary study in rats in which altered organ to body weight ratios were observed, using an uncertainty factor of 1000 (100 for inter- and intraspecies variation and an extra factor of 10 to compensate for the lack of adequate studies on long-term exposure and reproductive effects, a lack of data on carcinogenicity by the oral route and a much higher intestinal absorption when taken on an empty stomach in drinking-water than when taken together with food)
Limit of detection	0.1 µg/litre by ICP/MS; 1 µg/litre by EAAS or ICP/optical emission spectroscopy; 15 µg/litre by ICP; 20 µg/litre by FAAS
Treatment achievability	20 µg/litre should be achievable by conventional treatment, e.g., coagulation. However, nickel is not usually a raw water contaminant.
Guideline derivation • allocation to water • weight • consumption	 10% of TDI 60-kg adult 2 litres/day

Toxicological review

IARC concluded that inhaled nickel compounds are carcinogenic to humans (Group 1) and metallic nickel is possibly carcinogenic (Group 2B). However, there is a lack of evidence of a carcinogenic risk from oral exposure to nickel. Dose-related increases in perinatal mortality were observed in a carefully conducted two-generation study in rats, but variations in response between successive litters make it difficult to draw firm conclusions from this study.

History of guideline development

The 1958, 1963 and 1971 WHO *International Standards for Drinking-water* did not refer to nickel. In the first edition of the *Guidelines for Drinking-water Quality*, published in 1984, it was concluded that the toxicological data available indicate that a guideline value for nickel in drinking-water was not required. A health-based guideline value of 0.02 mg/litre was derived in the 1993 Guidelines, which should provide sufficient protection for individuals who are sensitive to nickel. This guideline value was maintained in the addendum to the Guidelines published in 1998 because, on the basis of the available data, it was considered to provide sufficient protection for individuals who are sensitive to nickel. However, the guideline value was designated as provisional owing to uncertainties about the effect level for perinatal mortality.

Assessment date
The risk assessment was originally conducted in 1998. The Final Task Force Meeting in 2003 agreed that this risk assessment be brought forward to this edition of the *Guidelines for Drinking-water Quality*.

Principal reference
WHO (2003) *Nickel in drinking-water. Background document for preparation of WHO Guidelines for drinking-water quality*. Geneva, World Health Organization (WHO/SDE/WSH/03.04/55).

12.94 Nitrate and nitrite
Nitrate and nitrite are naturally occurring ions that are part of the nitrogen cycle. Nitrate is used mainly in inorganic fertilizers, and sodium nitrite is used as a food preservative, especially in cured meats. The nitrate concentration in groundwater and surface water is normally low but can reach high levels as a result of leaching or runoff from agricultural land or contamination from human or animal wastes as a consequence of the oxidation of ammonia and similar sources. Anaerobic conditions may result in the formation and persistence of nitrite. Chloramination may give rise to the formation of nitrite within the distribution system if the formation of chloramine is not sufficiently controlled. The formation of nitrite is as a consequence of microbial activity and may be intermittent. Nitrification in distribution systems can increase nitrite levels, usually by 0.2–1.5 mg/litre.

Guideline value for nitrate	50 mg/litre to protect against methaemoglobinaemia in bottle-fed infants (short-term exposure)
Guideline value / Provisional guideline value for nitrite	• 3 mg/litre for methaemoglobinaemia in infants (short-term exposure) • 0.2 mg/litre (provisional) (long-term exposure) The guideline value for chronic effects of nitrite is considered provisional owing to uncertainty surrounding the relevance of the observed adverse health effects for humans and the susceptibility of humans compared with animals. The occurrence of nitrite in distribution as a consequence of chloramine use will be intermittent, and average exposures over time should not exceed the provisional guideline value.
Guideline value for combined nitrate plus nitrite	The sum of the ratios of the concentrations of each to its guideline value should not exceed 1.
Occurrence	In most countries, nitrate levels in drinking-water derived from surface water do not exceed 10 mg/litre, although nitrate levels in well water often exceed 50 mg/litre; nitrite levels are normally lower, less than a few milligrams per litre.

Basis of guideline derivation	• nitrate (bottle-fed infants): in epidemiological studies, methaemoglobinaemia was not reported in infants in areas where drinking-water consistently contained less than 50 mg of nitrate per litre • nitrite (bottle-fed infants): nitrite is 10 times more potent than nitrate on a molar basis with respect to methaemoglobin formation • nitrite (long-term exposure): based on allocation to drinking-water of 10% of JECFA ADI of 0.06 mg/kg of body weight per day, based on nitrite-induced morphological changes in the adrenals, heart and lungs in laboratory animal studies
Limit of detection	0.1 mg/litre (nitrate) and 0.05 mg/litre (nitrite) by liquid chromatography; 0.01–1 mg/litre (nitrate) by spectrometric techniques; 0.005–0.01 mg/litre (nitrite) by a molecular absorption spectrometric method; 22 µg/litre (nitrate) and 35 µg/litre (nitrite) by ion chromatography
Treatment achievability	• nitrate: 5 mg/litre or lower should be achievable using biological denitrification (surface waters) or ion exchange (groundwaters) • nitrite: 0.1 mg/litre should be achievable using chlorination (to form nitrate)
Additional comments	• Nitrite can occur in distribution at higher concentrations when chloramination is used, but the occurrence is almost invariably sporadic. Methaemoglobinaemia is therefore the most important consideration, and the guideline derived for protection against methaemoglobinaemia would be the most appropriate under these circumstances, allowing for any nitrate that may also be present. • All water systems that practise chloramination should closely and regularly monitor their systems to verify disinfectant levels, microbiological quality and nitrite levels. If nitrification is detected (e.g., reduced disinfectant residuals and increased nitrite levels), steps should be taken to modify the treatment train or water chemistry in order to maintain a safe water quality. Efficient disinfection must never be compromised. • Methaemoglobinaemia in infants also appears to be associated with simultaneous exposure to microbial contaminants.

Toxicological review

The primary health concern regarding nitrate and nitrite is the formation of methaemoglobinaemia, so-called "blue-baby syndrome." Nitrate is reduced to nitrite in the stomach of infants, and nitrite is able to oxidize haemoglobin (Hb) to methaemoglobin (metHb), which is unable to transport oxygen around the body. The reduced oxygen transport becomes clinically manifest when metHb concentrations reach 10% or more of normal Hb concentrations; the condition, called methaemoglobinaemia, causes cyanosis and, at higher concentrations, asphyxia. The normal metHb level in infants under 3 months of age is less than 3%.

The Hb of young infants is more susceptible to metHb formation than that of older children and adults; this is believed to be the result of the large proportion of fetal

12. CHEMICAL FACT SHEETS

Hb, which is more easily oxidized to metHb, still present in the blood of infants. In addition, there is a deficiency in infants of metHb reductase, the enzyme responsible for the reduction of metHb to Hb. The reduction of nitrate to nitrite by gastric bacteria is also higher in infants because of low gastric acidity. The level of nitrate in breast milk is relatively low; when bottle-fed, however, these young infants are at risk because of the potential for exposure to nitrate/nitrite in drinking-water and the relatively high intake of water in relation to body weight. The higher reduction of nitrate to nitrite in young infants is not very well quantified, but it appears that gastrointestinal infections exacerbate the conversion from nitrate to nitrite.

The weight of evidence is strongly against there being an association between nitrite and nitrate exposure in humans and the risk of cancer.

Studies with nitrite in laboratory rats have reported hypertrophy of the adrenal zona glomerulosa. The mechanism of induction of this effect and whether it occurs in other species is unclear. JECFA developed an ADI of 5 mg of potassium nitrite per kg of body weight based on the NOAEL in these studies.

History of guideline development

The 1958 WHO *International Standards for Drinking-water* referred to nitrates, stating that the ingestion of water containing nitrates in excess of 50–100 mg/litre (as nitrate) may give rise to methaemoglobinaemia in infants under 1 year of age. In the 1963 International Standards, this value was lowered to 45 mg/litre (as nitrate), which was retained in the 1971 International Standards. The 1971 International Standards first mentioned concern over the possibility of nitrosamine formation *in vivo*; as nitrosamines are a possible hazard to human health, the 1971 Standards stated that it may eventually become necessary to reduce the level of nitrates in water if it is found that this source makes a significant contribution to the hazard to human health arising from nitrosamines. In the first edition of the *Guidelines for Drinking-water Quality*, published in 1984, a guideline value of 10 mg/litre for nitrate-nitrogen was recommended. It was also recommended that the guideline value for nitrite must be correspondingly lower than that for nitrate, and it was noted that the nitrite-nitrogen level should be considerably lower than 1 mg/litre where drinking-water is correctly treated. The 1993 Guidelines concluded that extensive epidemiological data support the current guideline value for nitrate-nitrogen of 10 mg/litre, but stated that this value should be expressed not on the basis of nitrate-nitrogen but on the basis of nitrate itself, which is the chemical entity of concern to health. The guideline value for nitrate is therefore 50 mg/litre. This guideline value for methaemoglobinaemia in infants, an acute effect, was confirmed in the addendum to the Guidelines, published in 1998. It was also concluded in the 1993 Guidelines that a guideline value for nitrite should be proposed, although no suitable animal studies of methaemoglobinaemia were available. A provisional guideline value for nitrite of 3 mg/litre was therefore proposed by accepting a relative potency for nitrite and nitrate with respect to methaemoglobin formation of 10:1 (on a molar basis). In the addendum to the Guidelines, published in 1998, it was concluded that human data on nitrite reviewed by JECFA supported the current provisional guideline value of 3 mg/litre, based on induction of methaemoglobinaemia in infants. In addition, a guideline value of 0.2 mg/litre for nitrate ion associated with long-term exposure was derived in the addendum to the Guidelines, based on JECFA's ADI derived in 1995. However, because of the uncertainty surrounding the relevance of the observed adverse health effects for humans and the susceptibility of humans compared with animals, this guideline value was considered provisional. Because of the possibility of simultaneous occurrence of nitrite and nitrate in drinking-water, it was recommended in the 1993 and 1998 Guidelines that the sum of the ratios of the concentration of each to its guideline value should not exceed 1.

Assessment date
The risk assessment was originally conducted in 1998. The Final Task Force Meeting in 2003 agreed that this risk assessment be brought forward to this edition of the *Guidelines for Drinking-water Quality*.

Principal reference
WHO (2003) *Nitrate and nitrite in drinking-water. Background document for preparation of WHO Guidelines for drinking-water quality*. Geneva, World Health Organization (WHO/SDE/WSH/03.04/56).

12.95 Nitrilotriacetic acid (NTA)
Nitrilotriacetic acid (NTA) is used primarily in laundry detergents as a replacement for phosphates and in the treatment of boiler water to prevent accumulation of mineral scale.

Guideline value	0.2 mg/litre
Occurrence	Concentrations in drinking-water usually do not exceed a few micrograms per litre, although concentrations as high as 35 µg/litre have been measured.
TDI	10 µg/kg of body weight, based on nephritis and nephrosis in a 2-year study in rats and using an uncertainty factor of 1000 (100 for inter- and intraspecies variation and 10 for carcinogenic potential at high doses)
Limit of detection	0.2 µg/litre using GC with a nitrogen-specific detector
Treatment achievability	No data available
Guideline derivation • allocation to water • weight • consumption	 50% of TDI 60-kg adult 2 litres/day

Toxicological review

NTA is not metabolized in animals and is rapidly eliminated, although some may be briefly retained in bone. It is of low acute toxicity to animals, but it has been shown to produce kidney tumours in rodents following long-term exposure to doses higher than those required to produce nephrotoxicity. IARC has placed NTA in Group 2B. It is not genotoxic, and the reported induction of tumours is believed to be due to cytotoxicity resulting from the chelation of divalent cations such as zinc and calcium in the urinary tract, leading to the development of hyperplasia and subsequently neoplasia.

History of guideline development

The 1958 and 1963 WHO *International Standards for Drinking-water* did not refer to NTA. The 1971 International Standards stated that NTA should be controlled in drinking-water, but that insufficient information was available to enable a tentative limit to be established. In the first edition of the *Guidelines for Drinking-water Quality*, published in 1984, it was determined that no further action on NTA was required. A health-based guideline value of 0.2 mg/litre was established for NTA in the 1993 Guidelines.

Assessment date

The risk assessment was originally conducted in 1993. The Final Task Force Meeting in 2003 agreed that this risk assessment be brought forward to this edition of the *Guidelines for Drinking-water Quality*.

Principal reference

WHO (2003) *Nitrilotriacetic acid in drinking-water. Background document for preparation of WHO Guidelines for drinking-water quality.* Geneva, World Health Organization (WHO/SDE/WSH/03.04/30).

12.96 Parathion

Parathion (CAS No. 56-38-2) is a non-systemic insecticide that is used in many countries throughout the world. It is used as a fumigant and acaricide and as a pre-harvest soil and foliage treatment on a wide variety of crops, both outdoors and in greenhouses. Parathion released to the environment will adsorb strongly to the top layer of soil and is not likely to leach significantly. Parathion disappears from surface waters in about a week. The general population is not usually exposed to parathion from air or water. Parathion residues in food are the main source of exposure.

Parathion inhibits cholinesterase activity in all species tested. There has been no evidence of carcinogenicity in 2-year rat studies. JMPR concluded that parathion is not genotoxic.

A health-based value of 10 µg/litre can be calculated for parathion on the basis of an ADI of 0.004 mg/kg of body weight based on a NOAEL of 0.4 mg/kg body weight per day in a 2-year study in rats for retinal atrophy and inhibition of brain acetylcholinesterase at the higher dose, and using an uncertainty factor of 100. Lower NOAELs in animals, based only on inhibition of erythrocyte or brain acetylcholinesterase, were not considered relevant because of the availability of a NOAEL for erythrocyte acetylcholinesterase inhibition in humans, which was 0.1 mg/kg of body weight per day.

Intake of parathion from all sources is generally low and well below the ADI. As the health-based value is much higher than parathion concentrations likely to be found in drinking-water, the presence of parathion in drinking-water under usual conditions is unlikely to represent a hazard to human health. For this reason, the establishment of a guideline value for parathion is not deemed necessary.

History of guideline development

The 1958 and 1963 WHO *International Standards for Drinking-water* did not refer to parathion, but the 1971 International Standards suggested that pesticide residues that may occur in community water supplies make only a minimal contribution to the total daily intake of pesticides for the population served. Parathion was not evaluated in the first edition of the *Guidelines for Drinking-water Quality*, published in 1984, in the second edition, published in 1993, or in the addendum to the second edition, published in 1998.

Assessment date

The risk assessment was conducted in 2003.

Principal references

FAO/WHO (1996) *Pesticide residues in food – 1995 evaluations. Part II – Toxicological and environmental.* Geneva, World Health Organization, Joint FAO/WHO Meeting on Pesticide Residues (WHO/PCS/96.48).

WHO (2003) *Parathion in drinking-water. Background document for preparation of WHO Guidelines for drinking-water quality.* Geneva, World Health Organization (WHO/SDE/WSH/03.04/110).

12.97 Pendimethalin

Pendimethalin (CAS No. 40487-42-1) is a pre-emergence herbicide that is fairly immobile and persistent in soil. It is used in large amounts in Japan (5000 tonnes per year). It is lost through photodegradation, biodegradation and volatilization. The leaching potential of pendimethalin appears to be very low, but little is known about its more polar degradation products.

Guideline value	0.02 mg/litre
Occurrence	Rarely been found in drinking-water in the limited studies available (detection limit 0.01 µg/litre)
TDI	5 µg/kg of body weight, based on evidence of slight liver toxicity even at the lowest dose tested (5 mg/kg of body weight) in a long-term rat feeding study, with an uncertainty factor of 1000 (100 for inter- and intraspecies variation and 10 for a combination of the use of a LOAEL instead of a NOAEL and limitations of the database)
Limit of detection	0.01 µg/litre by GC/MS
Treatment achievability	1 µg/litre should be achievable using GAC
Guideline derivation • allocation to water • weight • consumption	10% of TDI 60-kg adult 2 litres/day

Toxicological review

In a short-term dietary study in rats, a variety of indications of hepatotoxicity as well as increased kidney weights in males were observed at the highest dose level. In a long-term dietary study, some toxic effects (hyperglycaemia in the mouse and hepatotoxicity in the rat) were present even at the lowest dose level. On the basis of available data, pendimethalin does not appear to have significant mutagenic activity. Long-term studies in mice and rats have not provided evidence of carcinogenicity; however, these studies have some important methodological limitations.

History of guideline development

The 1958 and 1963 WHO *International Standards for Drinking-water* did not refer to pendimethalin, but the 1971 International Standards suggested that pesticide residues that may occur in community water supplies make only a minimal contribution to the total daily intake of pesticides for the population served. Pendimethalin was not evaluated in the first edition of the *Guidelines for Drinking-water Quality*, published in 1984, but the 1993 Guidelines established a health-based guideline value of 0.02 mg/litre for pendimethalin in drinking-water.

Assessment date

The risk assessment was originally conducted in 1993. The Final Task Force Meeting in 2003 agreed that this risk assessment be brought forward to this edition of the *Guidelines for Drinking-water Quality*.

Principal reference

WHO (2003) *Pendimethalin in drinking-water. Background document for preparation of WHO Guidelines for drinking-water quality*. Geneva, World Health Organization (WHO/SDE/WSH/03.04/41).

12.98 Pentachlorophenol (PCP)

PCP (CAS No. 87-86-5) and other chlorophenols are used primarily for protecting wood from fungal growth. Food is usually the major source of exposure to PCP unless there is a specific local chlorophenol contamination of drinking-water or exposure from log homes treated with PCP.

Provisional guideline value	0.009 mg/litre The guideline value is considered provisional because of the variations in metabolism between experimental animals and humans.
Occurrence	Concentrations in water samples are usually below 10 µg/litre, although much higher concentrations in groundwater may be measured under certain conditions.
Basis of guideline derivation	Multistage modelling of tumour incidence in a US NTP bioassay without incorporation of a body surface area correction, recognizing that there are interspecies differences in metabolism between animals and humans, with an important metabolite formed in rats being only a minor metabolite in humans
Limit of detection	0.005–0.01 µg/litre by GC with ECD
Treatment achievability	0.4 µg/litre should be achievable using GAC
Additional comments	The concentration of PCP associated with a 10^{-5} upper-bound excess lifetime cancer risk is similar to the guideline value established in the second edition, so that guideline value is retained.

Toxicological review

IARC classified PCP in Group 2B (the agent is possibly carcinogenic to humans) on the basis of inadequate evidence of carcinogenicity in humans but sufficient evidence in experimental animals. There is suggestive, although inconclusive, evidence of the carcinogenicity of PCP from epidemiological studies of populations exposed to mixtures that include PCP. Conclusive evidence of carcinogenicity has been obtained in one animal species (mice). Although there are notable variations in metabolism between experimental animals and humans, it was considered prudent to treat PCP as a potential carcinogen.

History of guideline development

The 1958 and 1963 WHO *International Standards for Drinking-water* did not refer to PCP, but the 1971 International Standards suggested that pesticide residues that may occur in community water supplies make only a minimal contribution to the total daily intake of pesticides for the population served. In the first edition of the *Guidelines for Drinking-water Quality*, published in 1984, a health-based guideline value of 0.01 mg/litre was recommended for PCP. The 1993 Guidelines established a health-based guideline value of 0.009 mg/litre for PCP in drinking-water. This value was considered provisional because PCP was evaluated only at the Final Task Group Meeting on the basis of an EHC monograph (No. 71). The concentration of PCP associated with a 10^{-5} upper-bound excess lifetime cancer risk was found to be similar to the provisional guideline value established in 1993, and so that provisional guideline value was retained in the addendum to the Guidelines, published in 1998.

Assessment date
The risk assessment was originally conducted in 1998. The Final Task Force Meeting in 2003 agreed that this risk assessment be brought forward to this edition of the *Guidelines for Drinking-water Quality*.

Principal reference
WHO (2003) *Pentachlorophenol in drinking-water. Background document for preparation of WHO Guidelines for drinking-water quality*. Geneva, World Health Organization (WHO/SDE/WSH/03.04/62).

12.99 Permethrin
Permethrin (CAS No. 52645-53-1) is a contact insecticide effective against a broad range of pests in agriculture, forestry and public health. It is also a WHOPES-recommended larvicide used to control aquatic invertebrates in water mains. Permethrin is photodegraded both in water and on soil surfaces. Concentrations as high as 0.8 mg/litre have been recorded in surface water; levels in drinking-water have not been reported. In soil, permethrin is rapidly degraded by hydrolysis and microbial action under aerobic conditions. Exposure of the general population to permethrin is mainly via the diet.

Technical-grade permethrin is of low acute toxicity. The *cis* isomer is considerably more toxic than the *trans* isomer. IARC has classified permethrin in Group 3, as there are no human data and only limited data from animal studies. Permethrin is not genotoxic.

An ADI of 0.05 mg/kg of body weight for 2:3 and 1:3 *cis*:*trans*-permethrin has been derived by applying an uncertainty factor of 100 to a NOAEL of 100 mg/kg, equivalent to 5 mg/kg of body weight per day, from a 2-year dietary study in rats, based on clinical signs and changes in body and organ weights and blood chemistry, and a NOAEL of 5 mg/kg of body weight per day from a 1-year study in dogs, based on reduced body weight at 100 mg/kg of body weight per day. A health-based value of 20 µg/litre can be calculated for permethrin by allocating 1% of this ADI to drinking-water (because there is significant exposure to permethrin from the environment). However, because permethrin occurs at concentrations well below those at which toxic effects are observed, it is not considered necessary to derive a guideline value.

History of guideline development
The 1958 and 1963 WHO *International Standards for Drinking-water* did not refer to permethrin, but the 1971 International Standards suggested that pesticide residues that may occur in community water supplies make only a minimal contribution to

the total daily intake of pesticides for the population served. Permethrin was not evaluated in the first edition of the *Guidelines for Drinking-water Quality*, published in 1984, but the 1993 Guidelines established a health-based guideline value of 0.02 mg/litre for permethrin in drinking-water, based on an ADI established by JMPR in 1987 for 2:3 and 1:3 *cis*:*trans*-permethrin and recognizing the significant exposure to permethrin from the environment. It was noted that if permethrin is to be used as a larvicide for the control of mosquitos and other insects of health significance in drinking-water sources, the share of the ADI allocated to drinking-water may be increased.

Assessment date
The risk assessment was conducted in 2003.

Principal references
FAO/WHO (2000) *Pesticide residues in food – 1999. Evaluations – 1999. Part II – Toxicology*. Geneva, World Health Organization, Joint FAO/WHO Meeting on Pesticide Residues (WHO/PCS/00.4).

WHO (2003) *Permethrin in drinking-water. Background document for preparation of WHO Guidelines for drinking-water quality*. Geneva, World Health Organization (WHO/SDE/WSH/03.04/111).

12.100 pH
No health-based guideline value is proposed for pH. Although pH usually has no direct impact on consumers, it is one of the most important operational water quality parameters (see chapter 10).

History of guideline development
The 1958 WHO *International Standards for Drinking-water* suggested that pH less than 6.5 or greater than 9.2 would markedly impair the potability of the water. The 1963 and 1971 International Standards retained the pH range 6.5–9.2 as the allowable or permissible range. In the first edition of the *Guidelines for Drinking-water Quality*, published in 1984, a guideline value pH range of 6.5–8.5 was established for pH, based on aesthetic considerations. It was noted that the acceptable range of pH may be broader in the absence of a distribution system. No health-based guideline value was proposed for pH in the 1993 Guidelines. Although pH usually has no direct impact on consumers, it is one of the most important operational water quality parameters, the optimum pH required often being in the range 6.5–9.5.

Assessment date
The risk assessment was originally conducted in 1993. The Final Task Force Meeting in 2003 agreed that this risk assessment be brought forward to this edition of the *Guidelines for Drinking-water Quality*.

Principal reference
WHO (2003) *pH in drinking-water. Background document for preparation of WHO Guidelines for drinking-water quality*. Geneva, World Health Organization (WHO/SDE/WSH/03.04/12).

12.101 2-Phenylphenol and its sodium salt

2-Phenylphenol (CAS No. 90-43-7) is used as a disinfectant, bactericide and virucide. In agriculture, it is used in disinfecting fruits, vegetables and eggs. It is also used as a general surface disinfectant in hospitals, nursing homes, veterinary hospitals, poultry farms, dairy farms, commercial laundries, barbershops and food processing plants. 2-Phenylphenol is readily degraded in surface waters, with a half-life of about 1 week in river water.

2-Phenylphenol has been determined to be of low toxicity. Both 2-phenylphenol and its sodium salt are carcinogenic in male rats, and 2-phenylphenol is carcinogenic in male mice. However, urinary bladder tumours observed in male rats and liver tumours observed in male mice exposed to 2-phenylphenol appear to be threshold phenomena that are species- and sex-specific. JMPR has concluded that 2-phenylphenol is unlikely to represent a carcinogenic risk to humans. Although a working group convened by IARC has classified 2-phenylphenol, sodium salt, in Group 2B (possibly carcinogenic to humans) and 2-phenylphenol in Group 3 (not classifiable as to its carcinogenicity to humans), JMPR noted that the IARC classification is based on hazard identification, not risk assessment, and is furthermore limited to published literature, excluding unpublished studies on toxicity and carcinogenicity. JMPR also concluded that there are unresolved questions about the genotoxic potential of 2-phenylphenol.

A health-based value of 1 mg/litre can be calculated for 2-phenylphenol on the basis of an ADI of 0.4 mg/kg of body weight, based on a NOAEL of 39 mg/kg of body weight per day in a 2-year toxicity study for decreased body weight gain and hyperplasia of the urinary bladder and carcinogenicity of the urinary bladder in male rats, using an uncertainty factor of 100. Because of its low toxicity, however, the health-based value derived for 2-phenylphenol is much higher than 2-phenylphenol concentrations likely to be found in drinking-water. Under usual conditions, therefore, the presence of 2-phenylphenol in drinking-water is unlikely to represent a hazard to human health. For this reason, the establishment of a guideline value for 2-phenylphenol is not deemed necessary.

History of guideline development
The 1958 and 1963 WHO *International Standards for Drinking-water* did not refer to 2-phenylphenol, but the 1971 International Standards suggested that pesticide residues that may occur in community water supplies make only a minimal contribution to the total daily intake of pesticides for the population served. 2-Phenylphenol was not evaluated in the first edition of the *Guidelines for Drinking-water Quality*,

published in 1984, in the second edition, published in 1993, or in the addendum to the second edition, published in 1998.

Assessment date
The risk assessment was conducted in 2003.

Principal references
FAO/WHO (2000) *Pesticide residues in food – 1999 evaluations. Part II – Toxicological.* Geneva, World Health Organization, Joint FAO/WHO Meeting on Pesticide Residues (WHO/PCS/00.4).

WHO (2003) *2-Phenylphenol and its sodium salt in drinking-water. Background document for preparation of WHO Guidelines for drinking-water quality.* Geneva, World Health Organization (WHO/SDE/WSH/03.04/69).

12.102 Polynuclear aromatic hydrocarbons (PAHs)

PAHs form a class of diverse organic compounds each containing two or more fused aromatic rings of carbon and hydrogen atoms. Most PAHs enter the environment via the atmosphere from a variety of combustion processes and pyrolysis sources. Owing to their low solubility and high affinity for particulate matter, they are not usually found in water in notable concentrations. The main source of PAH contamination in drinking-water is usually the coal-tar coating of drinking-water distribution pipes, used to protect the pipes from corrosion. Fluoranthene is the most commonly detected PAH in drinking-water and is associated primarily with coal-tar linings of cast iron or ductile iron distribution pipes. PAHs have been detected in a variety of foods as a result of the deposition of airborne PAHs and in fish from contaminated waters. PAHs are also formed during some methods of food preparation, such as char-broiling, grilling, roasting, frying or baking. For the general population, the major routes of exposure to PAHs are from food and ambient and indoor air. The use of open fires for heating and cooking may increase PAH exposure, especially in developing countries. Where there are elevated levels of contamination by coal-tar coatings of water pipes, PAH intake from drinking-water could equal or even exceed that from food.

Guideline value for benzo[a]pyrene (BaP)	0.0007 mg/litre (0.7 µg/litre)
Occurrence	PAH levels in uncontaminated groundwater usually in range 0–5 ng/litre; concentrations in contaminated groundwater may exceed 10 µg/litre; typical concentration range for sum of selected PAHs in drinking-water is from about 1 ng/litre to 11 µg/litre

Basis of guideline derivation	Based on an oral carcinogenicity study in mice and calculated using a two-stage birth–death mutation model, which incorporates variable dosing patterns and time of killing; quantification of dose–response for tumours, on the basis of new studies in which the carcinogenicity of BaP was examined following oral administration in mice, but for which the number of dose groups was smaller, confirms this value
Limit of detection	0.01 µg/litre by GC/MS and reverse-phase HPLC with a fluorescence detector
Treatment achievability	0.05 µg/litre should be achievable using coagulation
Additional comments	• The presence of significant concentrations of BaP in drinking-water in the absence of very high concentrations of fluoranthene indicates the presence of coal-tar particles, which may arise from seriously deteriorating coal-tar pipe linings. • It is recommended that the use of coal-tar-based and similar materials for pipe linings and coatings on storage tanks be discontinued.

Toxicological review

Evidence that mixtures of PAHs are carcinogenic to humans comes primarily from occupational studies of workers following inhalation and dermal exposure. No data are available for humans for the oral route of exposure. There are few data on the oral toxicity of PAHs other than BaP, particularly in drinking-water. Relative potencies of carcinogenic PAHs have been determined by comparison of data from dermal and other studies. The order of potencies is consistent, and this scheme therefore provides a useful indicator of PAH potency relative to BaP.

A health-based value of 4 µg/litre can be calculated for fluoranthene on the basis of a NOAEL of 125 mg/kg of body weight per day for increased serum glutamate–pyruvate transaminase levels, kidney and liver pathology, and clinical and haematological changes in a 13-week oral gavage study in mice, using an uncertainty factor of 10 000 (100 for inter- and intraspecies variation, 10 for the use of a subchronic study and inadequate database and 10 because of clear evidence of cocarcinogenicity with BaP in mouse skin painting studies). However, this health-based value is significantly above the concentrations normally found in drinking-water. Under usual conditions, therefore, the presence of fluoranthene in drinking-water does not represent a hazard to human health. For this reason, the establishment of a guideline value for fluoranthene is not deemed necessary.

History of guideline development

The 1958 and 1963 WHO *International Standards for Drinking-water* did not refer to PAHs. The 1971 International Standards stated that some PAHs are known to be carcinogenic and that the concentrations of six representative PAH compounds (fluoranthene, 3,4-benzfluoranthene, 11,12-benzfluoranthene, 3,4-benzpyrene, 1,12-benzpyrene and indeno [1,2,3-cd] pyrene) should therefore not, in general, exceed 0.0002 mg/litre. In the first edition of the *Guidelines for Drinking-water Quality*, published in 1984, the only PAH for which there was sufficient substantiated toxicological evidence to set a guideline value was BaP. A health-based guideline value of 0.00001 mg/litre was recommended for BaP, while noting that the mathematical model appropriate to chemical carcinogens that was used in its derivation involved considerable uncertainty. It was also recommended that the control of PAHs in drinking-water should be based on the concept that the levels found in unpolluted groundwater should not be exceeded. The 1993 Guidelines concluded that there were insufficient data available to derive drinking-water guidelines for PAHs other than BaP. The guideline value for BaP, corresponding to an upper-bound excess lifetime cancer risk of 10^{-5}, was calculated to be 0.0007 mg/litre. This guideline value was retained in the addendum to the second edition of the Guidelines, published in 1998, as it was confirmed by new studies on the carcinogenicity of the compound. It was also recommended that the use of coal-tar-based and similar materials for pipe linings and coatings on storage tanks be discontinued. Although a health-based value for fluoranthene was calculated in the addendum, it was significantly above the concentrations found in drinking-water, and it was concluded that, under usual conditions, the presence of fluoranthene in drinking-water does not represent a hazard to human health; thus, the establishment of a guideline value for fluoranthene was not deemed necessary. As there are few data on the oral toxicity of other PAHs, particularly in drinking-water, relative potencies of carcinogenic PAHs were determined by comparison of data from dermal and other studies, which provides a useful indicator of PAH potency relative to BaP.

Assessment date

The risk assessment was originally conducted in 1998. The Final Task Force Meeting in 2003 agreed that this risk assessment be brought forward to this edition of the *Guidelines for Drinking-water Quality*.

Principal reference

WHO (2003) *Polynuclear aromatic hydrocarbons in drinking-water. Background document for preparation of WHO Guidelines for drinking-water quality*. Geneva, World Health Organization (WHO/SDE/WSH/03.04/59).

12.103 Propanil

Propanil (CAS No. 709-98-8) is a contact post-emergence herbicide used to control broad-leaved and grassy weeds, mainly in rice. It is a mobile compound with affinity for the water compartment. Propanil is not, however, persistent, being easily transformed under natural conditions to several metabolites. Two of these metabolites, 3,4-dichloroaniline and 3,3′,4,4′-tetrachloroazobenzene, are more toxic and more persistent than the parent compound. Although used in a number of countries, propanil has only occasionally been detected in groundwater.

Although a health-based value for propanil can be derived, this has not been done, because propanil is readily transformed into metabolites that are more toxic. Therefore, a guideline value for the parent compound is considered inappropriate, and there are inadequate data on the metabolites to allow the derivation of a guideline value for them. Authorities should consider the possible presence in water of more toxic environmental metabolites.

History of guideline development
The 1958 and 1963 WHO *International Standards for Drinking-water* did not refer to propanil, but the 1971 International Standards suggested that pesticide residues that may occur in community water supplies make only a minimal contribution to the total daily intake of pesticides for the population served. Propanil was not evaluated in the first edition of the *Guidelines for Drinking-water Quality*, published in 1984, but the 1993 Guidelines established a health-based guideline value of 0.02 mg/litre for propanil in drinking-water, noting that in applying this guideline, authorities should consider the possible presence of more toxic metabolites in water.

Assessment date
The risk assessment was conducted in 2003.

Principal reference
WHO (2003) *Propanil in drinking-water. Background document for preparation of WHO Guidelines for drinking-water quality*. Geneva, World Health Organization (WHO/SDE/WSH/03.04/112).

12.104 Pyriproxyfen
Pyriproxyfen (CAS No. 95737-68-1) is a broad-spectrum insect growth regulator with insecticidal activity against public health insect pests. It is a WHOPES-recommended insecticide for the control of mosquito larvae. In agriculture and horticulture, pyriproxyfen has registered uses for the control of scale, whitefly, bollworm, jassids, aphids and cutworms. Pyriproxyfen degrades rapidly in soil under aerobic conditions, with a half-life of 6.4–36 days. It disappeared from aerobic lake water–sediment systems with half-lives of 16 and 21 days. Pyriproxyfen appeared to be degraded much more slowly in anaerobic lake water–sediment systems. As pyriproxyfen is a new pesticide, few environmental data have been collected. Intake of pyriproxyfen from all sources is generally low and below the ADI.

Guideline value	0.3 mg/litre
Occurrence	No detectable concentrations found in surface water in the USA
ADI	0.1 mg/kg of body weight based on an overall NOAEL of 10 mg/kg of body weight per day for increased relative liver weight and increased total plasma cholesterol concentration in male dogs in two 1-year toxicity studies, using an uncertainty factor of 100
Limit of detection	No information found
Treatment achievability	No data available; 1 µg/litre should be achievable using GAC
Guideline derivation • allocation to water • weight • consumption	 10% of ADI 60-kg adult 2 litres/day

Toxicological review

JMPR concluded that pyriproxyfen was not carcinogenic or genotoxic. In short- and long-term studies of the effects of pyriproxyfen in mice, rats and dogs, the liver (increases in liver weight and changes in plasma lipid concentrations, particularly cholesterol) was the main toxicological target.

History of guideline development

The 1958 and 1963 WHO *International Standards for Drinking-water* did not refer to pyriproxyfen, but the 1971 International Standards suggested that pesticide residues that may occur in community water supplies make only a minimal contribution to the total daily intake of pesticides for the population served. Pyriproxyfen was not evaluated in the first edition of the *Guidelines for Drinking-water Quality*, published in 1984, in the second edition, published in 1993, or in the addendum to the second edition, published in 1998.

Assessment date

The risk assessment was conducted in 2003.

Principal references

FAO/WHO (2000) *Pesticide residues in food – 1999 evaluations. Part II – Toxicological.* Geneva, World Health Organization, Joint FAO/WHO Meeting on Pesticide Residues (WHO/PCS/00.4).

WHO (2003) *Pyriproxyfen in drinking-water. Background document for preparation of WHO Guidelines for drinking-water quality.* Geneva, World Health Organization (WHO/SDE/WSH/03.04/113).

12.105 Selenium

Selenium is present in the Earth's crust, often in association with sulfur-containing minerals. Selenium is an essential trace element, and foodstuffs such as cereals, meat and fish are the principal source of selenium in the general population. Levels in food also vary greatly according to geographical area of production.

Guideline value	0.01 mg/litre
Occurrence	Levels in drinking-water vary greatly in different geographical areas but are usually much less than 0.01 mg/litre.
NOAEL in humans	Estimated to be about 4 µg/kg of body weight per day, based on data in which a group of 142 persons with a mean daily intake of 4 µg/kg body weight showed no clinical or biochemical signs of selenium toxicity
Limit of detection	0.5 µg/litre by AAS with hydride generation
Treatment achievability	0.01 mg/litre should be achievable using coagulation for selenium(IV) removal; selenium(VI) is not removed by conventional treatment processes
Guideline derivation • allocation to wate • weight • consumption	10% of NOAEL 60-kg adult 2 litres/day

Toxicological review

Selenium is an essential element for humans, with a recommended daily intake of about 1 µg/kg of body weight for adults. Selenium compounds have been shown to be genotoxic in *in vitro* systems with metabolic activation, but not in humans. There was no evidence of teratogenic effects in monkeys. Long-term toxicity in rats is characterized by depression of growth and liver pathology. In humans, the toxic effects of long-term selenium exposure are manifested in nails, hair and liver. Data from China indicate that clinical and biochemical signs occur at a daily intake above 0.8 mg. Daily intakes of Venezuelan children with clinical signs were estimated to be about 0.7 mg on the basis of their blood levels and the Chinese data on the relationship between blood level and intake. Effects on synthesis of a liver protein were also seen in a small group of patients with rheumatoid arthritis given selenium at a rate of 0.25 mg/day in addition to selenium from food. No clinical or biochemical signs of selenium toxicity were reported in a group of 142 persons with a mean daily intake of 0.24 mg (maximum 0.72 mg) from food.

History of guideline development

The 1958 WHO *International Standards for Drinking-water* recommended a maximum allowable concentration of 0.05 mg/litre for selenium, based on health concerns. In the 1963 International Standards, this value was lowered to 0.01 mg/litre, which was retained in the 1971 International Standards as a tentative upper concentration limit, while recognizing that selenium is an essential trace element for some species. In the first edition of the *Guidelines for Drinking-water Quality*, published in 1984, the guideline value of 0.01 mg/litre was again retained, although it was noted that in areas of relatively higher or lower selenium dietary intake, the guideline value may have to be modified accordingly. The 1993 Guidelines proposed a health-based guideline value of 0.01 mg/litre on the basis of human studies.

Assessment date
The risk assessment was originally conducted in 1993. The Final Task Force Meeting in 2003 agreed that this risk assessment be brought forward to this edition of the *Guidelines for Drinking-water Quality*.

Principal reference
WHO (2003) *Selenium in drinking-water. Background document for preparation of WHO Guidelines for drinking-water quality*. Geneva, World Health Organization (WHO/SDE/WSH/03.04/13).

12.106 Silver
Silver occurs naturally mainly in the form of its very insoluble and immobile oxides, sulfides and some salts. It has occasionally been found in groundwater, surface water and drinking-water at concentrations above 5 µg/litre. Levels in drinking-water treated with silver for disinfection may be above 50 µg/litre. Recent estimates of daily intake are about 7 µg per person.

Only a small percentage of silver is absorbed. Retention rates in humans and laboratory animals range between 0 and 10%.

The only obvious sign of silver overload is argyria, a condition in which skin and hair are heavily discoloured by silver in the tissues. An oral NOAEL for argyria in humans for a total lifetime intake of 10 g of silver was estimated on the basis of human case reports and long-term animal experiments.

The low levels of silver in drinking-water, generally below 5 µg/litre, are not relevant to human health with respect to argyria. On the other hand, special situations exist where silver salts may be used to maintain the bacteriological quality of drinking-water. Higher levels of silver, up to 0.1 mg/litre (this concentration gives a total dose over 70 years of half the human NOAEL of 10 g), could be tolerated in such cases without risk to health.

There are no adequate data with which to derive a health-based guideline value for silver in drinking-water.

History of guideline development
The 1958, 1963 and 1971 WHO *International Standards for Drinking-water* did not refer to silver. In the first edition of the *Guidelines for Drinking-Water Quality*, published in 1984, it was not considered necessary to establish a guideline value for silver in drinking-water. No health-based guideline value for silver was proposed in the 1993 Guidelines. Where silver salts are used to maintain the bacteriological quality of

drinking-water, levels of silver up to 0.1 mg/litre can be tolerated without risk to health.

Assessment date
The risk assessment was originally conducted in 1993. The Final Task Force Meeting in 2003 agreed that this risk assessment be brought forward to this edition of the *Guidelines for Drinking-water Quality*.

Principal reference
WHO (2003) *Silver in drinking-water. Background document for preparation of WHO Guidelines for drinking-water quality*. Geneva, World Health Organization (WHO/SDE/WSH/03.04/14).

12.107 Simazine

Simazine (CAS No. 122-34-9) is a pre-emergence herbicide used on a number of crops as well as in non-crop areas. It is fairly resistant to physical and chemical dissipation processes in the soil. It is persistent and mobile in the environment.

Guideline value	0.002 mg/litre
Occurrence	Frequently detected in groundwater and surface water at concentrations of up to a few micrograms per litre
TDI	0.52 µg/kg of body weight, based on a NOAEL of 0.52 mg/kg of body weight from a long-term study in the rat (based on weight changes, effects on haematological parameters and an increase in mammary tumours) and an uncertainty factor of 1000 (100 for inter- and intraspecies variation and 10 for possible non-genotoxic carcinogenicity)
Limit of detection	0.01 µg/litre by GC/MS; 0.1–0.2 µg/litre by GC with flame thermionic detection
Treatment achievability	0.1 µg/litre should be achievable using GAC
Guideline derivation • allocation to water • weight • consumption	 10% of TDI 60-kg adult 2 litres/day

Toxicological review
Simazine does not appear to be genotoxic in mammalian systems. Recent studies have shown an increase in mammary tumours in the female rat but no effects in the mouse. IARC has classified simazine in Group 3.

History of guideline development
The 1958 and 1963 WHO *International Standards for Drinking-water* did not refer to simazine, but the 1971 International Standards suggested that pesticide residues that

may occur in community water supplies make only a minimal contribution to the total daily intake of pesticides for the population served. Simazine was not evaluated in the first edition of the *Guidelines for Drinking-water Quality*, published in 1984, but the 1993 Guidelines established a health-based guideline value of 0.002 mg/litre for simazine in drinking-water.

Assessment date
The risk assessment was originally conducted in 1993. The Final Task Force Meeting in 2003 agreed that this risk assessment be brought forward to this edition of the *Guidelines for Drinking-water Quality*.

Principal reference
WHO (2003) *Simazine in drinking-water. Background document for preparation of WHO Guidelines for drinking-water quality*. Geneva, World Health Organization (WHO/SDE/WSH/03.04/42).

12.108 Sodium

Sodium salts (e.g., sodium chloride) are found in virtually all food (the main source of daily exposure) and drinking-water. Although concentrations of sodium in potable water are typically less than 20 mg/litre, they can greatly exceed this in some countries. The levels of sodium salts in air are normally low in relation to those in food or water. It should be noted that some water softeners can add significantly to the sodium content of drinking-water.

No firm conclusions can be drawn concerning the possible association between sodium in drinking-water and the occurrence of hypertension. Therefore, no health-based guideline value is proposed. However, concentrations in excess of 200 mg/litre may give rise to unacceptable taste (see chapter 10).

History of guideline development
The 1958, 1963 and 1971 WHO *International Standards for Drinking-water* did not refer to sodium. In the first edition of the *Guidelines for Drinking-water Quality*, published in 1984, it was concluded that there was insufficient evidence to justify a guideline value for sodium in water based on health risk considerations, but it was noted that intake of sodium from drinking-water may be of greater significance in persons who require a sodium-restricted diet and bottle-fed infants. A guideline value of 200 mg/litre was established for sodium based on taste considerations. No health-based guideline value was proposed for sodium in the 1993 Guidelines, as no firm conclusions could be drawn concerning the possible association between sodium in drinking-water and the occurrence of hypertension. However, concentrations in excess of 200 mg/litre may give rise to unacceptable taste.

Assessment date
The risk assessment was originally conducted in 1993. The Final Task Force Meeting in 2003 agreed that this risk assessment be brought forward to this edition of the *Guidelines for Drinking-water Quality*.

Principal reference
WHO (2003) *Sodium in drinking-water. Background document for preparation of WHO Guidelines for drinking-water quality*. Geneva, World Health Organization (WHO/SDE/WSH/03.04/15).

12.109 Styrene

Styrene, which is used primarily for the production of plastics and resins, is found in trace amounts in surface water, drinking-water and food. In industrial areas, exposure via air can result in intake of a few hundred micrograms per day. Smoking may increase daily exposure by up to 10-fold.

Guideline value	0.02 mg/litre
Occurrence	Has been detected in drinking-water and surface water at concentrations below 1 µg/litre
TDI	7.7 µg/kg of body weight, based on a NOAEL of 7.7 mg/kg of body weight per day for decreased body weight observed in a 2-year drinking-water study in rats, and using an uncertainty factor of 1000 (100 for inter- and intraspecies variation and 10 for the carcinogenicity and genotoxicity of the reactive intermediate styrene-7,8-oxide)
Limit of detection	0.3 µg/litre by GC with photoionization detection and confirmation by MS
Treatment achievability	0.02 mg/litre may be achievable using GAC
Guideline derivation • allocation to water • weight • consumption	 10% of TDI 60-kg adult 2 litres/day
Additional comments	Styrene may affect the acceptability of drinking-water at the guideline value.

Toxicological review
Following oral or inhalation exposure, styrene is rapidly absorbed and widely distributed in the body, with a preference for lipid depots. It is metabolized to the active intermediate styrene-7,8-oxide, which is conjugated with glutathione or further metabolized. Metabolites are rapidly and almost completely excreted in urine. Styrene has a low acute toxicity. In short-term toxicity studies in rats, impairment of glutathione transferase activity and reduced glutathione concentrations were observed. In *in vitro* tests, styrene has been shown to be mutagenic in the presence of metabolic

activation only. In *in vitro* as well as in *in vivo* studies, chromosomal aberrations have been observed, mostly at high doses of styrene. The reactive intermediate styrene-7,8-oxide is a direct-acting mutagen. In long-term studies, orally administered styrene increased the incidence of lung tumours in mice at high dose levels but had no carcinogenic effect in rats. Styrene-7,8-oxide was carcinogenic in rats after oral administration. IARC has classified styrene in Group 2B. The available data suggest that the carcinogenicity of styrene is due to overloading of the detoxification mechanism for styrene-7,8-oxide (e.g., glutathione depletion).

History of guideline development
The 1958, 1963 and 1971 WHO *International Standards for Drinking-water* and the first edition of the *Guidelines for Drinking-water Quality*, published in 1984, did not refer to styrene. The 1993 Guidelines established a health-based guideline value of 0.02 mg/litre for styrene, noting that styrene may affect the acceptability of drinking-water at this concentration.

Assessment date
The risk assessment was originally conducted in 1993. The Final Task Force Meeting in 2003 agreed that this risk assessment be brought forward to this edition of the *Guidelines for Drinking-water Quality*.

Principal reference
WHO (2003) *Styrene in drinking-water. Background document for preparation of WHO Guidelines for drinking-water quality*. Geneva, World Health Organization (WHO/SDE/WSH/03.04/27).

12.110 Sulfate

Sulfates occur naturally in numerous minerals and are used commercially, principally in the chemical industry. They are discharged into water in industrial wastes and through atmospheric deposition; however, the highest levels usually occur in groundwater and are from natural sources. In general, the average daily intake of sulfate from drinking-water, air and food is approximately 500 mg, food being the major source. However, in areas with drinking-water supplies containing high levels of sulfate, drinking-water may constitute the principal source of intake.

The existing data do not identify a level of sulfate in drinking-water that is likely to cause adverse human health effects. The data from a liquid diet piglet study and from tap water studies with human volunteers indicate a laxative effect at concentrations of 1000–1200 mg/litre but no increase in diarrhoea, dehydration or weight loss.

No health-based guideline is proposed for sulfate. However, because of the gastrointestinal effects resulting from ingestion of drinking-water containing high sulfate levels, it is recommended that health authorities be notified of sources of drinking-water that contain sulfate concentrations in excess of 500 mg/litre. The presence of sulfate in drinking-water may also cause noticeable taste (see chapter 10) and may contribute to the corrosion of distribution systems.

History of guideline development

The 1958 WHO *International Standards for Drinking-water* suggested that concentrations of sulfate greater than 400 mg/litre would markedly impair the potability of the water. The 1963 and 1971 International Standards retained this value as a maximum allowable or permissible concentration. The first two editions of the International Standards also suggested that concentrations of magnesium plus sodium sulfate in excess of 1000 mg/litre would markedly impair drinking-water potability. In the first edition of the *Guidelines for Drinking-water Quality*, published in 1984, a guideline value of 400 mg/litre for sulfate was established, based on taste considerations. No health-based guideline value for sulfate was proposed in the 1993 Guidelines. However, because of the gastrointestinal effects resulting from ingestion of drinking-water containing high sulfate levels, it was recommended that health authorities be notified of sources of drinking-water that contain sulfate concentrations in excess of 500 mg/litre. The presence of sulfate in drinking-water may also cause noticeable taste at concentrations above 250 mg/litre and may contribute to the corrosion of distribution systems.

Assessment date

The risk assessment was conducted in 2003.

Principal reference

WHO (2003) *Sulfate in drinking-water. Background document for preparation of WHO Guidelines for drinking-water quality*. Geneva, World Health Organization (WHO/SDE/WSH/03.04/114).

12.111 2,4,5-T (2,4,5-Trichlorophenoxyacetic acid)

The half-lives for degradation of chlorophenoxy herbicides, including 2,4,5-T (CAS No. 93-76-5), in the environment are in the order of several days. Chlorophenoxy herbicides are not often found in food.

Guideline value	0.009 mg/litre
Occurrence	Chlorophenoxy herbicides not frequently found in drinking- water; when detected, concentrations are usually no greater than a few micrograms per litre
TDI	3 µg/kg of body weight, based on a NOAEL of 3 mg/kg of body weight for reduced body weight gain, increased liver and kidney weights and renal toxicity in a 2-year study in rats, with an uncertainty factor of 1000 (100 for inter- and intraspecies variation and 10 to take into consideration the suggested association between 2,4,5-T and soft tissue sarcoma and non- Hodgkin lymphoma in epidemiological studies)

Limit of detection	0.02 µg/litre by GC with an ECD
Treatment achievability	1 µg/litre should be achievable using GAC
Guideline derivation • allocation to water • weight • consumption	10% of TDI 60-kg adult 2 litres/day

Toxicological review

Chlorophenoxy herbicides, as a group, have been classified in Group 2B by IARC. However, the available data from studies in exposed populations and animals do not permit assessment of the carcinogenic potential to humans of any specific chlorophenoxy herbicide. Therefore, drinking-water guidelines for these compounds are based on a threshold approach for other toxic effects. The NOAEL for reproductive effects (reduced neonatal survival, decreased fertility, reduced relative liver weights and thymus weights in litters) of dioxin-free (<0.03 µg/kg) 2,4,5-T in a three-generation reproduction study in rats is the same as the NOAEL for reduced body weight gain, increased liver and kidney weights and renal toxicity in a toxicity study in which rats were fed 2,4,5-T (practically free from dioxin contamination) in the diet for 2 years.

History of guideline development

The 1958 and 1963 WHO *International Standards for Drinking-water* did not refer to chlorophenoxy herbicides, including 2,4,5-T, but the 1971 International Standards suggested that pesticide residues that may occur in community water supplies make only a minimal contribution to the total daily intake of pesticides for the population served. 2,4,5-T was not evaluated in the first edition of the *Guidelines for Drinking-water Quality*, published in 1984, but the 1993 Guidelines established a health-based guideline value of 0.009 mg/litre for 2,4,5-T.

Assessment date

The risk assessment was originally conducted in 1993. The Final Task Force Meeting in 2003 agreed that this risk assessment be brought forward to this edition of the *Guidelines for Drinking-water Quality*.

Principal reference

WHO (2003) *Chlorophenoxy herbicides (excluding 2,4-D and MCPA) in drinking-water. Background document for preparation of WHO Guidelines for drinking-water quality.* Geneva, World Health Organization (WHO/SDE/WSH/03.04/44).

12.112 Terbuthylazine (TBA)

TBA (CAS No. 5915-41-3), a herbicide that belongs to the chlorotriazine family, is used in both pre- and post-emergence treatment of a variety of agricultural crops and in forestry. Degradation of TBA in natural water depends on the presence of sediments and biological activity.

Guideline value	0.007 mg/litre
Occurrence	Concentrations in water seldom exceed 0.2 µg/litre, although higher concentrations have been observed.
TDI	2.2 µg/kg of body weight, based on a NOAEL of 0.22 mg/kg of body weight for decreased body weight gain at the next higher dose in a 2-year toxicity/carcinogenicity study in rats, with an uncertainty factor of 100 (for inter- and intraspecies variation)
Limit of detection	0.1 µg/litre by HPLC with UV detection
Treatment achievability	0.1 µg/litre should be achievable using GAC
Guideline derivation • allocation to water • weight • consumption	 10% of TDI 60-kg adult 2 litres/day

Toxicological review
There is no evidence that TBA is carcinogenic or mutagenic. In long-term dietary studies in rats, effects on red blood cell parameters in females, an increased incidence of non-neoplastic lesions in the liver, lung, thyroid and testis and a slight decrease in body weight gain were observed.

History of guideline development
The 1958 and 1963 WHO *International Standards for Drinking-water* did not refer to TBA, but the 1971 International Standards suggested that pesticide residues that may occur in community water supplies make only a minimal contribution to the total daily intake of pesticides for the population served. In the first edition of the *Guidelines for Drinking-water Quality*, published in 1984, no guideline value for triazine herbicides, which include TBA, was recommended after a detailed evaluation of the compounds. TBA was not evaluated in the second edition of the *Guidelines for Drinking-water Quality*, published in 1993. In the addendum to the second edition of the Guidelines, published in 1998, a health-based guideline value of 0.007 mg/litre was derived for TBA in drinking-water.

Assessment date
The risk assessment was originally conducted in 1998. The Final Task Force Meeting in 2003 agreed that this risk assessment be brought forward to this edition of the *Guidelines for Drinking-water Quality*.

Principal reference
WHO (2003) *Terbuthylazine in drinking-water. Background document for preparation of WHO Guidelines for drinking-water quality*. Geneva, World Health Organization (WHO/SDE/WSH/03.04/63).

12.113 Tetrachloroethene

Tetrachloroethene has been used primarily as a solvent in dry cleaning industries and to a lesser extent as a degreasing solvent. It is widespread in the environment and is found in trace amounts in water, aquatic organisms, air, foodstuffs and human tissue. The highest environmental levels of tetrachloroethene are found in the commercial dry cleaning and metal degreasing industries. Emissions can sometimes lead to high concentrations in groundwater. Tetrachloroethene in anaerobic groundwater may degrade to more toxic compounds, including vinyl chloride.

Guideline value	0.04 mg/litre
Occurrence	Concentrations in drinking-water are generally below 3 µg/litre, although much higher concentrations have been detected in well water (23 mg/litre) and in contaminated groundwater (1 mg/litre).
TDI	14 µg/kg of body weight, based on hepatotoxic effects observed in a 6-week gavage study in male mice and a 90-day drinking-water study in male and female rats, and taking into consideration carcinogenic potential (but not the short length of the study, in view of the database and considerations regarding the application of the dose via drinking-water in one of the two critical studies)
Limit of detection	0.2 µg/litre by GC with ECD; 4.1 µg/litre by GC/MS
Treatment achievability	0.001 mg/litre should be achievable using air stripping
Guideline derivation • allocation to water • weight • consumption	 10% of TDI 60-kg adult 2 litres/day

Toxicological review
At high concentrations, tetrachloroethene causes central nervous system depression. Lower concentrations of tetrachloroethene have been reported to damage the liver and the kidneys. IARC has classified tetrachloroethene in Group 2A. Tetrachloroethene has been reported to produce liver tumours in male and female mice, with some evidence of mononuclear cell leukaemia in male and female rats and kidney tumours in male rats. The overall evidence from studies conducted to assess the genotoxicity of tetrachloroethene, including induction of single-strand DNA breaks, mutation in germ cells and chromosomal aberrations *in vitro* and *in vivo*, indicates that tetrachloroethene is not genotoxic.

12. CHEMICAL FACT SHEETS

History of guideline development
The 1958, 1963 and 1971 WHO *International Standards for Drinking-water* did not refer to tetrachloroethene. In the first edition of the *Guidelines for Drinking-water Quality*, published in 1984, a tentative guideline value of 0.01 mg/litre was recommended; the guideline was designated as tentative because, although the carcinogenicity data did not justify a full guideline value, the compound was considered to have important health implications when present in drinking-water. The 1993 Guidelines established a health-based guideline value of 0.04 mg/litre for tetrachloroethene.

Assessment date
The risk assessment was originally conducted in 1993. The Final Task Force Meeting in 2003 agreed that this risk assessment be brought forward to this edition of the *Guidelines for Drinking-water Quality*.

Principal reference
WHO (2003) *Tetrachloroethene in drinking-water. Background document for preparation of WHO Guidelines for drinking-water quality*. Geneva, World Health Organization (WHO/SDE/WSH/03.04/23).

12.114 Toluene

Most toluene (in the form of benzene–toluene–xylene mixtures) is used in the blending of petrol. It is also used as a solvent and as a raw material in chemical production. The main exposure is via air. Exposure is increased by smoking and in traffic.

Guideline value	0.7 mg/litre
Occurrence	Concentrations of a few micrograms per litre have been found in surface water, groundwater and drinking-water; point emissions can lead to higher concentrations in groundwater (up to 1 mg/litre). It may also penetrate plastic pipes from contaminated soil.
TDI	223 µg/kg of body weight, based on a LOAEL of 312 mg/kg of body weight per day for marginal hepatotoxic effects observed in a 13-week gavage study in mice, correcting for 5 days per week dosing and using an uncertainty factor of 1000 (100 for inter- and intraspecies variation and 10 for the short duration of the study and use of a LOAEL instead of a NOAEL)
Limit of detection	0.13 µg/litre by GC with FID; 6 µg/litre by GC/MS
Treatment achievability	0.001 mg/litre should be achievable using air stripping
Guideline derivation • allocation to water • weight • consumption	 10% of TDI 60-kg adult 2 litres/day
Additional comments	The guideline value exceeds the lowest reported odour threshold for toluene in water.

Toxicological review
Toluene is absorbed completely from the gastrointestinal tract and rapidly distributed in the body, with a preference for adipose tissue. Toluene is rapidly metabolized and, following conjugation, excreted predominantly in urine. With occupational exposure to toluene by inhalation, impairment of the central nervous system and irritation of mucous membranes are observed. The acute oral toxicity is low. Toluene exerts embryotoxic and fetotoxic effects, but there is no clear evidence of teratogenic activity in laboratory animals and humans. In long-term inhalation studies in rats and mice, there is no evidence for carcinogenicity of toluene. Genotoxicity tests *in vitro* were negative, whereas *in vivo* assays showed conflicting results with respect to chromosomal aberrations. IARC has concluded that there is inadequate evidence for the carcinogenicity of toluene in both experimental animals and humans and classified it as Group 3 (not classifiable as to its carcinogenicity to humans).

History of guideline development
The 1958, 1963 and 1971 WHO *International Standards for Drinking-water* did not refer to toluene. In the first edition of the *Guidelines for Drinking-water Quality*, published in 1984, no guideline value was recommended after a detailed evaluation of the compound. The 1993 Guidelines established a health-based guideline value of 0.7 mg/litre for toluene, but noted that this value exceeds the lowest reported odour threshold for toluene in water (0.024 mg/litre).

Assessment date
The risk assessment was conducted in 2003.

Principal reference
WHO (2003) *Toluene in drinking-water. Background document for preparation of WHO Guidelines for drinking-water quality.* Geneva, World Health Organization (WHO/SDE/WSH/03.04/116).

12.115 Total dissolved solids (TDS)

TDS comprise inorganic salts (principally calcium, magnesium, potassium, sodium, bicarbonates, chlorides and sulfates) and small amounts of organic matter that are dissolved in water. TDS in drinking-water originate from natural sources, sewage, urban runoff and industrial wastewater. Salts used for road de-icing in some countries may also contribute to the TDS content of drinking-water. Concentrations of TDS in water vary considerably in different geological regions owing to differences in the solubilities of minerals.

Reliable data on possible health effects associated with the ingestion of TDS in drinking-water are not available, and no health-based guideline value is proposed. However, the presence of high levels of TDS in drinking-water may be objectionable to consumers (see chapter 10).

12. CHEMICAL FACT SHEETS

History of guideline development
The 1958 WHO *International Standards for Drinking-water* suggested that concentrations of total solids greater than 1500 mg/litre would markedly impair the potability of the water. The 1963 and 1971 International Standards retained this value as a maximum allowable or permissible concentration. In the first edition of the *Guidelines for Drinking-water Quality*, published in 1984, a guideline value of 1000 mg/litre was established for TDS, based on taste considerations. No health-based guideline value for TDS was proposed in the 1993 Guidelines, as reliable data on possible health effects associated with the ingestion of TDS in drinking-water were not available. However, the presence of high levels of TDS in drinking-water (greater than 1200 mg/litre) may be objectionable to consumers. Water with extremely low concentrations of TDS may also be unacceptable because of its flat, insipid taste.

Assessment date
The risk assessment was originally conducted in 1993. The Final Task Force Meeting in 2003 agreed that this risk assessment be brought forward to this edition of the *Guidelines for Drinking-water Quality*.

Principal reference
WHO (2003) *Total dissolved solids in drinking-water. Background document for preparation of WHO Guidelines for drinking-water quality*. Geneva, World Health Organization (WHO/SDE/WSH/03.04/16).

12.116 Trichloroacetic acid
Chlorinated acetic acids are formed from organic material during water chlorination.

Guideline value	0.2 mg/litre
Occurrence	Detected in US groundwater and surface water distribution systems at mean concentrations of 5.3 µg/litre (range <1.0–80 µg/litre) and 16 µg/litre (range <1.0–174 µg/litre), respectively; maximum concentration (200 µg/litre) measured in chlorinated water in Australia
TDI	32.5 µg/kg of body weight, based on a NOAEL of 32.5 mg/kg of body weight per day from a study in which decreased body weight, increased liver serum enzyme activity and liver histopathology were seen in rats exposed to trichloroacetate in drinking-water for 2 years, incorporating an uncertainty factor of 1000 (100 for inter- and intraspecies variation and 10 for database deficiencies, including the absence of a multigeneration reproductive study, the lack of a developmental study in a second species and the absence of full histopathological data in a second species)
Limit of detection	1 µg/litre by GC with ECD; 1 µg/litre by GC/MS

Treatment achievability	Trichloroacetic acid concentrations in drinking-water are generally below 0.1 mg/litre. Concentrations may be reduced by installing or optimizing coagulation to remove precursors and/or by controlling the pH during chlorination.
Guideline derivation • allocation to water • weight • consumption	 20% of TDI 60-kg adult 2 litres/day
Additional comments	A similar TDI for trichloroacetate was established by IPCS based on a NOAEL for hepatic toxicity in a long-term study in mice.

Toxicological review
Trichloroacetic acid has been shown to induce tumours in the liver of mice. It has given mixed results in *in vitro* assays for mutations and chromosomal aberrations and has been reported to cause chromosomal aberrations in *in vivo* studies. IARC has classified trichloroacetic acid in Group 3, not classifiable as to its carcinogenicity to humans. The weight of evidence indicates that trichloroacetic acid is not a genotoxic carcinogen.

History of guideline development
The 1958, 1963 and 1971 WHO *International Standards for Drinking-water* and the first edition of the *Guidelines for Drinking-water Quality*, published in 1984, did not refer to trichloroacetic acid. In the 1993 Guidelines, a provisional guideline value of 0.1 mg/litre was derived for trichloroacetic acid, with the provisional designation because of the limitations of the available toxicological database and because there were inadequate data to judge whether the guideline value was technically achievable. It was emphasized that difficulties in meeting the guideline value must never be a reason for compromising adequate disinfection.

Assessment date
The risk assessment was conducted in 2003.

Principal reference
WHO (2003) *Trichloroacetic acid in drinking-water. Background document for preparation of WHO Guidelines for drinking-water quality*. Geneva, World Health Organization (WHO/SDE/WSH/03.04/120).

12.117 Trichlorobenzenes (total)
Releases of trichlorobenzenes (TCBs) into the environment occur through their manufacture and use as industrial chemicals, chemical intermediates and solvents. TCBs are found in drinking-water, but rarely at levels above 1 µg/litre. General population exposure will primarily result from air and food.

The TCBs are of moderate acute toxicity. After short-term oral exposure, all three isomers show similar toxic effects, predominantly on the liver. Long-term toxicity and carcinogenicity studies via the oral route have not been carried out, but the data available suggest that all three isomers are non-genotoxic.

A health-based value of 20 μg/litre can be calculated for total TCBs on the basis of a TDI of 7.7 μg/kg of body weight, based on liver toxicity identified in a 13-week rat study, taking into consideration the short duration of the study. However, because TCBs occur at concentrations well below those at which toxic effects are observed, it is not considered necessary to derive a health-based guideline value. It should be noted that the health-based value exceeds the lowest reported odour threshold in water.

History of guideline development
The 1958, 1963 and 1971 WHO *International Standards for Drinking-water* did not refer to TCBs. In the first edition of the *Guidelines for Drinking-water Quality*, published in 1984, it was concluded that insufficient health data were available from which to derive a guideline value for 1,2,4-TCB. The 1993 Guidelines proposed a health-based guideline value of 0.02 mg/litre for total TCBs, because of the similarity in the toxicity of the three isomers, but noted that this value exceeds the lowest reported odour threshold in water (0.005 mg/litre for 1,2,4-TCB).

Assessment date
The risk assessment was conducted in 2003.

Principal reference
WHO (2003) *Trichlorobenzenes in drinking-water. Background document for preparation of WHO Guidelines for drinking-water quality*. Geneva, World Health Organization (WHO/SDE/WSH/03.04/117).

12.118 1,1,1-Trichloroethane

1,1,1-Trichloroethane is widely used as a cleaning solvent for electrical equipment, as a solvent for adhesives, coatings and textile dyes and as a coolant and lubricant. It is found mainly in the atmosphere, although it is mobile in soils and readily migrates to groundwaters. 1,1,1-Trichloroethane has been found in only a small proportion of surface waters and groundwaters, usually at concentrations of less than 20 μg/litre; higher concentrations (up to 150 μg/litre) have been observed in a few instances. There appears to be increasing exposure to 1,1,1-trichloroethane from other sources.

1,1,1-Trichloroethane is rapidly absorbed from the lungs and gastrointestinal tract, but only small amounts – about 6% in humans and 3% in experimental animals – are metabolized. Exposure to high concentrations can lead to hepatic steatosis (fatty liver) in both humans and laboratory animals. In a well conducted oral study in mice and rats, effects included reduced liver weight and changes in the kidney consistent with hyaline droplet neuropathy. IARC has placed 1,1,1-trichloroethane in Group 3. 1,1,1-Trichloroethane does not appear to be mutagenic.

A health-based value of 2 mg/litre can be calculated for 1,1,1-trichloroethane on the basis of a TDI of 0.6 mg/kg of body weight, based on changes in the kidney that were consistent with hyaline droplet nephropathy observed in a 13-week oral study in male rats, and taking into account the short duration of the study. However, because 1,1,1-trichloroethane occurs at concentrations well below those at which toxic effects are observed, it is not considered necessary to derive a guideline value.

History of guideline development
The 1958, 1963 and 1971 WHO *International Standards for Drinking-water* did not refer to 1,1,1-trichloroethane. In the first edition of the *Guidelines for Drinking-water Quality*, published in 1984, no guideline value was recommended after a detailed evaluation of the compound. The 1993 Guidelines proposed a provisional guideline value of 2 mg/litre for 1,1,1-trichloroethane. The value was provisional because it was based on an inhalation study rather than an oral study. It was strongly recommended that an adequate oral toxicity study be conducted to provide more acceptable data for the derivation of a guideline value.

Assessment date
The risk assessment was conducted in 2003.

Principal reference
WHO (2003) *1,1,1-Trichloroethane in drinking-water. Background document for preparation of WHO Guidelines for drinking-water quality*. Geneva, World Health Organization (WHO/SDE/WSH/03.04/65).

12.119 Trichloroethene

Trichloroethene is used mainly in dry cleaning and metal degreasing operations. Its use in industrialized countries has declined sharply since 1970. It is released mainly to the atmosphere but may be introduced into surface water and groundwater in industrial effluents. It is expected that exposure to trichloroethene from air will be greater than that from food or drinking-water. Trichloroethene in anaerobic groundwater may degrade to more toxic compounds, including vinyl chloride.

Provisional guideline value	0.07 mg/litre The guideline value is designated as provisional because of deficiencies in the toxicological database.
Occurrence	Found mostly in groundwater from which it is not lost to air; mean concentration of 2.1 µg/litre in a survey of drinking-water; also present in 24% of 158 non-random samples collected in a groundwater supply survey at a median level of 1 µg/litre and a maximum of 130 µg/litre

TDI	23.8 µg/kg of body weight (including allowance for 5 days per week dosing), based on a LOAEL of 100 mg/kg of body weight per day for minor effects on relative liver weight in a 6-week study in mice, using an uncertainty factor of 3000 (100 for intra- and interspecies variation, 10 for limited evidence of carcinogenicity and 3 in view of the short duration of the study and the use of a LOAEL rather than a NOAEL)
Limit of detection	0.037 µg/litre by capillary GC with ECD; 0.12 µg/litre by purge-and-trap packed column GC with ECD or microcoulometric detector; 0.2 µg/litre by purge-and-trap packed column GC/MS
Treatment achievability	0.02 mg/litre should be achievable using air stripping
Guideline derivation • allocation to water • weight • consumption	 10% of TDI 60-kg adult 2 litres/day

Toxicological review

The reactive epoxide trichloroethene oxide is an essential feature of the metabolic pathway. Trichloroethene has been classified by IARC in Group 3. It has been shown to induce lung and liver tumours in various strains of mice at toxic doses. However, there are no conclusive data to suggest that this chemical causes cancer in other species. Trichloroethene is a weakly active mutagen in bacteria and yeast.

History of guideline development

The 1958, 1963 and 1971 WHO *International Standards for Drinking-water* did not refer to trichloroethene. In the first edition of the *Guidelines for Drinking-water Quality*, published in 1984, a tentative guideline value of 0.03 mg/litre was recommended; the guideline was designated as tentative because, although carcinogenicity was observed in one species only, the compound occurs relatively frequently in drinking-water. The 1993 Guidelines established a provisional health-based guideline value of 0.07 mg/litre for trichloroethene. The value was provisional because an uncertainty factor of 3000 was used in its derivation.

Assessment date

The risk assessment was originally conducted in 1993. The Final Task Force Meeting in 2003 agreed that this risk assessment be brought forward to this edition of the *Guidelines for Drinking-water Quality*.

Principal reference

WHO (2003) *Trichloroethene in drinking-water. Background document for preparation of WHO Guidelines for drinking-water quality*. Geneva, World Health Organization (WHO/SDE/WSH/03.04/22).

12.120 Trifluralin

Trifluralin (CAS No. 1582-09-8) is a pre-emergence herbicide used in a number of crops. It has low water solubility and a high affinity for soil. However, biodegradation and photodegradation processes may give rise to polar metabolites that may contaminate drinking-water sources. Although this compound is used in many countries, relatively few data are available concerning contamination of drinking-water.

Guideline value	0.02 mg/litre
Occurrence	Not detected in the small number of drinking-water samples analysed; has been detected in surface water at concentrations above 0.5 µg/litre and rarely in groundwater
TDI	7.5 µg/kg of body weight, based on a NOAEL of 0.75 mg/kg of body weight for mild hepatic effects in a 1-year feeding study in dogs, with an uncertainty factor of 100 (for inter- and intraspecies variation)
Limit of detection	0.05 µg/litre by GC with nitrogen–phosphorus detection
Treatment achievability	1 µg/litre should be achievable using GAC
Guideline derivation • allocation to water • weight • consumption	 10% of TDI 60-kg adult 2 litres/day
Additional comments	Authorities should note that some impure technical grades of trifluralin could contain potent carcinogenic compounds and therefore should not be used.

Toxicological review

Trifluralin of high purity does not possess mutagenic properties. Technical trifluralin of low purity may contain nitroso contaminants and has been found to be mutagenic. No evidence of carcinogenicity was demonstrated in a number of long-term toxicity/carcinogenicity studies with pure (99%) test material. IARC recently evaluated technical-grade trifluralin and assigned it to Group 3.

History of guideline development

The 1958 and 1963 WHO *International Standards for Drinking-water* did not refer to trifluralin, but the 1971 International Standards suggested that pesticide residues that may occur in community water supplies make only a minimal contribution to the total daily intake of pesticides for the population served. Trifluralin was not evaluated in the first edition of the *Guidelines for Drinking-water Quality*, published in 1984, but the 1993 Guidelines established a health-based guideline value of 0.02 mg/litre for trifluralin in drinking-water, noting that authorities should be aware that some impure technical grades of trifluralin could contain potent carcinogenic compounds and therefore should not be used.

Assessment date
The risk assessment was originally conducted in 1993. The Final Task Force Meeting in 2003 agreed that this risk assessment be brought forward to this edition of the *Guidelines for Drinking-water Quality*.

Principal reference
WHO (2003) *Trifluralin in drinking-water. Background document for preparation of WHO Guidelines for drinking-water quality*. Geneva, World Health Organization (WHO/SDE/WSH/03.04/43).

12.121 Trihalomethanes (bromoform, bromodichloromethane, dibromochloromethane, chloroform)

Trihalomethanes (THMs) are generated principally as by-products of the chlorination of drinking-water, being formed from naturally occurring organic compounds. Hypochlorous acid oxidizes bromide ion to form hypobromous acid, which reacts with endogenous organic materials (e.g., humic or fulvic acids) to form brominated THMs. The amount of each THM formed depends on the temperature, pH and chlorine and bromide ion concentrations. It is assumed that most THMs present in water are ultimately transferred to air as a result of their volatility. For chloroform, for example, individuals may be exposed during showering to elevated concentrations from chlorinated tap water. Based on estimates of mean exposure from various media, the general population is exposed to chloroform principally in food, drinking-water and indoor air, in approximately equivalent amounts.

Guideline values	
Chloroform	0.2 mg/litre
Bromoform	0.1 mg/litre
Dibromochloromethane (DBCM)	0.1 mg/litre
Bromodichloromethane (BDCM)	0.06 mg/litre
Occurrence	THMs are rarely found in raw water but are often present in finished water; concentrations are generally below 100 µg/litre. In most circumstances, chloroform is the dominant compound.
TDIs	
Chloroform	13 µg/kg of body weight, based on slight hepatotoxicity (increases in hepatic serum enzymes and fatty cysts) observed in beagle dogs ingesting 15 mg of chloroform per kg of body weight per day in toothpaste for 7.5 years, incorporating an uncertainty factor of 1000 (100 for inter- and intraspecies variation and 10 for use of a LOAEL rather than a NOAEL and a subchronic study) and correcting for 6 days per week dosing

Bromoform	17.9 µg/kg of body weight, based on the absence of histopathological lesions in the liver in a well conducted and well documented 90-day study in rats, using an uncertainty factor of 1000 (100 for intra- and interspecies variation and 10 for possible carcinogenicity and short duration of exposure)
DBCM	21.4 µg/kg of body weight, based on absence of histopathological effects in the liver in a well conducted and well documented 90-day study in rats, using an uncertainty factor of 1000 (100 for intra- and interspecies variation and 10 for the short duration of the study); an additional uncertainty factor for potential carcinogenicity was not applied because of the questions regarding mouse liver tumours from corn oil vehicles and inconclusive evidence of genotoxicity
Basis of guideline derivation for BDCM	Application of the linearized multistage model for the observed increases in incidence of kidney tumours in male mice observed in an NTP bioassay, as these tumours yield the most protective value (guideline value is supported by a recently published feeding study in rats that was not available for full evaluation)
Limit of detection	0.1 µg/litre by GC with ECD; 2.2 µg/litre by GC/MS
Treatment achievability	Concentrations of chloroform, bromoform, BDCM and DBCM in drinking-water are generally <0.05 mg/litre. Concentrations can be reduced by changes to disinfection practice (reducing organic THM precursors) or using air stripping.
Guideline derivation • allocation to water	20% of TDI for bromoform and DBCM 50% of TDI for chloroform (based on estimates indicating that the general population is exposed to chloroform principally in food, drinking-water and indoor air in approximately equivalent amounts and that most of the chloroform in indoor air is present as a result of volatilization from drinking-water)
• weight	60-kg adult
• consumption	2 litres/day
Additional comments	For authorities wishing to establish a total THM standard to account for additive toxicity, the following fractionation approach could be taken: $$\frac{C_{bromoform}}{GV_{bromoform}} + \frac{C_{DBCM}}{GV_{DBCM}} + \frac{C_{BDCM}}{GV_{BDCM}} + \frac{C_{chloroform}}{GV_{chloroform}} \leq 1$$ where C = concentration and GV = guideline value. It is emphasized that adequate disinfection should never be compromised in attempting to meet guidelines for THMs.

Toxicological review

Chloroform

The weight of evidence for genotoxicity of chloroform is considered negative. The weight of evidence for liver tumours in mice is consistent with a threshold mechanism of induction. Although it is plausible that kidney tumours in rats may similarly be associated with a threshold mechanism, there are some limitations of the database in this regard. The most universally observed toxic effect of chloroform is damage to the centrilobular region of the liver. The severity of these effects per unit dose administered depends on the species, vehicle and method by which the chloroform is administered.

Bromoform

In an NTP bioassay, bromoform induced a small increase in relatively rare tumours of the large intestine in rats of both sexes but did not induce tumours in mice. Data from a variety of assays on the genotoxicity of bromoform are equivocal. IARC has classified bromoform in Group 3 (not classifiable as to its carcinogenicity to humans).

Dibromochloromethane

In an NTP bioassay, DBCM induced hepatic tumours in female and possibly in male mice but not in rats. The genotoxicity of DBCM has been studied in a number of assays, but the available data are considered inconclusive. IARC has classified DBCM in Group 3 (not classifiable as to its carcinogenicity to humans).

Bromodichloromethane

IARC has classified BDCM in Group 2B (possibly carcinogenic to humans). BDCM gave both positive and negative results in a variety of *in vitro* and *in vivo* genotoxicity assays. In an NTP bioassay, BDCM induced renal adenomas and adenocarcinomas in both sexes of rats and male mice, rare tumours of the large intestine (adenomatous polyps and adenocarcinomas) in both sexes of rats and hepatocellular adenomas and adenocarcinomas in female mice.

History of guideline development

The 1958, 1963 and 1971 WHO *International Standards for Drinking-water* did not refer to THMs. In the first edition of the *Guidelines for Drinking-water Quality*, published in 1984, no guideline values for THMs other than chloroform were recommended after a detailed evaluation of the compounds. A health-based guideline value of 0.03 mg/litre was established for chloroform only, as few data existed for the remaining THMs and, for most water supplies, chloroform was the most commonly encountered member of the group. It was noted that the guideline value for chloroform was obtained using a linear multistage extrapolation of data obtained from male rats, a mathematical model appropriate to chemical carcinogens that involves considerable uncertainty. It was also mentioned that although the available toxicological data were useful in establishing a guideline value for chloroform only, the concentrations of the other THMs should also be minimized. Limits ranging from 0.025 to 0.25 mg/litre, which represent a balance between the levels that can be achieved given certain circumstances and those that are desirable, have been set in several countries for the sum of bromoform, DBCM, BDCM and chloroform. In the 1993 Guidelines, no guideline value was set for total THMs, but guideline values were established separately for all four THMs. Authorities wishing to establish a total THM standard to account for additive toxicity could use a fractionation approach in which the sum of the ratios of each of the four THMs to their respective guideline values is less than 1. The 1993 Guidelines established health-based guideline values of 0.1 mg/litre for both bromoform and DBCM. Guideline values of 0.06 mg/litre for BDCM and 0.2 mg/litre for chloroform, associated with an upper-bound excess lifetime cancer risk of 10^{-5}, were also recommended. The guideline value of 0.2 mg/litre for chloroform was retained in the addendum to the second edition of the Guidelines, published in 1998, but was developed on the basis of a TDI for threshold effects.

Assessment date
The risk assessments were originally conducted in 1993 and 1998 (for chloroform). The Final Task Force Meeting in 2003 agreed that these risk assessments be brought forward to this edition of the *Guidelines for Drinking-water Quality*.

Principal reference
WHO (2003) *Trihalomethanes in drinking-water. Background document for preparation of WHO Guidelines for drinking-water quality*. Geneva, World Health Organization (WHO/SDE/WSH/03.04/64).

12.122 Uranium
Uranium is widespread in nature, occurring in granites and various other mineral deposits. Uranium is used mainly as fuel in nuclear power stations. Uranium is present in the environment as a result of leaching from natural deposits, release in mill tailings, emissions from the nuclear industry, the combustion of coal and other fuels and the use of phosphate fertilizers that contain uranium. Intake of uranium through air is low, and it appears that intake through food is between 1 and 4 µg/day. Intake through drinking-water is normally extremely low; however, in circumstances in which uranium is present in a drinking-water source, the majority of intake can be through drinking-water.

Provisional guideline value	0.015 mg/litre
	The guideline value is designated as provisional because of outstanding uncertainties regarding the toxicology and epidemiology of uranium as well as difficulties concerning its technical achievability in smaller supplies.
Occurrence	Levels in drinking-water are generally less than 1 µg/litre, although concentrations as high as 700 µg/litre have been measured in private supplies.

12. CHEMICAL FACT SHEETS

TDI	0.6 µg/kg of body weight per day, based on the application of an uncertainty factor of 100 (for inter- and intraspecies variation) to a LOAEL (equivalent to 60 µg of uranium per kg of body weight per day) for degenerative lesions in the proximal convoluted tubule of the kidney in male rats in a 91-day study in which uranyl nitrate hexahydrate was administered in drinking-water. It was considered unnecessary to apply an additional uncertainty factor for the use of a LOAEL instead of a NOAEL and the short length of the study because of the minimal degree of severity of the lesions and the short half-life of uranium in the kidney, with no indication that the severity of the renal lesions will be exacerbated following continued exposure. This is supported by data from epidemiological studies.
Limit of detection	0.01 µg/litre by ICP/MS; 0.1 µg/litre by solid fluorimetry with either laser excitation or UV light; 0.2 µg/litre by ICP using adsorption with chelating resin
Treatment achievability	1 µg/litre should be achievable using conventional treatment, e.g., coagulation or ion exchange
Guideline derivation • allocation to water • weight • consumption	 80% of TDI (because intake from other sources is low in most areas) 60-kg adult 2 litres/day
Additional comments	• The data on intake from food in most areas suggest that intake from food is low and support the higher allocation to drinking-water. In some regions, exposure from sources such as soil may be higher and should be taken into account in setting national or local standards. • The concentration of uranium in drinking-water associated with the onset of measurable tubular dysfunction remains uncertain, as does the clinical significance of the observed changes at low exposure levels. A guideline value of up to 30 µg/litre may be protective of kidney toxicity because of uncertainty regarding the clinical significance of changes observed in epidemiological studies. • Only chemical, not radiological, aspects of uranium toxicity have been addressed here. • A document on depleted uranium, which is a by-product of natural uranium, is available.

Toxicological review

There are insufficient data regarding the carcinogenicity of uranium in humans and experimental animals. Nephritis is the primary chemically induced effect of uranium in humans. Little information is available on the chronic health effects of exposure to environmental uranium in humans. A number of epidemiological studies of populations exposed to uranium in drinking-water have shown a correlation with alkaline phosphatase and β-microglobulin in urine along with modest alterations in proximal tubular function. However, the actual measurements were still within the normal physiological range.

History of guideline development
The 1958 and 1963 WHO *International Standards for Drinking-water* did not refer to uranium. The 1971 International Standards stated that uranium should be controlled in drinking-water, but that insufficient information was available to enable a tentative limit to be established. In the first edition of the *Guidelines for Drinking-water Quality*, published in 1984, it was concluded that no action was required for uranium. A health-based guideline value for uranium was not derived in the 1993 Guidelines, as adequate short- and long-term studies on the chemical toxicity of uranium were not available. Until such information became available, it was recommended that the limits for radiological characteristics of uranium be used. The equivalent for natural uranium, based on these limits, is approximately 0.14 mg/litre. In the addendum to the Guidelines, published in 1998, a health-based guideline value of 0.002 mg/litre was established. This guideline value was designated as provisional, because it may be difficult to achieve in areas with high natural uranium levels with the treatment technology available and because of limitations in the key study. It was noted that several human studies are under way that may provide helpful additional data.

Assessment date
The risk assessment was conducted in 2003.

Principal reference
WHO (2003) *Uranium in drinking-water. Background document for preparation of WHO Guidelines for drinking-water quality*. Geneva, World Health Organization (WHO/SDE/WSH/03.04/118).

12.123 Vinyl chloride
Vinyl chloride is used primarily for the production of PVC. Owing to its high volatility, vinyl chloride has rarely been detected in surface waters, except in contaminated areas. Unplasticized PVC is increasingly being used in some countries for water mains supplies. Migration of vinyl chloride monomer from unplasticized PVC is a possible source of vinyl chloride in drinking-water. It appears that inhalation is the most important route of vinyl chloride intake, although drinking-water may contribute a substantial portion of daily intake where PVC piping with a high residual content of vinyl chloride monomer is used in the distribution network. Vinyl chloride has been reported in groundwater as a degradation product of the chlorinated solvents trichloroethene and tetrachloroethene.

Guideline value	0.0003 mg/litre (0.3 µg/litre)
Occurrence	Rarely detected in surface waters, the concentrations measured generally not exceeding 10 µg/litre; much higher concentrations found in groundwater and well water in contaminated areas; concentrations up to 10 µg/litre detected in drinking-water
Basis for guideline derivation	Application of a linear extrapolation by drawing a straight line between the dose, determined using a pharmocokinetic model, resulting in tumours in 10% of animals in rat bioassays involving oral exposure and the origin (zero dose), determining the value associated with the upper-bound risk of 10^{-5} and assuming a doubling of the risk for exposure from birth
Limit of detection	0.01 µg/litre by GC with ECD or FID with MS for confirmation
Treatment achievability	0.001 mg/litre should be achievable using air stripping
Additional comments	• The results of the linear extrapolation are nearly identical to those derived using the linearized multistage model. • As vinyl chloride is a known human carcinogen, exposure to this compound should be avoided as far as practicable, and levels should be kept as low as technically feasible. • Vinyl chloride is primarily of concern as a potential contaminant from some grades of PVC pipe and is best controlled by specification of material quality.

Toxicological review

There is sufficient evidence of the carcinogenicity of vinyl chloride in humans from industrial populations exposed to high concentrations via the inhalation route, and IARC has classified vinyl chloride in Group 1. Studies of workers employed in the vinyl chloride industry have shown a marked exposure–response for all liver cancers, angiosarcomas and hepatocellular carcinoma, but no strong relationship between cumulative vinyl chloride exposure and other cancers. Animal data show vinyl chloride to be a multisite carcinogen. When administered orally or by inhalation to mice, rats and hamsters, it produced tumours in the mammary gland, lungs, Zymbal gland and skin, as well as angiosarcomas of the liver and other sites. Evidence indicates that vinyl chloride metabolites are genotoxic, interacting directly with DNA. DNA adducts formed by the reaction of DNA with a vinyl chloride metabolite have also been identified. Occupational exposure has resulted in chromosomal aberrations, micronuclei and sister chromatid exchanges; response levels were correlated with exposure levels.

History of guideline development

The 1958, 1963 and 1971 WHO *International Standards for Drinking-water* did not refer to vinyl chloride. In the first edition of the *Guidelines for Drinking-water Quality*, published in 1984, no guideline value was recommended, because the occurrence of vinyl chloride in water seemed to be associated primarily with the use of poorly polymerized PVC water pipes, a problem that was more appropriately controlled by product specification. The 1993 Guidelines calculated a guideline value of

0.005 mg/litre for vinyl chloride based on an upper-bound excess lifetime cancer risk of 10^{-5}.

Assessment date
The risk assessment was conducted in 2003.

Principal references
IPCS (1999) *Vinyl chloride*. Geneva, World Health Organization, International Programme on Chemical Safety (Environmental Health Criteria 215).
WHO (2003) *Vinyl chloride in drinking-water. Background document for preparation of WHO Guidelines for drinking-water quality*. Geneva, World Health Organization (WHO/SDE/WSH/03.04/119).

12.124 Xylenes

Xylenes are used in blending petrol, as a solvent and as a chemical intermediate. They are released to the environment largely via air. Exposure to xylenes is mainly from air, and exposure is increased by smoking.

Guideline value	0.5 mg/litre
Occurrence	Concentrations of up to 8 µg/litre have been reported in surface water, groundwater and drinking-water; levels of a few milligrams per litre were found in groundwater polluted by point emissions. Xylenes can also penetrate plastic pipe from contaminated soil.
TDI	179 µg/kg of body weight, based on a NOAEL of 250 mg/kg of body weight per day for decreased body weight in a 103- week gavage study in rats, correcting for 5 days per week dosing and using an uncertainty factor of 1000 (100 for inter- and intraspecies variation and 10 for the limited toxicological end-points)
Limit of detection	0.1 µg/litre by GC/MS; 1 µg/litre by GC with FID
Treatment achievability	0.005 mg/litre should be achievable using GAC or air stripping
Guideline derivation • allocation to water • weight • consumption	 10% of TDI 60-kg adult 2 litres/day
Additional comments	The guideline value exceeds the lowest reported odour threshold for xylenes in drinking-water.

Toxicological review
Xylenes are rapidly absorbed by inhalation. Data on oral exposure are lacking. Xylenes are rapidly distributed in the body, predominantly in adipose tissue. They are almost completely metabolized and excreted in urine. The acute oral toxicity of xylenes is low. No convincing evidence for teratogenicity has been found. Long-term carcino-

genicity studies have shown no evidence for carcinogenicity. *In vitro* as well as *in vivo* mutagenicity tests have proved negative.

History of guideline development
The 1958, 1963 and 1971 WHO *International Standards for Drinking-water* and the first edition of the *Guidelines for Drinking-water Quality*, published in 1984, did not refer to xylenes. The 1993 Guidelines proposed a health-based guideline value of 0.5 mg/litre for xylenes, noting that this value exceeds the lowest reported odour threshold for xylenes in drinking-water (0.02 mg/litre).

Assessment date
The risk assessment was originally conducted in 1993. The Final Task Force Meeting in 2003 agreed that this risk assessment be brought forward to this edition of the *Guidelines for Drinking-water Quality*.

Principal reference
WHO (2003) *Xylenes in drinking-water. Background document for preparation of WHO Guidelines for drinking-water quality.* Geneva, World Health Organization (WHO/SDE/WSH/03.04/25).

12.125 Zinc
Zinc is an essential trace element found in virtually all food and potable water in the form of salts or organic complexes. The diet is normally the principal source of zinc. Although levels of zinc in surface water and groundwater normally do not exceed 0.01 and 0.05 mg/litre, respectively, concentrations in tap water can be much higher as a result of dissolution of zinc from pipes.

In 1982, JECFA proposed a PMTDI for zinc of 1 mg/kg of body weight. The daily requirement for adult men is 15–20 mg/day. It was considered that, taking into account recent studies on humans, the derivation of a guideline value is not required at this time. However, drinking-water containing zinc at levels above 3 mg/litre may not be acceptable to consumers (see chapter 10).

History of guideline development
The 1958 WHO *International Standards for Drinking-water* suggested that concentrations of zinc greater than 15 mg/litre would markedly impair the potability of the water. The 1963 and 1971 International Standards retained this value as a maximum allowable or permissible concentration. In the first edition of the *Guidelines for Drinking-water Quality*, published in 1984, a guideline value of 5.0 mg/litre was established for zinc, based on taste considerations. The 1993 Guidelines concluded that, taking into account recent studies on humans, the derivation of a guideline value was not required at this time. However, drinking-water containing zinc at levels above 3 mg/litre may not be acceptable to consumers.

Assessment date
The risk assessment was originally conducted in 1993. The Final Task Force Meeting in 2003 agreed that this risk assessment be brought forward to this edition of the *Guidelines for Drinking-water Quality*.

Principal reference
WHO (2003) *Zinc in drinking-water. Background document for preparation of WHO Guidelines for drinking-water quality*. Geneva, World Health Organization (WHO/SDE/WSH/03.04/17).

ANNEX 1
Bibliography

Supporting documents

Ainsworth R, ed. (2004) *Safe piped water: Managing microbial water quality in piped distribution systems.* IWA Publishing, London, for the World Health Organization, Geneva.

Bartram J et al., eds. (2003) *Heterotrophic plate counts and drinking-water safety: the significance of HPCs for water quality and human health.* WHO Emerging Issues in Water and Infectious Disease Series. London, IWA Publishing.

Bartram J et al., eds. (2004) *Pathogenic mycobacteria in water: A guide to public health consequences, monitoring and management.* Geneva, World Health Organization.

Chorus I, Bartram J, eds. (1999) *Toxic cyanobacteria in water: A guide to their public health consequences, monitoring and management.* Published by E & FN Spon, London, on behalf of the World Health Organization, Geneva.

Davison A et al. (2004) *Water safety plans.* Geneva, World Health Organization.

Dufour A et al. (2003) *Assessing microbial safety of drinking water: Improving approaches and methods.* Geneva, Organisation for Economic Co-operation and Development/World Health Organization.

FAO/WHO (2003) *Hazard characterization for pathogens in food and water: guidelines.* Geneva, Food and Agriculture Organization of the United Nations/World Health Organization (Microbiological Risk Assessment Series No. 3). Available at http://www.who.int/foodsafety/publications/micro/en/pathogen.pdf.

Havelaar AH, Melse JM (2003) *Quantifying public health risks in the WHO Guidelines for drinking-water quality: A burden of disease approach.* Bilthoven, National Institute for Public Health and the Environment (RIVM Report 734301022/2003).

Howard G, Bartram J (2003) *Domestic water quantity, service level and health.* Geneva, World Health Organization.

LeChevallier MW, Au K-K (2004) *Water treatment and pathogen control: Process efficiency in achieving safe drinking-water.* Geneva, World Health Organization and IWA.

Sobsey M (2002) *Managing water in the home: Accelerated health gains from improved water supply.* Geneva, World Health Organization (WHO/SDE/WSH/02.07).

Sobsey MD, Pfaender FK (2002) *Evaluation of the H_2S method for detection of fecal contamination of drinking water*. Geneva, World Health Organization (WHO/SDE/WSH/02.08).

Thompson T et al. (2004) *Chemical safety of drinking-water: Assessing priorities for risk management*. Geneva, World Health Organization.

Wagner EG, Pinheiro RG (2001) *Upgrading water treatment plants*. Published by E & FN Spon, London, on behalf of the World Health Organization, Geneva.

WHO (in preparation) *The arsenic monograph*. Geneva, World Health Organization.

WHO (in preparation) *Desalination for safe drinking-water supply*. Geneva, World Health Organization.

WHO (in revision) *Guide to hygiene and sanitation in aviation*. Geneva, World Health Organization.

WHO (in revision) *Guide to ship sanitation*. Geneva, World Health Organization.

WHO (in preparation) *Health aspects of plumbing*. Geneva, World Health Organization.

WHO (in preparation) *Legionella and the prevention of legionellosis*. Geneva, World Health Organization.

WHO (in preparation) *Managing the safety of materials and chemicals used in the production and distribution of drinking-water*. Geneva, World Health Organization.

WHO (in preparation) *Protecting groundwaters for health – Managing the quality of drinking-water sources*. Geneva, World Health Organization.

WHO (in preparation) *Protecting surface waters for health – Managing the quality of drinking-water sources*. Geneva, World Health Organization.

WHO (in preparation) *Rapid assessment of drinking-water quality: a handbook for implementation*. Geneva, World Health Organization.

WHO (in preparation) *Safe drinking-water for travelers and emergencies*. Geneva, World Health Organization.

Cited references[1]

APHA (1998) *Standard methods for the examination of water and wastewater*, 20th ed. Washington, DC, American Public Health Association.

AS (1998) *Water quality – Sampling – Guidance on the design of sampling programs, sampling techniques and the preservation and handling of samples*. Australia and New Zealand Standards (AS/NZS 5667.1.1998).

Bartram J, Ballance R, eds. (1996) *Water quality monitoring: a practical guide to the design and implementation of freshwater quality studies and monitoring programmes*. Published by E & FN Spon, London, on behalf of the United Nations Educational, Scientific and Cultural Organization, the World Health Organization and the United Nations Environment Programme.

---

[1] In chapter 11, selected bibliographical references are included at the end of each microbial fact sheet. In chapter 12, principal references are provided at the end of each chemical fact sheet.

ANNEX 1. BIBLIOGRAPHY

Brikké F (2000) *Operation and maintenance of rural water supply and sanitation systems: a training package for managers and planners*. Delft, IRC International Water and Sanitation Centre; and Geneva, World Health Organization. Available at http://www.irc.nl/pdf/publ/ome.pdf.

Codex Alimentarius Commission (1985) *Code of practice for collecting, processing and marketing of natural mineral waters*. Rome, Food and Agriculture Organization of the United Nations and World Health Organization (CAC RCP 33). Available at ftp://ftp.fao.org/codex/standard/en/CXP_033e.pdf.

Codex Alimentarius Commission (1997) *Standard for natural mineral waters*. Rome, Food and Agriculture Organization of the United Nations and World Health Organization (CODEX STAN 108). Available at ftp://ftp.fao.org/codex/standard/en/CXS_108e.pdf.

Codex Alimentarius Commission (2001) *General standard for bottled/packaged waters (other than natural mineral waters)*. Rome, Food and Agriculture Organization of the United Nations and World Health Organization (CAC/RCP 48). Available at ftp://ftp.fao.org/codex/standard/en/CXP_048e.pdf.

Crump KS (1984) A new method for determining allowable daily intakes. *Fundamental and Applied Toxicology*, 4:854–871.

Dangendorf F et al. (2003) *The occurrence of pathogens in surface water*. Bonn, University of Bonn, World Health Organization Collaborating Centre (draft report).

Davis J, Lambert R (2002) *Engineering in emergencies: a practical guide for relief workers*, 2nd ed. London, Intermediate Technology Publications.

Evins C (2004) Small animals in drinking water systems. In: Ainsworth R, ed. *Safe, piped water: Managing microbial water quality in piped distribution systems*. IWA Publishing, London, for the World Health Organization, Geneva.

Farland W, Dourson ML (1992) Noncancer health endpoints: approaches to quantitative risk assessment. In: Cothern R, ed. *Comparative environmental risk assessment*. Boca Raton, FL, Lewis Publishers, pp. 87–106.

Guth DJ et al. (1991) *Evaluation of risk assessment methods for short-term inhalation exposure*. Presentation at the 84th Annual Meeting of the Air and Waste Management Association, Vancouver, British Columbia, 16–21 June 1991.

Haas CN, Rose JB, Gerba CP (1999) *Quantitative microbial risk assessment*. New York, NY, Wiley.

Havelaar AH et al. (2000) Balancing the risks of drinking water disinfection: Disability adjusted life-years on the scale. *Environmental Health Perspectives*, 108:315–321.

Hertzberg RC (1989) Fitting a model to categorical response data with application to special extrapolation to toxicity. *Health Physics*, 57(Suppl. 1):404–409.

Hertzberg RC, Miller M (1985) A statistical model for species extrapolation using categorical response data. *Toxicology and Industrial Health*, 1:43–57.

House SF, Reed RA (1997) *Emergency water sources: Guidelines for selection and treatment*. Loughborough, Water, Engineering and Development Centre.

Howard G et al. (2002) *Healthy villages: A guide for communities and community health workers*. Geneva, World Health Organization.

IAEA (1996) *International basic safety standards for protection against ionizing radiation and for the safety of radiation sources*. Vienna, International Atomic Energy Agency.

IAEA (1997) *Generic assessment procedures for determining protective actions during a reactor accident*. Vienna, International Atomic Energy Agency (IAEA-TecDoc-955).

IAEA (1998) *Diagnosis and treatment of radiation injuries*. Vienna, International Atomic Energy Agency/World Health Organization (Safety Reports Series No. 2).

IAEA (1999) *Generic procedures for monitoring in a nuclear or radiological emergency*. Vienna, International Atomic Energy Agency (IAEA-TecDoc-1092).

IAEA (2002) *Safety requirements on preparedness and response for a nuclear or radiological emergency*. Vienna, International Atomic Energy Agency (Safety Standards Series No. GS-R-2).

ICRP (1989) *Optimization and decision-making in radiological protection*. Annals of the ICRP, 20(1).

ICRP (1991) 1990 recommendations of the ICRP. *Annals of the ICRP*, 21(1.3). Oxford, Pergamon Press (International Commission on Radiological Protection Publication 60).

ICRP (1992) *Report of the Task Group on Reference Man*. New York, NY, Pergamon Press (International Commission on Radiological Protection No. 23).

ICRP (1996) *Age-dependent doses to members of the public from intake of radionuclides: Part 5. Compilation of ingestion and inhalation dose coefficients*. Oxford, Pergamon Press (International Commission on Radiological Protection Publication 72).

ICRP (2000) *Protection of the public in situations of prolonged radiation exposure*. Oxford, Pergamon Press (International Commission on Radiological Protection Publication 82).

IPCS (1994) *Assessing human health risks of chemicals: derivation of guidance values for health-based exposure limits*. Geneva, World Health Organization, International Programme on Chemical Safety (Environmental Health Criteria 170).

IPCS (2000) *Disinfectants and disinfectant by-products*. Geneva, World Health Organization, International Programme on Chemical Safety (Environmental Health Criteria 216).

IPCS (2001) *Guidance document for the use of data in development of chemical-specific adjustment factors (CSAFs) for interspecies differences and human variability in dose/concentration–response assessment*. Geneva, World Health Organization, International Programme on Chemical Safety (February 2001 draft).

ISO (1991a) *Water quality – Measurement of gross beta activity in non-saline water – Thick source method*. Geneva, International Organization for Standardization (International Standard 9695).

ISO (1991b) *Water quality – Measurement of gross alpha activity in non-saline water – Thick source method*. Geneva, International Organization for Standardization (International Standard 9696).

Jochimsen EM et al. (1998) Liver failure and death after exposure to microcystins at a hemodialysis center in Brazil. *New England Journal of Medicine*, 338(13):873–878.

Lloyd B, Bartram J (1991) Surveillance solutions to microbiological problems in water quality control in developing countries. *Water Science and Technology*, 24(2):61–75.

NCRP (1989) *Control of radon in houses. Recommendations of the National Council on Radiation Protection and Measurements*. Bethesda, MD, National Council on Radiation Protection and Measurements (NCRP Report No. 103).

Pouria S et al. (1998) Fatal microcystin intoxication in haemodialysis unit in Caruaru, Brazil. *Lancet*, 352:21–26.

Pylon (1989) *Instruction manual for using Pylon Model 110A and 300A Lucas cells with the Pylon Model AB-5*. Ottawa, Ontario, Pylon Electronic Development Company Ltd., 43 pp.

Pylon (2003) *Water degassing with Pylon WG-1001 to measure Rn in Lucas cells*. Available at http://www.pylonelectronics.com/nukeinst/sections/2.htm.

Renwick AG (1993) Data-derived safety factors for the evaluation of food additives and environmental contaminants. *Food Additives and Contaminants*, 10:275–305.

Rooney RM et al. (in press) Water safety on ships. A review of outbreaks of waterborne disease associated with ships. Accepted for publication in *Public Health Reports*.

Sawyer R, Simpson-Hébert M, Wood S (1998) *PHAST step-by-step guide: A participatory approach for the control of diarrhoeal disease*. Geneva, World Health Organization (unpublished document WHO/EOS/98.3). Available at http://www.who.int/water_sanitation_health/hygiene/envsan/phastep/en/.

Simpson-Hébert M, Sawyer R, Clarke L (1996) *The Participatory Hygiene and Sanitation Transformation (PHAST) initiative: a new approach to working with communities*. Geneva, World Health Organization, United Nations Development Programme/World Bank Water and Sanitation Program (WHO/EOS/96.11). Available at http://www.who.int/water_sanitation_health/hygiene/envsan/phast/en/.

UNSCEAR (2000) *Sources, effects and risks of ionizing radiation. UNSCEAR 2000 report to the General Assembly*. New York, NY, United Nations Scientific Committee on the Effects of Atomic Radiation.

US NAS (1999) *Risk assessment of radon in drinking water*. National Academy of Sciences, Committee on Risk Assessment of Exposure to Radon in Drinking Water. Washington, DC, National Academy Press, 296 pp.

US NRC (1999) *Health effects of exposure to radon; BEIR VI*. US National Research Council. Washington, DC, National Academy Press.

WHO (1976) *Surveillance of drinking-water quality*. Geneva, World Health Organization.

WHO (1983) Article 14.2. In: *International Health Regulations (1969)*, 3rd annotated ed. (updated and reprinted in 1992 and 1995). Geneva, World Health Organization. Available at http://policy.who.int/cgi-bin/om_isapi.dll?advquery=aircraft&hitsperheading=on&infobase=ihreg&rankhits=50&record={96}&softpage=Doc_Frame_Pg42&x=37&y=12&zz=.

WHO (1988) *Derived intervention levels for radionuclides in food.* Geneva, World Health Organization.

WHO (1997) *Guidelines for drinking-water quality*, 2nd ed. Vol. 3. *Surveillance and control of community supplies.* Geneva, World Health Organization. Available at http://www.who.int/water_sanitation_health/dwq/guidelines4/en/.

WHO (2003a) *Emerging issues in water and infectious disease.* Geneva, World Health Organization.

WHO (2003b) *Report of the WHO workshop: Nutrient minerals in drinking water and the potential health consequences of long-term consumption of demineralized and remineralized and altered mineral content drinking waters.* Rome, 11–13 November 2003 (SDE/WSH/04.01).

WHO/UNICEF Joint Monitoring Programme for Water Supply and Sanitation (2000) *Global water supply and sanitation assessment 2000 report.* Geneva, World Health Organization, Water Supply and Sanitation Collaborative Council and United Nations Children Fund.

Wisner B, Adams J (2003) *Environmental health in emergencies and disasters.* Geneva, World Health Organization.

World Health Assembly (1991) *Elimination of dracunculiasis: resolution of the 44th World Health Assembly.* Geneva, World Health Organization (Resolution No. WHA 44.5).

ANNEX 2

Contributors to the development of the third edition of the *Guidelines on drinking-water quality*

Mr M. Abbaszadegan, (21: iv), American Water Works Services Inc., Belleville, IL, USA
Dr M. Abdulraheem, (9), United Nations Environment Programme, Manama, Bahrain
Dr H. Abouzaid, (1, 7, 9, 15, 23, 25, 27), WHO, Regional Office for the Eastern Mediterranean, Cairo, Egypt
Mr R. Abrams, (19), WHO, Regional Office for the Western Pacific, Manila, Philippines
Mr J. Adams, (5), (formerly of Oxfam, Oxford, UK)
Dr Z. Adeel, (15), The United Nations University, Tokyo, Japan
Mr M. Adriaanse, (5), United Nations Environment Programme, The Hague, Netherlands
Mr R. Aertgeerts, (7, 15, 23, 25, 27), European Centre for Environment and Health, Rome, Italy
Dr R. Ainsworth, (12, 20, 23, 25), Water Science and Technology, Bucklebury, UK
Dr A. Aitio, (26), WHO, Geneva, Switzerland
Ms M. Al Alili, (9), Abu Dhabi Water and Electricity Authority, Abu Dhabi, United Arab Emirates
Dr F. Al Awadhi, (9), United Nations Environment Programme, Bahrain, and Regional Organization for the Protection of the Marine Environment, Kuwait
Dr M.M.Z. Al-Ghali, (21), Ministry of Health, Damascus, Syria
Dr B. Ali, (27), Kwame Nkrumah University of Science and Technology, Kumasi, Ghana
Dr M. Ali, (27), Water, Engineering and Development Centre, Loughborough University, Loughborough, UK
Dr A. Ali Alawadhi, (9), Ministry of Electricity and Water, Manama, Bahrain
Mr M. Al Jabri, (9), Ministry of Regional Municipalities, Environment and Water Resources, Muscat, Oman
Dr A. Allen, (27), University of York, Ireland
Dr M. Allen, (14), American Water Works Association, Denver, CO, USA
Mr H. Al Motairy, (9), Ministry of Defence and Aviation, Jeddah, Saudi Arabia

Ms E. Al Nakhi, (9), Abu Dhabi Water and Electricity Authority, Abu Dhabi, United Arab Emirates

Dr M. Al Rashed, (9), Kuwait Institute for Scientific Research, Safat, Kuwait

Mr M. Al Sofi, (9), House of Sofia, Al Khobar, Saudi Arabia

Dr M. Al Sulaiti, (9), Qatar Electricity and Water Corporation, Doha, Qatar

Dr S. Ambu, (11), Ministry of Health, Kuala Lumpur, Malaysia

American Chemistry Council, (19), Washington, DC, USA

Ms Y. Andersson, (6), Swedish Institute for Infectious Disease Control, Solna, Sweden

Dr M. Ando, (15), Ministry of Health, Labour and Welfare, Tokyo, Japan

Dr M. Asami, (11, 15), National Institute of Public Health, Tokyo, Japan

Dr N. Ashbolt, (6, 8, 13, 14, 23, 28), University of New South Wales, Sydney, Australia

Ms K. Asora, (10), Samoa Water Supply, Apia, Samoa

Dr K.-K. Au, (24), Greeley and Hansen, Limited Liability Company, Chicago, USA

Dr S. Azevedo, (29), Federal University of Rio de Janeiro, Rio de Janeiro, Brazil

Dr L. Backer, (19), National Center for Environmental Health, Atlanta, USA

Mr D. Bahadur Shrestha, (15), Department of Water Supply and Sewerage, Kathmandu, Nepal

Dr K. Bailey, (5), WRc-NSF Ltd, Marlow, UK (now retired)

Dr H. Bakir, (9), Centre for Environmental Health Activities, Amman, Jordan

Dr G. Ball, (3), NSF International, Ann Arbor, MI, USA

Dr M. Balonov, (20), International Atomic Energy Agency, Vienna, Austria

Mr R. Bannerman, (27), Water Resources Consultancy Service, Accra, Ghana

Dr J. Bartram, (1, 2, 3, 4, 5, 6, 7, 8, 9, 12, 13, 14, 15, 16, 18, 19: xiii–lii, liv–lxviii, 21: i–v, 22, 23, 24, 25, 29), WHO, Geneva, Switzerland

Dr A. Basaran, (10, 11, 12, 15, 25), WHO, Regional Office for the Western Pacific, Manila, Philippines

Dr A. Bathija, (19: xxvi), US Environmental Protection Agency, Washington, DC, USA

Mr U. Bayar, (11), State Inspectorate for Health, Ulaanbaatar, Mongolia

Mr G. Bellen, (2), NSF International, Ann Arbor, MI, USA

Dr R. Belmar, (15), Ministry of Health of Chile, Santiago, Chile

Dr R. Bentham, (16), Department of Environmental Health, Adelaide, Australia

Dr K. Bentley, (4), Centre for Environmental Health, Woden, Australia

Mrs U. Bera, (10), Ministry of Health, Suva, Fiji

Dr P. Berger, (21: iv, 27), US Environmental Protection Agency, Washington, DC, USA

Dr U. Blumenthal, (6, 28), London School of Hygiene and Tropical Medicine, London, UK

Dr A. Boehncke, (19: vii), Fraunhofer Institute of Toxicology and Experimental Medicine, Hanover, Germany

Ms E. Bolt, (27), International Research Centre on Water and Sanitation, Delft, Netherlands

Dr L. Bonadonna, (14, 21: i), Istituto Superiore di Sanità, Rome, Italy

ANNEX 2. CONTRIBUTORS TO THE DEVELOPMENT OF THE THIRD EDITION

Dr X. Bonnefoy, (19: xii, liii, lxix), WHO European Center for Environment and Health, Bonn, Germany (formerly of WHO Regional Office for Europe, Copenhagen, Denmark)
Mr L. Bontoux, (6), European Commission, Brussels, Belgium
Ms T. Boonyakarnkul, (8, 12, 15, 22, 25), Ministry of Public Health, Nonthaburi, Thailand
Professor K. Botzenhart, (5, 16, 21: iii), Tuebingen University, Tuebingen, Germany
Dr L. Bowling, (29), Department of Land and Water Conservation, Parramatta, Australia
Dr E. Briand, (16), Centre Scientifique et Technique du Bâtiment, Marne-la Vallée, France
Dr S. Bumaa, (11), Health Inspection Services, Ulaanbaatar, Mongolia
Mr M. Burch, (8, 29), Australian Water Quality Centre, Salisbury, Australia
Dr T. Burns, (19), The Vinyl Institute, Inc., Arlington, VA, USA
Professor D. Bursill, (8), Australian Water Quality Centre, Salisbury, Australia
Dr J. Butler, (21: iii), Centers for Disease Control and Prevention, Atlanta, GA, USA
Dr P. Byleveld, (10), New South Wales Department of Health, Gladesville, Australia
Mr P. Callan, (7, 8, 13, 15, 17, 19: xiii–lii, liv–lxviii, 22, 25), National Health and Medical Research Council, Canberra, Australia
Professor G. Cangelosi, (18), Seattle Biomedical Research Institute, Seattle, USA
Professor W. Carmichael, (29), Wright State University, Ohio, USA
Mr R. Carr, (23), WHO, Geneva, Switzerland
Dr R. Carter, (27), Cranfield University, Silsoe, UK
Dr C. Castell-Exner, (27), The German Technical and Scientific Association for Gas and Water, Bonn, Germany
Dr M. Cavalieri, (29), Local Agency for Electricity and Water Supply, Rome, Italy
Dr R. Chalmers, (26), Public Health Laboratory Service, Swansea, UK
Dr K. Chambers, (23), WRc-NSF Ltd, Swindon, UK
Professor P. Chambon, (1, 4, 19: i–xii), University of Lyon, Lyon, France,
Mr C.K.R. Chan, (11), Shatin Treatment Works, Shatin, Hong Kong, Special Administrative Region of China
Mr S. Chantaphone, (11), Ministry of Health, Ventiane, Lao People's Democratic Republic
Dr D. Chapman, (29), Cork, Ireland
Mr G.P.R. Chaney, (7), International Association of Plumbing and Mechanical Officials, Ontario, CA, USA
Ms L. Channan, (10), South Pacific Applied Geoscience Commission, Suva, Fiji
Professor W. Chee Woon, (11), University of Malaya, Kuala Lumpur, Malaysia
Dr T. Chi Ho, (11), Health Department, Macao, Macao, People's Republic of China
Dr N. Chiu, (15, 19: xlvi), US Environmental Protection Agency, Washington, DC, USA
Dr Y.-G. Cho, (11), Waterworks Gwangju, Gwangju City, Republic of Korea

Dr I. Chorus, (2, 5, 7, 8, 15, 20, 22, 25, 27, 29), Umweltbundesamt, Berlin, Germany

Dr W.T. Chung, (11), Department of Health, Wan Chai, Hong Kong, Special Administrative Region of China

Dr J. Clancy, (23), Clancy Environmental Consultants, St. Albans, VT, USA

Dr J. Clark-Curtiss, (18), Washington University, St. Louis, MO, USA

Dr E. Clayton, (21: ii), US Armed Forces Research Institute of Medical Sciences, Bangkok, Thailand

Professor G. Codd, (29), University of Dundee, Dundee, UK

Dr O. Conerly, (19), US Environmental Protection Agency, Washington, DC, USA

Dr M. Cooper, (19), Envirorad Services Pty Ltd, Victoria, Australia

Dr C. Corvalan, (26), WHO, Geneva, Switzerland

Dr A.L. Corwin, (21: ii), US Armed Forces Research Institute of Medical Sciences, Jakarta, Indonesia

Dr J. Cotruvo, (3, 5, 7, 9, 14, 18, 22, 23, 25), Joseph Cotruvo & Associates, Limited Liability Company, and NSF International, Washington, DC, USA

Professor D. Crawford-Brown, (26), University of North Carolina, Chapel Hill, NC, USA

Dr J. Creasy, (23), WRc-NSF Ltd, Swindon, UK

Dr S. Crespi, (16), Policlinica Miramar, Palma, Spain

Dr G. Cronberg, (29), Lund University, Lund, Sweden

Dr D. Cunliffe, (8, 13, 19, 20, 21: iv, 22, 23, 25, 27), Environmental Health Service, Adelaide, Australia

Dr F. Dagendorf, (16), Institute for Hygiene and Public Health, Bonn, Germany

Dr J.L. Daroussin, (20), European Commission, Luxembourg

Dr H. Darpito, (19, 20, 22), Ministry of Health, Jakarta Pusat, Indonesia

Dr A. Davison, (13, 25), Water Futures, Dundas Valley, NSW, Australia (formerly of the Ministry of Energy and Utilities, Parramatta, NSW, Australia)

Dr F. de Buttet, (19, 20), Gisenec-Unesen, Paris, France

Dr M.-A. DeGroote, (18), University of Colorado, Denver, CO, USA

Dr G. de Hollander, (26), National Institute for Public Health and the Environment (RIVM), Bilthoven, Netherlands

Dr D. Deere, (6, 8, 12, 13, 23, 25, 27), Water Futures, Dundas Valley, Australia (formerly of South East Water Ltd, Moorabbin, Australia)

Mr W. Delai, (10), Ministry of Health, Suva, Fiji

Dr J.M. Delattre, (14, 21: i), Institut Pasteur de Lille, Lille, France

Dr S. Dethoudom, (11), Water Supply Authority, Ventiane, Lao People's Democratic Republic

Professor B. De Villiers, (27), Potchefstroom University for CHE, Potchefstroom, South Africa

Mr I. Deyab, (9), Environment Public Authority, Safat, Kuwait

Professor H. Dieter, (19: xxii), Federal Environment Agency, Berlin, Germany

Dr P. Dillon, (27), Commonwealth Scientific and Industrial Research Organisation, Land and Water, Glen Osmond, Australia

ANNEX 2. CONTRIBUTORS TO THE DEVELOPMENT OF THE THIRD EDITION

Dr J. Donohue, (7, 19: xxxvi), US Environmental Protection Agency, Washington, DC, USA
Dr J. Doss, (19, 20), International Bottled Water Association, Alexandria, USA
Dr V. Drasar, (16), OHS-National Legionella Reference Laboratory, Vyskov, Czech Republic
Dr M. Drikas, (19, 29), Australian Water Quality Center, Salisbury, Australia
Dr J. Du, (19: lii), US Environmental Protection Agency, Washington, DC, USA
Dr A. Dufour, (6, 8, 14, 16, 27), US Environmental Protection Agency, Cincinnati, OH, USA
Dr S. Edberg, (14), Yale University, New Haven, CT, USA
Dr N. Edmonds, (19: xxxi), Health Canada, Ottawa, Canada
Dr J. Eisenberg, (6, 28), University of California, Berkeley, CA, USA
Dr M. El Desouky, (9), Kuwait Institute for Scientific Research, Safat, Kuwait
Dr H. El Habr, (9), United Nations Environment Programme, Managa, Bahrain
Professor F. El Zaatari, (18), Baylor College of Medicine, Houston, TX, USA
Dr M. Ema, (19: xlii, xlix), National Institute of Health Sciences, Tokyo, Japan
Mr P. Emile, (10), Ministry of Health, Rarotonga, Cook Islands
Dr R. Enderlein, (29), United Nations Economic Commission for Europe, Geneva, Switzerland
Dr T. Endo, (5, 7, 14, 15, 19, 22), Ministry of Health, Labor and Welfare, Tokyo, Japan
Mr H. Enevoldsen, (9), Intergovernmental Oceanographic Commission of UNESCO, IOC Science and Communication Centre on Harmful Algae, Copenhagen, Denmark
Dr S. Enkhsetseg, (15), Ministry of Health, Ulaanbaatar, Mongolia
Dr O. Espinoza, (19: xii, liii, lxix), WHO Regional Office for Europe, Copenhagen, Denmark
Mr S. Esrey, (6), deceased (formerly of UNICEF, New York, USA)
Mr G. Ethier, (4), International Council of Metals and the Environment, Ottawa, Canada
Dr C. Evins, (23), Drinking Water Inspectorate, London, UK
Dr M. Exner, (14, 16, 22), Universität Bonn, Bonn, Germany
Professor I. Falconer, (29), University of Adelaide, Adelaide, Australia
Dr J. Falkinham, (18), Fralin Biotechnology Center, Blacksburg, VA, USA
Dr M. Farrimond, (23), UK Water Industry Research, London, UK
Dr J. Fastner, (15, 29), Federal Environmental Agency, Berlin, Germany
Professor B. Fattal, (6), Hebrew University of Jerusalem, Jerusalem, Israel
Mr J. Fawell, (4, 5, 7, 15, 17, 19: vi, xii–lxix, 20, 22, 29), independent consultant, High Wycombe, UK
Ms F. Feagai, (10), Princess Margaret Hospital, Funafuti, Tuvalu
Dr T. Fengthong, (15), Ministry of Health, Ventiane, Lao People's Democratic Republic
Dr I. Feuerpfeil, (21: iv), Umweltbundesamt, Bad Elster, Germany

Dr L. Fewtrell, (6, 12), Center for Research into Environment & Health, University of Wales, Aberystwyth, UK

Mr B. Fields, (16), Centers for Disease Control and Prevention, Atlanta, GA, USA

Mr J. Filiomea, (10), Ministry of Health and Medical Service, Honiara, Solomon Islands

Dr J. Fitch, (20), South Australian Health Commission, Adelaide, Australia

Dr J. Fitzgerald, (29), South Australian Health Commission, Adelaide, Australia

Dr J. Fleisher, (6), State University of New York, Downstate Medical Center, New York, NY, USA

Dr L. Forbes, (23), Leith Forbes & Associates Pty Ltd, Victoria, Australia

Dr T. Ford, (18), Montana State University, Bozeman, MT, USA

Dr R. Franceys, (27), Cranfield University, Silsoe, UK

Ms P. Franz, (10), Paulau Environment Quality Protection Agency, Koror, Republic of Palau

Dr I. Fraser, (19, 20), Department of Health, London, UK

Dr C. Fricker, (14, 21: iv), CRF Consulting, Reading, UK

Dr A. Friday, (22), Ministry of Health, Kampala, Uganda

Dr E. Funari, (7), Istituto Superiore di Sanità, Rome, Italy

Dr H. Galal-Gorchev, (1, 2, 4, 5, 19: i–xii, liii, lxix), US Environmental Protection Agency, Washington, DC (formerly of WHO, Geneva, Switzerland)

Dr P. Gale, (8), WRc-NSF Ltd, Marlow, UK

Dr Y. Ganou, (22), Ministry of Health, Ougadougo, Burkino Faso

Dr M. Gardner, (19), WRc-NSF Ltd, Marlow, UK

Dr A.E.H. Gassim, (22), Ministry of Health, Makkah, Saudi Arabia

Dr R. Gaunt, (4), International Council of Metals and the Environment, Ottawa, Canada

Dr A.-M. Gebhart, (3), NSF International, Ann Arbor, MI, USA

Dr B. Genthe, (27), Division Environment, Pretoria, South Africa

Dr C. Gerba, (14, 28), Arizona University, Tucson, AZ, USA

Dr T. Gerschel, (19), European Copper Institute, Brussels, Belgium

Dr H. Gezairy, (9), WHO, Regional Office for the Eastern Mediterranean, Cairo, Egypt

Ms M. Giddings, (15, 19: xiii–lii, liv–lxviii, 20, 22, 29), Health Canada, Ottawa, Canada

Professor W. Giger, (27), Swiss Federal Institute for Environmental Science and Technology, Dübendorf, Switzerland

Dr N. Gjolme, (29), National Institute for Public Health, Oslo, Norway

Dr A. Glasmacher, (14), Universität Bonn, Bonn, Germany

Dr A. Godfree, (23, 25, 27), United Utilities Water, Warrington, UK

Mr S. Godfrey, (10, 12), Water, Engineering and Development Centre, Loughborough University, Loughborough, UK

Dr M.I. Gonzalez, (19, 20, 22), National Institute of Hygiene, Epidemiology and Microbiology, Havana, Cuba

Ms F. Gore, (22), WHO, Geneva, Switzerland

ANNEX 2. CONTRIBUTORS TO THE DEVELOPMENT OF THE THIRD EDITION

Dr P. Gosling, (21: i), Department of Health, London, UK
Dr P. Gowin, (9), International Atomic Energy Agency, Vienna, Austria
Professor W. Grabow, (5, 6, 8, 13, 19, 20, 21: ii, 22, 25), retired (formerly of University of Pretoria, Pretoria, South Africa)
Mr W. Graham, (4), CropLife International, Brussels, Belgium
Dr P. Grandjean, (19, 20), Institute of Public Health, Odense, Denmark
Dr S. Grant-Trusdale, (19: xxxiv), Health Canada, Ottawa, Canada
Dr R. Gregory, (29), WRc-NSF Ltd, Swindon, UK
Professor A. Grohmann, (19, 27), independent, Berlin, Germany
Dr S. Gupta, (19: v), Health Canada, Ottawa, Canada
Professor C. Haas, (6, 28), Drexel University, Philadelphia, PA, USA
Dr W. Haas, (18), Robert Koch Institute, Berlin, Germany
Ms L. Haller, (12), WHO, Geneva, Switzerland
Mr F. Hannecart, (10), Noumea City Hygiene Service, Noumea, New Caledonia
Dr K.-I. Harada, (29), Meijo University, Nagoya, Japan
Dr M. Hardiman, (20), WHO, Geneva, Switzerland
Mr H. Hashizume, (5, 9, 15, 17, 19: xiii–lii, liv–lxviii, 22), Ministry of the Environment, Tokyo, Japan (formerly of WHO, Geneva, Switzerland)
Dr A. Havelaar, (1, 2, 5, 6, 7, 8, 20, 21: i–v, 22, 25, 26, 28), National Institute for Public Health and the Environment (RIVM), Bilthoven, Netherlands
Mr T. Hayakawa, (1, 5), Ministry of Health & Welfare, Tokyo, Japan
Mr J. Hayes, (16), Institute for Healthcare Management, High Wycombe, UK
Mr P. Hecq, (22), European Commission, Brussels, Belgium
Mr P. Heinsbroek, (10, 11, 15), WHO, Regional Office for the South Pacific, Manila, Philippines
Dr R. Heinze, (29), Umweltbundesamt, Bad Elster, Germany
Mr E. Hellan, (10), Pohnpei Environment Protection Agency, Kolonia, Federated States of Micronesia
Dr R. Helmer, (1, 4, 19: xii, liii, lxix, 22), retired (formerly of WHO, Geneva, Switzerland)
Dr P. Henriksen, (29), National Environmental Research Institute, Roskilde, Denmark
Dr N. Hepworth, (27), Lancaster, UK
Professor J. Hermon-Taylor, (18), St George's Hospital Medical School, London, UK
Mr A. Hicking, (10), Marshall Islands Environment Protection Agency, Majuro, Marshall Islands
Dr G. Hoetzel, (29), La Trobe University, Victoria, Australia
Dr A. Hogue, (6), US Department of Agriculture, Washington, DC, USA (formerly of WHO, Geneva, Switzerland)
Dr D. Holt, (23), Thames Water Utilities Ltd, Reading, UK
Mr M. Hori, (7), Ministry of Health and Welfare, Tokyo, Japan
Professor H. Höring, (2), Umweltbundesamt, Bad Elster, Germany
Ms M. Hoshino, (15), UNICEF, Tokyo, Japan

Dr G. Howard, (2, 5, 7, 8, 12, 13, 15, 19, 20, 22, 23, 25), DFID Bangladesh, Dhaka, Bangladesh (formerly of Water Engineering and Development Centre, Loughborough University, Loughborough, UK)

Dr P. Howsam, (27), Cranfield University, Silsoe, UK

Professor S. Hrudey, (8, 29), University of Alberta, Edmonton, Canada

Mr J. Hueb, (20, 21: v, 23), WHO, Geneva, Switzerland

Dr J. Hulka, (19, 20), National Radiation Protection Institute, Prague, Czech Republic

Dr N. Hung Long, (15), Ministry of Health, Han Noi, Viet Nam

Dr P. Hunter, (14, 23), University of East Anglia, Norwich, UK

Dr K. Hussain, (9), Ministry of Health, Manama, Bahrain

Mr O.D. Hydes, (4, 5, 7), independent consultant, West Sussex, UK (formerly of Drinking Water Inspectorate, London, UK)

Dr A. Iannucci, (3), NSF International, Ann Arbor, MI, USA

Mr S. Iddings, (11, 15), WHO, Phnom Penh, Cambodia

Dr M. Ince, (12, 25), independent consultant, Loughborough, UK (formerly of Water, Engineering and Development Centre, Loughborough University, Loughborough, UK)

International Bottled Water Association, (19), Alexandria, VA, USA

Mr K. Ishii, (15), Japan Water Works Association, Tokyo, Japan

Mr J. Ishiwata, (11), Ministry of Health, Labour and Welfare, Tokyo, Japan

Mr P. Jackson, (2, 5, 7, 15, 19: xiii–lii, liv–lxviii, 22, 25), WRc-NSF Ltd, Marlow, UK

Dr J. Jacob, (21: v), (formerly of Umweltbundesamt, Bad Elster, Germany)

Dr M. Janda, (21: i), Health and Welfare Agency, Berkeley, CA, USA

Mr A. Jensen, (1, 2), DHI Water and Environment, Horsholm, Denmark

Dr R. Johnson, (19), Rohm and Haas Company, USA

Dr D. Jonas, (7), Industry Council for Development, Ramsgate, UK

Dr G. Jones, (29), Commonwealth Scientific and Industrial Research Organisation, Brisbane, Australia

Mr C. Jörgensen, (5), DHI Water and Environment, Horsholm, Denmark

Dr C. Joseph, (16), Communicable Disease Surveillance Control, London, UK

Mr H. Kai-Chye, (10), Canberra, Australia

Ms R. Kalmet, (10), Mines and Water Resources, Port Vila, Vanuatu

Mr I. Karnjanareka, (15), Ministry of Public Health, Nonthaburi, Thailand

Dr D. Kay, (6), University of Wales, Abeystwyth, UK

Dr H. Kerndorff, (27), Umweltbundesamt, Berlin, Germany

Dr S. Khamdan, (9), Ministry of State for Municipalities and Environment Affairs, Manama, Bahrain

Mr P. Khanna, (21: ii), National Environmental Engineering Institute, Nagpur, India

Mr M. Kidanu, (22), WHO, Regional Office for Africa, Harare, Zimbabwe

Dr J. Kielhorn, (4, 19: vii, xv, lxvii), Fraunhofer Institute of Toxicology and Experimental Medicine, Hanover, Germany

ANNEX 2. CONTRIBUTORS TO THE DEVELOPMENT OF THE THIRD EDITION

Dr R. Kirby, (8, 25), Industry Council for Development, Ramsgate, UK
Dr G. Klein, (1), WHO, Bonn, Germany (formerly of Umweltbundesamt, Bad Elster, Germany)
Dr J. Komarkova, (29), Hydrobiological Institute of the Czech Academy of Sciences, Ceské Budejovice, Czech Republic
Dr H. Komulainen, (22), National Public Health Institute, Kuopio, Finland
Dr F. Kondo, (29), Aichi Prefectural Institute of Public Health, Nagoya, Japan
Dr M. Koopmans, (26), National Institute for Public Health and the Environment (RIVM), Bilthoven, Netherlands
Dr F. Kozisek, (19), National Institute of Public Health, Prague, Czech Republic
Dr A. Kozma-Törökne, (29), National Institute for Public Health, Budapest, Hungary
Dr T. Kuiper-Goodman, (29), Health Canada, Ottawa, Canada
Dr S. Kunikane, (7, 15, 17, 22), Ministry of Health, Labour and Welfare, Tokyo, Japan
Dr T. Kwader, (27), URS Corporation, Tallahassee, FL, USA
Miss K. Kwee-Chu, (11), Ministry of Health, Kuala Lumpur, Malaysia
Mr P. Lafitaga, (10), Department of Health, Pago-pago, American Samoa
Dr B. Lang, (4), Novartis Crop Protection AG, Basel, Switzerland
Dr J. Langford, (8), Water Services Association, Melbourne, Australia
Dr J. Latorre, (25), Universidad del Valle, Cali, Colombia
Dr L. Lawton, (29), Robert Gordon University of Aberdeen, Aberdeen, UK
Dr M. LeChevallier, (7, 8, 14, 18, 23, 24), American Water Works Service Company, Inc., Voorhees, NJ, USA
Dr H. Leclerc, (14, 19, 20), University of Lille, Lille, France
Dr J. Lee, (5, 16, 21: iii), Queen's Medical Centre, Nottingham, UK
Mr F. Leitz, (9), Water Treatment Engineering and Research Group, Denver, CO, USA
Professor Le The Thu, (11), Institute of Hygiene and Public Health, Ho Chi Minh City, Viet Nam
Dr Y. Levi, (23), Laboratoire Santé Publique – Environnement, Université Paris XI, Chatenay-Malabry, France
Dr D. Levy, (19, 20), Centers for Disease Control and Prevention, Atlanta, GA, USA
Dr N. Lightfoot, (14), UK Public Health Laboratory Service, Newcastle-upon-Tyne, UK
Dr P. Literathy, (2, 5, 29), Kuwait Institute for Scientific Research, Safat, Kuwait (formerly of Water Resource Research Centre VITUKI, Budapest, Hungary)
Mr S. Loau, (15), Preventive Health Services, Apia, Samoa
Dr J.F. Luna, (26), Secretariat of Health, Mexico City, Mexico
Dr U. Lund, (4, 7, 19: i–xii, liii, lxix), DHI Water and Environment, Horsholm, Denmark
Dr Y. Magara, (1, 4, 5, 7, 14, 15, 19: xiii–lii, liv–lxviii, 21: iv, 22), Hokkaido University, Sapporo, Japan
Mr T. Magno, (10), WHO Representative's Office in Papua New Guinea, Port Moresby, Papua New Guinea

Dr B. Magtibay, (11, 22), Bureau of International Health Cooperation, Manila, Philippines
Dr I. Mäkeläinen, (20), Radiation and Nuclear Safety Authority, Helsinki, Finland
Dr D. Mangino, (3), NSF International, Ann Arbor, MI, USA
Dr A. Marandi, (19), University of Tartu, Tartu, Estonia
Dr T. Mariee, (21: iii), Queen Elizabeth II Health Science Centre, Halifax, Canada
Mr A. Marquez, (10),Guam Environmental Protection Agency, Barrigada, Guam
Dr B. Marsalek, (29), Institute of Botany, Brno, Czech Republic
Professor M. Martin, (27), Bangladesh University of Engineering and Technology, Dhaka, Bangladesh
Dr R. Mascarenhas, (19: xxiii, xxiv, xxx, lvi, lxii, lxiii), Metcalf and Eddy, Devizes, UK
Dr D. McFadden, (3), NSF International, Ann Arbor, MI, USA
Dr M. McLaughlin, (27), Commonwealth Scientific and Industrial Research Organisation, Land and Water, Glen Osmond, Australia
Dr B. McRae, (8), Australian Water Association, Artarmon, Australia
Dr D. Medeiros, (26), Health Canada, Ottawa, Canada
Dr G. Medema, (5, 7, 8, 21: iv), KIWA N.V, Nieuwegein, Netherlands
Ms M.E. Meek, (4), Health Canada, Ottawa, Canada
Dr J. Meheus, (4), International Water Supply Association, Antwerpen, Belgium
Ms G. Melix, (10), Papeete, French Polynesia
Dr J.M. Melse, (26), National Institute for Public Health and the Environment (RIVM), Utrecht, Netherlands
Dr T. Meredith, (22), WHO, Geneva, Switzerland
Mr T. Metutera, (10), Public Utilities Board, Tarawa, Kiribati
Dr E. Meyer, (3), Umweltbundesamt, Berlin, Germany
Dr S. Miller, (27), US Department of Agriculture – Agricultural Research Service (USDA-ARS), Tucson, AZ, USA
Dr B. Mintz, (19: xii, liii, lxix), Centers for Disease Control and Prevention (CDC), Atlanta, GA, USA
Mr M.Z. bin Mohd Talha, (11), Ministry of Health, Kuala Lumpur, Malaysia
Ms M.N. Mons, (7), KIWA Research and Consultancy, Nieuwegein, Netherlands
Professor M.R. Moore, (19), National Research Centre for Environmental Toxicology, Queensland, University of Queensland, Queensland, Australia
Dr G. Morace, (21: ii), Istituto Superiore di Sanità, Rome, Italy
Dr A. Moreau, (14, 19, 20), Danone Water Technology Centre, Evian, France
Dr R. Morris, (5, 7, 23), IWA, London, UK
Dr D. Mossel, (14), Eijkman Foundation, Utrecht, Netherlands
Ms G. Motturi, (2, 5), WHO, Regional Office for Europe, Copenhagen, Denmark
Dr G. Moy, (4, 29), WHO, Geneva, Switzerland
Dr L. Mur, (29), University of Amsterdam, Amsterdam, Netherlands
Ms S. Murcott, (19, 20), Massachusetts Institute of Technology, Massachusetts, USA
Dr P. Murphy, (26), US Environmental Protection Agency, Edison, NJ, USA

ANNEX 2. CONTRIBUTORS TO THE DEVELOPMENT OF THE THIRD EDITION

Dr S. Murphy, (3), NSF International, Ann Arbor, MI, USA
Dr F.J. Murray, (4), International Life Sciences Institute, San José, CA, USA
Mr M.W. Muru, (10), Health Protection, Waigani, Papua New Guinea
Mr C. Mwesigye, (7), WHO, Kampala, Uganda
Dr D. Nabarro, (22), WHO, Geneva, Switzerland
Dr G.B. Nair, (21: v), National Institute of Cholera and Enteric Diseases, Calcutta, India
Pr. K. Nath, (19), Institution of Public Health Engineering, Calcutta, India
Mr P. Navuth, (11), Ministry of Industry, Mines and Energy, Phnom Penh, Cambodia
Mr M. Neal, (19), Invista, Teesside, UK
Dr A. Neller, (8), University of Sunshine Coast, Maroochydore, Australia
Mr J. Newbold, (16), Health and Safety Executive, Bootle, UK
Dr E. Ngoni Mudege, (27), Institute of Water and Sanitation Development, Harare, Zimbabwe
Dr C. Nhachi, (15), WHO, Regional Office for Africa, Harare, Zimbabwe
Dr G. Nichols, (18), Health Protection Agency, London, UK
Dr T. Nishimura, (15, 19: xix, xlii, xlix, lvii), Ministry of Health, Labour and Welfare, Tokyo, Japan
Ms S. Nofal, (9), WHO, Cairo, Egypt
Dr C. Nokes, (25), Environmental Science and Research Ltd, Christchurch, New Zealand
Dr L. Ofanoa, (10), Ministry of Health, Nuku'olofa, Tonga
Dr H. Ogawa, (23), WHO, Regional Office for the Western Pacific, Manila, Philippines
Dr E. Ohanian, (4, 7, 19: i–lii; liv–lxviii, 22), US Environmental Protection Agency, Washington, DC, USA
Dr Y. Okumura, (20), Nagasaki University, Japan
Ms J. Orme-Zavaleta, (1), US Environmental Protection Agency, Washington, DC, USA
Dr Y. Ortega, (21: iv), University of Arizona, Tucson, AZ, USA
Dr J. Padisák, (29), University of Veszprém, Veszprém, Hungary
Dr F. Pamminger, (23), Yarra Valley Water, Melbourne, Australia
Mr I. Papadopoulos, (5, 7), European Commission, Athens, Greece (formerly of European Commission, Brussels, Belgium)
Dr C.N. Paramasivan, (18), Indian Council of Medical Research, Chennai, India
Mr R. Paramasivan, (21: ii), National Environmental Engineering Research Institute, Nagpur, India
Mr D. Parish, (10), Pacific Water Association, Suva, Fiji
Dr C. Pastoris, (21: iii), Istituto Superiore di Sanità, Rome, Italy
Dr E. Pawlitzky, (29), Umweltbundesamt, Berlin, Germany
Dr P. Payment, (6, 14, 23), National Institute of Scientific Research, University of Quebec, Montreal, Canada

Dr S. Pedley, (5, 21: ii, 29), Robens Centre, University of Surrey, Guildford, UK

Dr H.K. bin Pengiran Haji Ismail, (11), Ministry of Health, Bandar Seri Begawan Negara 1210, Brunei Darussalam

Dr G. Peralta, (10, 11, 15), University of the Philippines, Quezon City, Philippines

Mr A. Percival, (8), Consumer Health Forum, Cobargo, Australia

Dr K. Petersson Grawé, (4, 19: iii), National Food Administration, Uppsala, Sweden

Dr M.S. Pillay, (11, 15, 27), Ministry of Health, Kuala Lumpur, Malaysia

Mr S. Pita Helu, (15), Tonga Water Board, Nuku'alofa, Tonga

Dr J. Plouffe, (21: iii), University Hospitals, Columbus, OH, USA

Dr A. Pozniak, (18), Chelsea & Westminster Hospital, London, UK

Dr E. Pozio, (21: iv), Istituto Superiore di Sanità, Rome, Italy

Dr A. Pozniak, (18), Chelsea & Westminster Hospital, London, UK

Mr M. Pretrick, (10), National Government, Palikir, Federated States of Micronesia

Dr C. Price, (19, 20), American Chemical Council, Arlington, VA, USA

Mr F. Properzi, (22), WHO, Geneva, Switzerland

Dr V. Puklova, (19), National Institute of Public Health, Prague, Czech Republic

Mr T. Pule, (23, 25, 27), WHO, Regional Office for Africa, Harare, Zimbabwe

Dr D. Purkiss, (3), NSF International, Ann Arbor, MI, USA

Dr A. Pruess-Ustun, (6, 26), WHO, Geneva, Switzerland

Dr L. Quiggle, (3), NSF International, Ann Arbor, MI, USA

Dr P.P. Raingsey, (11), Ministry of Health, Phnom Penh, Cambodia

Dr C. Ramsay, (22), Scottish Centre for Infection and Environmental Health, Glasgow, UK

Dr P. Ravest-Weber, (3), NSF International, Ann Arbor, MI, USA

Dr D. Reasoner, (14, 23), National Risk Management Research Laboratory, US Environmental Protection Agency, Cincinnati, OH, USA

Dr G. Rees, (7, 18), Askham Bryan College, York, UK (formerly University of Surrey, Guildford, UK)

Dr S. Regli, (21: iv, 23), US Environmental Protection Agency, Washington, DC, USA

Dr R. Reilly, (26), Scottish Centre for Infection and Environmental Health, Glasgow, UK

Dr M. Repacholi, (20, 22), WHO, Geneva, Switzerland

Mrs M. Richold, (4), European Centre for Ecotoxicology and Toxicology, Sharnbrook, UK

Dr J. Ridgway, (3, 5), WRc-NSF Ltd, Marlow, UK

Ms J. Riego de Dios, (11, 15), National Center for Disease Prevention and Control, Manila, Philippines

Mrs U. Ringelband, (5), Umweltbundesamt, Berlin, Germany

Dr M. Rivett, (27), University of Birmingham, Birmingham, UK

Mr W. Robertson, (3, 7, 8, 14, 23, 26), Health Canada, Ottawa, Canada

Dr C. Robinson, (20), International Atomic Energy Agency, Vienna, Austria

Dr J. Rocourt, (28), WHO, Geneva, Switzerland

ANNEX 2. CONTRIBUTORS TO THE DEVELOPMENT OF THE THIRD EDITION

Ms R. Rooney, (12, 16), WHO, Geneva, Switzerland
Dr J. Rose, (14, 21: iv, 28), University of South Florida, St. Petersburg, FL, USA
Dr K. Rotert, (19, 20), US Environmental Protection Agency, Washington, DC, USA
Dr J. Rothel, (18), Cellestis Limited, Victoria, Australia
Mr H. Salas, (2), WHO, Regional Office for the Americas, Washington, DC, USA
Mr A. Salem, (9), Abu Dhabi Water and Electricity Authority, Abu Dhabi, United Arab Emirates
Dr P. Samnang, (15), Ministry of Health, Phnom Penh, Cambodia
Dr M. Santamaria, (16), WHO, Geneva, Switzerland
Mr M. Saray, (11), Ministry of Rural Development, Phnom Penh, Cambodia
Mr D. Sartory, (5, 21: i), Severn Trent Water Ltd, Shelton, UK
Dr M. Savkin, (20), Institut Biophysics, Moscow, Russian Federation
Dr S. Schaub, (6, 16, 21: iv, 28), US Environmental Protection Agency, Washington, DC, USA
Professor R. Schertenleib, (27), Swiss Federal Institute for Environmental Science and Technology, Dübendorf, Switzerland
Dr J. Schijven, (27), National Institute for Public Health and the Environment (RIVM), Bilthoven, Netherlands
Mrs G. Schlag, (5), Umweltbundesamt, Berlin, Germany
Mr O. Schmoll, (8, 15, 27), Umweltbundesamt, Berlin, Germany
Professor L. Schwartzbrod, (19, 20), WHO Collaborating Center for Microorganisms in Wastewater, Nancy, France
Mr P. Scott, (8), Melbourne Water, Melbourne, Australia
Professor K.-P. Seiler, (27), National Research Center for Environment and Health, Institut für Hydrologie, Neuherberg, Germany
Dr Y.-C. Seo, (11), Sangji University, Wonju, Republic of Korea
Dr I. Shalaru, (19, 20, 22), Ministry of Health, Chisinau, Republic of Moldova
Dr D. Sharp, (10, 15), WHO Representative Office in South Pacific, Suva, Fiji
Ms S. Shaw, (7, 21: iv), US Environmental Protection Agency, Washington, DC, USA
Ms F. Shoaie, (9), Department of the Environment, Tehran, Islamic Republic of Iran
Dr Y. Shun-Zhang, (29), Institute of Public Health, Shanghai, People's Republic of China
Dr E. Sievers, (19, 20), Kiel, Germany
Dr D. Simazaki, (15), National Institute of Public Health, Tokyo, Japan
Professor I. Simmers, (27), Vrije University, Amsterdam, Netherlands
Mr T. Simons, (1, 4), European Commission, Brussels, Belgium
Dr M. Sinclair, (8), Monash University Medical School, Prahran, Australia
Dr K. Sivonen, (29), University of Helsinki, Helsinki, Finland
Dr B. Skinner, (12), Water, Engineering and Development Centre, Loughborough University, Loughborough, UK
Dr O. Skulberg, (29), Norwegian Institute for Public Health, Oslo, Norway

Professor H.V. Smith, (5, 21: iv), Scottish Parasite Diagnostic Laboratory, Stobhill Hospital, Glasgow, UK

Dr M. Smith, (23), Water, Engineering and Development Centre, Loughborough University, Loughborough, UK

Dr M. Snozzi, (7), Swiss Federal Institute for Environmental Science and Technology, Dübendorf, Switzerland

Professor M. Sobsey, (7, 8, 12, 13, 19, 20, 22, 25, 28), University of North Carolina, Chapel Hill, USA

Professor J.A. Sokal, (19, 20, 22), Institute of Occupational Medicine and Environmental Health, Sosnowiec, Poland

Dr F. Solsona, (15, 23, 25, 27), retired (formerly of WHO, Regional Office for the Americas/Centro Panamericano de Ingeneria Sanitaria Ciencias del Ambiente [CEPIS], Lima, Peru)

Dr G.J.A. Speijers, (4, 19: iv), National Institute for Public Health and the Environment (RIVM), Bilthoven, Netherlands

Dr D. Srinivasan, (10), University of the South Pacific, Suva, Fiji

Dr G. Stanfield, (5, 7, 21: ii, 25), WRc-NSF Ltd, Marlow, UK

Dr T.A. Stenstrom, (6, 16, 22), Swedish Institute for Infectious Disease Control, Solna, Sweden

Dr M. Stevens, (8, 12, 13, 14, 15, 19, 20, 23, 25), Melbourne Water Corporation, Melbourne, Australia

Dr T. Stinear, (18), Institut Pasteur, Paris, France

Dr M. Storey, (23), University of New South Wales, Sydney, Australia

Mr M. Strauss, (6), Swiss Federal Institute for Environmental Science and Technology, Dübendorf, Switzerland

Dr K. Subramanian, (7), Health Canada, Ottawa, Canada

Dr S. Surman, (16), Health Protection Agency, London, UK

Mr T. Taeuea, (10), Ministry of Health, Tarawa, Kiribati

Mr P. Talota, (10), Ministry of Health and Medical Service, Honiara, Solomon Islands

Mr C. Tan, (11), Ministry of Environment, Singapore

Mr B. Tanner, (2), NSF International, Brussels, Belgium

Mr H. Tano, (4), Ministry of Health and Welfare, Tokyo, Japan

Professor I. Tartakovsky, (16), Gamaleya Research Institute for Epidemiology and Microbiology, Moscow, Russian Federation

Dr A. Tayeh, (20), WHO, Geneva, Switzerland

Dr M. Taylor, (8, 19, 20, 22, 27), Ministry of Health, Wellington, New Zealand

Dr R. Taylor, (8, 10, 15), Health Surveillance and Disease Control, Rockhampton, Australia

Mr J. Teio, (10), Department of Health, Waigani, Papua New Guinea

Dr P.F.M. Teunis, (7, 8, 28), National Institute for Public Health and the Environment (RIVM), Bilthoven, Netherlands

Dr B.H. Thomas, (4), independent consultant (formerly Health Canada, Ottawa, Canada)

Mr T. Thompson, (7, 12, 15, 17, 22, 23, 25, 27), WHO, Regional Office for South-East Asia, New Delhi, India

Dr F. Tiefenbrunner, (21: iii), Institute of Hygiene and Social Medicine, Innsbruck, Austria

Mr Tiew King Nyau, (11), Public Utilities Board, Singapore

Dr D. Till, (8, 28), Consultant Public Health Microbiologist, Wellington, New Zealand

Mr T. Tipi, (10), Health Department, Apia, Samoa

Mr T.Q. Toan, (11), National Institute of Occupational and Environmental Health, Hanoi, Viet Nam

Dr P. Toft, (1, 4, 7, 15, 19: xiii–lii, liv–lxviii, 22), independent consultant, Qualicum Beach, Canada

Mr V. Tovu, (10), Ministry of Health, Port Vila, Vanuatu

Mr L. Tu'itupou, (10), Ministry of Health, Nuku'olafo, Tonga

Professor J. Tuomisto, (4, 19: x), National Public Health Institute, Kuopio, Finland

Dr I. Turai, (20, 22), WHO, Geneva, Switzerland

Dr R. Uauy, (4, 15, 19, 20), Instituto de Nutrición y Technologia de los Alimentos, Santiago, Chile

Mr S. Unisuga, (2), Ministry of Health and Welfare, Tokyo, Japan

Dr H. Utkilen, (29), National Institute for Public Health, Oslo, Norway

Dr J. van Den Berg, (3), KIWA N.V., Nieuwegein, Netherlands

Dr D. van der Kooij, (14, 21: i, 23), KIWA N.V., Nieuwegein, Netherlands

Ms K. VandeVelde, (19), International Antimony Oxide Industry Association, Campine, Beerse, Belgium

Dr A.M. van Dijk-Looijaard, (4), KIWA N.V, Nieuwegein, Netherlands

Dr F.X.R. van Leeuwen, (4), National Institute of Public Health and the Environment (RIVM), Bilthoven, Netherlands (formerly of WHO European Centre for Environment and Health, Netherlands)

Dr J. Vapnek, (29), Food and Agriculture Organization of the United Nations, Rome, Italy

Mr A. Versteegh, (16), National Institute for Public Health and the Environment (RIVM), Bilthoven, Netherlands

Ms C. Vickers, (15, 19: xiii–lii, liv–lxviii),WHO, Geneva, Switzerland

Dr V. Vincent, (18), Institut Pasteur, Paris, France

The Vinyl Institute, (19), Arlington, VA, USA

Dr D. Vitanage, (23), Sydney Water, Sydney, Australia

Dr U. von Gunten, (19), Swiss Federal Institute for Environmental Science and Technology Dübendorf, Switzerland

Professor F. Von Reyn, (18), Dartmouth Hitchcock Medical Centre, Hanover, NH, USA

Professor M. Von Sperling, (6), Federal University of Minas Gerais, Belo Horizonte, Brazil
Dr T. Vourtsanis, (23), Sydney Water, Sydney, Australia
Dr P. Waggit, (27), Environment Australia, Darwin, Australia
Dr I. Wagner, (3), Technologie Zentrum Wasser, Karlsruhe, Germany
Mr M. Waite, (6), Drinking Water Inspectorate, London, UK
Mr M. Waring, (19, 20), Department of Health, London, UK
Ms M. Whittaker, (3), NSF International, Ann Arbor, MI, USA
Dr B. Wilkins, (20), National Radiological Protection Board, UK
Dr J. Wilson, (3), NSF International, Ann Arbor, MI, USA
Dr R. Wolter, (27), Umweltbundesamt, Berlin, Germany
Dr D. Wong, (19: xxvii, xxxiii, lxviii), US Environmental Protection Agency, Washington, DC, USA
Dr A. Wrixon, (20), International Atomic Energy Agency, Vienna, Austria
Professor Y. Xu, (27), University of the Western Cape, Bellville, South Africa
Mr T. Yamamura, (7), (formerly of WHO, Geneva, Switzerland)
Dr S. Yamashita, (20), Nagasaki University, Japan
Dr C. Yayan, (15), Institute of Environmental Health Monitoring, Beijing, People's Republic of China
Dr B. Yessekin, (27), The Regional Environmental Centre for Central Asia, Almaty, Kazakhstan
Dr Z. Yinfa, (11), Ministry of Health, Beijing, People's Republic of China
Mr N. Yoshiguti, (1), Ministry of Health and Welfare, Tokyo, Japan
Dr M. Younes, (1, 7, 20), WHO, Geneva, Switzerland
Mr J. Youngson, (10), Crown Public Health, Christchurch, New Zealand
Dr V. Yu, (21: iii), Pittsburgh University, Pittsburgh, PA, USA
Professor Q. Yuhui, (11), Institute of Environmental Health Monitoring, Beijing, People's Republic of China
Mrs N. Zainuddin, (11), Ministry of Health, Kuala Lumpur, Malaysia

1. *Expert Consultation on Rolling Revision of WHO Guidelines for Drinking-water Quality, Geneva, 13–15 December 1995*
2. *Expert Consultation on Protection and Control of Water Quality for the Updating of the WHO Guidelines for Drinking-water Quality, Bad Elster, Germany, 17–19 June 1996*
3. *Expert Consultation on Safety of Materials and Chemicals Used in Production and Distribution of Drinking-water, Ann Arbor, Michigan, USA, 23–24 January 1997*
4. *Expert Consultation on Rolling Revision of the Guidelines for Drinking-water Quality: Report of Working Group Meeting on Chemical Substances for the Updating of WHO Guidelines for Drinking-water Quality, Geneva, Switzerland, 22–26 April 1997*

ANNEX 2. CONTRIBUTORS TO THE DEVELOPMENT OF THE THIRD EDITION

5. *Expert Consultation on Rolling Revision of the Guidelines for Drinking-water Quality: Aspects of Protection and Control and of Microbiological Quality, Medmenham, UK, 17–21 March 1998*
6. *Expert Consultation on Harmonized Risk Assessment for Water-related Microbiological Hazards, Stockholm, Sweden, 12–16 September 1999*
7. *Drinking-water Quality Committee Meeting, Berlin, Germany, 5–9 June 2000*
8. *Expert Consultation on Effective Approaches to Regulating Microbial Drinking-water Quality, Adelaide, Australia, 14–18 May 2001*
9. *Consultation on Planning of Water Quality Guidelines for Desalination, Bahrain, 28–31 May 2001*
10. *Workshop on Drinking-water Quality Surveillance and Safety, Nadi, Fiji, 29 October–1 November 2001*
11. *Workshop on Drinking-water Quality Surveillance and Safety, Kuala Lumpur, Malaysia, 12–15 November 2001*
12. *Expert Consultation on Preparation of Supporting Documents for the Updating of Microbial Aspects of WHO Guidelines for Drinking-water Quality, Loughborough, UK, 18–23 November 2001*
13. *WHO Meeting: Guidelines on Drinking-water Quality, Micro Working Group, Melbourne, Australia, 13–14 April 2002*
14. *Meeting on HPC Bacteria in Drinking-water, Geneva, Switzerland, 25–26 April 2002*
15. *Global Meeting on the Revision of WHO Guidelines for Drinking-water Quality, Tokyo, Japan, 23–29 May 2002*
16. *Meeting on Prevention and Control of Legionnaires' Disease, London, 18–20 June 2002*
17. *Chemical Safety of Drinking-water: Assessing Priorities for Risk Management, Nyon, Switzerland, 26–30 August 2002*
18. *Expert Consultation on Mycobacterium Avium Complex, Guildford, UK, 18–20 September 2002*
19. *Contributors to the chemical substantiation document on:*
    i. *Aluminium*
    ii. *Boron*
    iii. *Nickel*
    iv. *Nitrate and Nitrite*
    v. *Cyanobacterial Toxins: Microcystin-LR*
    vi. *Edetic Acid (EDTA)*
    vii. *Polynuclear aromatic hydrocarbons*
    viii. *Cyanazine*
    ix. *1,2-Dichloropropane (1,2-DCP)*
    x. *Pentachlorophenol*
    xi. *Terbuthylazine (TBA)*
    xii. *Trihalomethanes*

xiii. 1,1,1-Trichloroethane
xiv. 1,2-Dibromoethane
xv. 1,2-Dichloroethane
xvi. Di(2-ethylhexyl)adipate
xvii. 2-Phenylphenol
xviii. 2,4-Dichlorophenoxyacetic acid
xix. Acrylamide
xx. Aldicarb
xxi. Aldrin and Dieldrin
xxii. Antimony
xxiii. Arsenic
xxiv. Barium
xxv. Bentazone
xxvi. Bromate
xxvii. Brominated Acetic Acids
xxviii. Cadmium
xxix. Carbofuran
xxx. Carbon Tetrachloride
xxxi. Monochloramine
xxxii. Chlordane
xxxiii. Monochloroacetic acid
xxxiv. Chlorite and Chlorate
xxxv. Chlorpyrifos
xxxvi. Copper
xxxvii. DDT and its Derivatives
xxxviii. Dimethoate
xxxix. Diquat
xl. Endosulfan
xli. Endrin
xlii. Epichlorohydrin
xliii. Fenitrothion
xliv. Fluoride
xlv. Glyphosate and AMPA
xlvi. Halogenated Acetonitriles
xlvii. Heptachlor and Heptachlor Epoxide
xlviii. Hexachlorobenzene
xlix. Hexachlorobutadiene
l. Lindane
li. Malathion
lii. Manganese
liii. Methoxychlor
liv. Methyl Parathion

ANNEX 2. CONTRIBUTORS TO THE DEVELOPMENT OF THE THIRD EDITION

 lv. *Monochlorobenzene*
 lvi. *MX*
 lvii. *Dialkyltins*
 lviii. *Parathion*
 lix. *Permethrin*
 lx. *Propanil*
 lxi. *Pyriproxyfen*
 lxii. *Sulfate*
 lxiii. *Inorganic Tin*
 lxiv. *Toluene*
 lxv. *Trichlorobenzenes*
 lxvi. *Uranium*
 lxvii. *Vinyl Chloride*
 lxviii. *Trichloroacetic Acid*
 lxix. *Dichloroacetic Acid*

20. *Provision of comments on drafts of the Guidelines for Drinking-water Quality (3rd edition)*
21. *Contributor to Guidelines for Drinking-water Quality (2nd edition), Addendum, Microbiological Agents in Drinking-water*
 i. *Aeromonas*
 ii. *Enteric Hepatitis Viruses*
 iii. *Legionella*
 iv. *Protozoan Parasites (Cryptosporidium, Giardia, Cyclospora)*
 v. *Vibrio cholerae*
22. *Participant in Final Task Force Meeting for 3rd Edition of Guidelines on Drinking-water Quality, Geneva, Switzerland, 31 March – 4 April 2003*
23. *Contributor to the background document "Safe, Piped Water: Managing Microbial Water Quality in Piped Distribution Systems."*
24. *Contributor to the background document "Water Treatment and Pathogen Control: Process Efficiency in Achieving Safe Drinking-water."*
25. *Contributor to the background document "Water Safety Plans."*
26. *Contributor to the background document "Quantifying Public Health Risk in the WHO Guidelines for Drinking-Water Quality: A Burden of Disease Approach."*
27. *Contributor to the background document "Protecting Groundwaters for Health – Managing the Quality of Drinking-water Sources."*
28. *Contributor to the background document "Hazard Characterization for Pathogens in Food and Water: Guidelines."*
29. *Contributor to the background document "Toxic Cyanobacteria in Water."*

ANNEX 3
Default assumptions

A3.1 Drinking-water consumption and body weight

Global data on the consumption of drinking-water are limited. In studies carried out in Canada, the Netherlands, the United Kingdom and the USA, the average daily per capita consumption was usually found to be less than 2 litres, but there was considerable variation between individuals. As water intake will vary with climate, physical activity and culture, the above studies, which were conducted in temperate zones, can give only a limited view of consumption patterns throughout the world. At temperatures above 25 °C, for example, there is a sharp rise in fluid intake, largely to meet the demands of an increased sweat rate (ICRP, 1992; see also Howard & Bartram, 2003).

In developing guidelines for microbial hazard, per capita daily consumption of 1 litre of unboiled water was assumed.

In developing the guideline values for potentially hazardous chemicals, a daily per capita consumption of 2 litres by a person weighing 60 kg was generally assumed. The guideline values set for drinking-water using this assumption do, on average, err on the side of caution. However, such an assumption may underestimate the consumption of water per unit weight, and thus exposure, for those living in hot climates, as well as for infants and children, who consume more fluid per unit weight than adults. The higher intakes, and hence exposure, for infants and children apply for only a limited time, but this period may coincide with greater sensitivity to some toxic agents and less for others. Irreversible effects that occur at a young age will have more social and public health significance than those that are delayed. Where it was judged that this segment of the population was at a particularly high risk from exposure to certain chemicals, the guideline value was derived on the basis of a 10-kg child consuming 1 litre per day or a 5-kg bottle-fed infant consuming 0.75 litre per day. The corresponding daily fluid intakes are higher than for adults on a body weight basis.

A3.2 Inhalation and dermal absorption

The contribution of drinking-water to daily exposure includes some indirect routes – such as inhalation of particles and droplets containing microbes and volatile substances, and dermal contact during bathing or showering – as well as direct ingestion.

In most cases, available data are insufficient to permit reliable estimates of exposure by inhalation and dermal absorption of contaminants present in drinking-water. It was not always possible, therefore, to address intake from these routes specifically in the derivation of the guideline values. However, that portion of the total tolerable daily intake (TDI) allocated to drinking-water is generally sufficient to allow for these additional routes of intake (see section 8.2.2). Should there be reason to believe that potential inhalation of volatile compounds and dermal exposure from various indoor water uses (such as showering) are not adequately addressed, authorities could consider taking this into account in setting national standards or guidelines.

ANNEX 4
Chemical summary tables

Table A4.1 Chemicals excluded from guideline value derivation

Chemical	Reason for exclusion
Amitraz	Degrades rapidly in the environment and is not expected to occur at measurable concentrations in drinking-water supplies
Beryllium	Unlikely to occur in drinking-water
Chlorobenzilate	Unlikely to occur in drinking-water
Chlorothalonil	Unlikely to occur in drinking-water
Cypermethrin	Unlikely to occur in drinking-water
Diazinon	Unlikely to occur in drinking-water
Dinoseb	Unlikely to occur in drinking-water
Ethylene thiourea	Unlikely to occur in drinking-water
Fenamiphos	Unlikely to occur in drinking-water
Formothion	Unlikely to occur in drinking-water
Hexachlorocyclohexanes (mixed isomers)	Unlikely to occur in drinking-water
MCPB	Unlikely to occur in drinking-water
Methamidophos	Unlikely to occur in drinking-water
Methomyl	Unlikely to occur in drinking-water
Mirex	Unlikely to occur in drinking-water
Monocrotophos	Has been withdrawn from use in many countries and is unlikely to occur in drinking-water
Oxamyl	Unlikely to occur in drinking-water
Phorate	Unlikely to occur in drinking-water
Propoxur	Unlikely to occur in drinking-water
Pyridate	Not persistent and only rarely found in drinking-water
Quintozene	Unlikely to occur in drinking-water
Toxaphene	Unlikely to occur in drinking-water
Triazophos	Unlikely to occur in drinking-water
Tributyltin oxide	Unlikely to occur in drinking-water
Trichlorfon	Unlikely to occur in drinking-water

ANNEX 4. CHEMICAL SUMMARY TABLES

Table A4.2 Chemicals for which guideline values have not been established

Chemical	Reason for not establishing a guideline value
Aluminium	Owing to limitations in the animal data as a model for humans and the uncertainty surrounding the human data, a health-based guideline value cannot be derived; however, practicable levels based on optimization of the coagulation process in drinking-water plants using aluminium-based coagulants are derived: 0.1 mg/litre or less in large water treatment facilities, and 0.2 mg/litre or less in small facilities
Ammonia	Occurs in drinking-water at concentrations well below those at which toxic effects may occur
Asbestos	No consistent evidence that ingested asbestos is hazardous to health
Bentazone	Occurs in drinking-water at concentrations well below those at which toxic effects may occur
Bromochloroacetate	Available data inadequate to permit derivation of health-based guideline value
Bromochloroacetonitrile	Available data inadequate to permit derivation of health-based guideline value
Chloride	Not of health concern at levels found in drinking-water[a]
Chlorine dioxide	Guideline value not established because of the rapid breakdown of chlorine dioxide and because the chlorite provisional guideline value is adequately protective for potential toxicity from chlorine dioxide
Chloroacetones	Available data inadequate to permit derivation of health-based guideline values for any of the chloroacetones
Chlorophenol, 2-	Available data inadequate to permit derivation of health-based guideline value
Chloropicrin	Available data inadequate to permit derivation of health-based guideline value
Dialkyltins	Available data inadequate to permit derivation of health-based guideline values for any of the dialkyltins
Dibromoacetate	Available data inadequate to permit derivation of health-based guideline value
Dichloramine	Available data inadequate to permit derivation of health-based guideline value
Dichlorobenzene, 1,3-	Toxicological data are insufficient to permit derivation of health-based guideline value
Dichloroethane, 1,1-	Very limited database on toxicity and carcinogenicity
Dichlorophenol, 2,4-	Available data inadequate to permit derivation of health-based guideline value
Dichloropropane, 1,3-	Data insufficient to permit derivation of health-based guideline value
Di(2-ethylhexyl)adipate	Occurs in drinking-water at concentrations well below those at which toxic effects may occur
Diquat	Rarely found in drinking-water, but may be used as an aquatic herbicide for the control of free-floating and submerged aquatic weeds in ponds, lakes and irrigation ditches
Endosulfan	Occurs in drinking-water at concentrations well below those at which toxic effects may occur
Fenitrothion	Occurs in drinking-water at concentrations well below those at which toxic effects may occur
Fluoranthene	Occurs in drinking-water at concentrations well below those at which toxic effects may occur
Glyphosate and AMPA	Occurs in drinking-water at concentrations well below those at which toxic effects may occur
Hardness	Not of health concern at levels found in drinking-water[a]
Heptachlor and heptachlor epoxide	Occurs in drinking-water at concentrations well below those at which toxic effects may occur

continued

Table A4.2 *Continued*

Chemical	Reason for not establishing a guideline value
Hexachlorobenzene	Occurs in drinking-water at concentrations well below those at which toxic effects may occur
Hydrogen sulfide	Not of health concern at levels found in drinking-water[a]
Inorganic tin	Occurs in drinking-water at concentrations well below those at which toxic effects may occur
Iodine	Available data inadequate to permit derivation of health-based guideline value, and lifetime exposure to iodine through water disinfection is unlikely
Iron	Not of health concern at concentrations normally observed in drinking-water, and taste and appearance of water are affected below the health-based value
Malathion	Occurs in drinking-water at concentrations well below those at which toxic effects may occur
Methyl parathion	Occurs in drinking-water at concentrations well below those at which toxic effects may occur
Monobromoacetate	Available data inadequate to permit derivation of health-based guideline value
Monochlorobenzene	Occurs in drinking-water at concentrations well below those at which toxic effects may occur, and health-based value would far exceed lowest reported taste and odour threshold
MX	Occurs in drinking-water at concentrations well below those at which toxic effects may occur
Parathion	Occurs in drinking-water at concentrations well below those at which toxic effects may occur
Permethrin	Occurs in drinking-water at concentrations well below those at which toxic effects may occur
pH	Not of health concern at levels found in drinking-water[b]
Phenylphenol, 2- and its sodium salt	Occurs in drinking-water at concentrations well below those at which toxic effects may occur
Propanil	Readily transformed into metabolites that are more toxic; a guideline value for the parent compound is considered inappropriate, and there are inadequate data to enable the derivation of guideline values for the metabolites
Silver	Available data inadequate to permit derivation of health-based guideline value
Sodium	Not of health concern at levels found in drinking-water[a]
Sulfate	Not of health concern at levels found in drinking-water[a]
Total dissolved solids (TDS)	Not of health concern at levels found in drinking-water[a]
Trichloramine	Available data inadequate to permit derivation of health-based guideline value
Trichloroacetonitrile	Available data inadequate to permit derivation of health-based guideline value
Trichlorobenzenes (total)	Occurs in drinking-water at concentrations well below those at which toxic effects may occur, and health-based value would exceed lowest reported odour threshold
Trichloroethane, 1,1,1-	Occurs in drinking-water at concentrations well below those at which toxic effects may occur
Zinc	Not of health concern at concentrations normally observed in drinking-water[a]

[a] May affect acceptability of drinking-water (see chapter 10).
[b] An important operational water quality parameter.

ANNEX 4. CHEMICAL SUMMARY TABLES

Table A4.3 Guideline values for chemicals that are of health significance in drinking-water

Chemical	Guideline value[a] (mg/litre)	Remarks
Acrylamide	0.0005[b]	
Alachlor	0.02[b]	
Aldicarb	0.01	Applies to aldicarb sulfoxide and aldicarb sulfone
Aldrin and dieldrin	0.00003	For combined aldrin plus dieldrin
Antimony	0.02	
Arsenic	0.01 (P)	
Atrazine	0.002	
Barium	0.7	
Benzene	0.01[b]	
Benzo[a]pyrene	0.0007[b]	
Boron	0.5 (T)	
Bromate	0.01[b] (A, T)	
Bromodichloromethane	0.06[b]	
Bromoform	0.1	
Cadmium	0.003	
Carbofuran	0.007	
Carbon tetrachloride	0.004	
Chloral hydrate (trichloroacetaldehyde)	0.01 (P)	
Chlorate	0.7 (D)	
Chlordane	0.0002	
Chlorine	5 (C)	For effective disinfection, there should be a residual concentration of free chlorine of ≥0.5 mg/litre after at least 30 min contact time at pH <8.0
Chlorite	0.7 (D)	
Chloroform	0.2	
Chlorotoluron	0.03	
Chlorpyrifos	0.03	
Chromium	0.05 (P)	For total chromium
Copper	2	Staining of laundry and sanitary ware may occur below guideline value
Cyanazine	0.0006	
Cyanide	0.07	
Cyanogen chloride	0.07	For cyanide as total cyanogenic compounds
2,4-D (2,4-dichlorophenoxyacetic acid)	0.03	Applies to free acid
2,4-DB	0.09	
DDT and metabolites	0.001	
Di(2-ethylhexyl)phthalate	0.008	
Dibromoacetonitrile	0.07	
Dibromochloromethane	0.1	
1,2-Dibromo-3-chloropropane	0.001[b]	
1,2-Dibromoethane	0.0004[b] (P)	
Dichloroacetate	0.05 (T, D)	
Dichloroacetonitrile	0.02 (P)	
Dichlorobenzene, 1,2-	1 (C)	

continued

Table A4.3 *Continued*

Chemical	Guideline value (mg/litre)	Remarks
Dichlorobenzene, 1,4-	0.3 (C)	
Dichloroethane, 1,2-	0.03[b]	
Dichloroethene, 1,1-	0.03	
Dichloroethene, 1,2-	0.05	
Dichloromethane	0.02	
1,2-Dichloropropane (1,2-DCP)	0.04 (P)	
1,3-Dichloropropene	0.02[b]	
Dichlorprop	0.1	
Dimethoate	0.006	
Edetic acid (EDTA)	0.6	Applies to the free acid
Endrin	0.0006	
Epichlorohydrin	0.0004 (P)	
Ethylbenzene	0.3 (C)	
Fenoprop	0.009	
Fluoride	1.5	Volume of water consumed and intake from other sources should be considered when setting national standards
Formaldehyde	0.9	
Hexachlorobutadiene	0.0006	
Isoproturon	0.009	
Lead	0.01	
Lindane	0.002	
Manganese	0.4 (C)	
MCPA	0.002	
Mecoprop	0.01	
Mercury	0.001	For total mercury (inorganic plus organic)
Methoxychlor	0.02	
Metolachlor	0.01	
Microcystin-LR	0.001 (P)	For total microcystin-LR (free plus cell-bound)
Molinate	0.006	
Molybdenum	0.07	
Monochloramine	3	
Monochloroacetate	0.02	
Nickel	0.02 (P)	
Nitrate (as NO_3^-)	50	Short-term exposure
Nitrilotriacetic acid (NTA)	0.2	
Nitrite (as NO_2^-)	3	Short-term exposure
	0.2 (P)	Long-term exposure
Pendimethalin	0.02	
Pentachlorophenol	0.009[b] (P)	
Pyriproxyfen	0.3	
Selenium	0.01	
Simazine	0.002	
Styrene	0.02 (C)	
2,4,5-T	0.009	
Terbuthylazine	0.007	
Tetrachloroethene	0.04	
Toluene	0.7 (C)	

ANNEX 4. CHEMICAL SUMMARY TABLES

Table A4.3 *Continued*

Chemical	Guideline value (mg/litre)	Remarks
Trichloroacetate	0.2	
Trichloroethene	0.07 (P)	
Trichlorophenol, 2,4,6-	0.2[b] (C)	
Trifluralin	0.02	
Trihalomethanes		The sum of the ratio of the concentration of each to its respective guideline value should not exceed 1
Uranium	0.015 (P, T)	Only chemical aspects of uranium addressed
Vinyl chloride	0.0003[b]	
Xylenes	0.5 (C)	

[a] P = provisional guideline value, as there is evidence of a hazard, but the available information on health effects is limited; T = provisional guideline value because calculated guideline value is below the level that can be achieved through practical treatment methods, source protection, etc.; A = provisional guideline value because calculated guideline value is below the achievable quantification level; D = provisional guideline value because disinfection is likely to result in the guideline value being exceeded; C = concentrations of the substance at or below the health-based guideline value may affect the appearance, taste or odour of the water, leading to consumer complaints.

[b] For substances that are considered to be carcinogenic, the guideline value is the concentration in drinking-water associated with an upper-bound excess lifetime cancer risk of 10^{-5} (one additional cancer per 100 000 of the population ingesting drinking-water containing the substance at the guideline value for 70 years). Concentrations associated with upper-bound estimated excess lifetime cancer risks of 10^{-4} and 10^{-6} can be calculated by multiplying and dividing, respectively, the guideline value by 10.

Index

Page numbers in **bold** indicate main discussions.

Acanthamoeba 122, 123, 125, **259–261**
Acceptability **7**, 23, **210–220**
  biologically derived contaminants
    211–213
  chemical contaminants 146, 156,
    213–219
  desalinated water 112–113
  in emergency and disaster situations
    106
Acceptable daily intake (ADI) 150
  derivation of guideline values 152
  uncertainty factors 150–151
Access to water (accessibility) 90, **91–92**
  definition of reasonable 91
  equitability 105
Acinetobacter 102, 124, **222–224**, 286
Acrylamide **296–297**
  analysis 162
  guideline value 194, 296, 491
Actinomycetes 212
Activated alumina 179
Activated carbon
  adsorption **176–177**
  granular (GAC) 176, 177
  powdered (PAC) 176
Additives 30
Adenoviruses 122, **248–250**, 295
Adequacy of supply, surveillance **90–93**
ADI *see* Acceptable daily intake
Advanced oxidation processes 173
Aeration processes **175**
Aeromonas 102, 124, **224–225**, 286
Aerosols 123
Affordability 90, **92**
Aggressivity, desalinated water 112
Aggressivity index 183

Agricultural activities, chemicals from 147
  analysis 159, 161
  guideline values **187–188**, 189, 190, 191
  treatment achievabilities 169–170
AIDS 124, 270
Air
  chemical intake 152
  radon intake 206–207
Air stripping 175
Aircraft **116–117**
Airports **116–117**
Alachlor **297–298**
  analysis 161
  guideline value 191, 298, 491
  treatment achievability 169, 298
Aldicarb **298–300**
  analysis 161
  guideline value 191, 299, 491
  treatment achievability 169, 299
Aldrin **300–301**
  analysis 161
  guideline value 191, 300, 491
  treatment achievability 169, 300
Algae 213
  blue-green *see* Cyanobacteria
  harmful events 111, 213
  toxins 111
Alkalinity 217
  corrosion and 181, 184
  see also pH
Alkylbenzenes 217
Alpha radiation activity
  measurement 207–208
  screening levels 204, 205, 206
Alumina, activated 179
Aluminium 193, 213, **301–303**, 489

494

INDEX

Alzheimer disease (AD) 302
Americium-241 202
Aminomethylphosphonic acid (AMPA) 190, **379–380**, 489
Amitraz 189, 488
Ammonia 190, **303–304**, 489
　taste and odour 213
　treatment to remove 220
Amoebae 63
　Legionella ingestion 234
　persistence in water 125
　see also Acanthamoeba; Entamoeba histolytica; Naegleria fowleri
Amoebiasis 266
Amoebic meningoencephalitis, primary (PAM) 123, 272, 273
AMPA 190, **379–380**, 489
Analytical methods
　chemicals **157–166**
　radionuclides **207–208**
Ancylostoma 124
Animals
　in drinking-water 212–213
　toxicity studies 148
　uncertainty factors 151
Anion exchange 177
Anthrax 225
Antimony **304–306**
　analysis 159
　guideline value 194, 305, 491
Appearance 7, 210, **211–220**
　biologically derived contaminants 211–213
　chemical contaminants 213–219
　treatments for improving 219–220
Argyria 434
Arsenic 6, **306–308**
　analysis 159
　in drinking-water sources 146, 306
　guideline value 186, 306, 491
　priority 35–36
　treatment achievability 167, 307
Asbestos 193, **308**, 489
Asbestos–cement pipes 183
Ascariasis *(Ascaris)* 124, 276
Asellus aquaticus 212
Aspergillus 102
Assessing Microbial Safety of Drinking Water: Improving Approaches and Methods 18, 59
Astroviruses **250–251**
Atomic absorption spectrometry (AAS) 159–164
Atomic emission spectrometry (AES) 164

Atrazine **308–309**
　analysis 161
　guideline value 191, 309, 491
　treatment achievability 169, 309
Audit **86–87**, 94
Avoidance, water **79**

Bacillus 221, **225–226**
Bacillus cereus 221, 225, 226
Bacillus thuringiensis israelensis 190
Backflow 62, 63
　large buildings 101
Bacteria 221
　indicator and index **282–289**
　pathogenic 122, **222–247**
　persistence in water 125
　treatment effects 138–141
Bacteriophages 142, **289–294**
　Bacteroides fragilis **292–294**
　coliphages **289–292**
Bacteroides fragilis phages **292–294**
Balantidium coli (balantidiasis) 124, **261–262**
Barium **310–311**
　analysis 159
　guideline value 186, 310, 491
BDCM *see* Bromodichloromethane
Becquerel (Bq) 201
Benchmark dose (BMD) 152, **153**
Bentazone 190, **311–312**, 489
Benzene **312–313**
　analysis 160
　guideline value 188, 312, 491
　treatment achievability 168, 312
3,4-Benzfluoranthene 429
11,12-Benzfluoranthene 429
Benzo[*a*]pyrene 428–429, 430
　analysis 162
　guideline value 194, 428, 491
1,12-Benzpyrene 429
3,4-Benzpyrene 429
Beryllium 187, 488
Beta-Poisson dose–response relation 129
Beta radiation activity 205
　measurement 207–208
　screening levels 204, 205, 206
Bilharziasis 123
Biofilms 4–5, 63
　atypical mycobacteria 235, 236
　coliform bacteria 283
　desalinated water 113
　Klebsiella 233
　Legionella 234, 235
Biological denitrification 179

495

Biological nitrification 179
Biologically derived contaminants **211–213**
Bleach, household 107
Blooms, cyanobacterial 195, 213, 281
"Blue-baby syndrome"
  (methaemoglobinaemia) 6, 418–420
Blue-green algae *see* Cyanobacteria
Body weight 150
  assumptions 486
Boil water orders **79**
Boiling of water
  bottle-fed infants 114
  emergencies and disasters 79, 107
  travellers 110
Borehole water supplies 65–66
Boron **313–314**
  analysis 159
  guideline value 186, 313, 491
Bottle-fed infants 114, 418, 419
Bottled water 113–115
  international standards 114–115
  potential health benefits 114
  travellers 110, 111
Brackish water 111
Brass corrosion **182–183**
Bromate 179, **315–316**
  analysis 162
  guideline value 194, 315, 491
  strategies for reducing 180
Brominated acetic acids **316–317**
Bromochloroacetate 193, 316–317, 489
Bromochloroacetonitrile 193, **380–382**, 489
Bromodichloromethane (BDCM) **451–454**
  analysis 162, 452
  guideline value 194, 451, 491
Bromoform **451–454**
  analysis 162
  guideline value 194, 451, 491
Buildings
  large **99–104**, 235
  plumbing systems 17–18
Burkholderia pseudomallei 122, 221, **226–227**
Burns injuries 103

Cadmium **317–319**
  analysis 159
  guideline value 188, 317, 491
  treatment achievability 168, 317
Caesium-134 (^{134}Cs), 202
Caesium-137 (^{137}Cs), 202
Calcium, taste threshold 215
Calcium carbonate
  corrosion control 181, 182, 183, 184

scale 183–184, 215–216
  see also Hardness
Calcium hypochlorite 107, 171
Calcium sulfate 218
Caliciviruses **251–253**
Campylobacter **228–229**
  performance target setting 132
  risk characterization 129, 130
  in source waters 137
Campylobacter coli 122, 228
Campylobacter jejuni 122, 228
Campylobacter pylori see Helicobacter pylori
Cancer
  radiation-induced 200
  radon-related risk 207
  tolerable risk 46–47
  see also Carcinogens
Carbofuran 161, **319–320**
  guideline value 191, 319, 491
  treatment achievability 169, 319
Carbon, activated *see* Activated carbon
Carbon-14 (^{14}C), 202
Carbon tetrachloride **320–321**
  analysis 160
  guideline value 188, 320, 491
  treatment achievability 168, 320
Carcinogens
  derivation of guideline values 149
  genotoxic 148–149, 154
  guideline values 154
  IARC classification 149
  non-genotoxic 149
  tolerable risk 46–47
  uncertainty factors 151
Cascade aeration 175
Catchments 53, 54, 56–59
  control measures 58–59
  hazard identification 56–58
  mapping, emergency and disaster situations **108**
  new systems 52–53
  roles and responsibilities 11, 12–13, 14
  see also Source waters
Categorical regression 152, **153–154**
Cation exchange 177
Cement, corrosion **183**
Cercariae 123
Certification 16–17, 42
  agencies **16–17**
  chemicals in water 43
  desalination systems 112
Chemical Safety of Drinking-water: Assessing Priorities for Risk Management 18, 36

INDEX

Chemical-specific adjustment factors (CSAF) 152, 154
Chemicals 6–7, **145–196**
 acceptability aspects 146, 156, **213–219**
 agricultural activities *see* Agricultural activities, chemicals from
 allocation of intake 151–152
 alternative routes of exposure 43–44, 146
 analytical methods 157–166
  achievabilities 157–158, 159, 160–163
  ranking of complexity 158
 categorization by source 147
 desalination systems 111–112
 emergencies involving 79, 108–109
 guideline values *see* Guideline values
 health-based targets 41, 42–43
 health hazards 6–7, **145–147**
 IARC classification 149
 industrial sources and human dwellings *see* Industrial sources and human dwellings, chemicals from
 information sources 36, 148, 156
 inorganic
  analytical methods 158, 159
  guideline values 185, 186
 mixtures 156
 naturally occurring *see* Naturally occurring chemicals
 non-guideline 156
 non-threshold 148–149
  derivation of guideline values **154**
  provisional guideline values 155–156
 organic, analytical methods 158, 160–161
 priority setting **35–36**
 on ships 118
 "short-listing" 36
 summary tables **488–493**
 threshold 148, **149–154**
  alternative approaches 152–154
  derivation of guideline values 149–152
 treatment **166–184**
  achievabilities 166–171
  for corrosion control 180–184
  process control measures 179–180
  processes 171–179
  used in treatment/materials in contact with water 147
  analysis 159, 162
  guideline values **188–190**, 193–194
  see also Disinfection by-products
 water quality
  emergency and disaster situations **108–109**

targets 42–43
verification **30–31**, 72, **73**
Children
 consumption assumptions 486
 hygiene education 103–104
 radionuclide guidance levels 204
 see also Infants
Chironomus larvae 212
Chloral hydrate (trichloroacetaldehyde) **321–322**
 analysis 162
 guideline value 194, 322, 491
Chloramination 63–64, 172
 by-products 179, 180, 192
 nitrite formation 417, 418
Chloramines 172
 dialysis water 103
 see also Monochloramine
Chlorate 179, **326–329**
 analysis 162
 guideline value 194, 326, 491
Chlordane **323–324**
 analysis 161
 guideline value 191, 323, 491
 treatment achievability 169, 323
Chloride 185, **324–325**, 489
 acceptability 213–214, 324
 corrosion and 181, 182, 184
Chlorinated acetic acids 145, 179, 349–350, 412–413, 445–446
Chlorinated anisoles 214
Chlorinated ketones 179
Chlorination 61, **171–172**
 breakpoint 171
 by-products 145, 179–180, 192, 451
 in emergencies 79
 marginal 171
 microbial reduction 140
 for travellers 110
Chlorine 5, 171, **325–326**
 acceptable levels 214
 analysis 162
 gas, liquefied 171
 guideline value 194, 325, 491
 residual
  emergency and disaster situations 107, 108
  monitoring 69, 82
 treatment *see* Chlorination
Chlorine dioxide 326
 by-products 179, 180, 192, 326
 see also Chlorate; Chlorite
 guideline value 193, 328, 489
 microbial reduction 140

497

toxicity 327
water treatment 173
Chlorite 179, **326–329**
 analysis 162
 guideline value 194, 326, 491
3-Chloro-4-dichloromethyl-5-hydroxy-
 2(5H) furanone (MX) 193,
 414–415, 490
Chloroacetones 193, **329**, 489
Chlorobenzilate 189, 488
Chloroform 145, **451–454**
 analysis 162, 452
 guideline value 194, 451, 491
2-Chlorophenol 193, 214, **329–331**,
 489
Chlorophenols 214, **329–331**
Chlorophenoxy herbicides 341, 342–343,
 361–362, 374–375, 439–440
Chloropicrin 193, **331–332**, 489
Chlorothalonil 189, 488
Chlorotoluron **332–333**
 analysis 161
 guideline value 191, 332, 491
 treatment achievability 169, 332
Chlorpyrifos 190, **333–334**
 analysis 163
 guideline value 195, 333, 491
Cholera 244–245
Chromatography 164–165
Chromium **334–335**
 analysis 159
 guideline value 186, 334, 491
Chydorus sphaericus 212
Citrobacter 282, 284
Clarification 138–139
 drinking-water for travellers 110
 emergency and disaster situations 105,
 107
Clostridium perfringens 142, **288–289**
Closure, drinking-water supply **79**
Cloudiness 211
Co-precipitation method, radionuclide
 analysis 208
Coagulation (chemical) 60, **175–176**
 before disinfection 179–180
 microbial reduction 138–139
Coal-tar linings, pipes 428, 430
Coastal water 111
Code of good practice 33–34
*Code of Practice for Collecting, Processing and
 Marketing of Natural Mineral Waters*
 115
Codex Alimentarius Commission (CAC)
 114–115

Coliform bacteria
 detection methods 144
 thermotolerant 142, 143, 282, **284–285**
 total **282–284**
Coliphages **289–292**
 F-RNA 290–291
 somatic 290, 291
Colitis, amoebic 266
Collection, water
 emergency and disaster situations
 106
 household use 71
Colorimetric methods 158
Colour 211, 214
Communication **27–28**
 emergency and disaster situations
 106
 surveillance information **95–97**
 water safety plans **82–83**
Community
 communication 28, **96**
 involvement in setting standards 34
 organizations 12, 96
Community drinking-water systems
 64–67
 control measures 65–67
 development of water safety plans
 (WSPs) 85
 ensuring operation and maintenance
 94
 grading schemes 97, 98
 hazard identification 64–65
 management **81–82**
 operational monitoring **71**, 82
 roles and responsibilities 11–12, **14–15**
 surveillance 87, **88–89**
 verification testing **74–75**
Concise International Chemical Assessment
 Documents (CICADs) 36
Concrete, dissolution **183**
Confidence intervals 153
Conjunctivitis, adenovirus 248, 249
Consumers
 acceptability to *see* Acceptability
 interaction with **96**
 right of access to information 83, 96
 roles and responsibilities **15–16**
Consumption, drinking-water, daily per
 capita 90
 assumptions 486
 performance target setting and 128,
 133–134
Contact, transmission via 221
Contact lenses 238, 260–261

INDEX

Continuity of supply 90, **92–93**
Control measures 26, 49, **68**
   assessment and planning 55–56
   defined 55
   monitoring performance *see* Operational monitoring
   operational and critical limits 70
   prioritizing hazards 53–55
   validation *see* Validation
Cooling towers 100, 234
Copper **335–337**
   acceptability 214–215
   analysis 159
   corrosion **182**
   guideline value 194, 336, 491
   impingement attack 182
   pitting 182
Corrosion **180–184**, 217
   control strategies 184
   galvanic 182
   indices 183–184
   inhibitors 181, 184
   pitting 182
Costs
   treatment 166–167
   water supply 92
Coxsackieviruses 253–254
Crangonyx pseudogracilis 212
Critical limits **70**
Crustaceans 212
Cryptosporidiosis 259, 262–263
Cryptosporidium (parvum) 122, **262–264**
   disinfection 140–141
   oocysts 110, 262, 263
   performance target setting 131–132, 133–134
   risk characterization 130
   in source waters 137
Ct concept 61
Culex larvae 212
Cyanazine **337–338**
   analysis 161
   guideline value 191, 337, 491
   treatment achievability 169
Cyanide **339–340**
   analysis 159
   guideline value 188, 339, 491
Cyanobacteria 147, 192, 221, **279–281**
   acceptability 213
   blooms 195, 213, 281
   health concerns 4
   toxins *see* Cyanotoxins
   treatment 171, 195

Cyanogen chloride 162, 194, **340**, 491
Cyanotoxins 4, 280, 281
   classification 192
   guideline values **192–196**
   treatment 171, 195
   see also Microcystin-LR
Cyclops 212, 276, 277
Cyclospora cayetanensis 122, 259, **264–265**
Cyclosporiasis 264
Cylindrospermopsin 192, 280
Cypermethrin 189, 488
Cystic fibrosis 238

2,4-D (2,4-dichlorophenoxyacetic acid) **340–342**
   analysis 161
   guideline value 191, 341, 491
   treatment achievability 169, 341
DALYs *see* Disability-adjusted life years
Data
   fitness for purpose 75
   regional use **96–97**, 98
   system assessment and design 53–56
Day care centres **103–104**
2,4-DB 161, 191, **342–343**, 491
DBCP *see* 1,2-Dibromo-3-chloropropane
DBPs *see* Disinfection by-products
DCBs *see* Dichlorobenzenes
DDT and metabolites 190, **343–345**
   analysis 163
   guideline value 195, 344, 491
   treatment achievability 170, 344
"Dealkalization" 177
Dechlorination 171
DEHA *see* Di(2-ethylhexyl)adipate
DEHP *see* Di(2-ethylhexyl)phthalate
Demineralized water 114
Denitrification, biological 179
Dermal absorption
   assumptions 486–487
   chemicals 152
Desalination systems **111–113**, 178
Detergents, synthetic 218
Developing countries, urban areas **88**
"Deviations" 77
Devices
   certification *see* Certification
   medical, washing 103
Dezincification of brass 182
Di(2-ethylhexyl)adipate (DEHA) 187, **362–363**, 489
Dialkyltins 193, **345–346**, 489
Dialysis, renal 103

Diarrhoea
  cryptosporidiosis 262–263
  Escherichia coli 230
  Giardia 267
  rotavirus 258
  travellers' 109
Diatomaceous earth 139
Diazinon 189, 488
1,2-Dibromo-3-chloropropane (DBCP) **346–347**
  analysis 161
  guideline value 191, 346, 491
  treatment achievability 169, 346
Dibromoacetate 193, 316, 489
Dibromoacetonitrile 162, 194, **380–382**, 491
Dibromochloromethane (DCBM) **451–454**
  analysis 162
  guideline value 194, 451, 491
1,2-Dibromoethane (ethylene dibromide) **347–349**
  analysis 161
  guideline value 191, 347, 491
  treatment achievability 169, 348
Dichloramine 193, 411, 489
Dichloroacetate 162, 194, **349–350**, 491
1,1-Dichloroacetone 329
Dichloroacetonitrile 162, 194, **380–382**, 491
3,4-Dichloroaniline 430
1,2-Dichlorobenzene **350–352**
  acceptable levels 215
  analysis 160
  guideline value 188, 350, 491
  treatment achievability 168, 351
1,3-Dichlorobenzene 187, **350–352**, 489
1,4-Dichlorobenzene **350–352**
  acceptable levels 215
  analysis 160
  guideline value 188, 350, 492
  treatment achievability 168, 351
Dichlorobenzenes (DCBs) 215, **350–352**
1,1-Dichloroethane 187, **352**, 489
1,2-Dichloroethane **353–354**
  analysis 160
  guideline value 188, 353, 492
  treatment achievability 168, 353
1,1-Dichloroethene 160, 188, **354–355**, 492
1,2-Dichloroethene **355–356**
  analysis 160
  guideline value 188, 355, 492
  treatment achievability 168, 355
Dichloromethane 160, 188, **357–358**, 492
2,4-Dichlorophenol 193, 214, **329–331**, 489
2,4-Dichlorophenoxyacetic acid *see* 2,4-D

1,2-Dichloropropane (1,2-DCP) **358–359**
  analysis 161
  guideline value 191, 358, 492
  treatment achievability 169, 358
1,3-Dichloropropane 190, **359–360**, 489
1,3-Dichloropropene 161, 191, **360–361**, 492
Dichlorprop (2,4-DP) 161, 191, **361–362**, 492
Dieldrin **300–301**
  analysis 161
  guideline value 191, 300, 491
  treatment achievability 169, 300
Dimethoate **364–366**
  analysis 161
  guideline value 191, 365, 492
  treatment achievability 169, 365
Dinoseb 189, 488
1,4-Dioxane 168
Di(2-ethylhexyl)phthalate (DEHP) 160, 188, **363–364**, 491
Diquat 190, **366–367**, 489
Disability-adjusted life years (DALYs) **45–47**
  microbial hazards 129–130
  reference level of risk and 45
Disasters 63, **104–109**
  chemical and radiological guidelines 108–109
  microbial guidelines 107–108
  monitoring 106–107
  practical considerations 105–106
  sanitary inspections and catchment mapping 108
  testing kits and laboratories 109
  see also Emergencies
Disease burden
  health outcome targets and 134–135
  waterborne infections 129–130
Disinfectants 188–189
  analysis 162
  DBP formation and 180
  guideline values 193, 194
  residual, piped distribution systems 63
  see also specific disinfectants
Disinfection **5–6**, 61
  in emergency and disaster situations 105–106, 107
  indicator organisms 283, 284, 286
  limitations 5
  methods **171–173**
  microbial reduction 140–141
  non-chemical 180
  resistant organisms 142

on ships 120
for travellers 110
vendor supplies 15
Disinfection by-products (DBPs) 5, 145, **179–180**, 189, 192
analysis 162
desalinated water 111–112
guideline values 193, 194
strategies for reducing 179–180
see also specific chemicals
Displaced populations 104
Distilled water 114
Documentation **27–28**
incidents and emergencies 28, 77
supporting 18–21
water safety plans **82–83**
Domestic supplies *see* Household drinking-water supplies
Domestic Water Quantity, Service Level and Health 18
Dose, infectious 129
Dose–response assessment, microbial pathogens 127, 128–129
Dracunculus Eradication Programme 276
Dracunculus medinensis (guinea worm) 123, 124, 221, **276–277**
intermediate host 212
significance in drinking-water 122, 277
Dreissena polymorpha 212
Droughts 104
Dysentery
amoebic 266
bacillary 240–241

Earthquakes 104
Echinococcus 124
Echoviruses 253
Edetic acid (EDTA) **367–368**
analysis 160
guideline value 188, 367, 492
treatment achievability 168, 367
EDTA *see* Edetic acid
Education programmes 12, 71, 89
establishing 94
schools and day care centres 103–104
Electrode, ion-selective 158
Electron capture detection (ECD) 165
Electrothermal atomic absorption spectrometry (EAAS) 164
ELISA (enzyme-linked immunosorbent assay) 165–166
Emergencies 76, **104–109**
chemical and radiological guidelines 108–109

documentation and reporting 28, 77
follow-up investigation 77
microbial guidelines 107–108
monitoring 106–107
practical considerations 105–106
radionuclide releases 198
response plans 76–77, **78–79**
sanitary inspections and catchment mapping 108
testing kits and laboratories 109
see also Disasters; Incidents
Emerging diseases 259
Empty bed contact time (EBCT) 177
Encephalitis, granulomatous amoebic (GAE) 260, 261
Encephalitozoon 270, 271
Endosulfan 190, **368–369**, 489
Endrin **369–370**
analysis 161
guideline value 191, 369, 492
treatment achievability 169, 369
Entamoeba histolytica 122, **265–267**
Enteric fever 239
Enteric pathogens, in source waters 136–137
Enteric viruses 247–248, **294–295**
coliphages as indicator 290–291
indicator value 294
in source waters 137
Enterobacter 282, 284
Enterococci, intestinal **287–288**
Enterococcus spp. 287
Enterocolitis, *Staphylococcus aureus* 242
Enterocytozoon 270
Enteroviruses 122, 142, **253–254**, 295
Environmental Health Criteria monographs (EHCs) 36
Environmental Protection Agency, US (US EPA) 36
Enzyme-linked immunosorbent assay (ELISA) 165–166
Epichlorohydrin (ECH) 162, 194, **370–372**, 492
Equitability, access to water 105
Escherichia coli 282
detection methods 144
emergency and disaster situations 108
enterohaemorrhagic (EHEC) 122, 229–230
enteroinvasive (EIEC) 229, 230
enteropathogenic (EPEC) 229, 230
enterotoxigenic (ETEC) 229, 230
guideline values 143

as indicator of faecal pollution 29, 142, **284–285**
  pathogenic 122, **229–231**
  phages (coliphages) 289–292
  piped distribution systems 63
  in source waters 137
  see also Coliform bacteria
Ethylbenzene **372–373**
  analysis 160
  guideline value 188, 372, 492
  odour and taste thresholds 215
  treatment achievability 168, 372
Ethylene dibromide *see* 1,2-Dibromoethane
Ethylene thiourea 189, 488
Evaluation of the H_2S Method for Detection of Fecal Contamination of Drinking Water 19
Evaporation method, radionuclide analysis 207–208
Exposure assessment, microbial pathogens 127, 128
Eye infections
  Acanthamoeba 260
  adenovirus 248, 249

Faecal–oral route of transmission 122, 221
Faecal contamination 3–4
  control measures 5, 59
  in emergencies 79, 107
  indicator organisms *see* Faecal indicator organisms
  large buildings 100
  on ships 117
Faecal indicator organisms 29, **281–295**
  community supplies 82
  criteria 281–282
  desalinated water 112
  emergency and disaster situations 107, 108
  guideline values 143
  methods of detection **143–144**
  operational monitoring 69
  presence/absence (P/A) testing 72
  in source waters 136–137
  verification testing 72, 74, 142
Fasciola 124, 276, **278–279**
Fascioliasis 278–279
Fasciolopsis 124
Fenamiphos 189, 488
Fenitrothion 190, **373–374**, 489
Fenoprop 161, 191, **374–375**, 492
Field test kits 109, 158
Filtration 60–61, **173–175**
  after coagulation 176
  direct 173
  drinking-water for travellers 110
  dual-media or multimedia 174
  granular high-rate 139
  horizontal 173, 174
  membrane 139
  microbial reduction 139–140
  precoat 139
  pressure 173, 174
  rapid gravity 173–174
  roughing 138, 174
  slow sand 139, 173, 174–175
First-flush diverters 66
Fit for purpose 75
Flame atomic absorption spectrometry (FAAS) 159
Flame ionization detection (FID) 165
Flavobacterium 124, 286
Flocculation 60, 138–139, 175–176
Floods 104
Flotation, dissolved air 138, 176
Flow diagrams 52
Fluoranthene 193, 428, 489
  health-based values 429, 430
Fluoride **375–377**
  analysis 159
  desalinated water 113
  guideline value 186, 376, 492
  health concerns 6, 376–377
  priority 35–36
  treatment achievability 167, 376
Fluorosis 376–377
Food
  acceptable daily intakes (ADIs) 150
  intake of chemicals 152
  production and processing **115–116**
  safety, travellers 109–110
Food and Agriculture Organization (FAO) 114
Food poisoning
  Bacillus cereus 225, 226
  Campylobacter 228
  Salmonella 239, 240
  Staphylococcus aureus 242
Formaldehyde 162, 194, **377–378**, 492
Formothion 189, 488
Framework for safe drinking water 2–3, **22–36**
  health-based targets 24–25
  key components 22
  management plans, documentation and communication 27–28
  operational monitoring 26–27
  requirements 22–29

risk assessment 44
supporting information 22–23
surveillance of drinking-water quality 28–29
system assessment and design 25–26
Fulvic acids 214
Fungi 212

β-Galactosidase 282, 283
Galvanized iron 183
Gammarus pulex 212
Gas chromatography (GC) 165
Gas chromatography/mass spectrometry (GC/MS) 165
Gastroenteritis
  adenovirus 248–249
  astrovirus 250
  calicivirus 252
  Campylobacter 228
  rotavirus 258
  Salmonella 239
  Yersinia 246
Genotoxic carcinogens 148–149
Geosmin 212, 213
Geothermal waters 272, 273
Giardia (intestinalis) 122, **267–268**
  disinfection 140–141
  in source waters 137
Giardiasis 267
β-Glucuronidase 284
Glyphosate 190, **379–380**, 489
Gnat larvae 212
Grading schemes, safety of drinking-water 29, 53–55, 97, 98
Granular activated carbon (GAC) 176, 177
Granulomatous amoebic encephalitis (GAE) 260, 261
Gray (Gy) 201
Groundwaters
  Acinetobacter 222–223
  arsenic contamination 146
  control measures 58, 59, 65–66
  hazard identification 56, 57
  pathogen occurrence 136–137
  radon 206
  system assessment and design 53, 54
Guide to Ship Sanitation 118
Guideline values (GVs) 1–2, 6–7, 25, 30
  acceptability and 156
  applying 30–31
  chemicals by source category **184–196**
  chemicals excluded 488
  chemicals of health significance 491–493
  chemicals without established 489–490

  derivation 47, **147–156**
    approaches 148–149
    data quality 154–155
    non-threshold chemicals (non-TDI-based) 154–155
    significant figures 152
    threshold chemicals (TDI-based) 149–154
  see also Tolerable daily intake
  in emergencies 108–109
  health-based targets based on 41
  mixtures of chemicals and 156
  provisional 31, 148, **155–156**
    high uncertainty and 151
    use and designation 155
  radionuclides **202–204**
  radon 207
  summary tables **488–493**
  treatment achievability 166–171
  verification of microbial quality 143
Guillain–Barré syndrome 228
Guinea worm *see Dracunculus medinensis*

Haemolytic uraemic syndrome (HUS) 229–230
Hafnia 282
Halogenated acetonitriles **380–382**
Hardness 185, **382–383**, 489
  acceptability 215–216
  corrosion and 182, 184
  treatment to reduce 220
Hazard 52
  identification 127
  prioritization, for control 53–55
Hazard Characterization for Pathogens in Food and Water: Guidelines 19
Hazardous events 52, 127
Health-based targets **24–25, 37–47**
  benefits 38
  establishing 43–47
  microbial hazards **126–135**
  role and purpose 37–39
  types 39–43
Health care facilities
  drinking-water quality 102–103
  health risk assessment 100
Health education 89, 103–104
  see also Education programmes
Health outcome targets 24–25, 40, **43**
  waterborne infections **134–135**
Health promotion 89
Health risks 3–7
  aircraft and airports 116
  chemicals 6–7, **145–147**

large buildings 100
microbial *see* Microbial hazards
radiological 7, 198, 200–201
ships 117–118
travellers 109
Helicobacter pylori 221, **231–232**
Helminths 4, 221, **275–279**
 significance in drinking-water 122, 124
Hepatitis A virus (HAV) 122, 125, **254–256**
Hepatitis E virus (HEV) 122, **256–257**
Heptachlor 190, **383–384**, 489
Heptachlor epoxide 190, **383–384**, 489
Heterotrophic micro-organisms 69, 286
Heterotrophic plate counts (HPC) 5, **285–286**
Heterotrophic Plate Counts and Drinking-water Safety 19
Hexachlorobenzene (HCB) 187, **385–386**, 490
Hexachlorobutadiene (HCBD) **386–387**
 analysis 160
 guideline value 188, 386, 492
 treatment achievability 168, 386
Hexachlorocyclohexanes 189, 488
High-income countries, rotavirus performance targets 131–132
High-performance liquid chromatography (HPLC) 165
Holistic approach 3
Hookworm infections 276
Hospital-acquired (nosocomial) infections
 Acinetobacter 222, 223
 Klebsiella 232, 233
 Pseudomonas aeruginosa 238
Hospitals
 drinking-water quality 102–103
 health risk assessment 100
Hot water systems 100, 234–235
Hotels 100
Household drinking-water supplies
 collection, transportation and storage of water 71
 control measures 65–67
 hazard identification 64–65
 management **81–82**
 operational monitoring **71**
 quantity of water collected and used 90–91
 roles and responsibilities 11–12, 15–16
 surveillance **89**
 system assessment **64–67**
 treatment 141
 water safety plans (WSPs) 48–49, 85

Human dwellings, chemicals originating from *see* Industrial sources and human dwellings, chemicals from
Humic acids 214
Hydrocarbons, low molecular weight 217
Hydrogen peroxide 173, 180
Hydrogen sulfide 185, **387–388**, 490
 acceptable levels 216
 treatment to remove 220
Hydroquinone 118
Hydroxyl radicals 173
Hygiene
 education programmes *see* Education programmes
 service level and 90, 91
Hypertension 436
Hypochlorite 107, 171
Hypochlorous acid 171

Ice 110, 113
Immunity
 acquired 125, 130–131
 variations in 121, 125
Immunocompromised persons 102, 124
 Aeromonas infections 224
 atypical mycobacteria infections 236
 disease burden estimates 130
 isosporiasis 269
 Klebsiella infections 232
 Pseudomonas aeruginosa 238
 toxoplasmosis 274
 travellers 111
 Tsukamurella infections 243
Impingement attack 182, 183
Improvement, drinking-water systems **67–68**
Incidents 76
 audit 86–87
 documentation and reporting 28, 77
 follow-up investigation 77
 predictable 77
 response plans 76–77, 78
 unplanned events 77–78
 see also Emergencies
Indeno [1,2,3-cd] pyrene 429
Index organisms **281–295**
Indicator organisms 29, **281–295**
Inductively coupled plasma/atomic emission spectrometry (ICP/AES) 164
Inductively coupled plasma/mass spectrometry (ICP/MS) 164
Industrial effluents 214

Industrial sources and human dwellings, chemicals from
  analysis 159, 160
  guideline values **185–187**, 188
  treatment achievability 168
Infants
  bottle-fed 114, 418, 419
  consumption assumptions 486
  see also Children
Infections, waterborne 4, **121–124**, 221
  asymptomatic 125–126
  emergency and disaster situations 79, 104, 106
  health-based targets 39, 43
  health outcome targets 134–135
  public health aspects 10–11, **125–126**
  risk characterization 127, 129–131
  routes of transmission 221
  ships 117
  see also Pathogens
Infiltration
  bankside 138
  contamination via 62, 63
Information channels, establishing 94
Ingress
  non-piped distribution systems 65
  piped distribution systems 62, 63
Inhalation
  assumptions 486–487
  chemicals 152
  micro-organisms 123, 221
  radionuclides 197
  radon 206–207
Inorganic tin 193, **388–389**
Insecticides, aquatic 190
Intakes
  control measures 59
  hazard identification 57–58
Intermittent water supply 63, 92–93, 101
International Agency for Research on Cancer (IARC) 149
International Atomic Energy Agency (IAEA) 201–202
International Commission on Radiological Protection (ICRP) 197, 198, 201–202
International Health Regulations 116
International Organization for Standardization (ISO) standards 75, 76, 144, 208
International standards 2
Interspecies variation 151

Intestinal enterococci **287–288**
Invertebrate animals 212–213
Iodine **389–390**
  guideline value 193, 389, 490
  treatment, for travellers 110, 111
Iodine-131 202
Ion chromatography 164–165
Ion exchange 139, **177**
Ion-selective electrode 158
Iron 193, **390–391**, 490
  acceptable levels 216, 390
  corrosion **181**
  galvanized 183
  priority 35–36
Iron bacteria 213, 216
Isoproturon **391–392**
  analysis 161
  guideline value 191, 391, 492
  treatment achievability 169, 391
Isospora belli 221, **268–270**
Isosporiasis 269

Jar tests 176
Joint FAO/WHO Expert Committee on Food Additives (JECFA) 36, 150
Joint FAO/WHO Meetings on Pesticide Residues (JMPR) 36, 150

Keratitis, *Acanthamoeba* 260–261
Keratoconjunctivitis, epidemic ("shipyard eye") 248, 249
Kits, testing **109**, 158
Klebsiella **232–233**
  as indicator organism 282, 284, 286
  pathogenicity 124, 232

Laboratories, in emergencies and disasters **109**
Lactose fermentation 282, 283, 284
Lakes 137
Land use 12–13
Langelier index (LI) 184
Large buildings **99–104**, 235
  drinking-water quality 102–104
  health risk assessment 100
  independent surveillance and supporting programmes 102
  management 101
  monitoring 101–102
  system assessment 100–101
Larson ratio 184
Larvae 212
Larvicides, aquatic 190

Latrines, contamination from 186
Laws, national drinking-water 31–32
Lead 6, **392–394**
  analysis 159
  corrosion **181–182**
  guideline value 194, 392, 492
  priority 35–36
  sampling locations 73
Lead-210 202
Legionella spp. 4, 123, 221, **233–235**
  control measures 64. 234–235
  health care facilities 103
  large building systems 100, 235
  persistence 125
  significance in drinking-water 122, 234–235
Legionellosis 100, 123, 233–234
Legionnaires' disease 123, 233–234
Likelihood categories 54–55
Lime softening 139, 179
Lindane **394–396**
  analysis 161, 395
  guideline value 191, 395, 492
  treatment achievability 169, 395
Liver flukes *see Fasciola*
LOAEL *see* Lowest-observed-adverse-effect level
Local authorities **11–12**
Low-income countries, rotavirus performance targets 131–132
Lowest-observed-adverse-effect level (LOAEL) 149, **150**
  uncertainty factors 151
Lung cancer, radon-related risk 207

Magnesium 215
Malathion 190, **396–397**, 490
Management
  aircraft and airports 117
  community and household supplies 81–82
  large buildings 101
  piped distribution systems **76–81**
  plans **27–28**, 49
  roles and responsibilities **8–18**
  ships 119–120
Managing Water in the Home 19, 66–67
Manganese **397–399**
  acceptability 216, 398
  analysis 159
  guideline value 186, 398, 492
  priority 36
  treatment to remove 167, 220
Mass spectrometry (MS) 164, 165

MCPA (4-(2-methyl-4-chlorophenoxy)acetic acid) **399–400**
  analysis 161
  guideline value 191, 399, 492
  treatment achievability 169, 399
MCPB 189, 488
MCPP *see* Mecoprop
Mean, arithmetic *vs* geometric 131
Mecoprop **400–401**
  analysis 161
  guideline value 191, 401, 492
  treatment achievability 169, 401
Medical devices, cleaning 103
Melioidosis 226–227
Membrane processes, water treatment **178**, 180
Meningoencephalitis, primary amoebic (PAM) 123, 272, 273
Mercury **402–403**
  analysis 159
  guideline value 188, 402, 492
  treatment achievability 168, 402
Meringue dezincification 182–183
Methaemoglobinaemia 6, 418–420
Methamidophos 189, 488
Methomyl 189, 488
Methoprene 190
Methoxychlor **403–404**
  analysis 161
  guideline value 191, 403, 492
  treatment achievability 169, 403
4-(2-Methyl-4-chlorophenoxy)acetic acid *see* MCPA
2-(2-Methyl-chlorophenoxy) propionic acid *see* Mecoprop
2-Methyl isoborneol 212, 213
Methyl parathion 190, **404–405**, 490
Methylene chloride *see* Dichloromethane
Methylmercury 402
Metolachlor **405–407**
  analysis 161
  guideline value 191, 406, 492
  treatment achievability 169, 406
Micro-organisms, indicator and index **281–295**
Microbial aspects **3–5, 121–144**
Microbial growth
  bottled water 114
  desalinated water 113
Microbial hazards 3–4, **121–126**
  health-based target setting 126–135
  identification 127
  water quality targets 43, 126
Microbial pathogens *see* Pathogens

Microbial quality
  assessing priorities 35
  emergency and disaster situations 79, **107–108**
  grading schemes based on 97, 98
  health care facilities 102–103
  verification **29–30, 72, 142–143**
Microcystin-LR 195–196, **407–408**, 492
Microcystins 103, 192, 196, 280
Microfiltration 139, 178
Microsporidia 221, 259, **270–272**
Microstraining 138
Millennium Development Goals 33
Mineral waters, natural 114–115
  see also Bottled water
Mining activities 186
Minister of health 33
Ministries, government 33, 34
Mirex 189, 488
Molinate 161, 191, **408–409**, 492
Molluscs 212
Molybdenum 159, 186, **410–411**, 492
Monitoring
  dissolved radionuclides **204–205**
  emergency and disaster situations 106–107
  operational see Operational monitoring
  plans, preparing **80**
  see also Sanitary inspection; Surveillance
Monobromoacetate 193, 316–317, 490
Monochloramine **411–412**
  acceptability 216–217
  analysis 162
  by-products 179, 180
  disinfection activity 140, 172
  guideline value 194, 411, 492
Monochloroacetate 162, 194, **412–413**, 492
Monochlorobenzene (MCB) 187, 217, **413–414**, 490
Monocrotophos 189, 488
Moraxella 286
Mudslides 104
Multiagency approach, collaborative 8
Multiple-barrier concept 3, 5, 56
MX (3-chloro-4-dichloromethyl-5-hydroxy-2(5H)-furanone) 193, **414–415**, 490
Mycobacterium (mycobacteria) **235–237**
  atypical (non-tuberculous) 122, 124, 221
  health care facilities 102
Mycobacterium avium complex 235, 236
Mycobacterium kansasii 235, 236

Naegleria fowleri 123, 125, 221, **272–273**
  control measures 64, 273
  significance in drinking-water 122, 273
Nais worms 212
Nanofiltration 140, 178
National Academy of Sciences (NAS) (USA) 207
National drinking-water policy **31–34**
National performance targets **133–134**
National priorities, supply improvement 93
National standards and regulations **31–32**
  chemical contaminants 146
  developing 2, **32–34**
Natural disasters 63, 104
Naturally occurring chemicals 147
  analysis 159
  guideline values **184–185**, 186
  treatment achievability 167
  see also Chemicals
Necator 124
Nematodes 212, 276
New drinking-water supply systems
  assessment and design **52–53**
  source verification 74
Nickel **415–417**
  analysis 159, 416
  guideline value 194, 416, 492
  leaching **183**
Nitrate 6, **417–420**
  agricultural sources 187
  analysis 159, 418
  guideline value 191, 417, 492
  treatment achievability 169, 418
Nitrification, biological 179
Nitrilotriacetic acid (NTA) **420–421**
  analysis 160, 420
  guideline value 188, 420, 492
  treatment achievability 168
Nitrite 6, **417–420**
  analysis 159, 418
  desalinated water 113
  guideline value 191, 417, 492
  treatment achievability 169, 418
Nitrosamines 419
No-observed-adverse-effect level (NOAEL) 149, **150**
  uncertainty factors 151
  vs benchmark dose 153
NOAEL see No-observed-adverse-effect level
Non-piped water systems **64–67**
  control measures 65–67
  hazard identification 64–65

operational monitoring **71**
  roles and responsibilities 16
  treatment 141
Norms, drinking-water 10
Noroviruses (Norwalk-like viruses) 122, 251
Nosema 270
Nosocomial infections *see* Hospital-acquired infections
Nuisance organisms 4–5
Nursing care homes 100

Octanol/water partition coefficient 177
Odour 7, 210, **211–220**
  biologically derived contaminants 211–213
  chemical contaminants 213–219
  treatments for removing 219–220
Oils, petroleum 186, 217
Operational limits **70**
Operational monitoring 26–27, 49, **68–71**
  aircraft and airports 116–117
  community supplies 71, 82
  defined 68
  large buildings 101–102
  parameters 68–70
  ships 119
Organic matter 214
Organisms, visible 211, 212–213
Organotins 345–346
Orthophosphate 181, 182
Orthoreoviruses **257–259**, 295
Osmosis 178
  reverse 140, 178
Oxamyl 189, 488
Oxidation processes, advanced 173
Oxygen
  dissolved 215
  transfer 175
Ozonation **172**
  by-products 179, 180, 192
  microbial reduction 141
Ozone 172, 173

Packaged drinking-water **113–115**
  international standards 114–115
  safety 113–114
  see also Bottled water
Parasites 420
  persistence in water 125
  secondary hosts 212
  waterborne 122, 124
  see also Helminths; Protozoa
Parathion 190, **421–422**, 490

Particulate matter 211, 219
Pathogenic Mycobacteria in Water 19
Pathogens 121–124
  alternative routes of transmission 5, 43–44, 122
  bacterial **222–247**
  dose–response assessment 127, 128–129
  exposure assessment 127, 128
  fact sheets **221–279**
  health-based targets 39
  helminth **275–279**
  occurrence 135, **136–137**
  performance targets 41–42, 131–134
  persistence and growth in water **124–125**
  protozoan **259–275**
  special properties 142
  transmission pathways 123
  treatment **137–141**
  viral **247–259**
  see also Infections, waterborne
Pendimethalin **422–423**
  analysis 161
  guideline value 191, 423, 492
Pentachlorophenol (PCP) **424–425**
  analysis 160, 424
  guideline value 188, 424, 492
  treatment achievability 168, 424
Performance targets 25, 40, **41–42**, 126
  national/local adaptation 133–134
  pathogens in raw water 131–132, 133
  risk-based development **131–134**
Perlite 139
Permethrin 190, **425–426**, 490
Pesticides 187
  used in water for public health 147
  analysis 161, 163
  guideline values **190–192**, 195
  treatment achievability 170
  see also Agricultural activities, chemicals from; *specific compounds*
Petroleum oils 186, 217
pH 185, **426–427**, 490
  chemical coagulation 175–176
  community supplies 82
  corrosion and 181, 182, 184
  DBP formation and 179–180
  emergency and disaster situations 108
  optimum range 217, 426
  saturation 184
Phages *see* Bacteriophages
Pharyngoconjunctival fever 248
2-Phenylphenol (and its sodium salt) 190, **427–428**, 490
Phorate 189, 488

INDEX

Piped distribution systems **61–64**
  assessment and design 54
  control measures 63–64
  hazard identification 62–63
  intermittent supply 63
  large buildings 100, 101
  management procedures **76–81**
  microbial hazards 123
  operational monitoring parameters 69
  on ships 118, 119
  verification testing 74
Pipes 17–18
  bursts 62
  cement lining 183
  coal-tar linings 428, 430
  contaminants 193, 194
  corrosion 181, 182, 183
  lead 181
Pitting corrosion 182
Platyhelminthes 276
Pleistophora 270
Plumatella 212
Plumbing **17–18**
  household 16
  on ships 118
Plumbosolvency 181–182
Plutonium-239 (^{239}Pu) 202
Pneumonia, *Burkholderia pseudomallei* 226
Poisson distribution 129
Policy
  development, wider 10
  national drinking-water **31–34**
Poliovirus 253, 295
Polonium-210 (^{210}Po) 202
Polyacrylamides 296
Polynuclear aromatic hydrocarbons (PAHs) **428–430**
Polyphosphates 181
Polyvinylchloride (PVC) 456
Pontiac fever 233, 234
Pools, stagnant 101
Port authority 118, 119
Potassium-40 (^{40}K) 205
  measurement **208**
Potassium bromate 315
Powdered activated carbon (PAC) 176
Presence/absence (P/A) testing 72
Pressure, water 62, 63
  large buildings 101
  measurement, operational monitoring 69
Pretreatment 60, 138
Prevention, disease 6
Preventive integrated management approach 8

Priorities
  assessing chemical 35–36
  assessing microbial 35
  identifying **34–36**
  setting 34
Problem formulation, microbial hazards 127
Propanil 190, **430–431**, 490
Propoxur 189, 488
Protozoa 221
  cysts and oocysts, removal 61
  pathogenic 122, **259–275**
  resistance to treatment 142
  treatment effects 138–141
Pseudomonas 286
Pseudomonas aeruginosa 102, 122, 124, **237–239**
Public awareness, establishing 94
Public health
  authorities, roles and responsibilities **10–11**, 13
  policy context 44
  surveillance 10–11
  waterborne infections and **125–126**
Purge-and-trap packed-column GC method 165
Purge-and-trap packed-column GC/MS method 165
Pylon technique 208
Pyridate 189, 488
Pyriproxyfen 190, **431–432**
  analysis 163
  guideline value 195, 432, 492
  treatment achievability 170, 432

QMRA *see* Quantitative microbial risk assessment
Quality assurance **75–76**
Quality control **8–9, 75–76**
Quantifying Public Health Risk in the WHO Guidelines for Drinking-water Quality 19, 47
Quantitative microbial risk assessment (QMRA) 43, 126–131
  dose–response assessment 128–129
  exposure assessment 128
  problem formulation and hazard identification 127
  risk characterization 129–131
Quantitative risk assessment 43
Quantitative service indicators 74–75
Quantity of supply
  assessment of adequacy **90–91**
  emergency and disaster situations 105
Quintozene 189, 488

Radiation
    absorbed dose 201
    background exposures 198
    committed effective dose 201, 205
    dose **201–202**
    effective dose 201
    equivalent dose 201
    exposure through drinking-water **200**
    health risks 7, 198, 200–201
    reference dose level (RDL) 198, 202
    sources **198–201**
Radioactivity
    measurement 207–208
    screening 204
    units **201–202**
Radiological aspects **7, 197–209**
Radionuclides 7, 197–209
    activity concentration 201, 202
    analytical methods 207–208
    dose coefficients 201–202
    emergency and disaster situations **108–109**
    guidance levels **202–204**
    monitoring and assessment for dissolved **204–205**
    remedial measures 205
    reporting of results 209
    sampling 209
    screening for 204, 206
    sources 200
    strategy for assessing drinking-water 205, 206
Radium-226 (^{226}Ra) 202
Radium-228 (^{228}Ra) 202
Radon (^{222}Rn) 197, **206–207**
    in air and water 206
    guidance levels 207
    measurement **208**
    risk 207
    sampling 209
Rainfall 29–30
Rainwater
    collection systems 65, 66, 141
    consumption 114
Records *see* Documentation
"Red water" 181, 216
Reference dose level (RDL) 198, 202
Reference level of risk 44–45, 47, 132–133
Regional level
    performance target setting **133–134**
    supply improvement 93
    use of data for priority setting **96–97**, 98
"Regrowth" 5

Regulations, national *see* National standards and regulations
Reoviridae 257
Reporting
    incidents and emergencies 28, 77
    radioactivity analysis **209**
    surveillance information **95–97**
Reservoirs 54
    control measures 58–59, 64
    hazard identification 57–58
    occurrence of pathogens 137
Resource protection **56–59**, 81
    control measures 58–59
    hazard identification 56–58
Respiratory infections, adenoviral 248
Reverse osmosis 140, 178
Risk
    defined 52
    judgement of tolerable 2, 37
    reference level **44–45**, 47, 132–133
    scoring 53–55
Risk–benefit approach 2, 45
Risk assessment 53–55
    in framework for safe drinking water **44**
    quantitative 43
    quantitative microbial *see* Quantitative microbial risk assessment
Risk characterization, waterborne infection 127, 129–131
Rivers, occurrence of pathogens 136, 137
Roles and responsibilities, management **8–18**
Rotaviruses (HRVs) 122, **257–259**
    performance target setting 131–132, 133, 134, 135
    risk characterization 129, 130–131
Roughing filters 138, 174
Routes of transmission 123

Safe, Piped Water: Managing Microbial Water Quality in Piped Distribution Systems 19–20
Salmonella (salmonellae) 122, 137, **239–240**
Salmonella Enteritidis 239
Salmonella Paratyphi 239
Salmonella typhi 122, 239
Salmonella Typhimurium 239, 240
Sample numbers, minimum 74
Sampling
    community-managed supplies 89
    frequencies 72, 73, 75
    ISO standards 75
    locations 73
    radioactive contaminants **209**

Sanitary code 33–34
Sanitary inspection 86
    community-managed supplies 71, 74, 75, 89
    emergency and disaster situations **108**
    use of data 97, 98
Sapovirus (Sapporo-like viruses) 122, 251
Scale, calcium carbonate 183–184, 215–216
Schistosoma spp. 122, 221
Schistosomiasis 123, 276
"Schmutzdecke" 174
Schools 100, **103–104**
Screening, radionuclides in drinking-water **204**, 206
Scum 215
Seasonal discontinuity of supply 93
Seawater 111, 112
Sedimentation 60, 138–139, 176
Selenium 6, **432–434**
    analysis 159, 433
    guideline value 186, 433, 492
    priority setting and 35–36
    treatment achievability 167, 433
Septata 270
Septic tanks 186
Serratia 124, 282, 286
Service indicators, quantitative 74–75
Service levels 90–91
Severity categories 54–55
Shigella 122, **240–241**
Shigellosis 240–241
Ships **117–120**
    health risks 117–118
    management 119–120
    operational monitoring 119
    surveillance 120
    system risk assessment 118
"Shipyard eye" 248, 249
Sievert (Sv) 201
Significant figures 152
Silicates 181
Silver **434–435**
    guideline value 193, 490
    treatment, for travellers 110
Simazine **435–436**
    analysis 161
    guideline value 191, 435, 492
    treatment achievability 170, 435
Single-hit principle 128–129
Skin absorption *see* Dermal absorption
Snails 123, 212
Sodium 185, **436–437**, 490
    taste threshold 217–218, 436
Sodium bromate 315

Sodium hypochlorite 107, 171
Sodium sulfate 218
Softening 177
    lime 139, 179
    precipitation 179
Solids, total dissolved (TDS) 185, 218, **444–445**, 490
Solubility, water 177
Source protection **56–59**, 66
Source waters
    chemical contaminants 147
    community and household systems 71, 82
    control measures 58–59
    desalination systems 111
    emergency and disaster situations 105
    hazard identification 56–58
    microbial hazards 123
    naturally occurring chemicals 185
    new systems 52–53
    operational monitoring 69, 71
    pathogen occurrence 135, 136–137
    seasonal fluctuation 93
    verification **73–74**
    see also Catchments
Spas 234, 273
Specified technology targets 25, 40, **41**
Spirometra 124
Springs 65, 141
Stagnant pools 101
Standard for Bottled/Packaged Waters 115
Standard for Natural Mineral Waters 114–115
Standard operating procedures (SOPs) 81
    incident responses 77, 78
Standards
    bottled drinking-water 114–115
    certification 17
    drinking-water 10
    national *see* National standards and regulations
Staphylococcus aureus **242–243**
Stomach cancer, radon-related risk 207
Storage
    after disinfection 61
    emergency and disaster situations 106
    home 71
    large buildings 101
    off-stream/bankside 138
    on ships 119
    systems
        control measures 58–59, 64, 66
        surveillance 89
Streams, occurrence of pathogens 136, 137

Streptococci, faecal 142, 287
Strongyloidiasis *(Strongyloides)* 124, 276
Strontium-90 (^{90}Sr) 202
Styrene **437–438**
  analysis 160, 437
  guideline value 188, 437, 492
  odour threshold 218
  treatment achievability 168, 437
Styrene-7,8-oxide 437, 438
Sulfate 185, **438–439**, 490
  acceptable level 218
  corrosion control 181, 184
  notifiable level 438–439
Superchlorination/dechlorination 171
Suppliers, drinking-water
  audit-based surveillance 87
  independence of surveillance 8–9
  legal functions and responsibilities 31–32
  management plans *see* Water safety plans
  roles and responsibilities 9, **13–14**
Supply, drinking-water
  adequacy **90–93**
  emergency and disaster situations 105–106
  improved technologies 92
  intermittent 63, 92–93, 101
  planning and implementing improvement 93–94
  unimproved technologies 92
Supporting programmes **80–81**
  aircraft and airports 117
  large buildings 102
  ships 120
Surface waters
  control measures 58, 66
  emergency and disaster situations 105
  hazard identification 56–57
  Helicobacter pylori 231
  pathogen occurrence 136–137
  system assessment and design 53, 54
  verification 73
Surveillance **8–9**, **28–29**, **84–98**
  adapted to specific circumstances 88–89
  adequacy of supply 90–93
  agencies 9, 32, 85
  aircraft and airports 117
  approaches 85–87
    audit-based 86–87
    direct assessment 87
  community drinking-water supplies 87, 88–89
  definition 9, 84
  large buildings 102
  planning and implementation 93–95

public health 10–11
reporting and communicating 95–97
ships 120
stages of development 94–95
urban areas in developing countries 88
see also Monitoring
Swimming pools 249, 272, 273
System assessment and design **25–26**, 49, **51–68**
  aircraft and airports 116
  collecting and evaluating available data 53–56
  large buildings 100–101
  ships 118
  treatment 59–61
Systems, drinking-water
  large buildings 99, 100
  maintaining control **68–71**
  new 52–53, 74
  non-piped *see* Non-piped water systems
  operational monitoring *see* Operational monitoring
  piped *see* Piped distribution systems
  resource and source protection 56–59
  on ships 118
  upgrade and improvement 67–68, 94
  validation *see* Validation
  verification *see* Verification

2,4,5-T (2,4,5-trichlorophenoxy acetic acid) **439–440**
  analysis 161
  guideline value 191, 439, 492
  treatment achievability 170, 440
Taenia solium 124
Tankers, water 15
Tanks, storage 64
Taps 101
Targets
  health-based *see* Health-based targets
  health outcome 24–25, 40, 43
  incremental improvements towards 2
  performance *see* Performance targets
  specified technology 25, 40, 41
  water quality *see* Water quality targets
Taste 7, 210, **211–220**
  biologically derived contaminants 211–213
  chemical contaminants 213–219
  treatments for removing 219–220
TBA *see* Terbuthylazine
TDI *see* Tolerable daily intake
Team, water safety planning 51
Temephos 190

INDEX

Temperature, water
  acceptable levels **220**
  Legionella growth/survival 100, 234–235
  Naegleria survival 272, 273
Terbuthylazine (TBA) **440–442**
  analysis 161
  guideline value 191, 441, 492
  treatment achievability 170, 441
Testing kits **109**, 158
3,3′,4,4′-Tetrachloroazobenzene 430
Tetrachloroethene **442–443**
  analysis 160, 442
  guideline value 188, 442, 492
  treatment achievability 168, 442
Thermotolerant coliform bacteria 142, 143, 282, **284–285**
THMs *see* Trihalomethanes
Thorium-228 202
Thorium-230 202
Thorium-232 202
Tin, inorganic 193, **388–389**, 490
Titration, volumetric 158
Tolerable daily intake (TDI) 149, **150**
  allocation to drinking-water 151–152
  alternative approaches 152–154
  calculation of guideline values 149–150, 152
  uncertainty factors 150–151
Toluene **443–444**
  acceptability 218
  analysis 160, 443
  guideline value 188, 443, 492
  treatment achievability 168, 443
Total coliform bacteria **282–284**
Total dissolved solids (TDS) 185, 218, **444–445**, 490
Toxaphene 189, 488
Toxic Cyanobacteria in Water 20
Toxic shock syndrome 242
Toxicity studies, animal 148
Toxocara 124
Toxoplasma gondii 122, **274–275**
Toxoplasmosis 274, 275
2,4,5-TP *see* Fenoprop
Trachipleistophora 270
Transportation, household water 71
Travellers **109–111**
Treatment **59–61, 166–184**
  achievability **166–171**
  chemicals used in *see under* Chemicals
  community sources 71
  control measures 60–61
  for corrosion control 180–184
  desalinated water 112

  emergency and disaster situations 105, 107
  hazard identification 59–60
  household 71, 89, 141
  indicator organisms 282, 286
  membrane processes 178, 180
  operational monitoring parameters 69
  pathogen removal **137–141**
  performance target setting and 131–132, 133–134
  processes 138–141, 171–179
    control measures 179–180
    ranking of complexity/costs 166–167
    validation 67
    see also specific treatments
  for ships 119
  system assessment and design 53, 54
  taste, odour and appearance problems **219–220**
  for travellers 110
  water quality targets 42
  see also Disinfection
Triazophos 189, 488
Tributyltin oxide (TBTO) 189, 488
Trichloramine 193, 411, 490
Trichlorfon 189, 488
Trichloroacetaldehyde *see* Chloral hydrate
Trichloroacetic acid 145, **445–446**
  analysis 162, 445
  guideline value 194, 445, 493
Trichloroacetonitrile 193, **380–382**, 490
Trichlorobenzenes (TCBs) 187, 218–219, **446–447**, 490
1,1,1-Trichloroethane 187, **447–448**, 490
Trichloroethene **448–449**
  analysis 160, 449
  guideline value 188, 448, 493
  treatment achievability 168, 449
Trichloronitromethane *see* Chloropicrin
2,4,6-Trichlorophenol **329–331**
  acceptable levels 214
  analysis 162
  guideline value 194, 330, 493
2,4,5-Trichlorophenoxy acetic acid *see* 2,4,5-T
2,4,5-Trichlorophenoxy propionic acid *see* Fenoprop
Trichuriasis *(Trichuris)* 124, 276
Trifluralin **450–451**
  analysis 161
  guideline value 191, 450, 493
  treatment achievability 170, 450

Trihalomethanes (THMs) 145, 179, **451–454**
  analysis 162
  guideline values 194, 451, 493
  strategies for reducing 179–180
Trimethylbenzene 217
Tritium (^{3}H) 202
True colour units (TCU) 214
Tsukamurella 221, **243–244**
Tubewells 65
Turbidity 5, 219
  community supplies 82
  emergency and disaster situations 108
  operational monitoring 69
Turner diagram 184
Typhoid fever 239, 240

Ultrafiltration 139, 178
Ultraviolet (UV) absorption 159
Ultraviolet (UV) irradiation 141, 173, 180
Uncertainty factors (UF) 149, **150–151**
  data-derived 154
United Nations Scientific Committee on the Effects of Atomic Radiation (UNSCEAR) 198–199, 207
Unplanned events **77–78**
Upgrading, drinking-water systems **67–68**, 94
Upgrading Water Treatment Plants 20
Uranium 6, **454–456**
  analysis 159, 455
  guideline value 186, 454, 493
  priority setting and 35–36
  treatment achievability 167, 455
Uranium-234 (^{234}U) 202
Uranium-238 (^{238}U) 202
Urban areas
  in developing countries **88**
  zoning 88
Uveitis, *Acanthamoeba* 260

Validation 26, 50–51, **67**, 136
Vendors, water **15**
Verification **29–31**, 51, **71–76**
  chemical quality 30–31, 72, 73
  community-managed supplies 74–75
  microbial safety and quality 29–30, 72, **142–143**, 284
  piped distribution systems 74
  quality assurance and quality control 75–76
  water sources 73–74
Vessels
  emergency and disaster situations 106
  packaged drinking-water 113

Vibrio **244–246**
Vibrio cholerae 122, 125, **244–246**
Vinyl chloride **456–458**
  analysis 162
  guideline value 194, 457, 493
Vinylidene chloride *see* 1,1-Dichloroethene
Viruses 221
  enteric *see* Enteric viruses
  indicator and index **289–295**
  pathogenic 122, **247–259**
  persistence in water 125
  treatment effects 138–141
Visible organisms 211, 212–213
Vittaforma 270
Volumetric titration 158

Warm water systems 100
Wastewater, domestic, chemicals in 186
Water avoidance orders 79
Water extraction systems, control measures 58–59
Water quality 90
  health care facilities 102–103
  monitoring *see* Monitoring
  sources, in disaster situations 105
  see also Guideline values
Water Quality Monitoring (Bartram & Ballance) 75–76
Water quality targets (WQTs) 25, 40, **42–43**, 126
Water resource management **12–13**
  see also Resource protection
Water Safety Plans 20, 48, 66
Water safety plans (WSPs) 4, 24, 26, **48–83**
  aircraft and airports 116
  approval and review 85
  audit 86, 94
  community and household supplies 85
  documentation and communication 82–83
  health care facilities 103
  key components 49
  large buildings 99, 102
  management 76–82
  model 66
  operational monitoring and maintaining control 68–71
  ships 120
  stages in development 50
  supporting programmes 80–81
  surveillance *see* Surveillance
  system assessment and design 51–68
  verification *see* Verification
Water sources *see* Source waters

Water suppliers *see* Suppliers, drinking-water
Water treatment *see* Treatment
Water Treatment and Pathogen Control 20, 61
Water vendors **15**
Waterborne infections *see* Infections, waterborne
Weight, body *see* Body weight
Wells 59, 65, 141
WHO Pesticide Evaluation Scheme (WHOPES) programme 148, 190
Winter vomiting disease 252
Wound infections, *Aeromonas* 224
WQTs *see* Water quality targets
WSPs *see* Water safety plans

Xanthomonas 286
Xylenes **458–459**
  acceptable level 219
  analysis 160, 458
  guideline value 188, 458, 493
  treatment achievability 168, 458

Yersinia **246–247**
Yersinia enterocolitica 122, 246, 247
Yersinia pseudotuberculosis 246, 247

Zinc 193, **459–460**, 490
  acceptable level 219, 459
  corrosion **183**
  dissolution from brass 182–183
Zoning, urban areas 88